Proceedings in Life Sciences

European Society for Comparative Physiology and Biochemistry
4th Conference, Bielefeld, September 8–11, 1982

Conference Organization

General Organizers

R. Gilles and H. Langer
Liège, Belgium/Bochum, FRG

Local Organizers

K. Immelman, E. Prove, and S. Sossinka
Bielefeld, FRG

Symposium Organizers
Intestinal Transport

M. Gilles-Baillien
Liège, Belgium

Hormones and Behaviour

J. Balthazart
Liège, Belgium

Under the Patronage of

The Deutsche Forschungsgemeinschaft

The Department for Scientific Research of the
Bundesland Nordrhein Westfalen

The Paul-Martini-Stiftung der
Medizinisch Pharmazeutischen Studiengesellschaft

The University of Bielefeld

The University of Liège

The European Society for Comparative Physiology and Biochemistry

Intestinal Transport

Fundamental and Comparative Aspects

Edited by
M. Gilles-Baillien and R. Gilles

With 155 Figures

Springer-Verlag
Berlin Heidelberg New York Tokyo 1983

Dr. M. GILLES-BAILLIEN, Scientific Editor
Laboratory of General and
Comparative Biochemistry
University of Liège
17, Place Delcour
4020 Liège, Belgium

Professor Dr. R. GILLES, Coordinating Editor
Laboratory of Animal Physiology
University of Liège
22, Quai Van Beneden
4020 Liège, Belgium

ISBN 3-540-12430-6 Springer-Verlag Berlin Heidelberg New York Tokyo
ISBN 0-387-12430-6 Springer-Verlag New York Tokyo Heidelberg Berlin

Library of Congress Cataloging in Publication Data. Main entry under title: Intestinal transport. (Proceedings in life sciences) Lectures from a symposium held at a conference of the European Society for Comparative Physiology and Biochemistry, Bielefeld, FRG, September 8–11, 1982. 1. Intestinal absorption–Congresses. 2. Biological transport–Congresses. I. Gilles-Baillien, M. (Michelle), 1939–. II. Gilles, R. II. European Society for Comparative Physiology and Biochemistry. IV. Series. QP156.I567 1983 599′.0132 83-4826

Offsetprinting and bookbinding: Brühlsche Universitätsdruckerei, Giessen

2131/3130-543210

Preface

The aim of this symposium was to provide a framework for fruitful discussion on intestinal transport, not only for advanced scientists but also for younger people starting in this field of research. Invited lectures, communications and poster presentations were focused on four central themes, all treating the properties of the sole intestinal epithelium, deliberately leaving aside problems dealing with more integrative functions of the whole intestine. The importance of motility or blood circulation, for instance, is certainly capital in the overall intestinal function, but these aspects by themselves deserve another meeting.

This volume has compiled the manuscripts of the invited lectures which substantially comprised the four sessions of the Symposium.

Part 1 is designed to emphasize actual knowledge of the transport of water, inorganic as well as organic ions and molecules across the isolated intestinal epithelium. An enormous wave of investigations has emerged from studies performed with "Ussing chambers", which roused interest in studies on absorption mechanisms and subsequently on secretory processes. This has triggered off a trend to research on isolated cells as absorption and secretion are the main function of the different cell types constituting the intestinal epithelium. In this first session not only the importance of the parallel arrangement of these different cellular entities is stressed, but also the role played by the paracellular route. Moreover, though the interference of unstirred layers, of the mucus coating, and of the basement membrane are illustrated in this first part, there is undeniably a gap in research at this level which has to be filled before the role of these extracellular compartments in transport processes across the whole epithelium can be fully assessed.

Part 2 reflects the present major interest in sophisticated studies performed with vesicles obtained from purified brush-border and basolateral membranes and designed to elucidate transport or carrier mechanisms at these two levels. With this material appears growing information on the function of "carriers" at the molecular level; the use of biochemical as well as biophysical technical methods allows an approach to the complex relationship between the different membrane components organized to perform exchanges of matter and energy across these membranes.

Part 3 deals with some aspects of the regulation and control of intestinal transport. The enormous amount of information presented by clinical studies of human pathology on the intestinal function and on the therapeutical solutions envisaged, has been and still is one of the major stimuli to the many possibilities and means of regulation of intestinal transport. Pioneer work in this field is presented in the third section.

In part 4 our aim has been to arouse cross-reactions and mutual interest between fundamentalists and comparatists. Indeed, from the many concepts of intestinal transport processes which have appeared in the course of animal evolution, we believe that a more comprehensive knowledge could be acquired.

We hope that this volume will be a valuable tool not only for young or advanced scientists but also for students whishing to bring their knowledge of progress and gaps in the field of intestinal transport up to date.

Liège, April 1983 M. GILLES-BAILLIEN

Contents

Part 2. Brush Border and Basolateral Membranes

Part 3. Regulation of Intestinal Transport

Part 4. Comparative Aspects of Intestinal Transport

List of Contributors

You will find the addresses at the beginning of the respective contribution

Introductory Survey

Contributions and Stimulus to Intestinal Transport Studies

K.A. MUNDAY and J.A. POAT [1]

Introduction

In the mid-1970s, under the stimulus of Dr John Robinson of Lausanne, many of us collected together at a Falk Symposium on Intestinal Ion Transport (1975) in which a major session of the Symposium was devoted to comparative studies. From the comparative standpoint, a number of us working in the field owe John Robinson a great debt, because this was the first occasion when comparative physiologists collected together as a group to discuss intestinal transport in all its facets. That early effort was followed in 1980 by the Second Conference of the European Society for Comparative Physiology and Biochemistry with a session of comparative intestinal transport studies and was associated with the regular European Intestinal Transport Meeting. Now again we have a major section devoted to reviewing the current state of comparative studies assessing their implications for intestinal transport. The work and continuing stimulus is therefore present for a vigorous exchange of results and discussion using studies from a wide variety of experimental animal sources to advance our understanding in this field.

When one discusses transport studies, most attention is focused on the results and their significance. This introduction will attempt to summarise, with some detail, important methods that have been used in the study of intestinal transport, and to try to show how in the wake of each new advance in experimental methodology has come parallel development in our understanding of mechanisms. We are all familiar with Dennis Parsons' excellent review chapter, concerning methods for investigation of intestinal absorption (Parsons 1968). This chapter does not intend to repeat a precis of that excellent work. However, for the benefit of younger, less specialist ESCPB members, with a particular interest in this subject, I intend to select some methods which have been used by my own group at Southampton, in the study of the mechanism of angiotensin activity on the regulation of sodium and water intestinal fluid absorption. All the work that I shall be reporting has been done in collaboration with my senior colleagues, Dr Judith Poat and Dr Brian Parsons. We have been supported by innumerable postdoctoral and postgraduate colleagues, some of whom will be named in the literature.

---

1 Department of Physiology and Pharmacology, University of Southampton, Southampton, Great Britain

Intestinal Transport
(ed. by M. Gilles-Baillien and R. Gilles)

As Dennis Parsons showed, initially much information was gained on intestinal transport from balance studies using fistulae etc. Probably the biggest impact came in the study of intestinal transport from the use of the everted gut sac preparation devised by Wilson and Wiseman (1954) in Sheffield.

Length of intestine are inverted so that the brush border lining is open to the exterior medium. This allows for easier oxygenation and provides a neat device, collecting the transported material into the closed sac. The tissue is incubated in Krebs' bicarbonate buffer with added glucose as an essential constituent. It is maintained in a 95% O_2:5% CO_2 atmosphere.

Using this method, David Smyth and his colleagues in Sheffield were able to measure the transport of sugars across the intestine, and this pioneer work was paralleled by the work of Matthews and colleagues on amino acid transport studies.

There are major experimental disadvantages with a closed intestinal sac prepared in this manner. Oxygenation inside the sac is not always fully effective and as materials are transported into the closed sac, so the pressure of the contents rises. This can lead, with inadequate oxygenation, to damage of the intestinal epithelial layers, and unsatisfactory transport measurements. As a result the technique of cannulation of everted sacs emerged whereby the hydrostatic pressure could be maintained within the sac to give a steady state with well-oxygenated fluids perfusing through the sacs. These latter techniques have been very widely used.

Our work at Southampton initially employed the in vitro everted closed sac preparation, and the first studies on the action of angiotensin in stimulating sodium and water transport through the gut wall was carried out in the later 1960s using this preparation (Crocker and Munday 1970, Davies et al. 1972). Angiotensin II at very low (physiological) levels was added serosally to the inside of the closed sac preparations, prepared from previously nephrectomised and adrenalectomised rats to increase the tissue sensitivity. At these very low angiotensin levels (10^{-10} M) we showed a stimulation of sodium and water transport in all intestinal areas studied. These results are summarised in Table 1.

Table 1. The effect of angiotensin (10^{-10} M) on rat intestinal everted sac fluid transport

	Mucosal fluid transfer $ml\,g^{-1}$ wet wt h^{-1}	P
Jejunum		
Control	0.87 ± 0.07	
+ angiotensin	1.34 ± 0.09	< 0.001
Ileum		
Control	0.71 ± 0.09	
+ angiotensin	1.19 ± 0.01	< 0.01
Colon		
Control	0.27 ± 0.04	
+ angiotensin	0.46 ± 0.06	< 0.01

Everted sacs were prepared from rats which had been adrenalextomised/nephrectomised 48 h previously. Results are expressed as mean ± S.E.M., n = 5–7

It was argued that results obtained from an in vitro everted portion of intestine could not possibly mimic a physiological situation. In an attempt to increase the validity of results from everted intestinal isolated preparations the use of non-everted sac preparations in vitro and in vivo were being investigated. In these latter respects, pioneer work was carried out by both Fisher and Parsons. The three in vivo preparations most used were closed sacs, perfused sacs, and the more complex perfused sac with perfused vasculature. Again, each of these preparations has its own advantages and disadvantages but each has led to significant increases in our understanding of transport mechanism in normal functioning animals.

Our group in the study of angiotensin action chose to use a closed in vivo sac preparation. The nerves and all vasculature to this preparation are intact, and we were able to show that fluid transport was increased by low levels of angiotensin. At the concentrations which stimulate transport, angiotensin was without effect on the circulation or blood pressure, to the region of the intestine from which the loop was prepared (Bolton et al. 1975).

Figure 1 shows the experimental set-up used in our preparation. A closed 15-cm sac of jejunum is washed out and filled with Krebs' bicarbonate buffer, containing a non-absorbable marker. We generally used (^{14}C)-polyethylene glycol (PEG) but have also used (^{3}H)-inulin. These non-absorbable markers allow changes in radioactivity within the closed sac to be measure in samples withdrawn from the sac, and so give an indication of the amount of fluid absorbed. Fluid measurements by this method correlate well with direct weight measurements of the sac at various time intervals. The experiments were divided into two time periods, the first being a control and the second the experimental period. Saline was infused through a femoral vein cannula during the first period, and saline and/or drug during the second. Thus each animal acts as its own experimental control. Full details of the method have been published (Bolton et al. 1975). Blood pressure can be simultaneously recorded from a carotid cannula and Tables 2 and 3 illustrate the type of results we obtained using this preparation. Table 2 shows the stimulatory effect of infusions of angiotensin II at 0.59 ng kg^{-1} min^{-1} during the second period. This dose of hormone raises circulating levels of angiotensin II as measured by radioimmunoassay from around 20 pg ml^{-1} to 40 pg ml^{-1}, which is well within the circulating variations of physiological levels, and falls away to normal levels within a minute on cessation of infusion.

Table 2. The effect of angiotensin (A II) (0.59 ng kg^{-1} min^{-1}) on fluid transport by rat jejunum and on circulating levels of angiotensin

	Mucosal fluid transfer ml 30 min^{-1} g^{-1} wet wt	Plasma A II pg ml^{-1}
1st period (saline infusion)	0.62 ± 0.06	20.1 ± 3.9
2nd period (saline)	0.78 ± 0.17	13.5 ± 1.0
2nd period (angiotensin)	1.02 ± 0.08	41.9 ± 2.6

Fluid transport was measured over two consecutive 30-min periods and the results expressed as Mean ± S.E.M. from 5 animals.

Plasma angiotensin was measured by radioimmunoassay for experiments with a second period, no blood samples was taken during the first period. The observations are from 5 animals

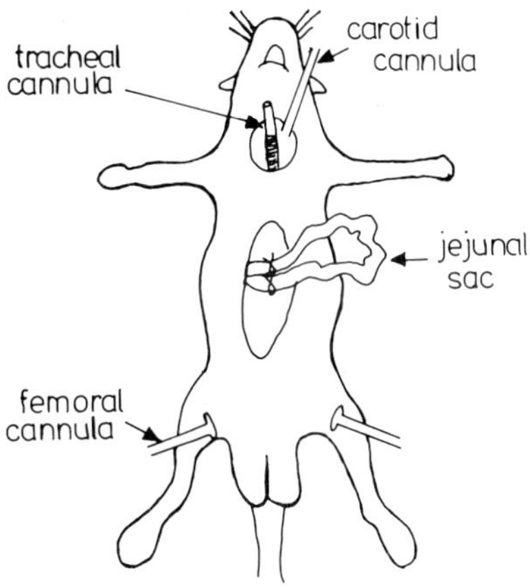

Fig. 1

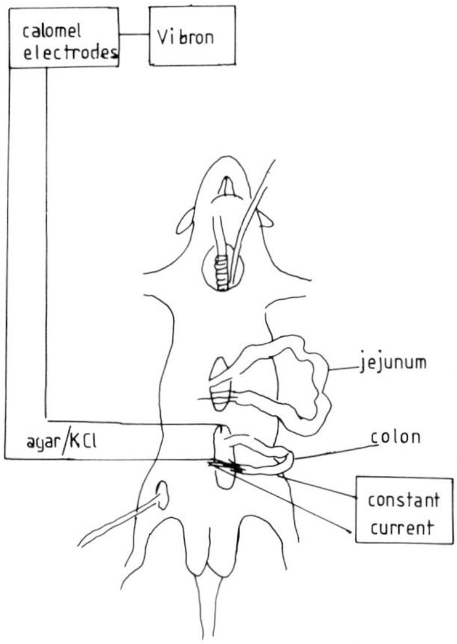

Fig. 2

Fig. 1. In vivo rat jejunal sac preparation. A 15-cm loop of jejunum was isolated, washed, ligatured and filled with Krebs' bicarbonate buffer, pH 7.4, containing 50,000 dpm (^{3}H)-inulin or (^{14}C)-polyethylene glycol and returned to the animal. Samples were withdrawn at various time intervals. The following cannulations were made; a carotid cannula for the measurement of blood pressure via a transducer and Servoscribe pen recorder, and two femoral vein cannulae for the infusion of saline or saline plus hormone

Fig. 2. In vivo rat jejunal, distal colon sac preparation. Rat jejunum was prepared for the measurement of fluid transport as described in Fig. 1, additionally a 3-cm distal colon sac was washed and filled with Krebs' bicarbonate, a length of portex PP30 tubing containing 3M potassium chloride in 4% agar gel was used as the mucosal p.d. electrode and ligatured into the distal end of the sac. An intraperitoneal agar-saline bridge was used as the serosal electrode, both electrodes were connected vial calomel half cell electrodes to a Vibron electrometer which was used to measure the potential difference across the intestine. Isc was measured with Ag/AgCl electrodes, one placed near the outer surface of the sac, and the other attached to the portex tubing. The Ag/AgCl electrodes were connected to a constant current box. This arrangement allows concomitant measures of fluid transport and electrical parameters in two distinct areas of the intestine which are sensitive to angiotensin

While these transport measurements are being recorded, it is also possible to make electrical measurements across the intestinal tissue, and the techniques used are illustrated in Fig. 2. The addition of calomel and Ag/AgCl electrodes to the basic in vivo loop preparation allows the measurement of short circuit current (s.c.c.) and potential differences (p.d.) across the tissue. The type of electrical measurements has been very widely used by Edmonds and co-workers for the study of the action of hormones such as aldosterone. Similarly, these techniques have been used to study the effects of acetylcholine and noradrenaline. Electrical measurements of this type primarily give information on the effects of hormones, neurotransmitters etc. on transport of fluid and ions, but have provided less information on the mechanisms. Table 3 demonstrates that the stimulation of sodium and fluid transport by angiotensin II is primarily via an electroneutral mechanism.

Table 3. The effect of angiotensin (0.59 ng kg^{-1} min^{-1}) on jejunal fluid transport and distal colon electrical parameters in the same animal

	Jejunal fluid transport ml g^{-1} wet wt 30 min^{-1}	S.c.c. μA cm^{-1}	p.d. mV	Resistance Ω cm^{-2}
1st period (saline)	0.58 ± 0.11	96.4 ± 8.7	20.1 ± 2.0	200.9 ± 15.3
2nd period (angiotensin)	0.94 ± 0.09	112.4 ± 13.9	22.5 ± 2.1	199.7 ± 15.0

The results are expressed as Mean ± S.E.M. from 4–6 rats, fluid transfer was measured during two consecutive time periods in the jejunum, and electrical measurements made in the same rats at the same time in the distal colon

Table 4 summarises the results of in vitro and in vivo preparations with respect to the angiotensin stimulation of sodium and fluid transport. It emphasises that the stimulation is via an electroneutral mechanism, that cyclic AMP is not involved, that cycloheximide blocking protein synthesis at the translation stage inhibits, whereas actinomycin D is without effect. The close similarity between the in vivo and in vitro findings offers substantial support to the assertion that the everted sac technique has relevance and meaning for the in vivo situation. This general conclusion has been confirmed by other workers.

Table 4. An in vivo/in vitro comparison of features in the angiotensin II response in the rat

In vitro	Area	In vivo
√	Jejunum	√
√	Ileum	–
√	Colon	√
	Effect	
	Low doses	
√	A II-stimulate transport	√
	High doses	
√	A II-inhibit transport	√
	Mechanism	
	Protein synthesis involvement	
√	Cycloheximide inhibits	√
√	Puromycin inhibits	–
√	Actinomycin D – no effect	√
√	Electroneutral	√
√	Cl^- necessary	–
√	Rapid	√
√	Ca^{2+}-dependent	–
√	cAMP not involved	–

– indicates no experiment

Another and now more usual in vitro preparation for the study of electrical changes associated with transport processes makes use of the Ussing chamber apparatus, originally developed by Ussing and Zerahn for studying transport across amphibian skin and bladder. This type of preparation was subsequently widely used by Schultz and his many co-workers and has led to major advances in our understanding of electrical features associated with the mechanisms of intestinal transport. Figure 3 is a summarising diagram of the basic features of the Ussing chamber apparatus. Using isotopes in this apparatus, it is possible to measure ion fluxes across isolated membrane preparations.

Table 5 illustrates some results we obtained with this type of apparatus when studying the effect of noradrenaline on rat jejunal, sodium and fluid transport. The results show that a change in J_{net} with noradrenaline, is due entirely to a change in J_{ms} and this is accompanied by a significant decrease in short circuit current. This effect of noradrenaline was first observed by Field and co-workers (Field and McColl 1973) in rabbit ileum and our results confirm these original findings. It is suggested that the fall in short circuit current after noradrenaline addition is associated with a bicarbonate process, but this is difficult to investigate.

Table 5. Effect of noradrenaline on sodium fluxes and short circuit current (s.s.c.) using stripped jejunum intestine of rat

	Flux of ^{22}Na ($\mu Eq\ h^{-1}\ cm^{-1}$)			
	J_{ms}	J_{sm}	J_{net}	I_{sc}
1st period	12.18 ± 1.29	0.27 ± 0.93	+ 2.91 ± 0.85	0.85 ± 0.14
2nd period + noradrenaline	15.25 ± 1.9	9.77 ± 1.14	+ 5.53 ± 0.3	0.60 ± 0.30
Δ	+ 3.07 ± 0.88	+ 0.50 ± 0.71	+ 2.64 ± 1.11	
p	< 0.05	N.S.	< 0.001	< 0.01

The experiments consisted of two consecutive 30 min flux periods, noradrenaline (5×10^{-4} M) was added 5 min prior to the second period. J_{ms} and J_{sm} represent unidirectional fluxes of Na, and J_{net} the difference.
Results are expressed as mean ± S.E.M., from 4–6 rats

By contrast, with angiotensin we have been unable to obtain any response using this experimental procedure at any dose on ion flux across prepared intestinal membranes. This result has caused discrepancy in our sequential interpretation of the mechanism of angiotensin action on gut epithelial cell. That the lack of response to angiotensin is not associated with inadequate methodology is confirmed by these noradrenaline effects to which we have just referred. Figure 4 is a model which we suggest could offer some explanation for our failure to show that angiotensin stimulates sodium and fluid transport in the Ussing chamber preparation. We suggest that there are presynaptic and postsynaptic noradrenaline receptor sites and then it may be that angiotensin II exercises its effect through a presynaptic angiotensin II

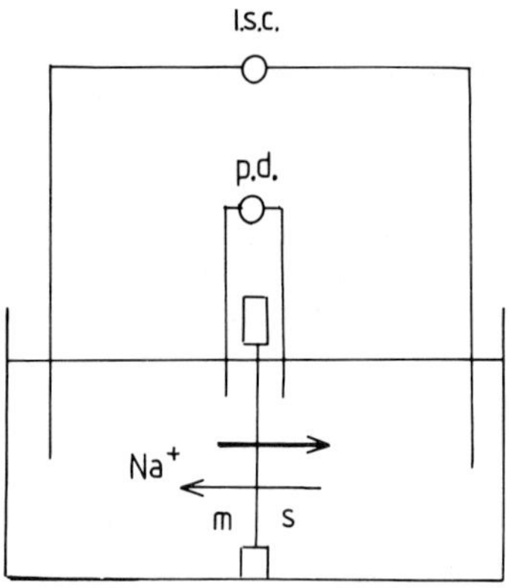

Fig. 3

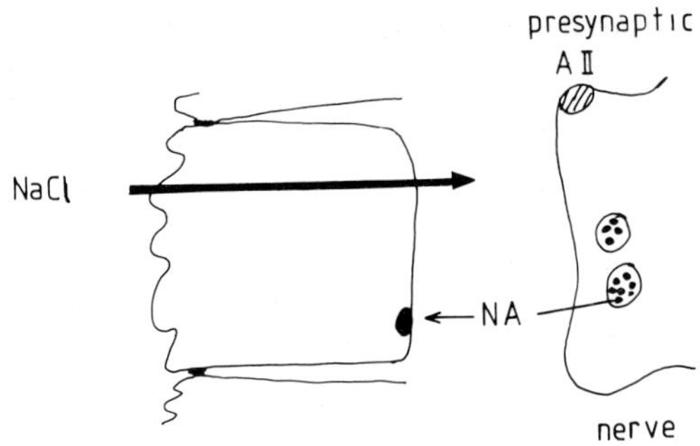

Fig. 4

Fig. 3. Diagrammatic representation of the Ussing chamber apparatus. Stripped rat jejunum (of the outer muscle only) was placed between the two chambers. Potassium chloride in agar formed the potential difference electrodes, these were connected by calomel half cells to a voltmeter for recording the potential difference. The I_{sc} electrodes were Ag/AgCl and were placed at some distance from the tissue. These were connected to a constant current box. Both halves of the chamber were filled with Krebs' bicarbonate buffer, pH 7.4 of identical composition and the buffer constantly bubbled with 95% O_2/5% CO_2 by a gas air lift arrangement. The chambers were surrounded by a water jacket which allowed the apparatus to be kept at 37°C for the duration of the experiment. ^{23}Na was added to either the mucosal or serosal surface and samples removed from the opposite site for γ-counting in a Beckman counter. The experiment was conducted in paired chambers to allow concurrent measurement of both fluxes

Fig. 4. A hypothetical model for angiotensin action on intestinal epithelial NaCl transport

receptor action stimulating in turn the release of noradrenaline. Noradrenaline then in its turn stimulates the sodium and fluid transport mechanism, as already shown in Table 5. In this way, the angiotensin II effect is an indirect action. The Ussing chamber intestinal preparation which was insensitive to angiotensin II was not highly innervated since the tissue was stripped and so may contain relatively little endogenous noradrenaline to be released by angiotensin II. Secondly, there is a small tissue to medium ratio in the chambers meaning that the oxygenation of the tissue could be very efficient, thus very rapidly inactivating any catecholamines that were present. We tested this speculation by studying the effects of tyramine upon the preparation and these results are summarised in Table 6. Treatment with tyramine, the sympathomimetic drug, which increases the release of noradrenaline from presynaptic terminals, resulted in an increase sodium flux paralleled by a slight increase in short circuit current, but both were not significant at the 5% level. We then carried out experiments with Ussing chamber preparations using pargyline – a substance which inhibits the monoamine oxidase inhibitor, and so leads to a rise in tissue noradrenaline concentration. We confirmed that this procedure was effective in causing a significant enhancement of the fall in potential difference across the tissue, but again the transport effects on sodium flux were equivocal. Our failure to advance our understanding of the mechanism of angiotensin action stimulating sodium and fluid transport with the Ussing chamber preparation was a disappointment, and this led us to look for other techniques of advancing our understanding of the angiotensin action.

Table 6. The effect of tyramine (5×10^{-4} M) on Na fluxes in rat ileum

	Flux of ^{22}Na (μEq h^{-1} cm^{-1})		
	J_{ms}	J_{sm}	J_{net}
1st period	11.12 ± 2.68	8.67 ± 2.22	+ 2.44 ± 1.12
2nd period (tyramine)	12.39 ± 2.74	9.48 ± 2.27	+ 2.92 ± 1.12
Δ	+ 1.28 ± 0.44	+ 0.81 ± 0.43	+ 0.43 ± 0.39
p	N.S.	N.S.	N.S.

The experiments consisted of two consecutive 30-min flux periods. Tyramine was added 5 min prior to the second period. J_{ms} and J_{sm} represent unidirectional fluxes of Na, and J_{net} the difference.
Results are expressed as mean ± S.E.M., required from 4–6 animals

Two in vitro preparations which have been very extensively used in recent years for molecular investigations are the preparation of cell scrapings, and the preparation of cells suspensions, from intestinal tissue. These preparations, initially developed for basic enzyme studies, led to the valuable membrane studies with vesicle preparations from basolateral and brush border membranes. Investigations using these preparations permit investigation of the different membrane functions in different locations of the enterocytes. Such preparations were pioneered by many groups, but particular mention should be made of the work of Murer und Kinne in Germany.

Our most recent work at Southampton has taken a variation of the Murer preparation and using differential centrifugation, we have produced crude basolateral membrane preparations. The procedure followed is based on that of Murer et al. (1974) and Scalera et al. (1980). The method initially produced epithelial cells and then membranes from these cells. The purity and specificity of the membranes are assessed by marker enzyme studies using alkaline phosphatase for the brush border, and ouabain sensitive Na, K, Mg, ATPase for the basolateral membrane.

We have used these preparations to study the possibility of postjunctional α-receptors located on intestinal basolateral membranes. Thus, the intestine is responsive to noradrenaline and it would be anticipated that receptor sites for the catecholamine would be located on the basolateral membranes. This possibility was investigated using radioligand binding assays. Adrenoceptors can be classified as α_1 (predominantly postsynaptically located) or α_2 (predominantly presynaptically, but some postsynaptically located). In this study we used two labelled ligands (^{3}H)-prazosin, an α_1 antagonist and (^{3}H)-clinidine, an α_2 agonist. Preliminary studies showed that sites labelled with (^{3}H)-prazosin were predominant in the preparation. These sites were further investigated by testing the ability of a variety of adrenoceptor agonists and antagonists to compete for the specific (^{3}H)-prazosin binding site, the results being expressed as Ki values. The binding data were compared with fluid transport responses and the same adrenoceptor agonists and antagonists tested for their ability to inhibit the stimulation produced by noradrenaline (antagonists) or mimic the response (agonists). The results are shown in Table 7. They suggest that the stimulation of fluid transport in rat jejunum by noradrenaline is predominantly an α_1-mediated mechanism, thus prazosin and indoramin are potent inhibitors whilst rawolscine and

Table 7. The effect of noradrenaline agonists and antagonists on rat jejunal fluid transport and (^{3}H)-prazosin binding to rat jejunal epithelial cell membranes

Agonists	EC_{50} Fluid transfer	K_i (^{3}H)-prazosin binding
Noradrenaline	4 μM	1.81 μM
Adrenaline	36 μM	22.22 μM
Clonidine	5 mM	8.99 μM
Phenylephrine	> 10 mM	84.87 μM
Methoxamine	> 10 mM	162.03 μM
Antagonists	**IC_{50}**	**K_i**
Prazosin	450 nM	3.45 nM
Phentolamine	68 μM	304.40 nM
WB 4101	170 μM	475.60 nM
Rawolscine	3090 μM	9.01 μM
Yohimbine	> 2 mM	6.15 μM

The results are expressed as EC_{50}, i.e., the amount of drug to give half maximal noradrenaline response (i.e., that caused by 1 mM) and IC_{50} the amount of drug to inhibit the stimulation to 1 mM noradrenaline by 50% for fluid transfer. Binding results are expressed as K_i values using the Cheng-Prussof equation

yohimbine are less potent. Furthermore there is a good correlation between the binding data and the physiological response.

These findings appear to be in some contradiction to recent work of Field and co-workers (Chang et al. 1982). Using rabbit ileum in Ussing chambers, they were able to show that clonidine was more effective than noradrenaline and that transport responses were blocked by yohimbine and rawolscine rather than prazosin. From these investigations we have gone on to study the kinetic parameters of binding in both species and both areas to see if a species tissue specificity might explain the result. Certainly the ileum and jejunum seem to differ in that in the jejunum of both animals we have α_1 binding site of high affinity and low capacity, whereas in the ileum the binding sites appear to have lower affinity but much increased capacity. Again, these results contrast with Starke's findings in the guinea-pig where he observes high affinity α_1 sites (Tanaka and Starke 1979). The conclusion from these preliminary binding data studies suggest that some of the differences we could be recording are due to different receptor populations in the two animals used in the investigation. It is tempting to suggest that α_1 adrenoceptors are involved in stimulation of fluid absorption whilst α_2 adrenoceptors are responsible for intestinal secretion.

Summary

1. This survey on a range of techniques used for transport studies shows that experimental technique is important in our understanding of the mechanism of the effect investigated. One must be conscious of the limitations of technique in the interpretation of results and the qualifications this might pose on the induction of general theories or models.

2. The synthesis of results from different animal species initially makes it difficult to suggest a unifying hypothesis. Most of the responses we investigate in transport studies are multifactorial and investigative methods may highlight only a particular facet of the mechanism, and so make generalisation difficult.

3. It may be possible to use animal groups for intestinal transport studies where, for example, they live at much lower temperatures than the mammal, and in this way the in vitro conditions become closer to the in vivo state. This can open up wider comparative possibilities for experimental design.

4. We hope this Symposium will generate ideas for younger comparative physiologists enabling them to take the work and ideas from mammalian transport studies and translate then to their own particular interests.

References

Bolton JE, Parsons BJ, Munday KA, York BG (1975) Effects of angiotensin II on fluid transport transmural potential difference and blood flow by rat jejunum in vivo. J Physiol (Lond) 253: 411–428

Chang EB, Field M, Miller RJ (1982) α_2-Adrenergic receptor regulation of ion transport in rabbit ileum. Am J Physiol 242:207–242

Crocker AD, Munday KA (1970) The effects of the renin-angiotensin system on mucosal water and sodium transfer in everted sacs of rat jejunum. J Physiol (Lond) 206:323–333

Davies NT, Munday KA, Parsons BJ (1972) Studies on the mechanism of action of angiotensin on fluid transport by the mucosa or rat distal colon. J Endocrinol 54:483

Field M, McColl I (1973) Ion transport in rabbit ileal mucosa. III. Effect of catecholamines. Am J Physiol 225:852–907

Murer H, Hopfer U, Kinne-Saffran E, Kinne R (1974) Glucose transport in isolated brush border and lateral basal plasma membrane vesicles from intestinal epithelial cells. Biochim Biophys Acta 345:170–179

Parsons DS (1968) Methods for investigation of intestinal absorption. In: Code C (ed) Handbook of Physiology – Alimentary canal III, Chap 64. Am Physiol Soc, Washington DC, p 1177

Robinson JWL (1975) Intestinal ion transport. Symp Titisee 1975. MTP Press

Scalera V, Storelli C, Storelli-Joss C, Haase W, Murer H (1980) A simple and fast method for the isolation of basolateral plasma membranes from rat small-intestinal epithelial cells. Biochem J 186:177–181

Tanaka T, Starke K (1979) Binding of (^{3}H)-clonidine in membranes of guinea-pig ileum. Naunyn-Schmiedeberg's Arch of Pharmac 309:207–215

Wilson TH, Wiseman G (1954) The use of sacs of everted small intestine for the study of the transference of substances from the mucosal to the serosal surface. J Physiol (Lond) 123:116–125

Part 1
From the Whole Epithelium to Isolated Cells

Routes of Water Flow Across the Intestine and Their Relationship to Isotonic Transport

R.J. NAFTALIN and S. TRIPATHI [1]

Roots and Routes of Isotonic Water Transport

There is still much discussion and uncertainty about the routes of water flow across loose epithelia like small intestine and gallbladder. Because of this uncertainty about the routes of fluid movement, the mechanism of isotonic transport also remains in doubt.

Work in the early 1960's by Curran (1960), Diamond (1979) reestablished that transporting epithelia, like the small intestine and gallbladder can transport fluid from the luminal (mucosal) to serosal side against a considerable osmotic pressure gradient. Between 50 und 200 mosmoles have been reported.

Curran (1960) suggested that the tissue Na^+-pump creats a hypertonic compartment within the small intestine, which is in contact with the mucosal solution, via tight channels. This hypertonic central compartment generates osmotic flow from the mucosal solution into the compartment. The hydraulic pressure built-up within the central compartment then forces fluid out across the leaky serosal border of the tissue, into the serosal bathing solution.

Diamond, working with a sac preparation of gallbladder, discovered that the transported fluid is isotonic with the mucosal solution over a wide concentration range, from half-diluted Ringer to 40 mosmoles hypertonic. He argued that isotonic transport implies that the fluid of the central compartment equilibrates rapidly and almost completely with the mucosal solution. The osmotic permeability of the mucosal border has to be very large, so that the residual osmotic gradient can generate a sufficiently large flow to match the observed transepithelial flow. Diamond and Bossert (1967) supported this view and suggested that isotonic transport might arise from hypertonicity within the lateral intercelluar spaces (LIS). Isotonic transport could be achieved more readily, if the Na^+-pump activity, which creates the osmotic pressure gradient, were confined to the proximal 10% of the length of the LIS; thereby permitting equilibration between the mucosal solution and the contents of the LIS in the distal 90% of the length of the LIS. This "standing gradient" mechanism implies that osmotic equilibration occurs via the transcellular route, as water enters the LIS via the baso-lateral cell membranes.

1 Department of Physiology, King's College, London, Strand, London WC2R 2LS, Great Britain

Intestinal Transport
(ed. by M. Gilles-Baillien and R. Gilles)

More recently, the standing-gradient model has been modified to take account of the possible secondary route of fluid entry via the tight-junction. The predicted effect of this second route would be to change the concentration profile within the LIS from a steep concentration decrease from the proximal to distal part of the LIS, as predicted by Diamond's standing gradient view of isotonic transport, to a more uniform concentration distribution along the length of the LIS, as predicted from the view that fluid enters the LIS via tight-junction (Sackin and Boulpaep 1975, Weinstein and Stephenson 1981).

However, as Hill (1980) has vigorously pointed out, dual access to the LIS is immaterial to the solution of the problem of isotonic transport. The isotonic constraint requires that the hydraulic conductivity of the mucosal border (transcellular and paracellular routes combined in parallel) has to be at least two orders of magnitude larger than the hydraulic conductivity of other cell membranes. The hydraulic permeability of gallbladder is approximately two orders of magnitude too small to accommodate isotonic equlibration via the mucosal route.

Diamond (1979) and co-workers have recently expressed the view that the osmotic permeability of loose epithelia is greatly underestimated because of solute polarization in unstirred layers adjacent to highly permeable membranes. Van Os et al. (1979) claim to have measured transient changes in water flow across rabbit ileum which are indicative of a very high hydraulic permeability.

On the other hand, Garson and Steward (1982) have shown recently, using an NMR method, which is not subject to unstirred layer effects, that the P_{os} of *Necturus* gallbladder cell membranes is similar to that of other cells.

Using X-ray electron microprobe microanalysis, it has been shown that the contents of the LIS of rabbit ileum are approximately 30–40 mosmoles hypertonic to the concentration of the mucosal solution (Gupta et al. 1978). Clearly, if this finding is correct, then the L_p of the mucosal border (tight-junction and transcellular route combined) must be considerably lower than that suggested by Diamond. A high L_p would not permit this substantial hypertonicity to be retained within the LIS. On the other hand, if the L_p of the mucosal border is substantially lower than that suggested by Diamond, then full equilibration between the LIS and the mucosal solution cannot occur and the fluid leaving the LIS must be hypertonic to the mucosal solution and hence, some other means of attainment of isotonicity must occur.

Hydraulic and Osmotic Flow Across Gallbladder and Small Intestine

Wright and his co-workers (Wright et al. 1972a, Smulders et al. 1972) showed that both hydraulic and osmotic flow across the gallbladder are asymmetric. The osmotic permeability of tissue, when measured with net mucosal-serosal flow, exceeds the permeability in the opposite direction. Mucosal-serosal flow is accompanied by a volume increase of the sub-mucosa and intercellular spaces. Serosal-mucosal flow is accompanied by tissue shrinkage.

Flow across the tissue induced by hydrostatic pressure is also asymmetric, however the hydraulic permeability in the serosal-mucosal direction greatly exceeds that of mucosal-serosal permeability.

The explanations offered for these phenomena are that the LIS behaves as a valve allowing osmotic flow in the mucosal-serosal direction, but not in the reverse direction.

With hydrostatic pressure, it is thought that serosal-mucosal flow goes via a shunt opened between the cells as a consequence of stretching (Hakim and Lifson 1969). A hydrostatic pressure applied to the mucosal side does not tend to cause cell separation.

A New Method of Continuous Measurement of Transmucosal and Transserosal Fluid Movement Across Rabbit Ileum

From the previous discussion, it can be seen that a precise means of determination of the mucosal border permeability is of crucial importance to determining the mechanism of fluid transport.

Gravimetric analysis, using serial measurement of weight changes in gallbladder can resolve changes within 5 min. With the intestine this kind of measurement has a time resolution of 20 min, which is insufficient to measure the transient changes in flow.

A method which uses a capacitance probe to monitor the change in volume of the fluid compartments bathing the tissue improves resolution by at least a hundredfold (Weidner 1976). However, this method has a serious drawback when applied to the small intestine, as it does not monitor the volume changes occurring within the tissue. Van Os et al. (1979) showed that within 30 min following exposure of rabbit gallbladder to a hypertonic serosal solution (100 mM sucrose), there is a 25% increase in tissue weight. However, as the overall fluid volume of the mucosal bathing solution does not change, this transferance of fluid from the mucosal bathing solution to the tissue compartment is not monitored by the capacitance probe. Hence, the measured transepithelial fluid movement is less than inflow across the mucosal border into the sub-mucosa.

The problem of the distensibility of the sub-mucosal space has been overcome as follows: the tissue volume changes are monitored simultaneously with transepithelial flow. The volume changes are measured using an optical lever. This consists of a light mirror, pivoted at its lower edge and resting against the mucosal surface of the tissue (see Fig. 1). Any volume change within the tissue rotates the mirror which deflects a laser beam. This deflection is monitored by photodetectors, electronically amplified and recorded. The resolution of volume change is 30 nl cm^{-2}. Fluid movement across the entire tissue is monitored simultaneously using the method of Weidner. We find it essential to immobilize the tissue by supporting the serosal surface with a fine stainless steel grid. This avoids artefacts due to tissue movement, which readily occur when the solution is changed, or a hydraulic pressure is applied unilaterally. The resolution of this method, when measuring flow across an tissue area of 10 cm^2 is approximately 10 nl cm^{-2}. The time resolution of both procedures is approximately two seconds (Naftalin and Tripathi 1982a).

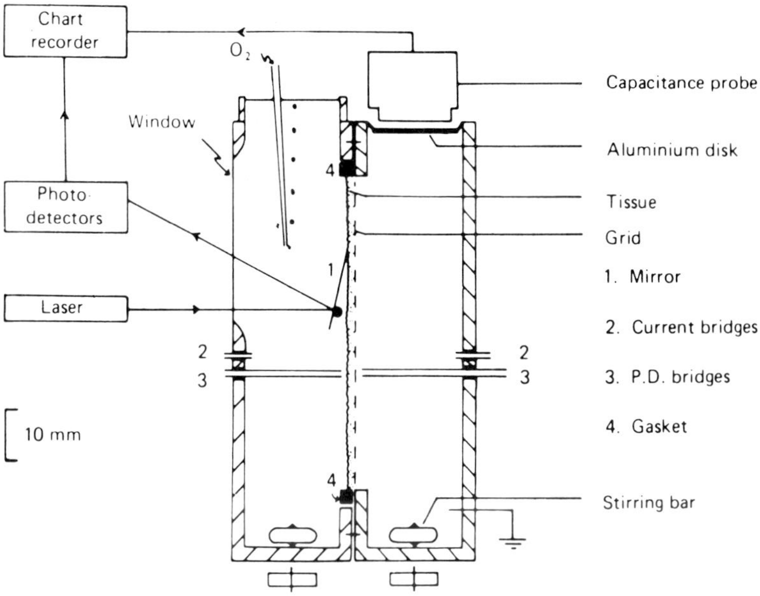

Fig. 1. The experimental set-up is shown schematically. One border of a sheet of rabbit ileum (exposed area 10 cm^2) is immobilized on a vertical grid by a small hydrostatic pressure head. Tissue volume changes are monitored with an optical lever. This consists of a mirror pivoted at its lower edge; its upper edge rests on the tissue. Any volume change within the tissue rotates the mirror which deflects a laser beam. This deflexion is detected by a pair of photodiodes and the difference signal is amplified and recorded. The resolution of tissue volume change is 30 nl cm^{-2}. Fluid movement across the immobilized border is detected as a change in the hight of the liquid column by a capacitance transducer (Dimeq TE200) (Weidner 1976). Evaporation is minimized and held constant by controlling the ambient temperature at 20 ± 0.5°C and covering the surface with a floating aluminium disk. For experiments at 35°C the capacitance probe is heated to 35°C with a constant temperature heating coil and the chamber waterjacket is perfused with water using a Haake thermocirculator

Determination of the Hydraulic Permeabilities of the Mucosal and Serosal Borders of Rabbit Ileum

When hypertonic sucrose (100 mM) is added to the serosal solution, following a delay, which lasts between 2 and 15 min, depending on the tissue thickness, fluid movement across the mucosal border increases simultaneously with a rise in streaming potential. The fluid entry rate continuously increases for about 20–30 min before reaching a steady state of around 25–30 μl cm^{-2} h^{-1}. As tissue volume increases, outflow across the serosal border rises to a steady rate of 10–15 μl cm^{-2} h^{-1}. Hence, the tissue continues to expand for at least 5 h at a rate of increase of 10–20 μl cm^{-2} h^{-1}.

A Direct Estimate of the L_p of the Serosal Border

When a hydrostatic pressure head is applied to the mucosal solution the pressure rise is instantaneously transmitted through the tissue to the surface resting on the grid. This is due to the low compressability of the tissue fluid and high flexibility of the tissue structures. Thus, when pressure is applied to the mucosal solution, a pressure gradient is present only in the region immediately adjacent to the support grid, i.e., across the serosal border. Consequently, an increase in pressure within the mucosal solution effectively squeezes fluid out of the sub-mucosa across the serosal border. As there is virtually no dissipation of the mucosal solution pressure head across the mucosal border, no flow across this border is directly induced by raised mucosal solution pressure.

Hence, by observing the effect of raised pressure on flow across the serosal border, it is possible to obtain a direct measure of the hydraulic permeability of the serosal border.

$$L_p = \Delta Jv/\Delta P.$$

The observed L_p across the serosal border of rabbit ileum at 20°C is 75×10^{-9} cm s^{-1} cm^{-1} H_2O (Naftalin and Tripathi 1982b).

Since the L_p of the serosal border is directly determined and flow is observed to be a linear function of pressure, over the range 0–100 mmHg, it follows that an interstitial pressure of 15–25 cm H_2O is required to generate a steady-state flow across the serosal border of 10–15 μl.

This interstitial pressure also exerts a considerable effect on water flow across the mucosal border.

Effect of Hydraulic Pressure on Fluid Movement Across the Mucosal Border

At 20°C, following addition of sucrose (100 mM) to the serosal bathing solution, inflow across the mucosal border is generated entirely by the osmotic pressure gradient across the mucosal layer due to sucrose entering the sub-mucosa. The flow induces a streaming potential, due to solvent drag of Na^+ via the cation-selective tight-junctions. Thus, when sufficient sucrose is added to the mucosal solution to nullify the streaming potential, the osmotic pressure gradient across the mucosal border is also reduced to zero.

Following a period of tissue expansion, after exposure to hypertonic serosal solution, the interstitial pressure rises, as is evident from the rise in outflow across the serosal border. Sufficient sucrose is then added to the mucosal solution to nullify the streaming potential, ca. 75 mM. The direction of flow across the mucosal border immediately reverses, but outflow across the serosal border continues at the same rate, so that the tissue volume decreases due to interstitial pressure-induced fluid loss to both the mucosal and serosal bathing solutions. As the interstitial pressure is determined and net outflow across the mucosal border is measured, it is possible to determine the L_p of the mucosal border.

L_p (mucosa) = J_v (mucosal exit)/interstitial pressure = 150–250 × 10^{-9} cm s^{-1} cm H_2O.

This lumped L_p of the shunt pathway, and the path of osmotic pressure-induced flow may be resolved into its composite parts by solution of the simultaneous equations.

(serosal sucrose) $$J_1 = L_1 \cdot (RT\Delta C - P) - L_2 \cdot P \qquad (1)$$

(mucosal sucrose = serosal sucrose) $$J_2 = -L_1 \cdot P - L_2 \cdot P \qquad (2)$$

where RT refers to the gas constant and temperature; ΔC, the osmotic gradient of sucrose, P, the interstitial pressure (cm H_2O), L_1 and L_2 to the hydraulic conductivities of the tight-junction and shunt pathways and J_1 the flow with osmotic gradient present and J_2 the flow after equilibration of the gradient.

The L_p of the mucosal osmotic pathway is L_1 = 10–15 × 10^{-9} cm s^{-1} cm H_2O^{-1} and that for the shunt pathway L_2 = 175–225 × 10^{-9} cm s^{-1} cm H_2O^{-1} (Naftalin and Tripathi 1982b).

When hypertonic sucrose is applied to the mucosal solution alone, immediately, there is a rapid and fairly constant exit across the mucosal border until the tissue shrinks beyond the resting volume. When the tissue volume decreases, beyond this point, the interstitial pressure falls rapidly and outflow across the serosal border decreases. Exit across the mucosal border also decreases.

Inflow across the serosal border occurs when the interstitial pressure becomes negative. In this phase, outflow across the mucosal border falls to a value close to the inflow across the mucosal border (5–10 μl cm^{-2} h^{-1} 100 $mosm^{-1}$ sucrose).

A New Interpretation of Osmotic and Hydraulic Permeability Asymmetries

These results allow us to reinterpret the mechanism of water permeability asymmetry of small intestine.

Water flow in the mucosal-serosal direction is generated by an osmotic pressure gradient across a single epithelial layer. Water flow in the serosal-mucosal direction is generated by a combined osmotic and negative hydrostatic pressure across the tissue and sub-mucosal layers respectively. The two layers are held together by negative interstitial pressure. The resistance to flow across the series barrier is greater than the sum of resistances across the two elements of the series barrier, because the osmotic pressure generated across the tight mucosal shunt pathway is dissipated across the wide mucosal shunt pathway. Hence the net resistance to serosal-mucosal transepithelial osmotic flow is much greater than the resistance to flow in the opposite direction.

The view that the paracellular pathway acts as a rectifier can be discounted, as the initial rate of exit of fluid from a distended sub-mucosa into a hypertonic mucosal solution is rapid, indicating that there is no asymmetry in the water permeability of the mucosal barrier itself. The asymmetry of hydraulic pressure-induced flow across unsupported tissue is almost certainly due to separation of the intercellular junctions, when the tissue's natural concavity is reversed by serosal pressure. When the tissue is supported, so that tissue stretching is prevented, no asymmetric permeabilities are observed.

Resolution of Mucosal and Serosal Pore Widths and Numbers by Osmotic Probes

The size of the mucosal and serosal border channels are measured by using osmotic probes with a wide range of molecular weight (60–150,000).

The change in flow per osmole after addition of solute is plotted against the appropriate hydrodynamic radius of each solute, obtained from literature values. The relationship between the changes in flow across the mucosal and serosal borders to the solute radii are plotted. The best fit of the data of flow across the serosal border to a modified Renkin equation (Bean 1972) shows that flow across the serosal border is consistent with flow via pores of 6.5–7.5 nm having an L_p of 75×10^{-9} cm s^{-1} H_2O.

The functional relationship between osmotic flow across the mucosal border and solute radius is inconsistent with flow via a single pore radius. The flows are consistent with flow via pores of three main radial sizes; 0.4, 0.7 and 6.5 nm, which have L_ps of $1–5 \times 10^{-9}$, $10–15 \times 10^{-9}$ and $100–150 \times 10^{-9}$ cm s^{-1} cm^{-1} H_2O respectively (Naftalin and Tripathi 1982b).

Thus, whilst flow across the serosal border is consistent with flow across a homogeneous matrix, the mucosal border differs from the serosal border in having heterogeneous pores; large pores which can exclude macromolecules above 4000 M.W. (the paracellular shunt); intermediate pores which can exclude solutes above sucrose (the LIS and tight-junctional route) and narrow pores which can exclude glycerol (the transcellular route) (Fig. 2).

Comparison of Effects of Probes Added to the Mucosal Solution on Streaming Potential and Mucosal Fluid Movement

A change in flow across the mucosal border is accompanied by a change in streaming potential. A plot of the change in streaming potential relative to the change in mucosal flow per osmole of solute added to the mucosal bathing solution varies with the solute radius. The ratio of ΔP.D./ΔJ (mucosa) increases as the probe radius increases to that of sucrose (0.52 nm). Above the size of sucrose the ratio of change in streaming P.D./J (mucosa) decreases. These findings are consistent with the view that a large electropositive streaming potential with water flow is generated via the intermediate (tight-junctional) channels. Flow via the narrow (transcellular) and wide (shunt) channels is not accompanied by a cation stream. This suggests that the transcellular channels exclude Na^+ and that the shunt channels are too wide to be ion-selective.

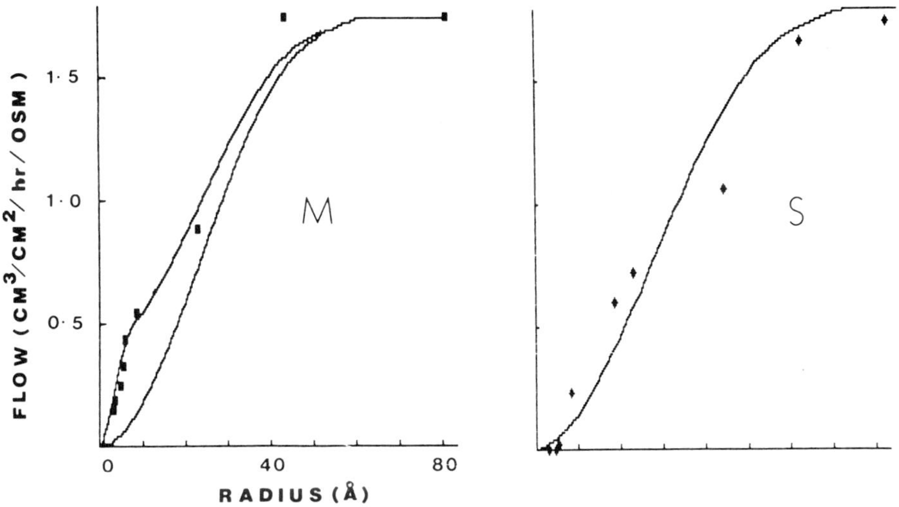

Fig. 2. A plot of fluid exit across the mucosal border *(M)* and serosal border *(S)* using osmotic probes of radial size as indicated on the abscissa. The lines are fitted using the modified Renkin equation (Bean 1972). *Line on right,* serosal exit shows the best fit of the data to a single pore size 6.5 nm radius. *Line on left* a single pore size fit and two pore sizes 0.7 and 6.5 nm

Experimental Observations on Fluid Transport Across Small Intestine at 35°C

The above experimental observations were obtained at 20°C. At this low temperature, no active transport is observed in rabbit ileum, so all the flows are passive. At 35°C, with Ringer containing 25 nM D-glucose on both sides and gassed with 95% O_2 5% CO_2 on the mucosal side, we find that a steady-state inflow of 25 μl cm^{-2} h^{-1} across the mucosal border and 9 μl cm^{-2} h^{-1} exit across the serosal border.

Effect of Solutes Added to the Mucosal Solution on Inflow

Sucrose and NaCl

In Fig. 3, the effect of 50 mM sucrose on fluid inflow across the mucosal border is shown. Immediately following addition, inflow falls from 26 μl cm^{-2} h^{-1} to an outflow of 25 μl cm^{-2} h^{-1}. This is followed by a rapid recovery of inflow (half-time for recovery = 80 s) to 12 μl cm^{-2} h^{-1}. During this time outflow across the serosal border remains roughly constant at 9–10 μl cm^{-2} h^{-1}. In the presence of ouabain (0.1 mM), following addition of sucrose to the mucosal solution, the rate of decrease of outflow is much slower than in uninhibited tissue (Naftalin and Tripathi 1982c).

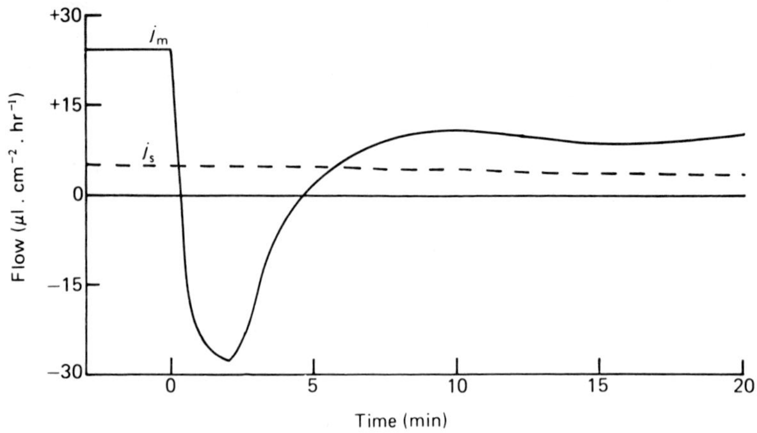

Fig. 3. Net mucosal (J_m) and serosal (J_s) flow were measured at 35°C, continuously as described in Fig. 1. This figure shows a typical record. Following a control period of absorption, the mucosal Ringer was rapidly made hypertonic at time zero. Positive flows are in the direction of mucosa to serosa; negative flows are in the opposite direction. Ringer contained D-glucose (25 mM)

Polyethylene Glycol 4000

Addition of 5 mM PEG 4000 to the mucosal solution is followed by much larger reduction in inflow per mM of solute added to the mucosal soltuion (Fig. 4). The change in inflow per mosmole is 500% larger than observed with sucrose. Recovery in inflow is not observed until PEG is removed from the mucosal solution.

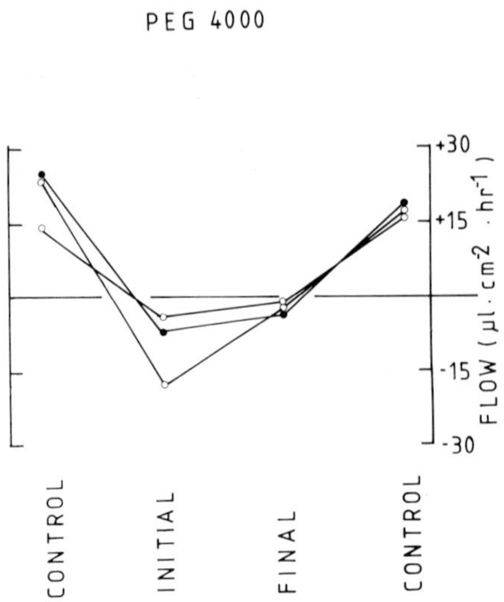

Fig. 4. Effect of addition of 5 mM polyethylene glycol (4000) to the mucosal bathing solution. Control flow is inflow across the mucosal border immediately before addition of PEG. Initial flow is flow across the mucosal border within 2 min of the change in mucosal solution. Final flow is the flow after 15 min of unchanging flow. The final control is the flow after removal of the PEG from the mucosal bathing solution

A New Model for Isotonic Transport

These data indicate that water flow across the mucosal border at 35°C is via channels of heterogeneous widths. Narrow transcellular channels can be demonstrated with solutes like glycerol.

The presence of intermediate channels can be deduced from the larger effects of solutes like NaCl and sucrose on flow and streaming potential across the mucosal border.

The evidence for wide channels is the effect of low concentrations of PEG 4000 on mucosal flow. No recovery in inflow is observed whilst PEG remains in the mucosal solution, indicating that solute polarization within the channels, which PEG affects, does not influence flow: i.e., the channels are too wide to reflect NaCl.

Recovery in inflow after addition of a hypertonic solution to the mucosal side arises from an increase in solute concentration within the LIS. This is due to water extraction from the LIS due to the hypertonicity of the mucosal solution. As the solute concentration within the LIS rises, outflow decreases. The rate of decrease of outflow depends on the ratio of the L_p of the mucosal border (transcellular and tight-junction routes in parallel) to the volume of the LIS. The volume of the LIS is determined within a narrow range (5%–10% of tissue volume). If the L_p were > 500 cm s^{-1} cm^{-1} H_2O, as Diamond predicted, then the half-time for recovery of inflow following addition of hypertonic sucrose, or NaCl to the mucosal solution would be less than 1 s. We observe the half-time of recovery is 80 s, which is only consistent with an L_p of the 10–30 cm s^{-1} cm^{-1} H_2O.

Solute Equilibration Within the Sub-Mucosa

Because the L_p of the tight-junction and mucosal border adjacent to the intracellular spaces, which permits concentration polarization of NaCl, is low, it follows that the solute concentration emerging from these spaces is hypertonic to the mucosal solution. However, the fluid pools in the submucosa, due to the hydraulic resistance of the serosal barrier. The hypertonicity of the submucosal fluid exerts a transcellular osmotic pressure, which generates a transcellular flow of water (the transcellular channels are very narrow) (Fig. 5). This transcellular water flow dilutes the submucosal fluid. With the transcellular L_p in the range $1–5 \times 10^{-9}$ cm s^{-1} cm^{-1} H_2O the fluid within the LIS is within 2% of the mucosal solution at steady-state over a wide range of mucosal solution concentrations (30–200 mM). Thus the concentration of the fluid emerging from the sub-mucosa into the serosal bathing solution, the transepithelial fluid flow, is nearly isotonic.

The wide mucosal shunt pathway serves to act as an escape route, limiting the size of the sub-mucosal compartment and oedema. Because the shunt is very wide, it could also allow solute equlibration between the mucosal and sub-mucosal compartments by diffusion in addition to the equilibration due to convective flows. However, in practice this does not seem to be a significant route for solute equilibration; isotonic transport can be supported by convective flows alone, without any significant net diffusion of solute across the mucosal or serosal borders.

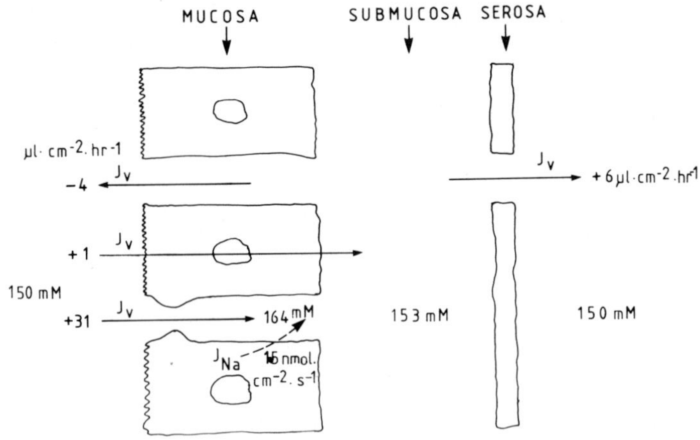

Fig. 5. This is a schematic diagram of our transport model for isotonic fluid movement in rabbit ileum. The figures next to the *arrows* J_v are the fluid flows $\mu l\ cm^{-2}\ h^{-1}$. J_{Na^+} is the pump rate n.osmoles $cm^{-2}\ s^{-1}$. The concentrations within the LIS and sub-mucosa are the steady-state concentrations within these compartments with the appropriate L_ps assinged as measured. The concentration of fluid emerging from the LIS is approximately 164 mM. Water, with solute removed, crosses via the transcellular route. The concentration of fluid emerging from the serosal solution is 2% above that in the mucosal bathing solution

Summary and Conclusions

Isotonic transport, defined as the near equality of the concentration of the fluid transported across an epithelium with the concentration of the solution bathing the mucosal surface of the epithelium, arises from active transport of salt into the lateral intercellular spaces.

The assumption that equilibration between the fluid in the lateral intercellular space and the mucosal solution is shown to be untenable. Fluid movement across the mucosal and serosal borders is measured continuously with a new high resolution method. The L_p of the tight-junction and mucosal border measured with this technique is low. Hence the concentration of fluid emerging from the LIS into the sub-mucosa must be substantially above that in the mucosal solution.

The sub-mucosa acts as fluid reservoir. Because there is a substantial resistance to fluid outflow across both the serosal and mucosal barriers, osmotic inflow across the mucosa leads to swelling of the sub-mucosa. This volume increase leads to an increase in interstitial pressure, which induces mass flow via the wide channels present in both the mucosal and serosal layers.

The tonicity of the sub-mucosal compartment is reduced by transcellular flow of water, induced by the transcellular osmotic gradient. The observed L_p values of the paracellular and tight-junctional routes are consistent with isotonic fluid movement.

Acknowledgement. The authors wish to thank the Medical Research Council for financial support.

References

Bean CP (1972) The physics of porous membrane-neutral pores. In: Eisenman G (ed) Membranes, vol 1. Dekker, New York

Curran PF (1960) Na, Cl, and water transport by rat ileum in vitro. J Gen Physiol 43:1137–1148

Diamond JM (1979) Osmotic water flow in leaky epithelia. J Membr Biol 51:195–216

Diamond JM, Bossert WH (1967) Standing gradient osmotic flow: a mechanism for coupling of salt and water transport in epithelia. J Gen Physiol 50:2061–2083

Garson MJ, Steward MC (1982) Water permeability of the epithelial cells of *Necturus* gallbladder. J Physiol (Lond) 326:44

Gupta B, Hall T, Naftalin RJ (1978) Microprobe measurement of Na, K and Cl concentration profiles in epithelial cells and intercellular spaces of rabbit ileum. Nature 272:70–73

Hakim AA, Lifson N (1969) Effects of pressure on water and solute transport by dog intestinal mucosa in vitro. Am J Physiol 216:276–284

Hill AE (1980) Salt-water coupling in leaky epithelia. J Membr Biol 56:177–182

Naftalin RJ, Tripathi S (1982a) A high resolution method for continuous measurement of trans-epithelial water movements across isolated sheets of rabbit ileum. J Physiol (Lond) 326:3–4

Naftalin RJ, Tripathi S (1982b) Determination of the hydraulic conductivities of the mucosal and serosal surfaces of the isolated rabbit ileum. J Physiol (Lond) 329:69

Naftalin RJ, Tripathi S (1982c) The effects of changing tonicity of the mucosal solution on fluid transport by isolated rabbit ileum. J Physiol (Lond) 332:112–113

Sackin H, Boulpaep E (1975) Models for coupling of salt and water transport. Proximal tubular reabsorption in *Necturus* kidney. J Gen Physiol 6:671–733

Smulders AP, Tormey J McD, Wright EM (1972) The effect of osmotically induced water flows on the permeability and ultrastructure of the rabbit gallbladder. J Membr Biol 7:164–197

Van Os CH, Weidner G, Wright EM (1979) Volume flows across gallbladder epithelium induced by small hydrostatic and osmotic gradients. J Membr Biol 49:1–20

Weidner G (1976) Method to detect volume flows in the nanoliter range. Rev Sci Instrum 47: 775–776

Weinstein AM, Stephenson JL (1981) Models of coupled salt and water transport across leaky epithelia. J Membr Biol 60:1–20

Wright EM, Smulders AP, Tormey J McD (1972) The role of the lateral intercellular spaces and solute polarization effects in the passive flow of water across rabbit gallbladder. J Membr Biol 7:198–219

Absorption of Inorganic Ions and Short-Chain Fatty Acids in the Colon of Mammals

W. v. ENGELHARDT and G. RECHKEMMER [1]

Introduction

Comprehensive reviews on transport of electrolytes across the colon epithelium have been published recently (Binder 1978, Phillips and Devroede 1979, Powell 1979, Schultz 1981a, Wrong et al. 1981). Therefore we shall try to concentrate mainly on topics that have been considered less extensively. We will emphasize comparative aspects of colonic function in mammals, and we will particularly discuss segmental differences in absorptive and secretory processes in the hindgut.

Anatomical Heterogeneity of the Hindgut

To obtain a better idea of the wide range of functions the enormous heterogeneity in anatomy and size of the hindgut in mammals has to be taken into account. Carnivores have a short and relatively simple large intestine, consisting of a small caecum and a non-sacculated, non-voluminous colon. Most of the herbivores and also many omnivores have a large and complex hindgut (Wrong et al. 1981). For satisfactory microbial digestion of fibre a sufficiently long retention time is required, either for the forestomach as in ruminants or for the lower gut. Hindgut fermenters have either the fermentation chamber in the caecum, like the rabbit and the guinea pig, or in the caecum and the colon, like equines. The relative size of the hindgut varies considerably in different species. Contents, as a percentage of body weight, range from about 0.5% in man and dog to 13% in elephant and horse (Fig. 1). The larger the hindgut, the better the conditions for extensive microbial fibre digestion. These differences in size also mean remarkable differences in epithelial surface areas. Therefore distinct species differences in absorption rates can be expected. Comparing the net water absorption in the hindgut of man with that of the pony (calculated on the basis of a body weight of 75 kg), the pony hindgut absorbs about 8 times more water in 24 h than does the hindgut of man (Fig. 2). In addition, the absorptive process in the pony hindgut varies along the colon, i.e., in the ventral colon 2 l of water are absorbed whereas in the voluminous dorsal colon substantial net secretion occurs, and in the small colon net water absorption takes place (Argenzio et al. 1974a).

1 Department of Physiology, School of Veterinary Medicine, D-3000 Hannover, Fed. Rep. of Germany

Intestinal Transport
(ed. by M. Gilles-Baillien and R. Gilles)

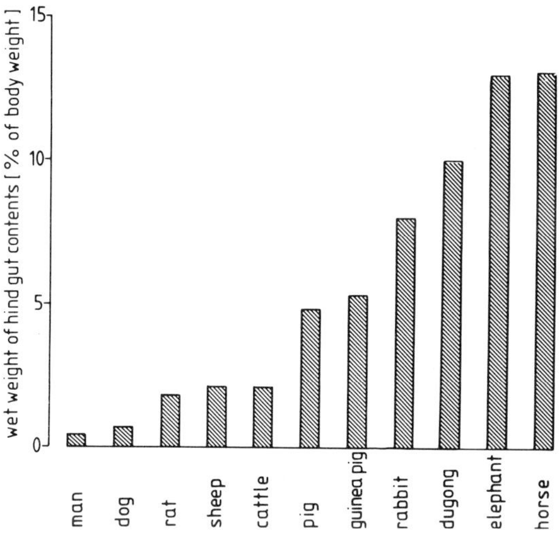

Fig. 1. Wet weight of hindgut contents as a percentage of body weight. (After Engelhardt and Rechkemmer 1983)

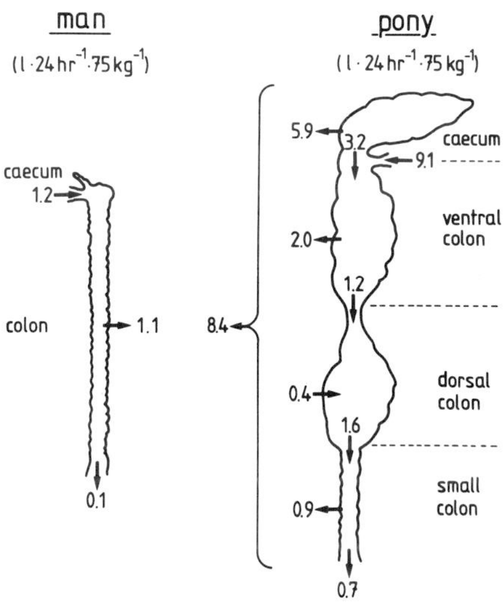

Fig. 2. Absorption and secretion of water in the hindgut of man and pony. (After Argenzio et al. 1974a)

Nomenclature of the Hindgut

In the past, textbooks as well as research articles have mostly treated the colon as a uniform, homogeneous organ. However, functions, transport capacities and also transport mechanisms can be rather different in the various sections along the hindgut. The hindgut has traditionally been divided into caecum, colon and rectum. The subdivisions of the colon are mainly defined, according to the anatomy of man, as an ascending, transverse and descending colon. This nomenclature is correct for man but not convenient for most animals because they do not have "ascending" or "descending" parts of the colon. Frequently, therefore, the terms proximal and distal colon are preferred. If a mid-section is present which is arranged transverse to the body, as in the guinea pig, the term transverse colon is appropriate. The rat and also the rabbit have no clear transverse colon; the region near the major flexure in the rat may be a short transverse colon (Fig. 3). The junction between the distal colon and the rectum is not well established; that is the reason why often distal colon and rectum were used for a confusing overlapping range in the large intestine. Alexander (1965) gave an unusual definition, he stated "the rectum of the rabbit and guinea pig was regarded as the part of the gut containing faecal pellets". In this case all of the distal colon and most of the transverse colon in the guinea pig would be rectum. To prevent further confusions a common nomenclature is urgently needed.

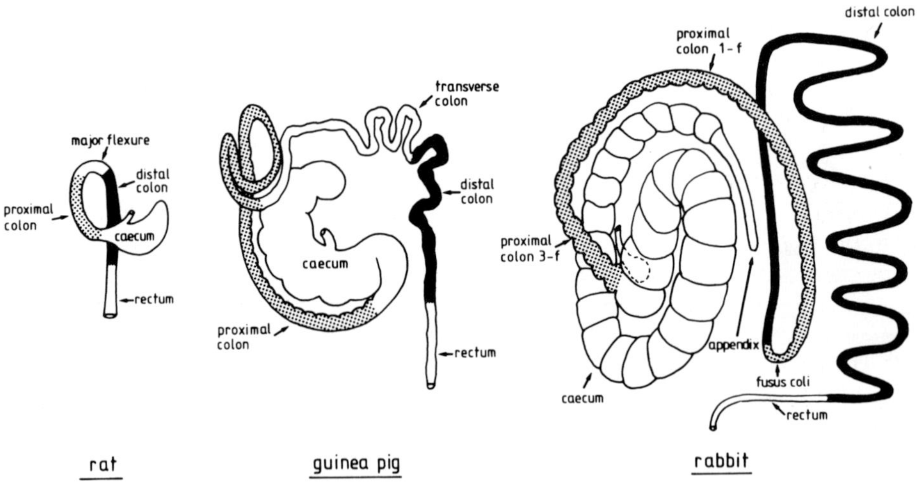

Fig. 3. Schematic drawing of the large intestine of rat, guinea pig and rabbit. Relations in size shown in the figure corresponds approximately to those in the adult animals. *1-f* region with one taenia, *3-f* three taeniae. (After Rechkemmer and Engelhardt 1982; we are grateful to Dr. Clauss, University Stuttgart-Hohenheim, for comments concerning the rabbit large intestine)

An example for the diversity of form and function along its proximo-distal axis is the rabbit colon. Based on macroscopic and also microscopic criteria Snipes et al. (1982) divided the proximal colon into three sections. The portion immediately distal to the caecum is endowed with three taeniae, the adjoining portion possesses only one taenia. The third portion, the fusus coli, is only 4 cm in length, is free of taeniae, but exhibits longitudinal folds on its inner surface. In the guinea pig the proximal colon is endowed with two taeniae and has a single row of haustra, whereas in the rat no taeniae nor haustra are present. The human colon has three taeniae and is haustrated in its whole length except in the sigmoid colon and in the rectum. In the first two portions of the rabbit colon the surface topography is characterized by wart-like protrusions, these are obviously effective enlargements of the surface area. The surface of the distal colon, in contrast to the proximal portions, shows macroscopically no surface specialization. Great differences in absorption and secretion in the sections of the rabbit colon have been shown by Clauss (1978).

Histology, Histochemistry and Ultrastructural Observations

The mucosa of the colon of adult mammals has no villi but exhibits numerous crypts. Goblet or vacuolated cells are present in the crypts but less occur at the surface where absorptive cells predominate. Thus in the colon not only segmental differences are present, even in the same colonic segment different cell populations occur in varying numbers. Wide paracellular spaces are often seen between epithelial cells at the surface of the rat colon (Specht 1977), and it is assumed that absorption takes place mainly at this surface region of the epithelium. So far only few comparative histological and histochemical studies have been done for the various colonic segments. In the guinea pig the number of crypts along the large intestine and also their length are increasing towards the rectum. In the rat, on the other hand, crypts are longer in the proximal and shorter in the distal colon (Kashgarian 1980).

Little attention has been given to the more or less continuous luminal mucin layer (LML) at the surface of the epithelial cells. The LML shows marked regional differences in its compactness, thickness, and histochemical composition (Table 1). In the proximal colon of mice, rat and guinea pig a LML is present, though spongy and not entirely continuous. Its thickness varies very much even in the same preparation. In the distal colon LML is thin but very compact and more homogeneous. The LML in the distal colon is so tight that even solid plant material obviously does not penetrate to the mucosa, whereas in the proximal colon particles sometimes even contact the mucosal surface.

Histochemically the mucin in the proximal colon of rats consists of neutral mucin and some sialo-mucin. In the distal colon the luminal mucin contains little neutral but mainly acid mucin, like sialo-mucin and sulfo-mucin (Sakata and Engelhardt 1981). The role of mucin in the large intestine is still a matter of debate.

Recent electronmicroscopic studies of zonulae occludentes between epithelial cells bordering the lumen of the guinea pig colon also indicate regional differences (Luciano et al. 1982). In the distal colon the zonulae occludentes have a compact appearance. The number of strands is higher in the distal colon, and the strands are more frequently interconnected. The possibility of paracellular movement of solutes, therefore, seems to be less likely in the distal colon compared to the proximal, at least in the guinea pig. In the proximal colon zonulae occludentes seem to be less compact, strands are arranged more often vertically, and fewer strands have to be passed by a penetrating substance (Fig. 4a,b).

Table 1. Characterization of the luminal mucin layer in mice, rat and guinea pig. (After Sakata and Engelhardt 1981)

	Proximal colon	Distal colon
Thickness		
Mice	182 ± 170 μm	34 ± 19 μm
Rat	151 ± 110 μm	16 ± 7 μm
Guinea pig	30 ± 29 μm	29 ± 20 μm
Histochemical composition	Mainly neutral glycoproteins	Mainly acidic glycoproteins
Compactness	Spongy	Compact

Results are expressed as means ± standard deviation

Fig. 4a

Fig. 4b

Solute Concentration and Production of Short-Chain Fatty Acids

In previous absorption studies the actual electrolyte concentration of normal colonic contents have rarely been taken into account. Estimations in a number of animals showed marked differences in the concentration of electrolytes between blood plasma and colonic fluid (Fig. 5). Mean concentration of sodium in colonic fluid is only half, and that of chloride is only one fifth that in the blood. Potassium concentration in the colon is about ten times, and anorganic phosphorus twenty times higher than in plasma. Short-chain fatty acids (SCFA) do exist in the blood only in very small quantities; SCFA are, however, the major anions in colonic fluid. SCFA are produced in the hindgut in cosiderable amounts by anaerobic microbial breakdown of polysaccharides. SCFA production in the hindgut may account for 7% of energy required for maintenance in dog and man with a small, simple large intestine. In animals with a voluminous fermentation chamber in the hindgut energy derived from SCFA comes up to 80% (Stevens et al. 1980, Cummings 1981). Despite of great differences in anatomy, size and production of SCFA the concentrations of SCFA in the hindgut contents are similar in most of the mammals studied so far. In most of these animals concentration in caecal contents is near 100 mmol/l. Furthermore, rather constant high SCFA concentrations are found throughout the entire length of the large bowel (Engelhardt and Rechkemmer 1983). Although SCFA are the predominant anions in digesta of the hindgut in mammals as well as in birds it is remarkable that studies of water and electrolyte transport in the colon have proceeded for more than 20 years, and yet, only recently attention has been given to the role of SCFA (Cummings 1981).

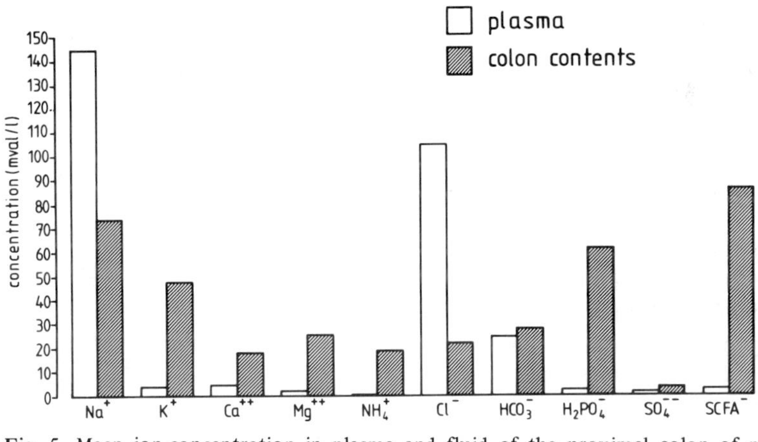

Fig. 5. Mean ion concentration in plasma and fluid of the proximal colon of mammals (data from various sources for several species)

◀ Fig. 4 a,b. Surface epithelium of the proximal (a) and distal (b) segments of the guinea pig colon. Freeze fracture replicas of the zonulae occludentes. In a the strands change their course from parallel to perpendicular to the luminal surface. Therefore, at very short distance many superimposed strands *(arrows)* are alternate by sites in which only 1 to 2 strands *(arrowheads)* separate the lumen from the intercellular space. In b strands are interconnected forming a network with polygonal meshes and a compact aspect. This network pattern is the same along the whole exposed zonula occludens. a × 76,800; b × 74,000. (Electron micrographs kindly provided by Profs. L. Luciano and E. Reale, Laboratory of Electron Microscopy, Hannover, Medical School)

Absorption of Electroytes and Short-Chain Fatty Acids

In the guinea pig colonic segments were perfused with a solution similar in chemical composition to that of normal colonic contents. Marked regional differences in solute net fluxes were observed (Table 2). Sodium and chloride absorption in the proximal colon were about four times that in the distal region. Luminal accumulation of bicarbonate was significantly higher in the proximal colon; this might be related to higher carbonic anhydrase activity in the proximal colon (Carter and Parsons 1972, Loennerholm 1977). Potassium was not absorbed in the proximal colon, whereas in the distal colon a remarkable absorption was observed. Acetate absorption was about 50% higher in the proximal than in the distal colon; for propionate and butyrate no significant differences were found.

Table 2. Net flux of solutes in the proximal and the distal colon of the guinea pig. Composition of the perfusion solution was (mM): Na 110, K 20, Mg 7.5, Ca 2.5, Cl 30, HCO_3 20, phosphate 30, acetate 60, propionate 10, butyrate 10; pH 6.1; adjusted to 300 mosmol l^{-1} by addition of mannitol. (After Rechkemmer and Engelhardt 1982)

Solute	Proximal colon (N = 8, n = 69) (mmol h^{-1} g^{-1} d.w.)	Distal colon (N = 4, n = 36) (mmol h^{-1} g^{-1} d.w.)
Sodium	4.0 ± 1.4	1.1 ± 0.5
Potassium	0.1 ± 0.3	0.9 ± 0.5
Chloride	2.0 ± 0.7	0.5 ± 0.4
Bicarbonate	– 0.7 ± 0.3	0.1 ± 0.5
Acetate	3.4 ± 1.3	2.3 ± 0.8
Propionate	0.6 ± 0.2	0.7 ± 0.2
Butyrate	0.7 ± 0.2	0.9 ± 0.3

Results are expressed as means ± standard deviation.
Negative values indicate secretion into the lumen, positive values absorption.
N = number of animals; n = number of samples

Our results for electrolyte absorption in the guinea pig agree with those of Edmonds and Thompson (1980) for the rat colon, and Devroede et al. (1971) for the colon of man, the absorptive capacity of the proximal colon being much higher than of the distal section. These findings were also supported by in vitro studies by Yau and Makhlouf (1975), who found that the distal colon of the rat absorbs significantly less sodium and that potassium was absorbed only in the distal but not in the proximal colon. Regional differences along the colon of the pony have been described by Argenzio et al. (1977). Clauss (1978) has demonstrated a significant diversity in absorption along the rabbit hindgut. Fromm and Hegel (1978) failed to reveal any difference between the proximal and the distal colon of the rat; however, these authors defined about 25%–35% of the large intestine, excluding the caecum, as rectum, and they found a significantly lower net absorption in this rectal section. Binder and Rawlins (1973), in the rat, as well as Argenzio and Whipp (1979) in the pig, did not find differences between the proximal and the distal colon.

SCFA are rapidly absorbed. Rough calculations indicate that 95%–99% of SCFA produced in the hindgut are absorbed. Their net movement is more rapid than net transport of sodium and chloride at equimolar concentrations. SCFA absorption is passive and increases linearly with the corresponding increase in concentration (Ruppin et al. 1980, McNeil 1982, Rechkemmer and Engelhardt 1982, Engelhardt and Rechkemmer 1983).

In order to compare the permeability of the mucosa for SCFA the clearance of the perfusate from SCFA was calculated; clearance is defined as net flux divided by the mean SCFA-concentration in the perfusate (Table 3). This gives the possibility to compare the permeability of different segments of the colon mucosa for SCFA. Findings show that (1) in the proximal colon chain length has only a minor effect on the absorption rate of SCFA; (2) in the distal colon clearance increases with increasing chain length, clearance is about doubled with each additional CH_2-group; and (3) acetate clearance is significantly higher in the proximal colon as compared to the distal colon. For butyrate the contrary was observed and for propionate no difference existed.

Table 3. Clearance of SCFA from the luminal solution of the proximal and distal colon of the guinea pig. The perfusion solution was the same as in Table 2. Clearance was calculated as netflux divided by the mean SCFA-concentration in the perfusate. (After Rechkemmer 1981)

Solute	Proximal colon (N = 6, n = 49) ($ml\,h^{-1}\,g^{-1}$ d.w.)	Distal colon (N = 5, n = 45) ($ml\,h^{-1}\,g^{-1}$ d.w.)
Acetate	48 ± 11	28 ± 14
Propionate	58 ± 12	65 ± 32
Butyrate	65 ± 12	102 ± 48

These findings indicate that in the proximal colon SCFA may be absorbed in the lipid-soluble as well as in the undissociated form. The larger the chain length, the better the lipid solubility and the higher the absorption rate of the unionized form. The less dense zonulae occludentes and the fewer strands in the proximal colon seem to enable a paracellular pathway in the proximal colon but not in the distal colon. The parallel channel model which Jackson et al. (1981) have suggested agrees with our observations in the proximal colon. In the distal colon a pH-partition model seems to give an appropriate explanation.

In normal colonic contents about 99% of SCFA are present as anions. Because the rapid transport of SCFA appears to occur mainly in the undissociated, lipid-soluble form an equimolar disappearance of H-ions would be required. Therefore a signficant increase in absorption rate should be expected if the pH in the luminal fluid is lowered. However, in the proximal colon of the guinea pig a decrease in the pH of the luminal solution within the range pH 5 and 9 had no, and in the distal colon only a small, effect on SCFA absorption rates. In the distal colon absorption rates increased indeed when the pH of the perfusate decreased; this increase was, however, much less than expected from the increase in undissociated SCFA (Rechkemmer 1981).

pH-Microclimate at the Surface of the Colonic Epithelium

The consequent question is: How can this unexpected independence of SCFA absorption from H-ion concentration in the perfusate be explained? The answer to this question is: The H-ion concentration at the surface of the luminal membrane is rather independent from the H-ion concentration in the bulk luminal fluid. With pH-microelectrodes H-ion activities were measured in the region near the luminal membrane; this region is called microclimate. In the proximal and in the distal colon of guinea pig and rat a rather stable neutral pH was measured (Rechkammer et al. 1979, McNeil and Ling 1980). The pH was constant near neutrality and independent from changes of the pH in the luminal solution (Fig. 6). The pH in the proximal colon (7.08 ± 0.14) was slightly higher than in the distal colon (6.91 ± 0.14) when the pH in the luminal solution was changed between pH 5 and 9 (Rechkemmer et al. 1979). An acid microclimate as proposed by Schanker (1959) for the rat colon was not confirmed in our experiments.

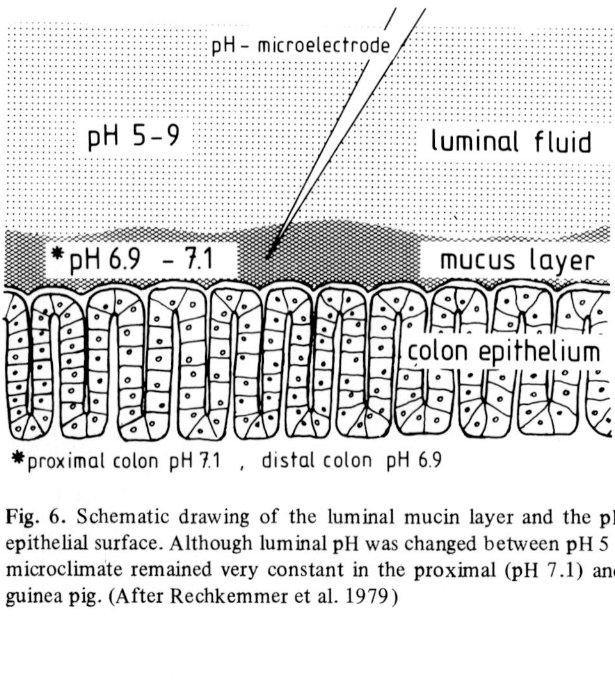

Fig. 6. Schematic drawing of the luminal mucin layer and the pH-microclimate at the luminal epithelial surface. Although luminal pH was changed between pH 5 and pH 9 H-ion activity in the microclimate remained very constant in the proximal (pH 7.1) and distal (pH 6.9) colon of the guinea pig. (After Rechkemmer et al. 1979)

Interrelationship Between SCFA Absorption and Bicarbonate Gain

A possibility to obtain H-ions for SCFA absorption would be a bicarbonate gain in the lumen. It has been shown in a number of studies that bicarbonate accumulates in the lumen of the colon in proportion to the amount of SCFA absorbed (horse: Argenzio et al. 1977; pig: Argenzio and Whipp 1979; sheep: Rübsamen and Engelhardt 1981; rat: Umesaki et al. 1979; guinea pig: Rechkemmer 1981; man: McNeil et al. 1978, Ruppin et al. 1980). Although the colon is exposed to high concentrations of

SCFA the pH in the lumen is maintained just below 7. Bicarbonate accumulation induced by SCFA might be considered an essential buffering system as well as a H-ion gain for SCFA absorption. Umesaki et al. (1979) (Fig. 4 in Engelhardt and Rechkemmer 1983) have demonstrated a close linear relationship between bicarbonate accumulation and SCFA absorption from the in situ perfused rat colon. The three SCFA were absorbed at the same rate, and the individual SCFA had a similar effect on bicarbonate accumulation. On a molar basis H-gain from bicarbonate accumulation seems to be equal to only one fifth of the number of SCFA molecules that were absorbed (Umesaki et al. 1979). No significant influence on chloride absorption was seen in these experiments with and without the presence of SCFA; the increased appearance of bicarbonate was a function of SCFA but not of chloride.

Interrelationship Between SCFA and Sodium Absorption

In the temporarily isolated colon of conscious goats (Argenzio et al. 1975) or conscious sheep (Rübsamen and Engelhardt 1981) a close relationship between SCFA absorption and sodium absorption was shown. A stimulatory effect of SCFA on sodium absorption has been confirmed in recent years in pig (Argenzio and Whipp 1979), rat (Parsons and Patterson 1965, Umesaki et al. 1979), man (Ruppin et al. 1980, Roediger and Moore 1981), and also in the caecum of domestic fowl (Rice and Skadhauge 1982). In a more detailed study in the guinea pig where colonic segments were perfused (Rechkemmer 1981) a significant relationship between sodium and SCFA absorption could be seen only in the proximal colon; no such influence was apparent in the distal colon (Fig. 7). Thus, the mechanism responsible for the inter-

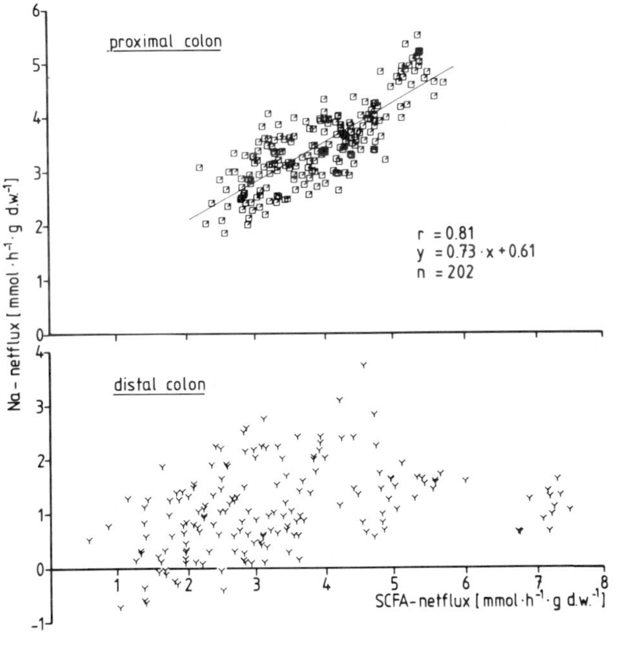

Fig. 7. Relationship between net absorption of sodium and of SCFA in the proximal and distal colon of the guinea pig. Findings are related to dry weight (d.w.) of stripped colonic mucosa. (After Engelhardt and Rechkemmer 1983)

relationship between sodium absorption and SCFA absorption seems to be located in the proximal part of the colon, at least in the guinea pig.

How can such an interrelationship between SCFA and sodium be explained?

1. The metabolism of SCFA in colonic epithelial cells could be an energy source for active sodium transport.
2. Sodium and/or SCFA may augment the Na-H exchange at the apical membrane.

SCFA as an Energy Source for Sodium Transport?

It is well documented that in the hindgut mucosa of guinea pig (Wirthensohn 1980), pony (Argenzio et al. 1974b), rabbit (Henning and Hird 1972), rat (Remesy and Demigne 1976), man (Roediger 1979, 1982) a substantial amount of SCFA are metabolized within the epithelial cells.

Sodium uptake into isolated colonocytes from the proximal colon of guinea pig was studied when acetate, propionate or butyrate were the only substrates for metabolism (Fig. 8). After an initial rapid ^{22}Na uptake, intracellular ^{22}Na concentration reached a plateau, indicating that the amount of sodium that had entered the cell was pumped out at the basolateral membrane in the presence of SCFA. ATP gained from SCFA metabolism may have been used for the active sodium transport. If no SCFA were present a gradual increase of ^{22}Na concentration in the cell was seen; if the sodium pump was inhibited by ouabain also a steady increase in intracellular ^{22}Na concentration occurred (Fig. 8) (Wirthensohn 1980).

Our findings in guinea pigs and also those of Roediger (1980) in human colonocytes seem to be evidence that SCFA metabolism could be an important supply of fuel for active Na transport in colonic epithelial cells. However, other arguments appear to disprove this hypothesis:

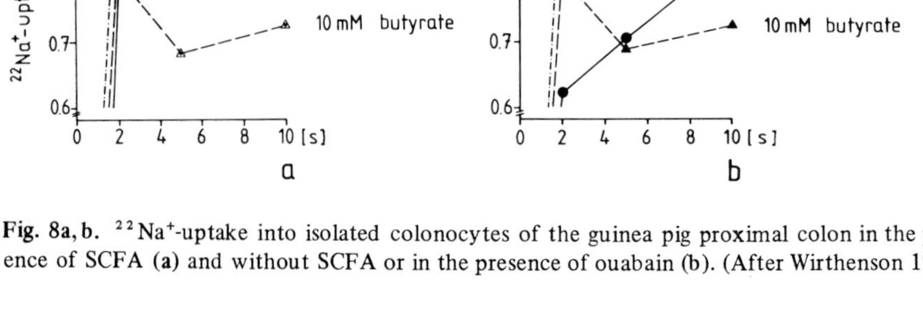

Fig. 8a, b. ^{22}Na^{+}-uptake into isolated colonocytes of the guinea pig proximal colon in the presence of SCFA (**a**) and without SCFA or in the presence of ouabain (**b**). (After Wirthenson 1980)

1. The three SCFA are metabolized in the colon epithelium at different rates (Wirthensohn 1980); all three SCFA have, however, a similar stimulatory effect on net sodium transport.

2. Rübsamen and Engelhardt (1981) achieved a decrease in net Na absorption by replacing sodium by lithium in the perfusate; this decrease in net sodium absorption diminished significantly also SCFA absorption, although SCFA concentrations were kept constant. This mutual interaction between sodium and SCFA absorption is the most conclusive argument against a metabolism hypothesis. It is believed that the Na-H exchange may play a fundamental role in this interrelationship, at least in the proximal colon of the guinea pig.

Na-H Exchange and Recycling of H as Undissociated SCFA

Na-H exchange is supposed to be located in the mucosal side of the membrane (Binder and Rawlins 1973). Thereby sodium is not only absorbed by an electrogenic system responsible for the generation of short circuit current but also by an electroneutral system of Na-H exchange which is dependent on exogenous substrates. The driving force for this Na-H exchange can be the removal of H-ions (Fig. 9). Epithelial transport of SCFA in the undissociated form, therefore, should result in an increased Na-H exchange, whereby the protons recycle from the cell across the luminal membrane. Conversely, these H-ions delivered from the exchange with Na can help to maintain a high rate of SCFA absorption in the proximal colon.

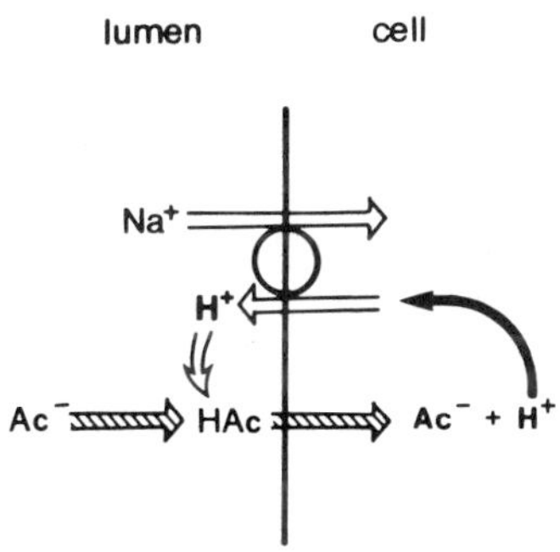

Fig. 9. Hypothesis for coupling between Na-H countertransport and SCFA absorption. H-ions are recycled by this mechanism

Experiments with the perfused total colon of conscious sheep where sodium was replaced by lithium indicated that net absorption of sodium and SCFA in the colon is closely linked with the availability of H-ions (Rübsamen and Engelhardt 1981). From these experiments in sheep is was estimated that 70% of the sodium may be transported by the electroneutral Na-H exchange and 30% by another system, probably electrogenic sodium transport. It is uncertain whether such an Na-H exchange exists also in the distal colon. Electrophysiological data, so far available, do not give clear indications.

Electrophysiological Data

Table 4 compiles some of the recent data from electrophysiological in vitro experiments for the rat and the rabbit. Great regional and marked species differences are apparent, and a considerable variability between respective data is seen. Research activities were focused nearly exclusively on the distal colon. In the rabbit only one, and in the rat only three studies are known for the proximal colon.

The confusing diversity of reported data does not allow a precise definition of the sodium transport mechanism. Most of the authors agree that in the distal colon of the rat a substantial proportion of the sodium absorption occurs as an electroneutral transport across the apical membrane. On the other hand, in the distal colon of the rabbit sodium transport seems to be entirely electrogenic. The very few electrophysiological experiments performed in the proximal colon of rat and rabbit do not allow a comparison.

The conductance of the rabbit distal colon is in general much lower than the conductance of the rat distal colon. Except for one paper (Sernka et al. 1982), in the proximal colon of rat and rabbit conductance was higher than in the distal colon. This may agree with our recent electronmicroscopic findings on the compactness of the tight junctions in the proximal and distal colon of the guinea pig (Luciano et al. 1982). In general conductance in the rat colon (proximal and distal segments) seems to be higher than in the rabbit.

In the rat distal colon the ψ^{ms} varies from 1.7 mV to 24.6 mV. This great variation is probably due to different techniques producing different degrees of edge-damage. For the rabbit distal colon much more consistent ψ^{ms} values are reported within a range from 10–22.7 mV. Yorio and Bentley (1977) found ψ^{ms} as high as –70 mV (Mucosal side negative) in unstripped preparations (these data are not included in Table 4).

In Table 5 some recent observations on the transport of potassium in colonic epithelium are shown. These conflicting results obtained by different groups do not allow any convincing conclusions on potassium transport across the colonic epithelium so far. In the in situ perfused guinea pig distal colon potassium is absorbed (Table 2) even at potassium concentrations as low as 5 mM in considerable amounts against the electrochemical gradient; this might indicate the presence of an active potassium absorption in this colonic segment.

In conclusion, the electrophysiological studies available so far make evident that:

1. Several parameters show an enormous variation.
2. Most studies were done on the distal and only a few in the proximal colon. Data indicate differences between the proximal and the distal colon.
3. Obviously marked species differences exist, at least between rats and rabbit.
4. Careful comparisons between colonic segments are urgently needed, with the same technique within and between species.

Table 4. Recent electrophysiological data for the proximal and the distal colon of rat and rabbit

	Proximal colon				Distal colon				Reference
	ψ^{ms}	G^t	I_{SC}	J_{Na}	ψ^{ms}	G^t	I_{SC}	J_{Na}	
1. Rat					− 8.9	9.9	3.2	6.5	Nell (1977)
						14.5	2.9	4.9	Racusen and Binder (1977)
					− 2.6	13.4	1.3	8.3	Binder (1978)
					− 3.1	14.5	1.7	5.4	Dharmsathaporn et al. (1980)
	− 1.8	24.2	1.6						Lee et al. (1980)
	− 2.6	18.9	1.9		− 4.9	12.0	2.3		Will et al. (1980)
					− 8.3	8.9	2.8		Favus et al. (1981)
					− 24.6	5.9	5.5	4.7	Goerg et al. (1981)
					− 1.7	19.0	1.1	6.7	Gazitua and Robinson (1982)
					− 15.0	7.2	3.9		Schulzke et al. (1982)
	− 5.7	3.6	0.7	5.0					Sernka et al. (1982)
					− 2.8	13.2	1.3	7.2	Zimmerman et al. (1982)
2. Rabbit					− 12.8	4.6	2.2	1.9	Frizzell (1977)
					− 13.0	4.9	2.2	3.0	Frizzell and Schultz (1978)
					− 20.0	3.5	2.6		Schultz (1979)
							1.1	1.2	Luger and Turnheim (1981)
	− 23.3	8.5	7.3	5.5	− 13.0	4.7	2.2	2.3	Clauss and Hoernicke (1982)
					− 15.4	6.1	3.5	5.1	McCabe et al. (1982)
					− 10.0	6.1	2.3		Thompson et al. (1982)
					− 22.7	3.4	2.9		Wills and Biagi (1982)

ψ^{ms} (transepithelial potential difference): mV;
G^t (tissue conductance): mS cm^{-2};
I_{SC} (short circuit current density): μmol cm^{-2};
J_{Na} (sodium netflux): μmol cm^{-2} h^{-1}

Table 5. Recently published papers concerning potassium transport in the colon

Reference	Animal	Segment	Method	Conclusions
Edmonds and Smith (1979)	Rat	Distal	In vivo perfusion	J_K^{sm} 80% through a paracellular K-selective pathway, small amount through a transcellular pool
Kliger et al. (1981)	Rat Rat	Proximal Distal	In vivo perfusion In vivo perfusion	Active secretion Active absorption
Frizzell et al. (1976)	Rabbit	Distal	In vitro SCC	Passive paracellular pathway, after addition of amphotericin B active K-secretion
Fromm and Schultz (1981)	Rabbit	Distal	In vitro SCC	Non-selective paracellular pathway "free solution shunt", concentration-dependent K-selective pathway
Moreto et al. (1981)	Rabbit	Distal	In vitro SCC	K-Cl-cotransport or K-HCO_3-cotransport, no electrogenic transport system, amiloride no effect, apical membrane permeability activated by calcium
Schultz (1981b)	Rabbit	Distal	In vitro SCC	Transepithelial movement of potassium restricted to paracellular route
McCabe et al. (1982)	Rabbit	Distal	In vitro SCC	Active secretion and active absorption possible, paracellular pathway not potassium selective
Wills and Biagi (1982)	Rabbit	Distal	In vitro SCC	Active secretion and active absorption possible, two separate transport systems linked to the metabolic activity of the cell

Summary and Conclusions

In most mammals the hindgut is not a homogeneous uniform organ. The anatomical complexity of the large intestine varies greatly between animals. Functions, transport capacities and transport mechanisms can be rather different in different sections of the colon.

In most mammals studied so far the absorptive capacity for electrolytes and SCFA is considerably higher in the proximal compared to the distal colon. SCFA are the major anions in colonic contents. They are rapidly absorbed. Absorption is rather independent from the pH in the luminal fluid. A constant pH microclimate at the surface of the colonic epithelium seems to be responsible for the pH independent absorption rates of SCFA. SCFA stimulate sodium absorption in the proximal but not in the distal colon, at least in the guinea pig. Bicarbonate accumulates in the lumen of the colon in proportion to SCFA-absorption.

Absorption data and electrophysiological observations give hints on differences in transport mechanisms between the proximal and the distal colon. Our findings on sodium transport in the guinea pig may fit into a model shown in Fig. 10. In the proximal colon a considerable amount of sodium seems to be absorbed by an electroneutral system, probably a Na-H-exchange. In the distal colon the electrogenic transport predominates, and an electrogenic potassium transport may also exist. The distal, not the proximal colon of the guinea pig, is amiloride sensitive. Amphotericin B increases net sodium transport in the distal, not in the proximal colon (Rechkemmer, unpublished.

The data so far available indicate that the distal sections of the hindgut are more tight than the proximal ones. However, in electrophysiological terms both epithelia may be called "leaky".

In Fig. 11 a model proposed by Argenzio et al. (1977) and Ruppin et al. (1980) has been modified for SCFA absorption in the proximal and in the distal colon

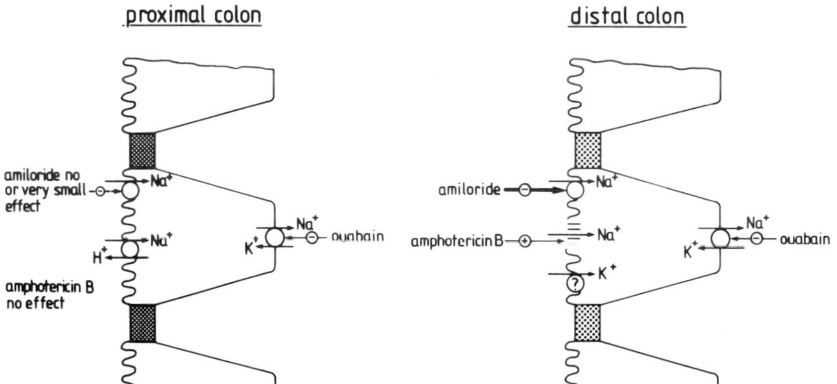

Fig. 10. Proposed model for sodium transport in the proximal and the distal colon of the guinea pig. ⊖ inhibition, ⊕ stimulation

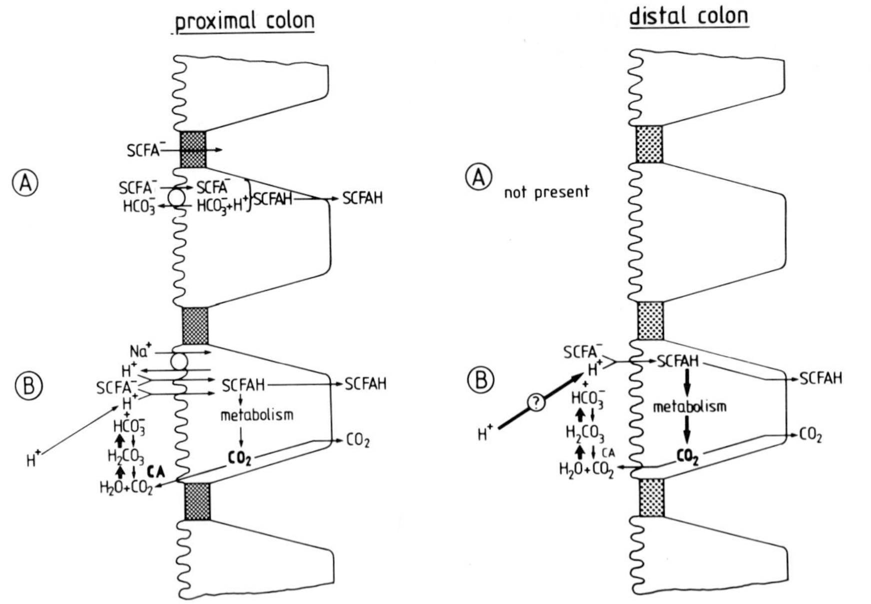

Fig. 11. Proposed model for absorption of SCFA in the ionized *(A)* or unionized *(B)* form in the proximal and in the distal colon

according to our findings. In the proximal colon some of the SCFA seem to be transported in the ionized form via a paracellular pathway or as a SCFA-bicarbonate exchange mechanism (A). SCFA are however more easily absorbed in the unionized form. H-ions needed for this rapid absorption of SCFA may be available from (1) the Na-H-exchange, (2) from the bicarbonate gain in the lumen and (3) from the bulk solution in the lumen (B).

For the proximal colon this model gives a satisfactory explanation for the H-ions needed for SCFA absorption. In the distal colon SCFA are not absorbed in ionized form; no interrelation between sodium transport and SCFA absorption was seen. The origin of most of the H-ions needed for SCFA absorption in the distal colon is not known.

Acknowledgement. The work from the authors' laboratory was supported by a grant of the Deutsche Forschungsgemeinschaft (DFG En 65/12).

References

Alexander F (1965) The concentration of electrolytes in the alimentary tract of the rabbit, guinea pig, dog and cat. Res Vet Sci 6:238–244

Argenzio RA, Lowe JE, Pickard DW, Stevens CE (1974a) Digesta passage and water exchange in the equine large intestine. Am J Physiol 226:1035–1042

Argenzio RA, Miller N, Engelhardt W v (1975) Effect of volatile fatty acids on water and ion absorption from the goat colon. Am J Physiol 229:997–1002

Argenzio RA, Southworth M, Lowe JE, Stevens CE (1977) Interrelationship of Na, HCO_3, and volatile fatty acid transport by equine large intestine. Am J Physiol 233:E469–E478

Argenzio RA, Southworth M, Stevens CE (1974b) Sites of organic acid production and absorption in the equine gastrointestinal tract. Am J Physiol 226:1043–1050

Argenzio RA, Whipp SC (1979) Inter-relationship of sodium, chloride, bicarbonate and acetate transport by the colon of the pig. J Physiol 295:365–381

Binder HJ (1978) Na and Cl transport across colonic mucosa in the rat. In: Hoffmann JF (ed) Membrane transport processes, vol 1. Raven Press, New York, pp 309–330

Binder HJ, Rawlins CL (1973) Electrolyte transport across isolated large intestinal mucosa. Am J Physiol 225:1232–1239

Carter MJ, Parsons DS (1972) The isoenzymes of carbonic anhydrase: kinetic properties with particular reference to the function in the intestinal tract. J Physiol 220:465–478

Clauss W (1978) Resorption und Sekretion von Wasser und Elektrolyten im Colon des Kaninchens im Zusammenhang mit der Bildung von Weichkot und Hartkot. Thesis, University Stuttgart-Hohenheim

Clauss W, Hoernicke H (1982) Ion transport and electrical properties in different parts of the rabbit colon in vitro. In: European Society for Comparative Physiology and Biochemistry (ed) Intestinal transport: fundamental and comparative aspects. Abstract volume of the 4th conference, pp 9–10

Cummings JH (1981) Short-chain fatty acids in the human colon. Gut 22:763–779

Devroede GJ, Phillips SF, Code CF, Lind JF (1971) Regional differences in rates of insorption of sodium and water from the human large intestine. Can J Physiol Pharmacol 49:1023–1029

Dharmsathaporn K, Racusen L, Dobbins JW (1980) Effect of somatostatin on ion transport in the rat colon. J Clin Invest 66:813–820

Edmonds CJ, Smith T (1979) Epithelial transport pathways of rat colon determined in vivo by impulse response analysis. J Physiol 296:471–485

Edmonds CJ, Thompson BD (1980) Absorption by the colon during prolonged infusions in conscious, unrestrained rats. J Physiol 302:399–409

Engelhardt W v, Rechkemmer G (1983) The physiological effects of short-chain fatty acids in the hind gut. In: Royal Society of New Zealand (ed) Dietary fibre in human and animal nutrition (in press)

Favus MJ, Kathpatlia SC, Coe FL (1981) Kinetic characteristics of calcium absorption and secretion by rat colon. Am J Physiol 240:G350–G354

Frizzell RA (1977) Active chloride secretion by rabbit colon: calcium-dependent stimulation by ionophore A23187. J Membr Biol 35:175–187

Frizzell RA, Koch JM, Schultz SG (1976) Ion transport by rabbit colon. I. Active and passive components. J Membr Biol 27:297–316

Frizzell RA, Schultz SG (1978) Effect of aldosterone on ion transport by rabbit colon in vitro. J Membr Biol 39:1–26

Fromm M, Hegel U (1978) Segmental heterogeneity of epithelial transport in rat large intestine. Pflügers Arch 378:71–83

Fromm M, Schultz SG (1981) Potassium transport across rabbit descending colon in vitro: evidence for single-file diffusion through a paracellular pathway. J Membr Biol 63:93–98

Gazitua S, Robinson JWL (1982) Ion fluxes and electrical characteristics of the short-circuited rat colon in vitro. Pflügers Arch 394:32–37

Goerg KJ, Wanitschke R, Soergel KH, Wood CM, Nell G (1981) The short-circuited everted sac of rat colon mucosa. Res Exp Med 178:247–256

Henning SJ, Hird FJR (1972) Transport of acetate and butyrate in the hind-gut of rabbits. Biochem J 130:791–796

Jackson MJ, Tai CY, Steane JE (1981) Weak electrolyte permeation in alimentary epithelia. Am J Physiol 240:G191–G198

Kashgarian M (1980) Changes in cell membrane surfaces associated with alterations of transepithelial ion movement. In: Boulpaep EL (ed) Current topics in membranes and transport, vol 13. Academic Press, pp 149–160

Kliger AS, Binder HJ, Bastl C, Hayslett JP (1981) Demonstration of active potassium transport in the mammalian colon. J Clin Invest 67:1189–1196

Lee DBN, Walling MW, Gafter U, Silis V, Coburn JW (1980) Calcium and inorganic phosphate transport in rat colon. J Clin Invest 65:1326–1331

Loennerholm G (1977) Carbonic anhydrase in the intestinal tract of the guinea-pig. Acta Physiol Scand 99:53–61

Luciano L, Rechkemmer G, Engelhardt W v, Reale E (1982) Die Zonulae occludentes des Colonepithels. Physiologische und morphologische Unterschiede zwischen proximalem und distalem Colon beim Meerschweinchen. Verh Anat Ges 77: (in press)

Luger A, Turnheim K (1981) Modification of cation permeability of rabbit descending colon by sulphydryl reagents. J Physiol 317:49–66

McCabe R, Cooke HJ, Sullivan LP (1982) Potassium transport by rabbit descending colon. Am J Physiol 242:C81–C86

McNeil NI (1982) Human large-intestinal absorption of short-chain fatty acids. In: Kasper H, Goebell H (eds) Colon and nutrition. MTP Press, Lancaster, pp 55–59

McNeil NI, Cummings JH, James WPT (1978) Rectal absorption of short chain fatty acids in the absence of chloride. Gut 20:400–403

McNeil NI, Ling KLE (1980) The mucosal surface pH of the large intestine. Gastroenterology 78: 1220

Moreto M, Planas JM, Naftalin RJ (1981) Effects of secretagogues on the K^+ permeability of mucosal and serosal borders of rabbit colonic mucosa. Biochim Biophys Acta 648:215–224

Nell G (1977) Elektrogener und neutraler Natriumtransport am Rattendarm. Analyse der gegenseitigen Abhängigkeit mit Hilfe von Inhibitoren. Habilitationsschrift, University Saarland, Homburg

Parsons DS, Patterson CF (1965) Fluid and solute transport across rat colonic mucosa. Quart J Exp Physiol 50:220–231

Phillips SF, Devroede GJ (1979) Function of the large intestine. Int Rev Physiol 19:263–290

Powell DW (1979) Transport in large intestine. In: Giebisch G, Tosteson DC, Ussing HH (eds) Membrane transport in biology, vol IV B. Springer, Berlin Heidelberg New York, pp 779–809

Racusen LC, Binder HJ (1977) Alteration in large intestinal electrolyte transport by vasoactive intestinal polypeptide in the rat. Gastroenterology 73:790–796

Rechkemmer G (1981) Resorption and Sekretion von Elektrolyten und kurzkettigen Fettsäuren im proximalen und distalen Colon des Meerschweinchens und die Bedeutung des wandnahen Mikroklimas. Thesis, University Stuttgart-Hohenheim

Rechkemmer G, Engelhardt W v (1982) Absorptive processes in different colonic segments of the guinea pig and the effects of short-chain fatty acids. In: Kasper H, Goebell H (eds) Colon and nutrition. MTP Press, Lancaster, pp 61–67

Rechkemmer G, Wahl M, Kuschinsky W, Engelhardt W v (1979) pH-microclimate at the surface of the intestine in guinea pig and rat. Pflügers Arch 382:R31

Remesy C, Demigne C (1976) Partition and absorption of volatile fatty acids in the alimentary canal of the rat. Ann Rech Vet 7:39–45

Rice GE, Skadhauge E (1982) Caecal water and electrolyte absorption and the effects of acetate and glucose in dehydrated, low-NaCl diet hens. J Comp Physiol B 147:61–64

Roediger WEW (1979) Nutrition of the colonic mucosa. In: Truelove SC, Willoughby CP (eds) Topics in gastroenterology, vol 7. Blackwell Sci Publ, Oxford, pp 281–287

Roediger WEW (1980) Role of anaerobic bacteria in the metabolic welfare of the colonic mucosa in man. Gut 21:793–798

Roediger WEW (1982) The effect of bacterial metabolites on nutrition and function of the colonic mucosa. Symbiosis between man and bacteria. In: Kasper H, Goebell H (eds) Colon and nutrition. MTP Press, Lancaster, pp 11–26

Roediger WEW, Moore A (1981) Effect of short-chain fatty acid on sodium absorption in isolated human colon perfused through the vascular bed. Dig Dis Sci 26:100–106

Rübsamen K, Engelhardt W v (1981) Absorption of Na, H ions and short-chain fatty acids from the sheep colon. Pflügers Arch 391:141–146

Ruppin H, Bar-Meir S, Soergel KH, Wood CM, Schmitt MG Jr (1980) Absorption of short-chain fatty acids by the colon. Gastroenterology 78:1500–1507

Sakata T, Engelhardt W v (1981) Luminal mucin in the large intestine of mice, rats and guinea pigs. Cell Tissue Res 219:629–635

Schanker LS (1959) Absorption of drugs from the rat colon. J Pharmacol Exp Ther 126:283–290

Schultz SG (1979) Application of equivalent electrical circuit models to study of sodium transport across epithelial tissues. Fed Proc 38:2024–2029

Schultz SG (1981a) Ion transport by mammalian large intestine. In: Johnson LR (ed) Physiology of the gastrointestinal tract, vol 2. Raven Press, New York, pp 991–1002

Schultz SG (1981b) Potassium transport by rabbit descending colon in vitro. Fed Proc 40:2408–2411

Schulzke JD, Fromm M, Lempart U, Hegel U (1982) Effect of stripping on electrical parameters of rat rectum in vitro. In: European Society for Comparative Physiology and Biochemistry, Intestinal Transport: Fundamental and comparative aspects. Abstr vol of the 4th conference, pp 38–39

Sernka TJ, Rood RP, Mah MY, Tseng CH (1982) Antiabsorptive effects of 16,16 dimethyl prostaglandin E_2 in isolated rat colon. Prostaglandins 23:411–432

Snipes RL, Clauss W, Weber A, Hörnicke H (1982) Structural and functional differences in various divisions of the rabbit colon. Cell Tissue Res 225:331–346

Specht W (1977) Morphology of the intestinal wall. In: Kramer M, Lauterbach F (eds) Intestinal permeation. Excerpta Medica, Amsterdam Oxford, pp 4–40

Stevens CE, Argenzio RA, Clemens ET (1980) Microbial digestion: Rumen versus large intestine. In: Ruckebusch Y, Thivend P (eds) Digestive physiology and metabolism in ruminants. MTP Press, Lancaster, pp 685–706

Thompson SM, Suzuki Y, Schultz SG (1982) The electrophysiology of rabbit descending colon. I. Instantaneous transepithelial current-voltage relations and the current-voltage relations of the Na-entry mechanism. J Membr Biol 66:41–54

Umesaki Y, Yajima T, Yokohura T, Mutai M (1979) Effect of organic acid absorption on bicarbonate transport in rat colon. Pflügers Arch 379:43–47

Will PC, Lebowitz JL, Hopfer U (1980) Induction of amiloride-sensitive sodium transport in the rat colon by mineralocorticoids. Am J Physiol 238:F261–F268

Wills NK, Biagi B (1982) Active potassium transport by rabbit descending colon epithelium. J Membr Biol 64:195–203

Wirthensohn K (1980) Der Stoffwechsel kurzkettiger Fettsäuren im Colonepithel des Meerschweinchens und seine Bedeutung für die Natriumresorption. Thesis, University Stuttgart-Hohenheim

Wrong OM, Edmonds CJ, Chadwick VS (1981) The large intestine. Its role in mammalian nutrition and homeostasis. MTP Press, Lancaster

Yau WM, Makhlouf GM (1975) Comparison of transport mechanisms in isolated ascending and descending rat colon. Am J Physiol 228:191–195

Yorio T, Bentley PJ (1977) Permeability of the rabbit colon in vitro. Am J Physiol 232:F5–F9

Zimmerman TW, Dobbins JW, Binder HJ (1982) Mechanism of cholinergic regulation of electrolyte transport in rat colon in vitro. Am J Physiol 242:G116–G123

Cellular Aspects of Amino-Acid Transport

M.W. SMITH, F.V. SEPÚLVEDA and J.Y.F. PATERSON [1]

Introduction

The object of the present review is to describe the characteristics of amino acid transport studied at the cellular level. Such studies have been carried out using enterocytes isolated from villi; they have also involved transport studies carried out on enterocytes cultured from the intestine of germ-free animals. More recently, studies have also been carried out using whole tissue, enterocytes involved in amino acid transport being identified and analysed subsequently using quantitative autoradiography.

The reasons why one should study amino acid transport at the cellular level, the increase in quality of questions that can be asked using these systems and the potential they have for finding out how the expression of amino acid transport might be regulated in the intestine, are emphasised. Attention is also paid to the limitations encountered when working at this level of organisation.

Transport into Isolated Enterocytes

Enterocytes can be obtained from intestinal villi using a variety of techniques, most of which impair subsequent viability (Kimmich 1975). Measurements of amino acid transport into such preparations are likely to include a variable fraction of diffusional entry and this limits their usefulness as models for research. In spite of this it has been possible to show, even with EDTA or citrate-treated preparations, a certain amount of Na-dependent amino acid accumulation in cell cytoplasm (Søgnen 1967, Reiser and Christiansen 1971). The best conditions obtained so far for preparing enterocytes appear to be those described for chick intestine by Kimmich (1975). In this method tissue is incubated with hyaluronidase only. It is then possible to use these preparations to measure the initial rate of amino acid uptake and study the Na-dependence and cross inhibition that occurs between different amino acids. The results of one such experiment are shown in Fig. 1.

About 60% of total valine uptake (from a concentration of 1 mM) is Na-dependent and about 90% of the uptake is inhibited in the presence of 25 mM leucine. The transport of valine by isolated chick enterocytes has been shown subsequently to be

1 A.R.C. Institute of Animal Physiology, Babraham, Cambridge CB2 4AT, Great Britain

Intestinal Transport
(ed. by M. Gilles-Baillien and R. Gilles)

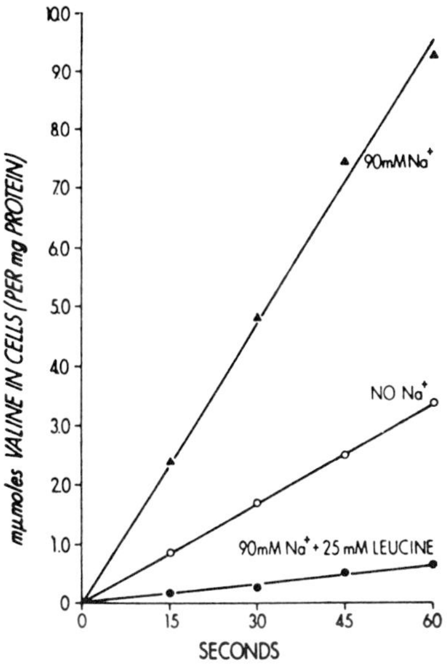

Fig. 1. Na-dependence and inhibition by leucine of the uptake of valine by chick enterocytes. Valine concentration was 1 mM throughout. (Kimmich 1975)

coupled to that of Na in a 1:1 ratio (Shapiro and Kimmich 1981). This work confirms our own results showing that several neutral amino acids are transported across the mucosal border of rabbit ileal enterocytes with a stoichiometry of 1 (Paterson et al. 1980a).

All the work quoted above uses a mixed population of enterocytes obtained from all parts of the villus. It is also possible, using a modification of the chelator technique, to obtain enterocytes from the villus and crypt according to age (Weiser 1973). Using this technique it has been possible to show that several neutral amino acids are taken up preferentially by villus tip enterocytes (Garvey et al. 1976). Whether or not this reflects a true functional difference in the capacity to take up amino acids is not certain in these experiments, since the treatment used to obtain crypt cells (67 min at 37°C in chelator) is far more traumatic than that used to obtain enterocytes from the tips of villi.

The recent establishment of different cell lines from rat intestine exhibiting some of the immunological and morphological characteristics of undifferentiated enterocytes, provides an alternative way to study the characteristics of amino acid transport in crypt cells (Inui et al. 1980). These cells take up alanine at 1 mM through a Na-dependent process; in contrast leucine uptake shows no dependence on the presence of Na. These characteristics are very similar to those reported recently for a wide variety of cells in culture (Gazzola et al. 1980, Bass et al. 1981, Handlogten et al. 1981, Sepúlveda and Pearson 1982).

What have been the real advantages of using isolated enterocytes up till now and what are the likely advances that might be made in the immediate future? The above results with mature villus enterocytes appear mainly to confirm findings obtained earlier using preparations of whole tissue (much the same may be said for a lot of the work published using membrane vesicles). Nothing, however, was known about the transport properties of crypt cells until they were isolated and this could prove an interesting topic to study in the future, particularly if it became possible to initiate differentiation in vitro. Isolating enterocytes also allows the experimenter to characterise amino acid transport across the basolateral as well as the brush border membrane, thereby relating amino acid uptake to other events taking place within the cell. Initial experiments carried out on isolated chick enterocytes have, for instance, already shown that influx of K does not change when Na influx is increased in the presence of alanine (Sepúlveda et al. 1982). It is anticipated that these and related lines of research will be followed using isolated enterocytes in the future.

Transport into Intact Tissue

To describe amino acid transport into whole tissue in an article concerned only with cells would seem a paradox. Yet it is true that when experimental conditions are chosen with care it becomes possible to characterise transport taking place across a single membrane of the enterocyte. The original idea for studying unidirectional movement across the brush border membrane of rabbit enterocytes came from the work of Curran et al. (1967). Isotope was presented to the mucosal surface only for periods of up to 1 min using flat sheets of intestine. Under these conditions only a small amount of amino acid has time to cross the basolateral membrane (Paterson et al. 1982a). An extracellular space marker is included with the amino acid to enable a correction to be made for the small amount of material adhering to the luminal surface (about 5% of total counts using 1 mM alanine in the presence of Na). Using this approach it has been possible to define the characteristics of neutral, basic and acidic amino acid entry into mammalian enterocytes (see the review by Munck 1981).

The most recent description to emerge of how basic and neutral amino acids enter enterocytes has been based on experiments combining rapid uptake measurements with computer analysis of kinetic data, emphasis being placed on the description of how selected amino acids inhibit as well as enter the transporting enterocytes. A summary of these results, obtained for three different neutral amino acids, is shown in Fig. 2.

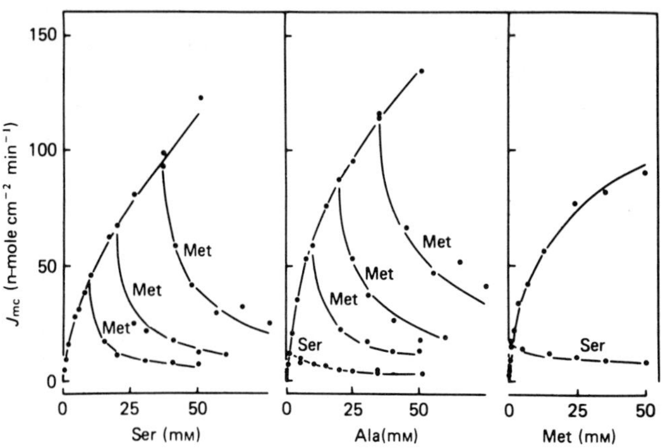

Fig. 2. Amino acid uptake by rabbit ileal enterocytes. Curves were derived by simultaneous analysis of all data points using a subroutine of the MINIM program assuming two transport systems of mediated entry. (Paterson et al. 1979)

Curves drawn in Fig. 2 have been based on a simultaneous analysis of 83 data points, assuming two carrier systems to mediate amino acid entry. Consideration of simpler solutions such as a single transport system or the presence of a large diffusional entry, were not acceptable (Paterson et al. 1979). Constants describing these systems are compared with those obtained using other techniques in Table 1.

Table 1. K_m values for amino acid transport into rabbit ileal enterocytes

	a Na present		b Serine inhibition experiments	c Na absent	d Electrical effect
	System 1	System 2			
Serine	6.3	(1303)	3.2	89	6.3
Alanine	4.6	77	0.7	75	5.0
Methionine	0.4	22	0.3	23	1.2

a, b, c and d refer to work published by Paterson et al. (1979), Sepúlveda and Smith (1978), Paterson et al. (1980b) and James and Smith (1980) respectively

There was generally good agreement between the high affinity constants determined by computer analysis (System 1) and those calculated from serine inhibition experiments and experiments where the concentration-dependence of amino acid-induced changes in short circuit current has been measured. When the kinetics for amino acid uptake were determined in the absence of Na, they could be described as taking place on single carrier having K_m-values similar to those determined for System 2. This has led to the proposal that System 1 is fully dependent on the presence of Na and that System 2 is Na-independent (Paterson et al. 1980b).

The notation used to describe these transport systems for amino acids was originally left vague because of the difficulty involved in reconciling all kinetic properties with known transport systems and an unwillingness to invent new ones unless that became necessary. Parallel studies of multi-transport systems for amino acid entry into rat intestinal mucosa suggested that the Na-dependent fraction enters on an A and an ASC system and that Na-independent entry is through an L system (Kilberg et al. 1981). It was further suggested that our results, obtained for Na-dependent transport in the rabbit ileum, could be explained by postulating the presence of an ASC system identical to that characterised in hepatocytes (Kilberg et al. 1981). This seems unlikely since the preferred substrates for System 1 in rabbit enterocytes are the hydrophobic bulky-residue neutral amino acids while for an ASC system the preferred substrates are alanine, serine, cysteine and α-amino-n-butyric acid (ANB). We have, nevertheless, carried out further experiments using cysteine to test directly for the presence of an ASC system in rabbit enterocytes. The results obtained are shown in Table 2.

The most effective inhibitors of cysteine uptake in the presence or absence of Na, are methionine and leucine; the three preferred substrates for an ASC system (ANB, alanine and serine) were less effective. The supposedly specific inhibitors of systems A and L (MeAIB and BCH respectively) were without effect on cysteine uptake.

Table 2. Cysteine uptake by rabbit ileal enterocytes

Inhibitor	Uptake ($nmol\ min^{-1}\ cm^{-2}$)	
	+ Na	– Na
None	9.5 ± 1.2	4.8 ± 0.8
ANB	5.0 ± 0.9**	2.6 ± 0.8**
Ala	7.9 ± 1.1*	3.2 ± 0.6**
Met	3.9 ± 1.1**	1.4 ± 0.4**
Leu	2.2 ± 0.5**	1.8 ± 0.4**
None	13.3 ± 2.6	5.4 ± 0.5
MeAIB	13.8 ± 2.5	6.1 ± 0.6
BCH	15.2 ± 3.2	5.1 ± 0.4
Gly	11.4 ± 1.6	5.8 ± 0.5
Ser	6.7 ± 1.0*	4.4 ± 0.4*

Results are means ± S.E. of experiments performed in 6 rabbits. Cysteine was used at a concentration of 0.5 mM in the presence of 10 mM DL-dithiothreitol. Inhibitor concentrations, 10 mM. * $P < 0.05$; ** $P < 0.01$ by paired t-test. ANB, α-amino-n-butyric acid; MeAIB, α-methylamino-isobutyric acid; BCH, 2-aminonorbornane-2-carboxylic acid

These characteristics are sufficiently different from those expected for A, ASC and L systems to justify their separate classification (Sepúlveda and Smith 1978). These differences, which have also been confirmed using brush border membrane vesicles (Stevens et al. 1982), could be connected with the specific function of enterocytes in transferring amino acids from lumen to blood.

An analysis similar to that described above for neutral amino acids has also been carried out for lysine using cross-inhibition experiments with alanine. A summary of these results is given in Table 3.

Table 3. Lysine uptake by rabbit ileal enterocytes

Kinetic constants	Lys transport/ala inhibitor		Ala transport/lys inhibitor	
	Basic/neutral system	Basic system	Basic/neutral system	System 2
K_m (mM)	1.0	108	14.2	115
K_i (mM)	14.3	∞	0.96	∞
J_{max} ($nmol\ min^{-1}\ cm^{-2}$)	12.8	194	41.3	323

All experiments performed in the absence of Na. Constants taken from Paterson et al. (1981)

Lysine enters the enterocyte on a high and low affinity transport system as reported previously by Munck and Schultz (1969). Using lysine to inhibit the Na-independent transport of alanine reveals the presence of a second component to alanine entry mediated by the basic/neutral amino acid carrier. The low affinity component of lysine uptake is not due to diffusion since it can be inhibited by the presence of arginine (King et al. 1981).

Kinetic analysis has some limitations which should not be overlooked. Several models can usually be made to fit most sets of experimental results and any decision as to which model to support should come from as many sources of independent evidence as possible.

None of the above results provides information about how amino acids distribute within enterocytes or how they leave the intestinal epithelium, neither do they give any information on the cellular site of transport.

Transport into Enterocytes Along the Villus Measured in situ

The facility to identify particular enterocytes transporting amino acids from lumen to blood, *while still present on the villus,* has been available since 1965. The technique used originally involved presenting radioactive amino acids to pieces of intestine in vitro, subsequent localisation of counts being made by carrying out autoradiography on frozen sections (Kinter and Wilson 1965). This technique has been modified more recently by fixing amino acids to cellular proteins with glutaraldehyde and the whole process of obtaining autoradiographs has been speeded up by physical intensification of Ag grain deposits (King et al. 1981). Measurement of Ag grain density has been made quantitative through the use of microdensitometry. The use of suitable standards then allows one to calculate tissue concentrations of amino acid (Syme and Smith 1982, Smith and Syme 1982). In the following section we describe different ways in which this technique can be applied to study problems associated with enterocyte differentiation of transport function as well as the distribution of amino acids within erythrocytes and amino acid efflux across the basal membranes.

Appearance of Amino Acid Transport Function During Enterocyte Development

The general picture of amino acid location in rabbit ileum following 45-s contact of the luminal surface with tritiated amino acid is shown in Fig. 3.

The uptake of methionine, alanine and lysine can be seen to be confined to the upper third of each villus (the average length of a rabbit ileal villus is 510 μm). Methionine penetrates the villus core to a significantly greater extent than does alanine or lysine during this short period of contact with isotope. A typical quantitative analysis of amino acid distribution in rabbit ileal villi is shown, for alanine, in Fig. 4.

The average time taken for a rabbit enterocyte to travel from base of crypt to tip of villus has been estimated, from thymidine injection experiments to be about 96 h (Cremaschi et al. 1982). Amino acid uptake across the brush border membrane is already detected by the time an enterocyte becomes 70 h old. The ability of enterocytes to perform this function then increases in an apparently biphasic fashion, cells being lost from the villus tip 26 h later. Alanine transport in the present case can be

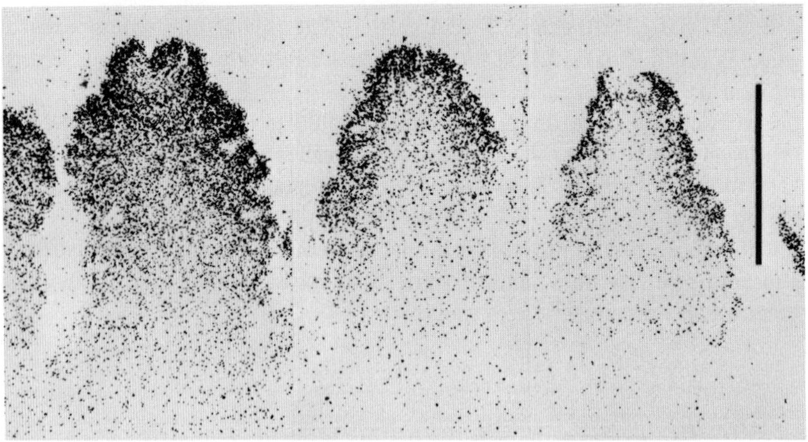

Fig. 3. Autoradiography of rabbit ileal villi showing, from the left, the distribution of transported methionine, alanine and lysine. The tissue had been exposed previously to 1 mM tritiated amino acid for 45 s in the presence of Na. Incubation was terminated with glutaraldehyde fixation as described previously. (King et al. 1981). Scale bar 100 μm

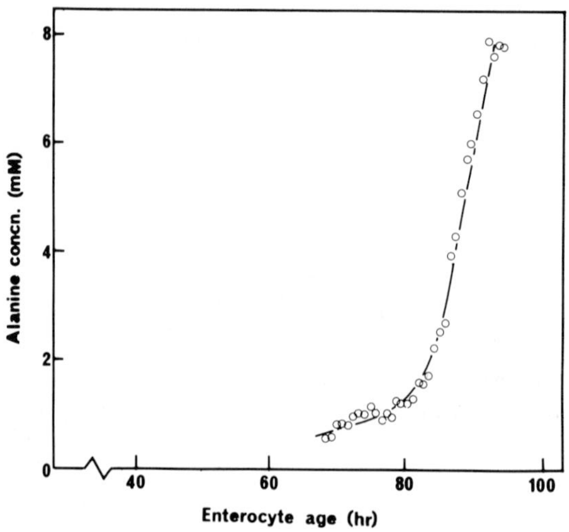

Fig. 4. Distribution of transported alanine along rabbit ileal villi. The tissue had been exposed to 1 mM tritiated alanine for 45 s. Distribution of optical density along the villus was assessed by microdensitometry and converted into tissue concentration by comparison with appropriate standards. *Abscissa* shows the age of cells calculated from the cell turnover time

seen to take place against a concentration gradient. Earlier work has shown alanine uptake in this area of the villus to be inhibited, in the absence of Na or presence of non-radioactive serine, to a degree similar to that predicted from constants calculated previously (Paterson et al. 1980b, see Table 1 of the present article for the relevant kinetic data). The possibility that the distribution of amino acid transport activity illustrated in Fig. 3 might be due to a diffusion artefact or to the effect of an unstirred layer has been ruled out by the finding that neither the rate of stirring of the mucosal solution (King et al. 1981) nor the time of contact with isotope (5–180 s; Paterson and Smith 1982) has any significant effect upon the distribution pattern for alanine.

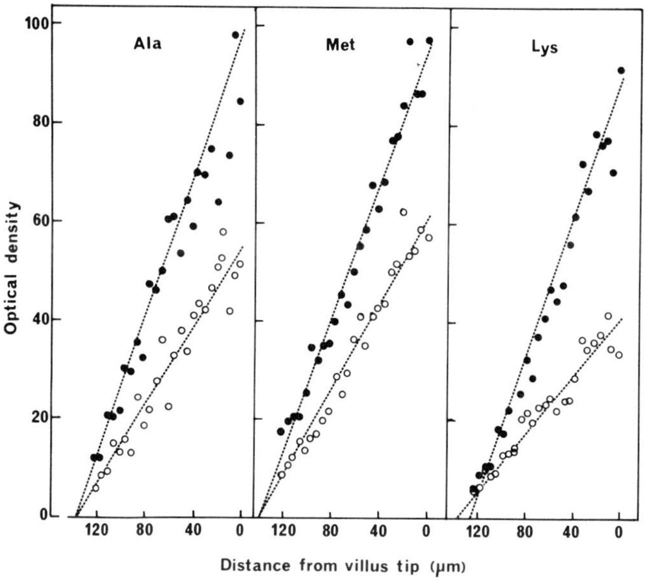

Fig. 5. Distribution of transported alanine, methionine and lysine along rabbit ileal villi. Incubations in the presence of 1 mM tritiated amino acid were stopped after 45 s *(open symbols)* or 180 s *(solid symbols)* by addition of glutaraldehyde. Lines through experimental points have been fitted by least squares regression analysis

Figure 5 compares the distribution pattern obtained for alanine in rabbit ileal tissue with that obtained for methionine and lysine for two different times of incubation. Lengthening the time of contact with isotope from 45 to 180 s increases the amount of uptake in the villus tip for all three amino acids. In no case does the increase in incubation time alter the proportion of enterocytes able to perform the transport function.

Having established that the method could be used quantitatively to measure the location of transporting enterocytes it then became possible to study how this standard pattern of differentiation could become modified under different circumstances. The results of some of these experiments have been summarized below.

Factors Affecting the Pattern of Enterocyte Differentiation

Effect of Intestinal Resection

Removing part of the intestinal tract is generally considered to provide one of the strongest stimuli for modification of intestinal function (Lipkin 1981). Taking away 60% of the proximal intestine in rats, for instance, leads to an approximate doubling of villus height and crypt depth and a 45% reduction in enterocyte turnover time in the remaining ileal segment (Menge et al. 1982a). The effect such treatment has on the cellular uptake of alanine is shown in Fig. 6.

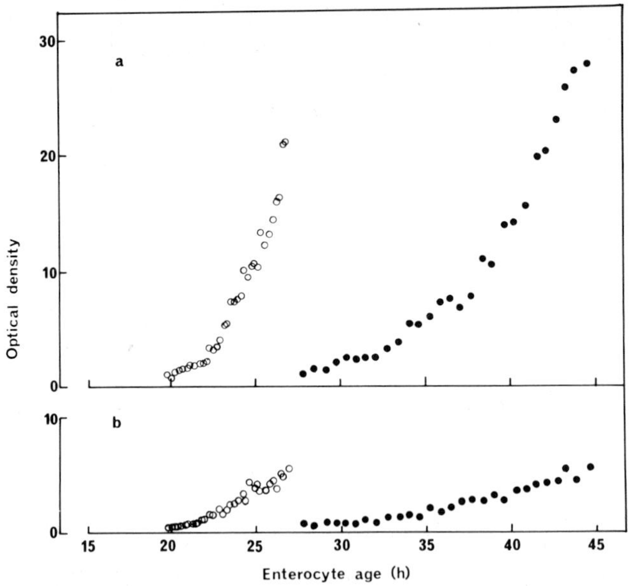

Fig. 6. Relation between enterocyte age and its ability to transport alanine in rat small intestinal villi. Pieces of control rat ileum (•) or ileal remnants after a 60% proximal resection (○) were exposed to 1 mM tritiated alanine in the presence (a) or absence(b) of Na. Enterocyte lifespan is 45 h for control tissue and 27 h for ileal remnants. (Menge et al. 1982b)

Results were obtained from female rats weighing 170–200 g, intestinal resection being performed three weeks before starting the experiment (Menge et al. 1982b). Entry of amino acid across the brush border membrane of rat enterocytes could be detected 30 h after cell birth under control conditions. This time was reduced to 20h as a result of intestinal resection. Once expressed, amino acid transport capacity continued to increase until the enterocyte reached the tips of villi, but the rate of increase after resection was approximately twice that found for control tissue. Similar results have been found using tritiated lysine (Menge et al. 1982b). Removing Na from the incubation medium caused a 60% reduction in alanine uptake, but the time at which transport into the enterocyte first became detectable remained less for tissue taken following intestinal resection. We conclude from these results that both systems 1 and 2 are subject to adaptational regulation in the rat (supposing the situation in rabbit ileum applies to rat jejunum).

Effect of Diet

Placing weanling rats on an isoenergetic diet containing 20 as opposed to 5% protein causes the villi to increase to 1.5 times the length found in animals fed a low protein diet (Syme 1982). In this case, however, it was found that the turnover time of enterocytes remained constant due to a 1.5 fold increase in the rate of mitosis (Syme and Smith 1982). The ability of pieces of jejunum taken from rats fed the different diets to concentrate valine is shown in Fig. 7.

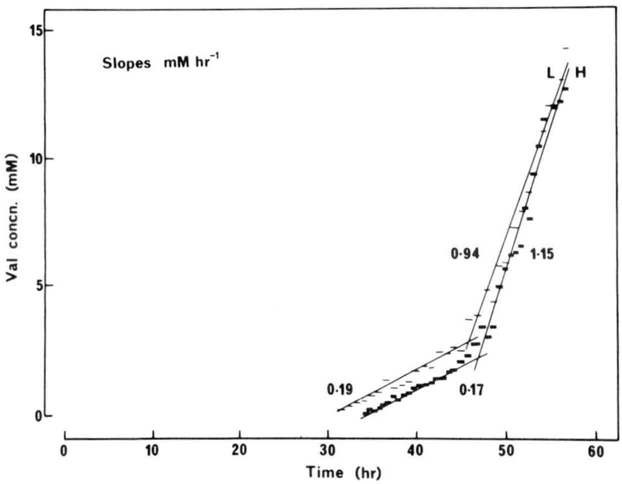

Fig. 7. Relation between enterocyte age and its ability to take up valine. Jejunal tissue from rats fed a 5 *(L)* or a 20% *(H)* protein diet were incubated in the presence of tritiated valine for 45 s and uptake assessed by autoradiography. The time base was calculated from the respective cell turnover times. (Syme and Smith 1982)

Tip enterocytes in the rat jejunum accumulate valine during 45 s exposure to isotope (12 mM within the enterocyte as against 1 mM in the bathing medium). This ability to transport valine starts 30 h after cell birth; a similar time to that shown in Fig. 6 for alanine uptake under control conditions. The effect of changing the diet on the time of onset and the rate of increase in transport capacity was insignificant (cf. slopes for the increase in optical density shown for the two conditions in Fig. 7). Compensation of transport capacity is achieved merely by maintaining a balance between the rate of mitosis and the length of the crypt/villus axis.

Differentiation of Enterocytes Associated with Peyer's Patches

Valine uptake has been measured into enterocytes covering Peyer's patches and compared with that found in surrounding villi. Autoradiographs and histological sections of this area of the intestine are shown in Fig. 8.

Radioactivity is seen to be concentrated in the dome region of the epithelium associated with the Peyer's patch (follicle-associated epithelium) and in the upper half of the surrounding villi. These villi seem to take up significantly greater amounts of valine than does the follicle-associated epithelium (Fig. 8a). The continuity of the epithelium lining the villi and that covering the follicular-lymphocytes is shown in Fig. 8b. Quantitation of these results is shown in Fig. 9.

Enterocyte turnover times are assumed to be equal for both villus and follicle-associated epithelium by analogy with results obtained previously in the mouse (Smith et al. 1980). Valine uptake is first detected in enterocytes 30 h after birth, a result identical to that found in villi taken from Peyer's patch-free regions of the rat

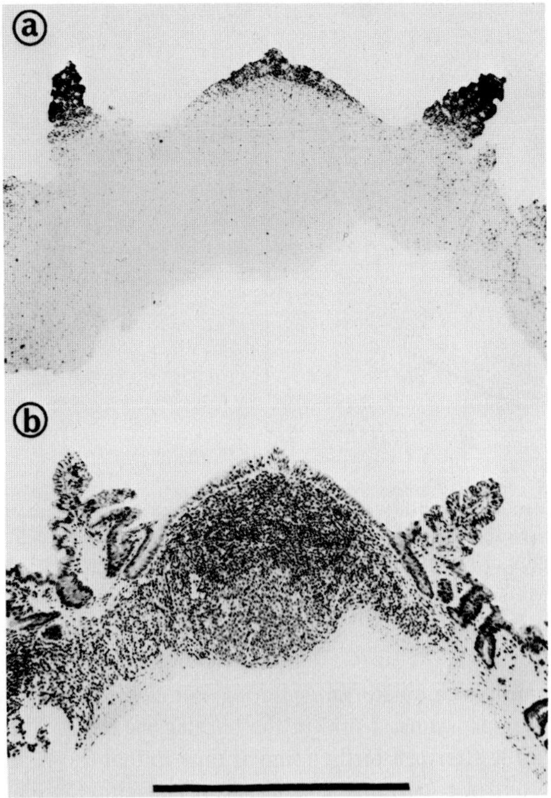

Fig. 8 a,b. Serial sections of a rat jejunal Peyer's patch showing the location of transported valine in an unstained section (a) or a hematoxylin and eosin stained section (b). Tissue was incubated for 45 s in the presence of tritiated valine. Scale bar, 500 μm. (Syme and Smith 1982)

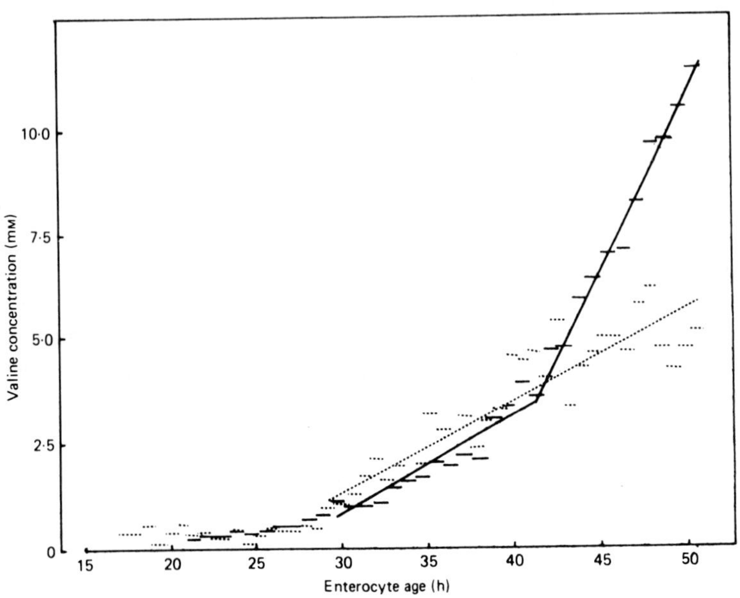

Fig. 9. Relation between age of enterocyte and its ability to take up tritiated valine during a 45 s incubation at 37°C. Enterocytes were studied in follicle-associated epithelium *(broken line)* and in adjacent villi *(continuous line)*. (Syme and Smith 1982)

jejunum (Syme and Smith 1982). Valine uptake increases as enterocytes grow older, both on the villus and over the follicle, but the final phase of rapid increase in transport capacity is absent in follicle-associated enterocytes. It is concluded from these results that the tissue over which an enterocyte migrates plays a part in the final expression of amino acid transport function. This might be due to lymphocyte inhibition of full differentiation by enterocytes; it could also be that the mesenchyme of the normal villus plays a regulatory role in organising the ontogeny of brush border intestinal function, but see the recent work of Kedinger et al. (1981)

Other situations where enterocyte differentiation of amino acid transport is obviously different from that found in adult mammals include that found in the new-born pig intestine (Smith 1981) and that found in the adult frog (Cheeseman and Smith, unpublished results). There is in both these cases a much wider distribution of enterocytes capable of transporting amino acids across brush border membranes. This could be connected with the slow rate of mitosis found in both instances.

Amino Acid Gradients Within Enterocytes

It is also possible to use quantitative autoradiography to determine the concentration of amino acids, both across enterocytes and into the centre of the villus core, after initial contact of the tissue with tritiated substrates (Paterson et al. 1982a). The concentration gradients for six amino acids across rabbit distal ileum are shown, for comparison, in Fig. 10.

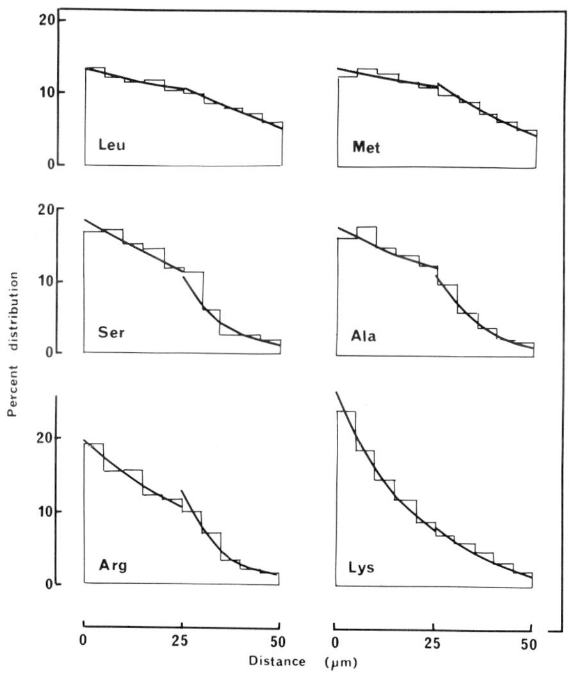

Fig. 10. Concentration gradients for different amino acids across rabbit distal ileum. Tritiated amino acid (1 mM) was presented to the mucosal surface for 45 s and the tissue fixed and analysed as described in the text. Distance is measured from the brush border to the centre of the villus core. The first 5 measurements are in the enterocyte. (Paterson et al. 1982a)

Contact with 1 mM amino acid was for 45 s. Enterocytes chosen for analysis were taken 50 μm from the tips of villi. The first five readings of optical density, which were all intracellular, show negligible gradients from brush border to basal membrane for leucine and methionine. More pronounced gradients existed for serine and alanine and for the basic amino acids arginine and lysine. All gradients became steeper on moving from the back of the enterocyte to the centre of the villus core (25–50 μm from the brush border membrane). Control experiments showed similar gradients of intracelluar concentration to be maintained for up to 3 min contact with isotope and for 45 s in the presence of 1 to 40 mM alanine, methionine and lysine. A similar gradient of intracellular concentration was obtained for ornithine, an amino acid not coded for in eukaryotic systems. It was concluded from these experiments that some physical property of the free amino acid rather than its selective incorporation into cellular protein was determining its ability to form concentration gradients within enterocytes. Further evidence that each amino acid remained essentially free within the enterocyte was provided in additional experiments where an initial 15-s uptake of tritiated alanine, methionine or lysine was followed by a 4 min period during which the mucosal surface was washed with amino acid-free bicarbonate saline. Such treatment caused minimal loss of total counts from the tissue showing negligible backflux of amino acid from cell to lumen. The results obtained from these experiments are shown in Fig. 11.

There were, as before, steep gradients of intracellular concentration for alanine and lysine, with no noticeable gradient for methionine, following initial contact of the tissue with isotope. These gradients were completely dissipated during the 4 min wash showing that virtually all of the intra-enterocyte amino acid eventually finds its way into the core of the villus. Amino acids showing steepest intracellular gradients

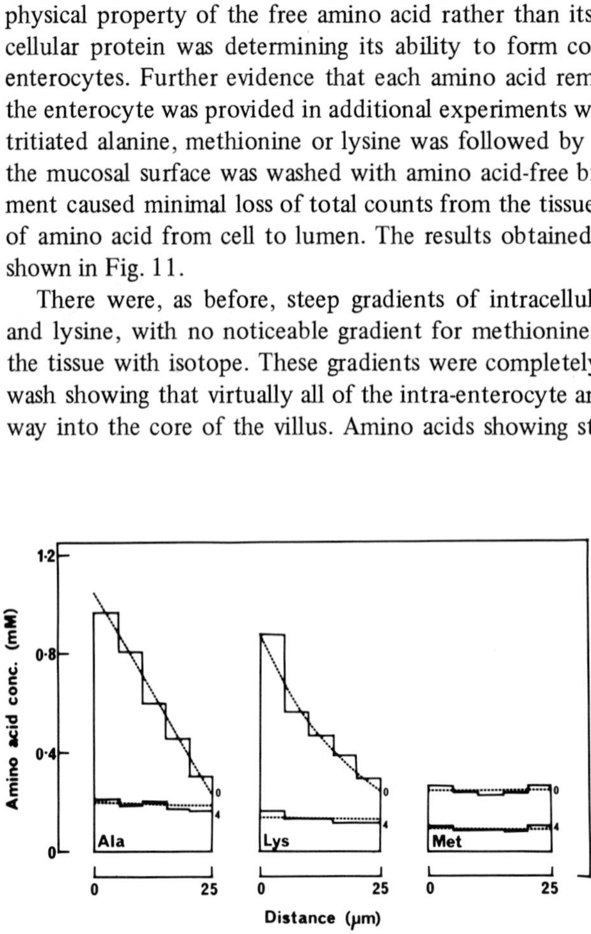

Fig. 11. Concentration gradients for amino acids within rabbit ileal enterocytes measured before and after a 4-min wash with bicarbonate saline. Tritiated amino acid (1 mM) was presented to the mucosal side for 15 s and samples taken for analysis (0). Adjacent samples were then superfused for 4 min before being taken for analysis (4). Superfusion had no effect on uptake (Ala, 2.21 and 2.17; Lys, 2.27 and 2.16; Met, 2.75 and 3.70 nmol cm^{-2} : 15 s uptake measured before and after a 4-min wash)

are those having least solubility in organic solvents. Lipid-like barriers to free diffusion within enterocytes (in the shape of intracellular membranes or other structures) could impede the free diffusion of amino acids during transport. An assumption that such intracellular structures might be permeated more readily by the hydrophobic amino acids could explain the observed difference in gradients seen between different substrates (Paterson et al. 1982a).

Amino Acid Efflux from Enterocytes

Concentrations of amino acids found within the villus core can be converted into measurements of total uptake and related to the intraenterocyte concentration of amino acid. Using this method the quantity of amino acid recovered from the villus core has been shown to be directly related to the intra-enterocyte concentration of amino acid determined at a site immediately adjacent to the basal membrane (Paterson et al. 1982b). Results showing this for methionine, alanine and lysine are shown in Fig. 12.

There was in all cases a straight line relationship between the quantity of amino acid recovered and the intra-enterocyte concentration of amino acid irrespective of whether amino acid concentration within the enterocyte had been changed by varying the time of contact with isotope or increasing the concentrattion of amino acid presented for uptake. This inability to saturate efflux mechanisms across the basal membrane confirms the work of Danisi et al. (1976) using compartmental analysis.

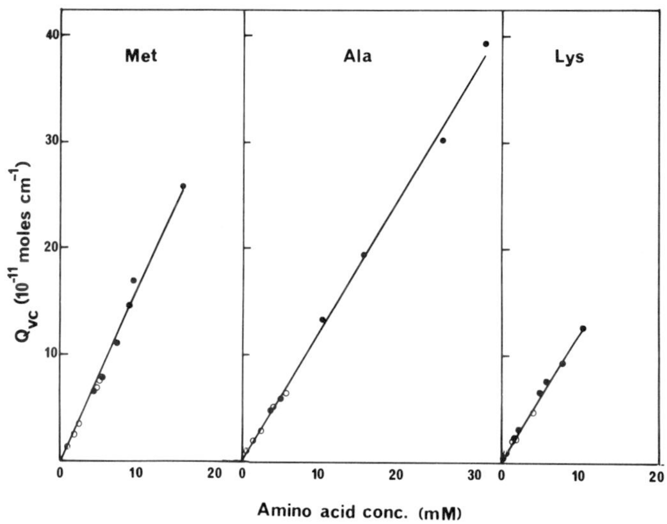

Fig. 12. Concentration dependence of amino acid efflux from rabbit ileal enterocytes. The net efflux (Q_{VC}) and intra-enterocyte concentration of methionine, alanine and lysine were determined as described previously (Paterson et al. 1982b). The original data came from both time course (–○–) and substrate concentration (–●–) studies. *Lines* through the experimental points have been fitted by least squares regression analysis. (Paterson et al. 1982b)

Table 4. Concentration dependence of amino acid efflux across basal membranes of rabbit ileal enterocytes

Amino acid	Amino acid permeability (10^{-5} cm^2)
Methionine	1.64 ± 0.11
Leucine	1.57 ± 0.06
Alanine	1.22 ± 0.05
Lysine	1.21 ± 0.06
Serine	1.10 ± 0.09
Arginine	1.06 ± 0.05

Values for amino acid permeability give the slopes of fitted regression lines ± S.E. of efflux concentration curves

The relative ease with which amino acids cross the basal membrane is given by the slope of lines shown in Fig. 12. These, together with others obtained using leucine, serine and arginine, have been summarised in Table 4. The permeability for methionine and leucine across the basal membrane was significantly greater than for alanine, lysine, serine and arginine.

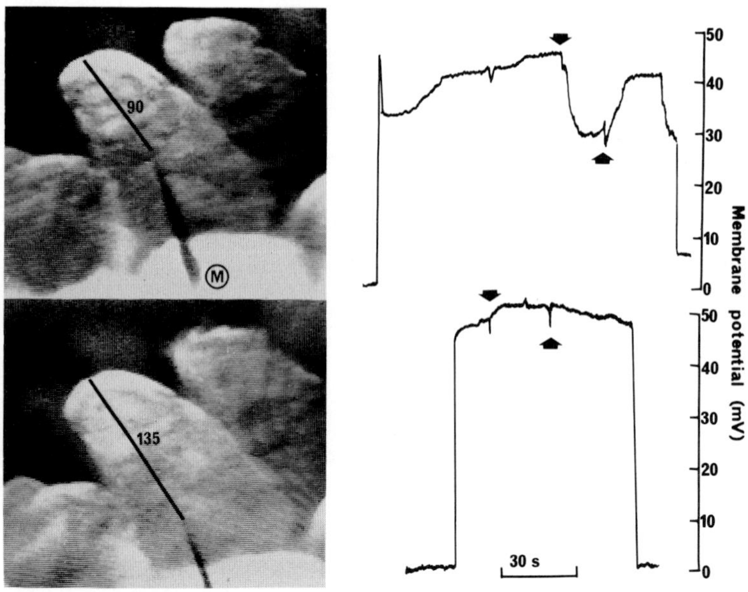

Fig. 13. A preparation of intestinal villi suitable for recording electrical events associated with amino acid entry into identified enterocytes. The method used to visualize individual villi was as published previously (James and Smith 1981). Villi come from rabbit distal ileum; *numbers* give distance in μm from microelectrode tip to tip of villus. *M* microelectrode barrel coated in Rotring black ink. Alanine effects on the two recorded membrane potentials are shown on the *right-hand side* of the figure

Villus Location of Enterocytes Showing Amino Acid-Induced Depolarisation of Membrane Potential

It is possible to mount stretched pieces of intestine on a plastic wedge and observe individual villi through a dissecting microscope at a final magnification of × 300 on a TV monitor (James and Smith 1981). Microelectrodes can then be inserted into known areas of the villus. Solutions containing amino acids are added to the bath to test directly for depolarisation of the brush border membrane potential. Two typical recordings from one such experiment are shown in Fig. 13.

The microelectrode in the first instance was located 90 μm from the tip of a rabbit ileal villus. The membrane potential recorded from this enterocyte fell in the presence of alanine (top trace). The same microelectrode was then placed in an enterocyte 135 μm from the villus tip. In this instance alanine was without effect on the recorded membrane potential (bottom trace). Earlier work using quantitative autoradiography showed the Na-dependent uptake of alanine to cease at about this distance from the villus tip. Amino acid-induced depolarisation of brush border membrane potential in one enterocyte could change the potential in another if both were connected electrically. Because of this one must conclude that any plot of enterocytes showing amino acid-induced depolarisation represents a *maximal* number of cells containing an operational Na-coupled entry system for alanine.

References

Bass R, Hedegaard HB, Dillehay L, Moffet J, Englesberg E (1981) The A, ASC and L systems for the transport of amino acids in Chinese hamster ovary cells. J Biol Chem 265:10259–10266

Cremaschi D, James PS, Meyer G, Peacock MA, Smith MW (1982) Membrane potentials of differentiating enterocytes. Biochim Biophys Acta 688:271–274

Curran PF, Schultz SG, Chez RA, Fuisz RE (1967) Kinetic relations of the Na-amino acid interaction at the mucosal border of intestine. J Gen Physiol 50:1261–1286

Danisi G, Tai Y-H, Curran PF (1976) Mucosal and serosal fluxes of alanine in rabbit ileum. Biochim Biophys Acta 455:200–213

Garvey TQ, Hyman PE, Isselbacher KJ (1976) γ-Glutamyl transpeptidase of rat intestine: Localization and possible role in amino acid transport. Gastroenterology 71:778–785

Gazzola GC, Dall'Asta V, Guidotti GG (1980) The transport of neutral amino acids in cultured human fibroblasts. J Biol Chem 255:929–936

Handlogten ME, Garcia-Cañero R, Lancaster KT, Christensen HN (1981) Surprising differences in substrate selectivity and other properties of systems A and ASC between rat hepatocytes and the hepatoma cell line HTC. J Biol Chem 256:7905–7909

Inui KI, Quaroni A, Tillotson LG, Isselbacher KJ (1980) Amino acid and hexose transport by cultured crypt cells from rat small intestine. Am J Physiol 239:C190–C196

James PS, Smith MW (1980) Amino acid-induced changes in intestinal short-circuit current. J Physiol (Lond) 303:79P

James PS, Smith MW (1981) A preparation of intestinal villi suitable for recording membrane potentials in enterocytes taken at different stages of differentiation. J Physiol (Lond) 319: 6–7P

Kedinger M, Simon PM, Grenier JF, Haffen K (1981) Role of epithelial-mesenchymal interactions in the ontogenesis of intestinal brush-border enzymes. Devel Biol 86:339–347

Kilberg MS, Lancaster KT, Christensen HN (1981) Neutral amino acid transport in isolated rat intestinal segments. Fed Proc 39: abstr 570

Kimmich GA (1975) Preparation and characterization of isolated intestinal epithelial cells and their use in studying intestinal transport. In: Korn E (ed) Methods in membrane biology, vol IV. Plenum, New York, pp 51–115

King IS, Sepúlveda FV, Smith MW (1981) Cellular distribution of neutral and basic amino acid transport systems in rabbit ileal mucosa. J Physiol (Lond) 319:355–368

Kinter WB, Wilson TH (1965) Autoradiography study of sugar and amino acid absorption by everted sacs of hamster intestine. J Cell Biol 25:19–39

Lipkin EM (1981) Proliferation and differentiation of gastrointestinal cells in normal and disease states. In: Physiology of the gastrointestinal tract, vol I. Raven Press, New York, pp 145–168

Menge H, Hopert R, Alexopoulos T, Riecken EO (1982a) Three-dimensional structure and cell kinetics at different sites of rat intestinal remnants during the early adaptive response to resection. Res Exp Med 181:77–94

Menge H, Sepúlveda FV, Smith MW (1982b) Cellular adaptation of amino acid transport following intestinal resection in the rat. J Physiol (Lond) 334:213–233

Munck BG (1981) Intestinal absorption of amino acids. In: Physiology of the gastrointestinal tract, vol II. Raven Press, New York, pp 1097–1122

Munck BG, Schultz SG (1969) Lysine transport across isolated rabbit ileum. J Gen Physiol 53: 157–182

Paterson JYF, Smith MW (1982) Testing the hypothesis that substrate availability determines the cellular distribution of amino acid uptake by rabbit ileal mucosa. J Physiol (Lond) 327: 96–97P

Paterson JYF, Sepúlveda FV, Smith MW (1979) Two-carrier influx of neutral amino acids into rabbit ileal mucosa. J Physiol (Lond) 292:339–350

Paterson JYF, Sepúlveda FV, Smith MW (1980a) Stoichiometry versus coupling ratio in the co-transport of Na and different neutral amino acids. Biochim Biophys Acta 603:288–297

Paterson JYF, Sepúlveda FV, Smith MW (1980b) A sodium-independent low affinity transport system for neutral amino acids in rabbit ileal mucosa. J Physiol (Lond) 298:333–346

Paterson JYF, Sepúlveda FV, Smith MW (1981) Distinguishing transport systems having overlapping specificities for neutral and basic amino acids in the rabbit ileum. J Physiol (Lond) 319:345–354

Paterson JYF, Sepúlveda FV, Smith MW (1982a) Distribution of transported amino acid within rabbit ileal mucosa. J Physiol (Lond) 331:523–535

Paterson JYF, Sepúlveda FV, Smith MW (1982b) Amino acid efflux from rabbit ileal enterocytes. J Physiol (Lond) 331:537–546

Reiser S, Christiansen PA (1971) The properties of the preferential uptake of L-leucine by isolated intestinal epithelial cells. Biochim Biophys Acta 225:123–139

Sepúlveda FV, Pearson JD (1982) Characterization of neutral amino acid uptake by cultured epithelial cells from pig kidney. J Cell Physiol 112:182–188

Sepúlveda FV, Smith MW (1978) Discrimination between different entry mechanisms for neutral amino acids in rabbit ileal mucosa. J Physiol (Lond) 282:73–90

Sepúlveda FV, Burton KA, Brown PD (1982) Relation between sodium-coupled amino acid and sugar transport and sodium/potassium pump activity in isolated intestinal epithelial cells. J Cell Physiol 111:303–308

Shapiro MP, Kimmich GA (1981) Na^+:L-valine coupling stoichiometry in isolated intestinal cells. Fed Proc 39:abstr 569

Smith MW (1981) Autoradiographic analysis of alanine uptake by newborn pig intestine. Experientia 37:868–869

Smith MW, Jarvis LG, King IS (1980) Cell proliferation in follicle-associated epithelium of mouse Peyer's patch. Am J Anat 159:157–166

Smith MW, Syme G (1982) Functional differentiation of enterocytes in the follicle-associated epithelium of rat Peyer's patch. J Cell Sci 55:147–156

Søgnen E (1967) A method for the preparation of suspensions of intestinal mucosal cells by means of calcium chelators. Acta Vet Scand 8:76–82

Stevens BR, Ross HJ, Wright EM (1982) Multiple transport pathways for neutral amino acids in rabbit jejunal brush border vesicles. J Membr Biol 66:213–225

Syme G (1982) The effect of protein-deficient isoenergetic diets on the growth of rat jejunal mucosa. Br. J Nutr 48:25–36

Syme G, Smith MW (1982) Intestinal adaptation to protein deficiency. Cell Biol Int Rep 6:573–578

Weiser MM (1973) Intestinal epithelial cell surface membrane glycoprotein synthesis. An indicator of cellular differentiation. J Biol Chem 248:2536–2541

Statistical Analysis of Solute Influx Kinetics

J.W.L. ROBINSON †, G. VAN MELLE [1], and S. JOHANSEN [2]

Introduction

Statistical analysis has seldom been applied to solute influx kinetics in the intestine, or indeed in other tissues. However, with the advent of numerical techniques readily implementable on desk-top computers, non-linear regression analysis can be employed to test the applicability of transport models to a given data set, or to examine which parameter in a given model is significantly influenced by different experimental conditions. In the present communication, we describe the analytical procedure that we have applied to the evaluation of solute influx in the small intestine and emphasize its wide scope and its usefulness in testing the statistical significance between different models and/or parameters.

The availability of large numbers of tissue rings excised from guinea-pig intestine is very advantageous for the sort of analysis that we have developed. In this species, we have shown that there is very little variation in kinetic behaviour over the entire jejunum-ileum, as regards amino-acid or monosaccharide influx (Robinson and van Melle 1982). We are therefore able to use the entire intestine for a single experiment, and our experience has shown that a maximum of 36 flasks, each bearing three replicate tissues, can be conveniently processed by two teams of two technicians working in parallel. Thus, since each tissue is treated and counted separately, we obtain fairly large data sets for introduction into our analysis.

The tissue rings were incubated for 2 min in a ^{14}C-labelled solution of the substrate. Each ring was then weighed and dissolved separately in 30% KOH for counting. No correction was applied for the entry of substrate into the extracellular space by the use of a so-called space marker, but a diffusional term (designated by K_D) is added to each model equation to effect this correction; the advantages and disadvantages of this approach have been discussed in earlier publications (Robinson et al. 1980, van Melle and Robinson 1981, Robinson and van Melle 1982). Therefore the data sets in general were fitted to a model comprising the sum of saturable and diffusional components:

$$V = \frac{V_{max} \cdot [S]}{[S] + K_m} + K_D \cdot [S] \qquad (1)$$

1 Secteur Mathématiques, Centre Hospitalier Universitaire Vaudois, Lausanne, Switzerland
2 Institut for Matematisk Statistik, Københavns Universitet, København, Denmark

Intestinal Transport
(ed. by M. Gilles-Baillien and R. Gilles)

Preliminary perusal of a series of data sets persuaded us that a recurrent observation was the proportionality between standard deviation and mean uptake at a given substrate concentration, indicating that the random errors are multiplicative rather than additive. For this reason, all our analyses have been performed on the log scale (van Melle and Robinson 1981).

The next stage of the preparation of our data set for final analysis involves a procedure rather similar to trimming, which will permit the identification of outliers. Following the recommendations of Anscombe and Tukey (1963), we removed all values that differed by more than three standard deviations from the mean of all observations at a given experimental condition (pooling animals and replicates). For this computation, the pooled standard deviation is used, since the log transformation has resulted in homogeneity of variance. This procedure never led to the removal of more than 2% of the points, and the proportion removed was generally much less.

The chosen model was then fitted to the data sets by minimising the sum of squares of errors (SSE) using the technique of Fletcher and Powell (1964) for non-linear regression models. This procedure was performed for the data for individual animals and for the data set obtained by pooling all animals. Using the analysis of variance approach, we can decompose the SSE into a pure error component (SSPE) which describes the variation around the means, and a "lack-of-fit" component (SSLF) which denotes the error around the regression. At this stage, comparison of the SSLF and SSPE with the aid of an F-test indicates whether the model describes the data satisfactorily. Since, by definition, the SSPE is the same whatever model is being considered, the quantity that is being minimised is the SSLF. It is worth pointing out at this juncture that if the F-test comparing SSLF and SSPE for a given model is not significant, then no other model can ever be found to give a *significantly* better fit to the data.

The objective of our work is to find the simplest model to describe a given set of data. We start with a complex model which we simplify by imposing restrictions on given parameters. This process automatically leads to an increase in the value of SSLF. We can then test whether the restriction is permitted by applying the following F-test:

$$F' = \left\{ [SS(R) - SS(F)] \,/\, [df(R) - df(F)] \right\} / \, [SS(F) \,/\, df(F)] \tag{2}$$

where SS(R) and SS(F) refer to the sums of squares associated with restricted and full models respectively, and df(R) and df(F) are the corresponding degrees of freedom. F′ then follows the F-distribution with [df(R) – df(F)] and df(F) degrees of freedom (Neter and Wasserman 1974). Note that SS(R) and SS(F) in Eq. (2) can refer to SSE or to SSLF. The use of SSLF gives a more sensitive test – and is indeed the one that we have chosen to use – but before it can be applied, it must first be demonstrated that the model fits the data adequately. Clearly, however, there is little sense in applying restrictions to a model which does not fit originally.

In the present contribution, we wish to illustrate these principles with three rather different examples. In the first case, we demonstrate that α-methyl-glucoside and galactose are absorbed by a single, shared transport site in the guinea-pig brush-border membrane. In the second, we have examined the applicability of several different models to describe the sodium-dependent uptake of β-methyl-glucoside in guinea-pig intestine. In the third, we explore the changes in kinetic parameters describing phenylalanine uptake by hypertrophic mucosa from self-filling blind loops of rat small intestine.

Example 1: Kinetics of Uptake for Different Monosaccharides

Studies on the kinetics of the uptake of each of the monosaccharides separately over a wide range of substrate concentrations (0.5–50 mM) revealed good fits to Eq. (1). At first sight, there seemed no reason to suspect the existence of a second transport site for sugars in guinea-pig small intestine, in contradiction to the claims of other authors (Honegger and Semenza 1973; Syme and Levin 1980, Debnam 1982) who worked with other mammals. Confirmation of this conclusion was obtained from the following tests:

1. Attempts to fit the data to a two-site model (plus a diffusion term) resulted in only a small decrease in the sums of squares which was not statistically significant. In the case of both sugars, the V_{max} for the second site turned out to be negative, which is clearly absurd.
2. Examination of the residuals around the curve fitted to Eq. (1) revealed no bias in their distribution at any given concentration.
3. Computation of the kinetic parameters was not dependent on the range of concentration used. When the full data set was sub-divided, along the lines proposed in an earlier article (Robinson and van Melle 1982), and the sub-sets were fitted to Eq. (1), then the same kinetic parameters always emerged.

Next, an experiment was devised to show that not only were the two sugars taken up by a single site, but also that they shared this site. For this purpose, both substrate and inhibitor concentrations were varied within the same experiment, and the data were fitted to the following equation for fully competitive inhibition for a single site:

$$V = \frac{V_{max} \cdot [S]}{[S] + K_m (1 + [I] / K_i)} + K_D \cdot [S] \qquad (3)$$

In order to avoid any bias inherent in the use of different animals, two half-experiments were performed with tissues extracted from the same guinea-pig, along the lines described previously (Robinson and van Melle 1982). For this purpose, mixtures of the two monosaccharides were prepared; 18 were labelled with ^{14}C-galactose (which then was "substrate" for this half-experiment) and 18 were labelled with ^{14}C-α-methyl-glucoside. Care was taken in the choice of concentrations for each condition that would aid in the determination of the parameters; a symmetrical plan, for instance, would not have been helpful. The experiment was repeated on six animals and the flasks were incubated in a random order.

The parameters emerging from each half-experiment when Eq. (3) is fitted to the data are listed in Table 1, together with the analysis of variance. The F-test appearing at the foot of the table shows an excellent fit to the model. We then pooled the data

Table 1. Estimates for individual half-experiments concerning the mutual inhibition between galactose and α-methyl-glucoside

[S] = galactose [I] = α-methyl-glucoside			[S] = α-methyl-glucoside [I] = galactose		
V_{max}^{gal}	0.347	± 0.062	$V_{max}^{\alpha mg}$	0.228	± 0.035
K_m^{gal}	18.1	± 4.2	$K_m^{\alpha mg}$	6.05	± 1.37
K_D^{gal}	0.00129	± 0.00037	$K_D^{\alpha mg}$	0.00227	± 0.00069
$K_i^{\alpha mg}$	11.0	± 1.7	K_i^{gal}	10.3	± 2.0
SSE	25.93558		SSE	21.33703	
SSPE	25.00529		SSPE	20.73513	
SSLF	0.93029		SSLF	0.60190	
$F_{14,304}$	0.81		$F_{14,303}$	0.63	

Eq. (3) is fitted to the experimental results (for 6 animals) illustrated in Figs. 1 and 2 by non-linear least-squares regression. The results tabulated are the estimates for each half-experiment analysed separately. V_{max} is expressed in μmol/100 mg · 2 min; K_D in μmol/100 mg · 2 min · mM; and K_m and K_i in mM

Table 2. Joint fit of the two half experiments concerning the mutual inhibitions between galactose and α-methyl-glucoside

V_m^{gal}	0.265	± 0.023
K_m^{gal}	12.91	± 1.17
K_D^{gal}	0.00215	± 0.00041
$V_m^{\alpha mg}$	0.272	± 0.029
$K_m^{\alpha mg}$	7.86	± 1.01
$K_D^{\alpha mg}$	0.00152	± 0.00060

	SSLF	df	MS
Restricted	1.73896	30	
Full	1.53219	28	0.05472
Error	0.20677	2	0.10338
		$F_{2,28}$ = 1.89	

A joint fit of Eq. (3) to the results of the two half-experiments illustrated in Figs. 1 and 2 under the constraint of equality of the respective K_m's and K_i's is performed. The lower panel gives the analysis of variance to test the validity of the constrained pooling procedure. Units as in Table 1

of the two half-experiments and effected a joint fit under the constraint that the two K_m's equalled the corresponding K_i's. The results are displayed in Table 2, together with the analysis of variance, performed according to Eq. (2) on the SSLF; this confirms that the restrictions are permissible. It may appear at first sight surprising that these constraints are permitted when the original estimates of K_m and K_i in the individual half-experiments differed considerably. The reason for this paradox is that the estimates of the parameters are highly correlated, even if the design used is almost optimal; the surface of the function is very flat around the minimum, and a fairly wide range of estimates will fit the data well with only a small change in the sum of squares. Figures 1 and 2 illustrate, for galactose and α-methyl-glucoside respectively, the three-dimensional curves drawn using the common estimates, together with the experimental points. On this type of representation, for a good fit, the experimental points should fall at the intersections of the solid lines. The fits are clearly most satisfactory.

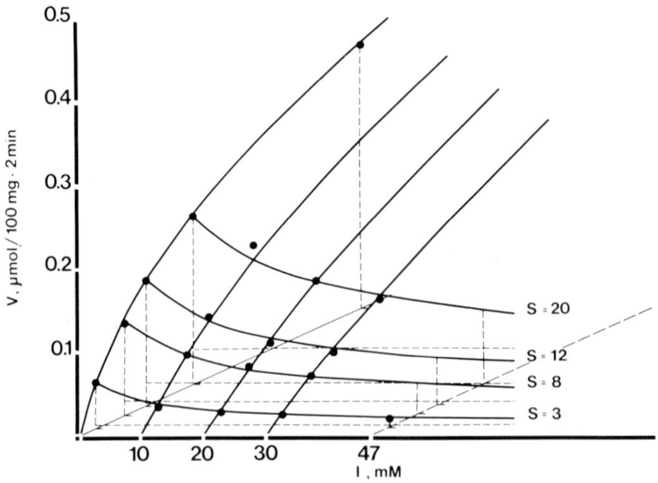

Fig. 1. Three-dimensional diagram illustrating the inhibition of galactose influx by α-methyl-glucoside. *Points* represent the antilogs of the logarithmic means, since the minimisation was performed on a logarithmic scale. *Curves* were drawn using the parameters obtained by the pooled analysis as listed in Table 2. *Horizontal dotted lines* indicate the asymptotes of each curve (= $K_D \cdot [S]$)

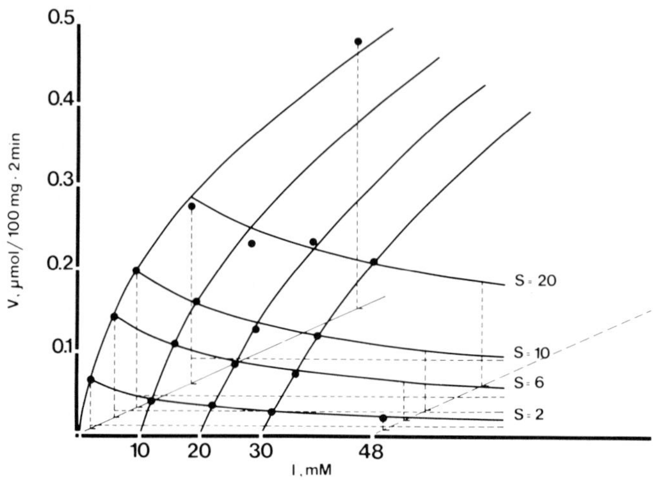

Fig. 2. Three-dimensional diagram illustrating the inhibition of α-methyl-glucoside influx into guinea-pig intestinal rings by galactose. Results of the other half-experiment corresponding to that illustrated in Fig. 1

This experimental design has another statistical advantage, namely that the analysis that we have just described can be applied to each animal separately. In Table 3, the values of the parameters emerging from the joint fit under the restriction that the K_m's are equal to the corresponding K_i's are listed, together with the F-values concerning the validity of the restriction. The restriction is accepted for all six animals. In five of the six geuinea-pigs, the estimates of each parameter were very similar, whereas the sixth animal gave absurd estimates; but perusal of the original data confirmed that the behaviour for this guinea-pig was rather erratic. It is nevertheless certain that one of the reasons that we get such clear-cut results from our analyses is the fact that variation between animals is rather small, and so we can treat the data as if they come from a "mean guinea-pig".

Table 3. Pooled data from half-experiments for individual animals (Interactions between galactose and α-methylglucoside)

Animal	1	2	3	4	5	6
V_m^{gal}	0.328	0.180	0.292	0.211	0.190	0.467
K_m^{gal}	14.8	11.6	9.85	10.7	10.4	21.2
K_D^{gal}	0.00244	0.00113	0.00301	0.00276	0.00284	0.000863
$V_m^{\alpha mg}$	0.378	0.227	0.231	0.209	0.175	0.506
$K_m^{\alpha mg}$	9.84	8.36	5.62	6.53	5.30	12.4
$K_D^{\alpha mg}$	0.000672	0.000778	0.00462	0.00232	0.00276	−0.00259
$F_{2,28}$	0.930	0.657	1.71	0.0418	2.00	0.999

Analysis performed as in Table 2 on data for each animal separately. The bottom line gives the results of the analysis of variance testing the validity of the constraint that the K_m's are equal to the corresponding K_i's. Units as in Table 1

From these results we can conclude: (1) the one-site model is acceptable for both sugars, as witnessed by the F-tests performed on the half-experiments, and (2) the monosaccharides share the same transport site, since the F-test performed on the joint fit permits the constraints that the two K_m's equal the corresponding K_i's.

Example 2: Kinetics of β-Methyl-Glucoside Influx at Different External Sodium Concentrations

The aim of this experiment was to examine the behaviour of β-methyl-glucoside influx in guinea-pig intestinal rings at different external sodium concentrations, and to attempt to find a kinetic model that would adequately describe the data. The experimental design involved 36 incubation media, covering all combinations of

6 sodium and 6 β-methyl-glucoside concentrations; the experiment was repeated on 6 guinea-pigs. The details of this experiment are to be published elsewhere (van Melle and Robinson, unpublished data) and only a summary will be presented in this communication.

The first step in our analysis of these data was to fit Eq. (1) to the values (on the log scale) at each individual sodium concentration. In every case, including the one when the external sodium concentration was zero, a good fit was obtained, both for each animal individually and for the pooled animals. We then demonstrated that a common K_D was acceptable at all sodium concentrations, and next tested whether a common V_{max} or a common K_m could be imposed. It turned out that the apparent V_{max} significantly decreased as external sodium concentration was reduced, whilst the apparent K_m concomitantly increased. The two observations, namely that a saturable function is obtained in the complete absence of external sodium, and that the apparent V_{max} is sodium-dependent, rule out the applicability of usual variations of the kinetic model shown in Fig. 3. In models permitting the transfer of binary and ternary forms of the carrier across the membrane with equal permeability coefficients ($P_2 = P_3 = P_4$), V_{max} is implicitly independent of sodium concentration. In variants in which only the transfer of the ternary complex occurs ($P_2 = P_4 = 0$), no mediated uptake should occur in the complete absence of sodium. In view of our experimental observations, we then tested a proposal of Alvarado and Lherminier (1982), who suggested that truly sodium-free conditions were unobtainable in guinea-pig intestinal rings in vitro, since there was sufficient sodium within the tissue to maintain a self-perpetuating reservoir in the immediate vicinity of the brush-border membrane. In this way, the persistence of a saturable component of sugar uptake in the nominal absence of sodium from the bulk medium could be explained. Since the apparent

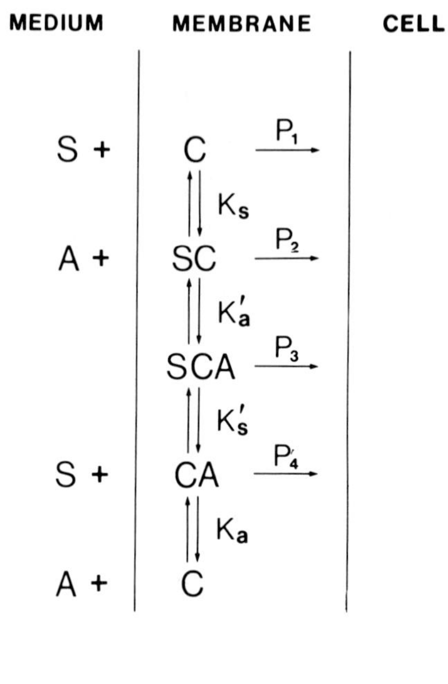

Fig. 3. General kinetic model describing coupled influx of sugar *(S)* and sodium ion *(A)* across the intestinal brush-border membrane. The diagram only illustrates the substrate-carrier interactions at the external face of the membrane and the translocation steps, P_1 to P_4, which are assumed to be rate-limiting

V_{max} is sodium-dependent, an "obligatory" model (meaning that only the ternary complex can cross the membrane) is necessary. We therefore modified the equations describing uptake by such "obligatory" models by the introduction of a factor, A_{min}, which represents the sodium concentration in the "reversoir". In the case of the formation of the ternary complex in a random manner, i.e., from either binary complex, the equation becomes:

$$V = \frac{V_{max} \dfrac{A^*}{K'_a + A^*}}{1 + \dfrac{K'_s}{[S]} \cdot \dfrac{K_a + A^*}{K'_a + A^*}} + K_D \cdot [S] \qquad (4)$$

$$\text{where } A^* = \begin{cases} [A] \text{ if } [A] > A_{min} \\ A_{min} \text{ otherwise} \end{cases}$$

If the ternary complex is formed compulsorily from the binary complex, SC, then the equation is modified by letting K'_s go to zero. These two models were then fitted to our data set: Eq. (4) was found to fit adequately ($F_{7.23}$ = 2.23 when comparing the SSLF for this model with that of the full model with common K_D imposed). The model involving the non-random formation of the ternary complex was rejected in a highly significant manner ($F_{8.23}$ = 35.8, $P \ll 0.001$). The parameter estimates of the random, obligatory model with sodium reversvoir incorporated, as given in Eq. (4), are listed in Table 4 for the pooled data set and for individual animals. The constancy of the value of A_{min} is particularly striking; in addition, its numerical value coincides with that predicted by Alvarado and Lherminier (1982). Figure 4 provides a three-dimensional diagram of the fit of the model to the experimental data, with A_{min} illustrated as a break in the curve.

Table 4. Parameter estimates for Na^+-dependent β-methyl-glucoside influx according to the model described by Eq. (4)

Animal	K_D	V_{max}	K'_a	K_s	K'_s	K_a	A_{min}
All	39.8 (2.6)	2190 (99)	4.23 (0.91)	13.4 (3.3)	2.41 (0.24)	23.5 (3.4)	4.80 (0.34)
1	44.5	1600	2.66	9.8	2.46	10.6	4.34
2	42.0	1700	1.99	26.3	1.37	38.2	4.95
3	42.3	2270	8.35	10.8	2.76	32.7	4.84
4	36.8	2170	3.32	13.8	2.40	19.1	3.54
5	45.7	2960	7.02	7.7	3.47	15.6	5.47
6	26.3	2850	3.33	24.0	2.50	32.0	5.27

Eq. (4) fitted to the data on Na^+-dependent β-methyl-glucoside influx for pooled animals and for each guinea-pig separately. Units of V_{max} are n-mol/g · 2 min and of K_D are n-mol/g · 2 min · mM; the other constants are expressed in mM. The values of K_a are obtained from the identity $K_s \cdot K'_a = K_a \cdot K'_s$

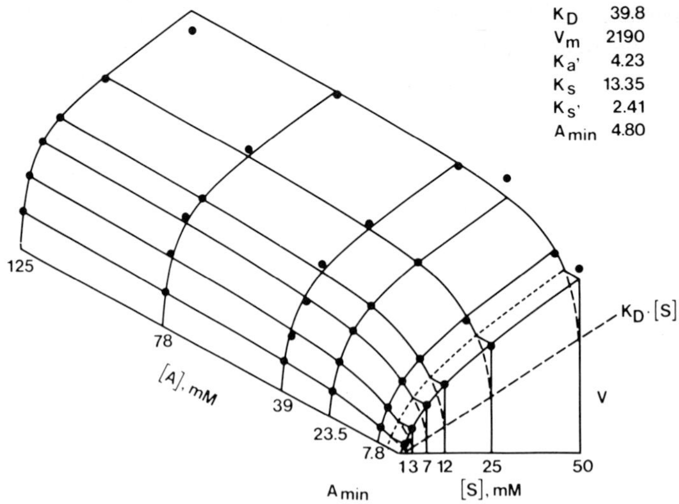

Fig. 4. Three-dimensional diagram illustrating the dependence of β-methyl-glucoside influx on the external sodium concentration, according to the model described by Eq. (4). *Abscissa in the plane* sugar concentration; *other abscissa* sodium. The experimental data (antilogs of the logarithmic means) are represented by *points; solid lines* describe the uptake according to Eq. (4), using the parameters listed in the *top righ-hand corner.* For sodium concentrations below A_{min}, the sugar uptake is assumed to be independent of sodium concentration, the transition being represented by *dotted line*

In conclusion, a relatively simple kinetic model has been found to describe the sodium-dependent uptake of monosaccharides across the intestinal brush-border membrane. The principal novel feature of this model is the concept of a sodium reservoir in the vicinity of the brush-border membrane which had been predicted, largely on theoretical grounds, by Alvarado and Lherminier (1982). The good fit of this model to the experimental data, incidentally, precludes the development of another model which will give a significantly better fit.

Example 3: Kinetics of Phenylalanine Influx into Samples from Blind-Loop Mucosa

The third example that we wish to present is of a somewhat different nature. Together with Dr. H. Menge from Berlin, we have been studying the changes that occur in the small intestinal mucosa when segments are transformed surgically into self-filling blind loops (Menge et al. 1979). The operative technique is simple: Rat jejunum is transected and the proximal opening is sewn up. The distal orifice is then joined to the upper jejunum by end-to-side anastomosis. The animals are then left for three weeks, during which time the blind loop undergoes considerable hypertrophy and is the site of markedly reduced functional capacity (Menge et al. 1979). In this context, we were interested in examining the kinetics of phenylalanine influx into samples from blind-loop mucosa. The results, providing a comparison between uptake by normal rat jejunal mucosa and influx into blind-loop tissue, are illutrated in Fig. 5. Clearly, phenylalanine transport is more rapid in the control series. In order to determine which kinetic parameter(s) are affected by the treatment, global fits to Eq. (1) are first performed. The results (Table 5) reveal that the V_{max} for phenylalanine influx is considerably smaller in the blind-loop mucosa whereas K_m is approximately the same in both series. In addition, the K_D parameter, describing diffusion into the tissue, is somewhat larger in the blind-loop mucosa then in control jejunum. In order to test the statistical significance of these differences, we proceed in a manner analogous to that described above. We pool the two data sets and then carry out a joint fit, under the restriction that one of the parameters is common to the two data sets. The analysis is then repeated for the other two parameters separately. We then apply an F-test to determine whether the imposition of a given restriction causes a significant rise in the SSLF. The results show that the two V_{max}'s are very significantly different from one another, whereas the difference between the two K_D's just attains significance at the 5% level.

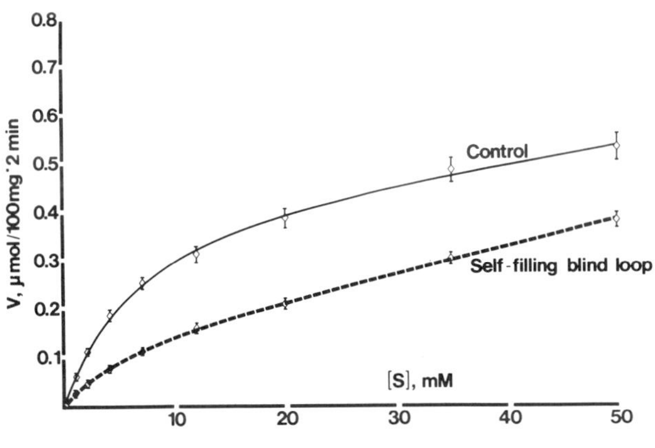

Fig. 5. Uptake of phenylalanine by control rat jejunal mucosa and by mucosa excised from self-filling jejunal blind loops. *Points* antilogs of the log mean uptakes; *lines* best fits, drawn with the parameters listed in Table 5

Table 5. Parameters describing phenylalanine influx into control rat jejunum and into mucosa from self-filling jejunal blind loops

	Control (n = 8)	Blind loop (n = 15)	Statistical evaluation
V_{max}, n-mol/100 mg · 2′	434 ± 28	130 ± 21	$F_{1,10}$ = 44.2
K_m, mM	6.5 ± 0'3	5.8 ± 1.1	$F_{1,10}$ = 0.38
K_D, n-mol/100 mg · 2′ · mM	2.9 ± 0.6	5.3 ± 0.5	$F_{1,10}$ = 5.19

Data obtained by incubating tissues from n animals in solutions at 8 different phenylalanine concentrations between 1 and 50 mM. Eq. (1) was then fitted to pooled data. Statistical analysis performed by effecting a joint fit under separate constraints that one kinetic parameter was common to both data sets

It is known that epithelial damage occurs in the mucosa of self-filling blind loops (Block et al. 1976). Thus the pronounced reduction in V_{max} which has been revealed in this analysis probably signifies that there is a decrease in the number of transport sites per unit weight of mucosa. A constant morphological finding in self-filling blind loops is increased proliferation rate in lengthened crypts (Block et al. 1976, Menge et al. 1979), so that our finding may simply reflect a reduction in the proportion of mature absorptive cells in the mucosa. The fact that K_m is unchanged suggests that there has been no alteration at the cellular level. The increased K_D presumably indicates the presence of breaks in a damaged epithelium.

This analytical procedure is clearly of general applicability and can be used to examine the identity of kinetic parameters whenever two sets of data are available. One warning should be expressed, however: The test is very sensitive, and so when data are obtained, for instance, in different animals and significant differences emerge, the possibility of animal variability alone being responsible for the difference must be considered.

Conclusions

It has been pointed out on several occasions that the conditions needed for the application of a least-squares analysis are very seldom met in kinetic experiments. In particular, the error structure is normally not well known. Endrenyi and Kwong (1979) give a convenient F_k-test to distinguish between constant error variance and constant relative errors. For our data, we have been able to work under the assumption of constant relative errors and hence, after a log transformation, to apply unweighted least squares. Another important point is that the presence of outliers will bias the parameter estimates in their direction. We believe that our data trimming takes care of this problem. Finally, it should be remembered that the theoretical derivation of the procedures and the tests applied in the present work is based on linear (in the parameters) models, but that, asymptotically, the estimates and tests in the non-linear case behave as their linear counterparts. Thus, even though the significance levels associated with the tests presented here are only approximate, the large sizes of our data sets allow the procedure to be quite satisfactory in the discrimination between rival models, as well as in model building. Furthermore, the parameter estimates resulting from the final fits are quite reliable, as witnessed by their respective standard errors.

There has recently been a tendency to avoid least-squares analysis in favour of robust and/or non-parametric methods, such as the median estimate from a direct linear plot (Cornish-Bowden 1979). There is little doubt that, in cases of pure enzyme kinetics where only two parameters need to be evaluated, such methods provide the best estimates when least-squares requirements are not met. However, such techniques are not readily applicable when large data sets are available, nor are they appropriate when several parameters are involved.

Acknowledgements. The experimental work described in this paper was largely supported by the Fonds National Suisse de la Recherche Scientifique. We are grateful to Dr H. Menge (Berlin) for his agreement to our quoting unpublished work performed in collaboration with him. We are indebted to Pierrette Ganguillet, Sylvianne Henriot, Sonya Jaquet and Dominique Mettraux for skilful technical assistance.

References

Alvarado F, Lherminier M (1982) Phenylalanine transport in guinea-pig jejunum. General mechanism for organic solute and sodium cotransport. J Physiol (Paris) 78:131–145

Anscombe FJ, Tukey JW (1963) The examination and analysis of residuals. Technometrics 5: 141–160

Bloch R, Menge H, Lrenz-Meyer H, Stöckert HG, Riecken EO (1976) Functional, biochemical and morphological alterations in the intestines of rats with an experimental blind-loop syndrome. Res Exp Med 166:67–78

Cornish-Bowden A (1979) Robust estimation in enzyme kinetics. In: Endrenyi L (ed) Kinetic data analysis. Plenum Press, New York London, pp 105–119

Debnam ES (1982) Effect of sodium concentration and plasma sugar concentration on hexose absorption by the rat jejunum in vivo. Further evidence for two transport mechanisms. Pflügers Arch 393:104–108

Endrenyi L, Kwong FHF (1979) Tests for the behaviour of experimental errors. In: Enrenyi L (ed) Kinetic data analysis. Plenum Press, New York London, pp 89–103

Fletcher R, Powell MJD (1964) A rapidly convergent descent method for minimization. Computer J 6:163–168

Honegger P, Semenza G (1973) Multiplicity of carriers for free glucalogues in hamster small intestine. Biochim Biophys Acta 318:390–410

Melle G van, Robinson JWL (1981) A systematic approach to the analysis of intestinal transport kinetics. J Physiol (Paris) 77:1011–1016

Menge H, Köhn R, Dietermann KH, Lorenz-Meyer H, Riecken EO, Robinson JWL (1979) Structural and functional alterations in the mucosa of self-filling blind loops in rats. Clin Sci 56: 121–131

Neter J, Wasserman W (1974) Applied linear statistical models. Richard D Irwin, Homewood Ill, p 88

Robinson JWL, van Melle G (1982) Single-site uptake of neutral amino acids into guinea-pig intestinal rings. J Physiol (Lond) 323:569–587

Robinson JWL, Antonioli JA, Johansen S (1980) The kinetics of sodium-dependent phenylalanine influx in the intestine of the dog: A comparison between ileum and colon. J Physiol (Paris) 76:637–645

Syme G, Levin RJ (1980) The validity of assessing changes in intestinal absorption mechanisms for dietary sugars with non-metabolizable analogues (glucalogues). Brit J Nutr 43:435–443

Intestinal Secretion of Organic Ions

F. LAUTERBACH[1]

Introduction

Active absorption of nutrients has been known for a long time. Active secretion of foreign compounds (xenobiotics) by the mucosal epithelium of the intestine, however, is a rather recent discovery. It has been demonstrated first for cardiotonic steroids. After i.v. administration a number of these uncharged compounds are concentrated in the intestinal fluid of rats and guinea pigs above the blood level (Lauterbach 1971a,b, 1975, 1977a; for review see Lauterbach 1981). In these as well as in other studies on the phenomenon of intestinal secretion the in vitro method of the isolated mucosa of guinea pig intestine appeared to be especially useful. Permeation by diffusion as well as absorptive and secretory transport processes are easily differentiated by comparing transepithelial fluxes across the tiny mucosal sheet devoid of additional tissue layers and mounted in a flux chamber. Furthermore, determination of tissue content at the end of the experiments allows conclusions on the respective role of the luminal and basolateral membranes of the enterocytes in the transport processes (Fig. 1) (Lauterbach 1977b).

Secretion of Quaternary Ammonium Compounds

General Characteristics in vitro

Intestinal secretion has likewise been demonstrated for a second group of xenobiotics, namely the cationic quaternary ammonium compounds. Since the intestinal permeation of these substances has been reviewed just recently (Lauterbach 1983) only the main points will be reported here.

1. In the isolated mucosa permeation of quaternary ammonium compounds from the blood side to the luminal side is higher than permeation in the absorptive direction. With N-methylscopolamine, for example, at 1 μM concentration a flux ratio of 12 was achieved, a value far exceeding that to be expected for passive distribu-

1 Institut für Pharmakologie und Toxikologie, Ruhr-Universität, D-4630 Bochum, Fed. Rep. of Germany

Intestinale Transport
(ed. by M. Gilles-Baillien and R. Gilles)

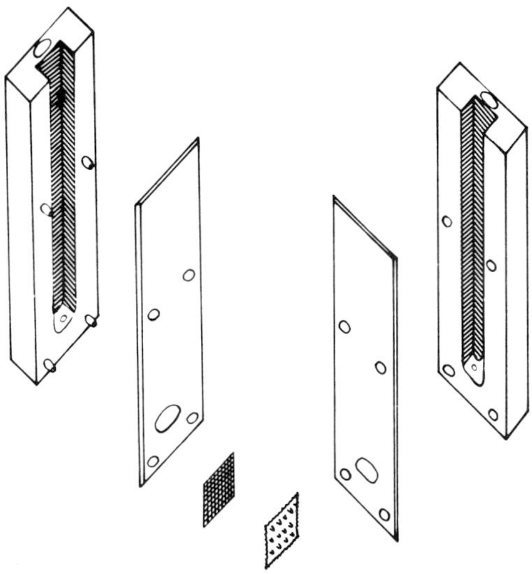

Fig. 1. Method of the isolated mucosa of guinea pig intestine. The isolated mucosa is captured on a nylon mesh and placed between two fenestrated polyvinyl chloride sheets, thus forming a separating membrane between two flux chambers. Window diameter 5 mm. The chambers are filled with 0.2 ml incubation solution and continuously aerated through drill holes in the chamber walls. (Lauterbach 1977b)

tion of a monovalent cation at the existing potential difference of 2–4 mV (Turnheim and Lauterbach 1977a).

2. Secretion rate decreased with increasing concentration indicating saturability of the transport mechanism.
3. Secretion was dependent on aerobic metabolism.
4. Tissue content was always found higher after blood side than after luminal side administration – in contrast to the behaviour observed with nutrients like sugars and amino acids which were taken up preferentially from the luminal side.

Substrate Specificity

Comparison of the secretion of six monoquaternary ammonium compounds in the isolated mucosa of guinea pig jejunum revealed a certain correlation between secretory rate and lipophilicity and/or size of the molecules. Whereas the small representatives (N-methylnicotinamide, pyridostigmine and tetraethylammonium) were secreted to only a small extent, the larger and more lipophilic tropanium derivatives (N-methylscopolamine, ipratropium and α-phenylcyclopentane acetic acid-N-isopropyl-nortropin ester methobromide) displayed severalfold higher secretory rates. Tissue content after blood side administration always exceeded that after administration on the luminal side. There was, however, no correlation between tissue content and secretory rate (Turnheim and Lauterbach 1977a, Pieper and Lauterbach 1979).

Transepithelial permeation of bisquaternary compounds (decamethonium, diquat, paraquat) seemed to be restricted almost entirely to paracellular pathways. Permeation of these compounds was strictly correlated with that of polyethylene glycol (M.W. 900), resulting in regression lines with ordinate intercepts not significantly different from zero.

Localization of the Secretory System in the Enterocyte

In countertransport experiments, uptake of (^{3}H)N-methylscopolamine from the blood side into mucosae preloaded with unlabelled N-methylscopolamine was enhanced. Likewise, efflux of labelled N-methylscopolamine from isolated mucosae into the blood side compartment was stimulated by addition of external N-methylscopolamine. Both results indicated the existence of a transport mechanism for quaternary ammonium compounds in the basolateral membrane capable of accelerated exchange diffusion (Turnheim et al. 1977).

Phenoxybenzamine, a β-haloalkylamine forming an alkylating intermediate aziridinium cation, inhibits secretion of N-methylscopolamine completely and reduced tissue content by 70%, when added to the blood side of the isolated mucosa. This has been regarded as additional evidence for the existence of a basolateral entrance system for quaternary ammonium compounds. In contrast, secretion of the more lipophilic congener, α-phenyl-cyclopentane acetic acid-N-isopropyl-nortropin ester methobromide was inhibited by only 30% under identical conditions. It was therefore concluded that quaternary ammonium compounds of sufficiently high lipophilicity might cross the basolateral membrane by diffusion as well (Pieper 1979, Pieper and Lauterbach 1979).

Existence of a transport mechanism in the luminal membrane was indicated by a number of further results. In efflux experiments, release of N-methylscopolamine from preloaded mucosae into the luminal solution was inhibited by cyanide (Turnheim et al. 1977). Anaerobiosis completely stopped secretion of all quaternary ammonium compounds, including the lipophilic cyclopentane derivative. Tissue content revealed a complex behaviour under these conditions. After luminal administration of quaternary ammonium compounds it was always increased, whereas it remained unchanged or was decreased after blood side administration (Turnheim and Lauterbach 1977a).

Hence, taking into consideration all results obtained so far, transepithelial secretion of quaternary ammonium compounds can be described unequivocally by two transport mechanisms in series. This system is similar to that inferred previously for the secretion of cardiac glycosides (Lauterbach 1971a, 1981). A first mechanism in the basolateral membrane, mediating uptake into the cell, probably dependent on metabolic energy and paralleled by a diffusional pathway, and a second mechanism in the luminal membrane extruding its substrates into the luminal solution.

Secretion in vivo

Intestinal secretion of quaternary ammonium compounds was substantiated in vivo. After i.v. administration, concentrations of N-methylscopolamine, N-methylnicotinamide and tetraethylammonium in the jejunal fluid exceeded plasma levels severalfold. As in the isolated mucosa, tetraethylammonium revealed the smallest secretory rate. With N-methylscopolamine a decrease of the concentration gradient with increasing doses has been shown, indicating saturability of the secretory system in vivo (Lauterbach 1970, Turnheim and Lauterbach 1972, 1977b).

Secretion of Organic Acids

Secretion in the Isolated Mucosa

The detection of the intestinal secretion of electroneutral cardiotonic steroids and cationic quaternary ammonium bases raised the question, whether the intestine would excrete likewise a third group of xenobiotics, organic acids. In order to minimize interference of non-ionic diffusion of the substrates, experiments were started with sulfonic acids.

The first acid investigated was sulfanilic acid. Its permeation across th isolated mucosa was strictly correlated to the simultaneous permeation of inulin in the direction lumen to blood as well as in the direction blood to lumen. There was no significant difference in the slopes of the two regression lines. Whereas, however, the regression line in the absorptive direction passed through the origin, the regression line for the secretory direction had a positive ordinate intercept. These results have been interpreted as demonstrating restriction of the absorptive flux of sulfanilic acid to paracellular, inulin permeable shunt pathways, and a small, but significant transcellular flux in the direction of secretion. As is visualized from Fig. 2, this difference would hardly have been judged to be significant by mere comparison of the mean values of the permeation in both directions. Hence, in accordance with previous experiences, estimation of and correction for simultaneous fluxes across paracellular shunt pathways turned out to be indispensible for a pertinent interpretation of the experimental data. The metabolic dependency of the secretory flux is seen under anaerobic conditions, where the difference between the two regression lines is abolished (Sund and Lauterbach 1978) (Fig. 2).

A better substrate for studying intestinal secretion of organic anions is β-naphthol orange, which was obtained by diazo coupling of (^{35}S) sulfanilic acid with β-naphthol. This acid displayed a significant difference between the mean value of permeation from lumen to blood and blood to lumen. Permeation from lumen to blood was again strictly correlated to the simultaneous permeation of inulin, pointing to restriction of the transepithelial absorptive flux to the paracellular route. In contrast, in the direction blood to lumen, no further correlation was seen, probably owing to the much greater proportion of the transcellular as compared to the paracellular flux. Under anaerobic conditions, however, permeation in both directions scatter around the same regression line (Fig. 2) (Lauterbach 1979). Thus, as observed with quaternary ammonium compounds, intestinal secretion seems to be favoured by an increased lipophilicity of the uncharged part of the molecule.

Intestinal secretion of anions is not restricted to sulfonic acids. The carbonic acid salicylic acid is even better secreted by the isolated mucosa of guinea pig jejunum than β-naphthol orange (Fig. 3). A number of results, however, raised serious doubts as to whether β-naphthol orange and salicylic acid share the same transport system. Amongst these, two deserve special mention.

The first one concerns the occurrence of organic anion secretion in different parts of the intestine. There is no parallel change of the secretory rates along the intestine. As compared with the jejunum, in the colonic mucosa the secretory rate of β-naphthol orange is eight times higher, wheras no net secretion of salicylic acid is seen anymore (Fig. 3) (Lauterbach et al. 1982).

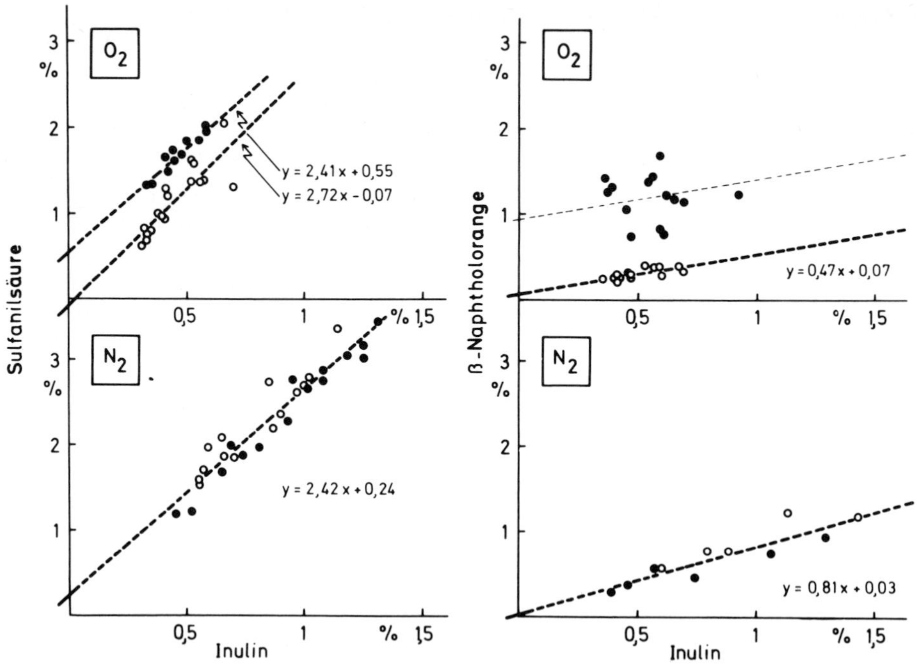

Fig. 2. Correlation between permeation of inulin and sulfanilic acid or β-naphthol orange through the isolated mucosa of guinea-pig jejunum. *Ordinates:* Permeation rate of (^{35}S)sulfanilic acid *(left)* or (^{35}S)β-naphthol orange *(right)*. *Abscissa:* Permeation rate of (^{3}H)inulin. Isolated mucosae of guinea-pig jejunum were incubated with 5 μM sulfanilic acid or 10 μM β-naphthol orange at 37°C. The acids were administered either on the luminal side *(circles)* or on the blood side *(dots)*. Permeation rate is expressed as concentration in the countercompartment after 45 min in percent of the concentration administered. Incubation under aerobic (O_2) and anaerobic (N_2) conditions. Each *point* represents one single experiment. Regression lines are given by *broken lines;* the corresponding equations are indicated *beneath the curves.* There is no significant correlation between permeation of inulin and β-naphthol orange after blood side administration under aerobic conditions; the *thin broken line* is a parallel shift of the corresponding regression line for the permeation lumen to blood intersecting the coordinates of the mean values of the permeation blood to lumen. (Lauterbach 1979)

The second observation concerns the influence of inhibitors. It turned out that the inhibitory effect of phenoxybenzamine is not as specific as might be expected from its ability to form an intermediate cation and from its introduction in transport studies as a tool for inhibiting renal secretion of quaternary ammonium compounds (Ross et al. 1968). As expected, in the isolated mucosa absorption of β-methyl-glucoside and secretion of the cardiotonic steroid digoxin were not influenced by phenoxybenzamine. Secretion of salicylic acid remained also unchanged, but secretion of β-naphthol orange was drastically inhibited (Fig. 4). Hence, there is reason to believe that more than one transport system for the secretion of organic anions exists in the intestine. A comparable situation is discussed at present for the kidney (Barany 1974, Hewitt et al. 1977, Greven 1981).

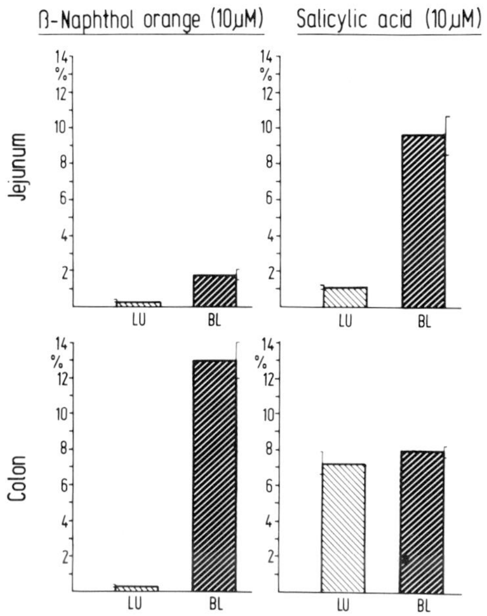

Fig. 3. Secretion of β-naphthol orange and salicylic acid in the isolated mucosa of guinea-pig jejunum and colon. Mucosae were incubated with 10 μM acid for 45 min at 37°C under aerobic conditions. Mean values ± S.E.M. of 6–16 experiments are depicted of the permeation lumen-to-blood *(LU)* and blood-to-lumen *(BL)*, expressed as concentration in the countercompartment in percent of the concentration administered. (Kilian, Lauterbach and Steinke, unpublished data)

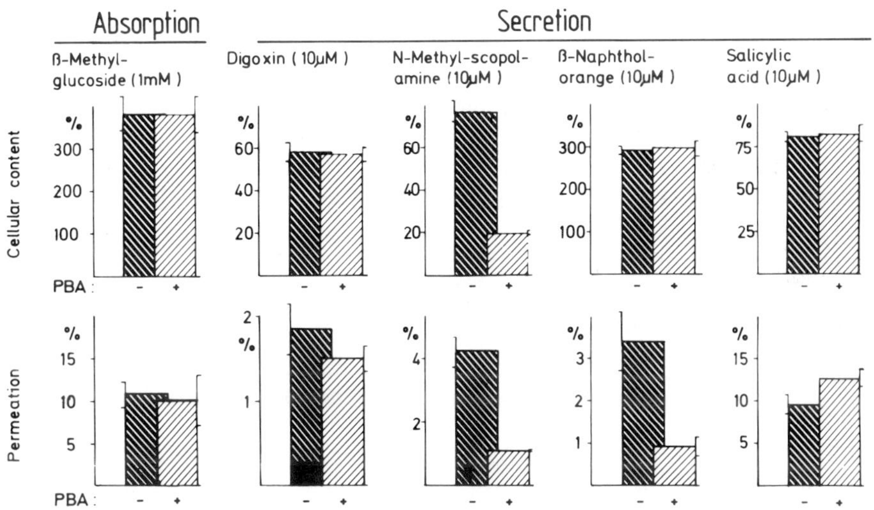

Fig. 4. Influence of phenoxybenzamine *(PBA)* on intestinal transport processes. Isolated mucosae of guinea-pig jejunum were preincubated either with 10^{-4} M PBA (+) on both sides (β-methyl-glucoside) or on the blood side only (all other substrates) or without PBA (–) for 120 min at 37°C. Substrates were administered in the concentrations indicated either on the luminal side (absorption) or on the blood side (secretion) for an additional 45-min period. *Upper part* Cellular content, defined as amount of substrate in the tissue, corrected for extracellular amount, expressed as concentration in the intracellular space in percent of the concentration administered. *Lower part* Permeation, expressed as concentration in the countercompartment in percent of the concentration administered. Means ± S.E.M. of 6–20 experiments. (Lauterbauch, Pieper and Steinke, unpublished data)

Concerning the inhibition of β-naphthol orange secretion by phenoxybenzamine, it is interesting to note that Holohan et al. (1979) observed a copurification of the renal cation (N-methylnicotinamide) and anion (p-aminohippurate) carrier protein although the two transports could be demonstrated to be distinct entities.

Secretion in vivo

Intestinal secretion of organic anions has been substantiated in vivo. After i.v. administration of β-naphthol orange the concentration ratio of unmetabolized, non protein-bound acid jejunal fluid/plasma approached 13 in the rat within 135 min, and more than 40 in guinea pig within 60 min (Lauterbach 1977a, Kilian and Lauterbach 1978).

Interference of Intestinal Secretion with Absorption

The secretion of xenobiotic substances must necessarily interfere with their absorption. Indeed, the discovery of intestinal secretion allows an explanation of peculiarities in the absorption of xenobiotics, which have been long known, though widely ignored, and which are of particular importance for the absorption kinetics of drugs. First, absorption of several drugs is characterized by a rapid initial uptake, followed by a standstill of absorption in spite of large amounts of unabsorbed drug remaining in the intestine. This phenomenon has been observed with quaternary ammonium compounds (Levine et al. 1955, Levine and Clark 1957a,b, Levine 1959, 1960, 1961, Turnheim and Lauterbach 1980) as well as with cardiac glycosides (Seidenstücker and Lauterbach 1976, Seidenstücker 1978, Lauterbach 1981). It could be demonstrated theoretically as well as experimentally that this phenomenon in fact reveals an equilibrium between absorption and secretion. The situation is most easily understood by regarding a computer simulation of the time course of absorption in a simplified two-compartment model (Fig. 5a). The model describes the permeation of a substrate across a barrier by a transport mechanism mediating permeation in both directions. It is capable of net secretory transport because of a lower K_m value at the trans-side than at the cis-side of the membrane. In addition, it assumes a parallel, diffusive pathway. As is visualized from Fig. 5, the time course of absorption depends on the intestinal concentration in a characteristic manner. Absorption of low doses is initially favoured by the kinetic advantage of being transported in the absorptive direction by the secretory system. But later, an equilibrium between absorption and secretion is attained resulting in a standstill of net absorption. Increasing doses, however, tend to saturate the secretory system which, therefore, becomes more and more unable to cope with the amounts absorbed. Therefore, no equilibrium is reached and absorption continues resulting in a higher absorption rate than that observed with lower doeses. In vivo experiments with N-methylscopolamine (Fig. 6) as well as with digoxin (Seidenstücker and Lauterbach 1976, Seidenstücker 1978, Lauterbach 1981) verified these predicted absorption kinetics. Validity of the interpretation of the halt in absorption as an equilibrium between absorption and secretion was proved by perturbation experiments. After the absorption had caesed, it could be restarted by

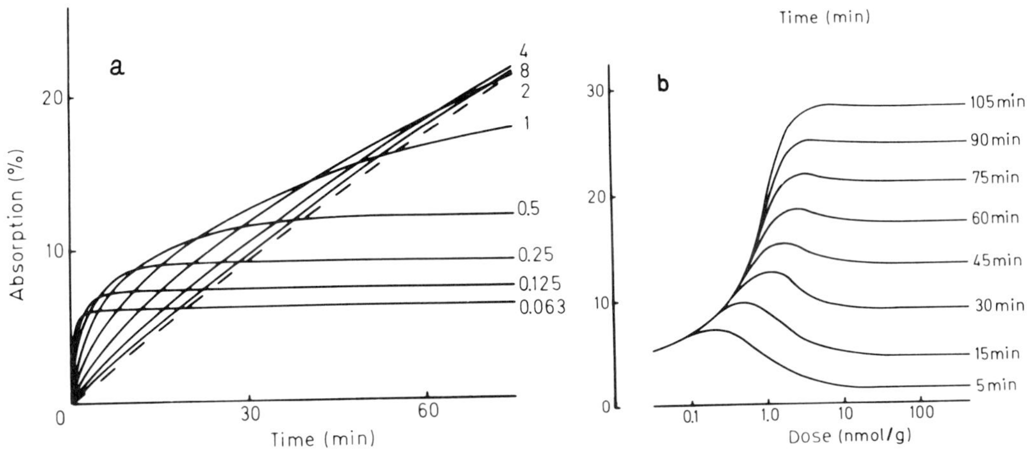

Fig. 5 a,b. Computer simulation of the absorption kinetics as a function of time and concentration in a simplified two-compartment system. The substrate permeates the separating membrane by a secretory transport system and a parallel, diffusional pathway. Absorption is assumed to proceed in an animal of 350 g from the intestinal lumen (volume 1 cm^3) across an epithelium (area 30 cm^2) into an apparent distribution volume (800 cm^3). The parameters used for calculation have the following values: $P = 1 \times 10^{-4}$ cm min^{-1}, $J_{max} = 0.3$ nmol min^{-1} cm^{-2}, $k_{12} = 8 \times 10^{-5}$ M, $K_{21} = 1.2 \times 10^{-8}$ M. **a** Time dependence of absorption at the indicated dose (nmol g^{-1}). *Broken line* represents the asymptote of the time course of absorption which is approached at very high intestinal substrate concentration. **b** Dose dependence of absorption at the indicated absorption periods. (Turnheim and Lauterbach 1980)

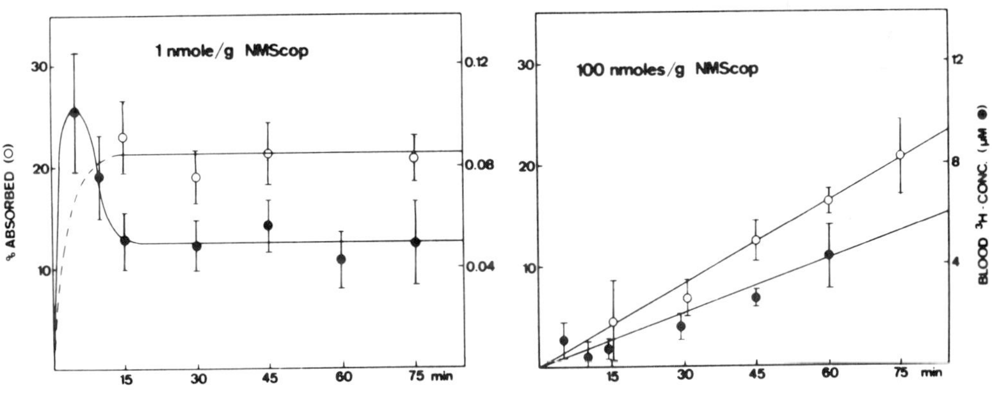

Fig. 6. Time course of absorption at different doses. 1 nmol g^{-1} *(left)* or 100 nmol g^{-1} *(right)* of (^{3}H)N-methylscopolamine were administered into an isolated jejunal loop of guinea pigs. *Ordinates* indicate absorption *(circles)* or blood levels of radioactivity *(dots)*. Means ± S.E. of four to seven experiments. (Turnheim and Lauterbach 1980)

reducing the concentration on the blood side to zero by transferring the intestinal content to another, untreated animal. It could be restarted also by i.v. administration of an excess of unlabelled substrate competing with the absorbed, labelled substrate for resecretion into the intestinal lumen (Seidenstücker and Lauterbach 1976, Seidenstücker 1978, Turnheim and Lauterbach 1980).

Second, absorption against a secretory system offers an explanation for another peculiarity, namely, an *increase* in absorption *rate,* determined after a defined absorption period, with rising doses. Theoretically this phenomenon can be ascribed to an increasing state of saturation of the secretory system (cf. Fig. 5b). Hypothetically, it could be deduced from data published by Levine and Pelikan already in 1961 (Lauterbach 1975). It was substantiated for the absorption of N-methylscopolamine in the guinea pig (Fig. 7) (Turnheim and Lauterbach 1980). Even the occurrence of a maximum in the absorption rate-dose relation could be demonstrated experimentally for ouabain in the rat (Seidenstücker 1978).

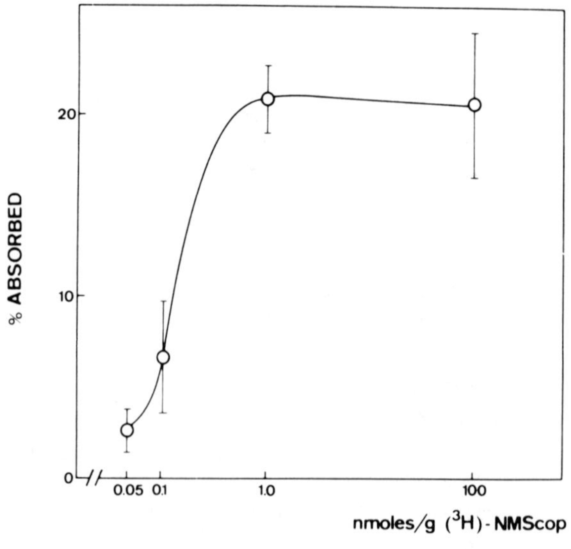

Fig. 7. Dose dependence of absorption rate: (^{3}H)N-methylscopolamine was administered in the doses indicated on the *abscissa* into an isolated jejunal loop of guinea pigs. Absorption rate within 75 min is indicated on the *ordinate.* Means ± S.E. of 4–7 experiments. (Turnheim and Lauterbach 1980)

Conclusions

Transport of organic substances through the mucosal epithelium of the intestinal tract is no one-way route. It turns out that absorptive *and* secretory transport systems occur in close spatial relationship, a situation which has long been realized for inorganic ions. At the present state of knowledge it seems that the substrates of the secretory systems belong to those classes of compounds which are no normal constituents or nutrients of the organism (xenobiotics). So far, intestinal secretion of neutral cardiotonic steroids, cationic quaternary ammonium compounds and anionic sulfonic and carbonic acids has been discovered. There are some indications that at least for organic acids there might be more than one transport system. Thus the intestine joins the kidney and the liver as a third excretory organ for the elimination of xenobiotic substances and their metabolites.

All available evidence leads to the conclusion that intestinal secretion of organic substances is brought about by two transport mechanisms in series located in the basolateral and the luminal membranes of the enterocytes. Though intestinal secretion is dependent on metabolism, the energization of the transport step itself remains unclear. Experiments with the cardiotonic steroid digoxin gave no indication for a sodium gradient driven process since the secretion was not influenced by the omission of sodium from the incubation medium (Lauterbach 1977a).

Besides its importance for the elimination of foreign compounds, intestinal secretion has far-reaching consequences for their absorption. As long as xenobiotics, e.g., drugs, are substrates of these systems, their absorption has to proceed *against* a secretory transport. Simultaneously, absorption is very likely mediated partly *by* these transport systems. As a consequence, absorption rate becomes a complex function of intestinal concentration and absorption time. In these cases, absorption rate can be determined only for one set of experimental conditions. Extrapolation to other conditions of concentration and time might lead to serious errors.

References

Bárány EH (1974) Selectivity of probenecid congeners for different organic acid transport systems in rabbit renal cortex. Acta Pharmacol Toxicol 35:309–316

Greven J (1981) Renal transport of drugs. In: Greger R, Lang F, Silbernagl S (eds) Renal transport of organic substances. Springer, Berlin Heidelberg New York, pp 262–277

Hewitt WR, Wagner PA, Bostwick EF, Hook JB (1977) Transport ontogeny and selective substrate stimulation as models for identification of multiple renal organic anion transport systems. J Pharmacol Exp Ther 202:711–723

Holohan PD, Pessah NI, Ross CR (1979) Reconstitution of N^1-methylnicotinamide and p-aminohippuric acid transport in phospholipid vesicles with a protein fraction isolated from dog kidney membranes. Pharmacology 16:343–356

Kilian U, Lauterbach F (1978) Intestinal secretion of xenobiotics. Gastroenterol Clin Biol 2:323

Lauterbach F (1970) Werden quaternäre Ammoniumverbindungen über einen enteralen Sekretionsmechanismus resorbiert? Naunyn-Schmiedeberg's Arch Pharmakol 266:388

Lauterbach F (1971a) Untersuchungen über den Mechanismus der Permeation cardiotoner Steroide durch die Mucosa des Dünndarmes – ein Beitrag zur Theorie der Resorption von Pharmaka. Habilitationsschrift, Bochum/Essen

Lauterbach F (1971b) Absorption of cardiac glycosides. Acta Pharmacol Toxicol 29: Suppl 4, 80

Lauterbach F (1975) Resorption und Sekretion von Arzneistoffen durch die Mukosaepithelien des Gastrointestinaltraktes. Arzneim Forsch 25:479–488

Lauterbach F (1977a) Intestinal secretion of organic ions and drugs. In: Kramer M, Lauterbach F (eds) Intestinal permeation. Excerpta Medica, Amsterdam Oxford, pp 173–194

Lauterbach F (1977b) Passive permeabilities of luminal and basolateral membranes in the isolated mucosal epithelium of guinea pig small intestine. Naunyn-Schmiedeberg's Arch Pharmacol 297:201–212

Lauterbach F (1979) Enterale Resorptions- und Sekretionsmechanismen. In: Rietbrock N, Schnieders B (eds) Bioverfügbarkeit von Arzneimitteln. Fischer, Stuttgart New York, pp 109–126

Lauterbach F (1981) Intestinal absorption and secretion of cardiac glycosides. In: Greeff K (ed) Cardiac glycosides. Springer, Berlin Heidelberg New York (Handbook of experimental pharmacology, vol 56/II, pp 105–139)

Lauterbach F (1982) Intestinal permeation of organic bases and quaternary ammonium compounds. In: Csáky TZ (ed) Pharmacology of intestinal permeation. Springer, Berlin Heidelberg New York (Handbook of experimental pharmacology, vol 70/II)

Lauterbach F, Giese M, Kilian U, Steinke W (1982) Comparative aspects of diffusion and transport of drugs across the mucosa of small intestine and colon. Naunyn-Schmiedeberg's Arch Pharmacol 321:R55

Levine RM (1959) The intestinal absorption of the quaternary derivatives of atropine and scopolamine. Arch Int Pharacodyn Ther 121:146–149

Levine RM, Clark BB (1957a) The physiological disposition of oxyphenonium bromide (antrenyl) and related compounds. J Pharmacol Exp Ther 121:63–70

Levine RM, Clark BB (1957b) A note on the intestinal absorption of penthienate (monodral). Arch Int Pharmacodyn Ther 112:458–462

Levine RM, Blair MR, Clark BB (1955) Factors influencing the intestinal absorption of certain monoquaternary anticholinergic compounds with special reference to benzomethamine (N-diethylaminoethyl-N′-methyl-benzilamide methobromide (MC-3199). J Pharmacol Exp Ther 114:78–86

Levine RR (1960) The physiological disposition of hexamethonium and related compounds. J Pharmacol Exp Ther 129:296–304

Levine RR (1961) The influence of the intraluminal intestinal milieu on absorption of an organic cation and an anionic agent. J Pharmacol Exp Ther 131:328–333

Levine RR, Pelikan EW (1961) The influence of experimental procedures and dose on the intestinal absorption of an onium compound, benzomethamine. J Pharmacol Exp Ther 131:319–327

Pieper B (1979) Influence of phenoxybenzamine on the intestinal secretion of various quaternary ammonium compounds by two transport mechanisms in series. Naunyn-Schmiedeberg's Arch Pharmacol 307:R5

Pieper B, Lauterbach F (1979) Specificity, localization and inhibition of intestinal transport systems for quaternary ammonium compounds. Gastroenterol Clin Biol 3:167–168

Ross CR, Pessah NI, Farah A (1968) Inhibitory effects of β-haloalkylamines on the renal transport of N-methylnicotinamide. J Pharmacol Exp Ther 160:375–380

Seidenstücker R (1978) Beziehungen zwischen enteraler Sekretion und Resorption herzwirksamer Glycoside. Ph D Thesis, Bochum

Seidenstücker R, Lauterbach F (1976) Mediation of intestinal absorption of cardiotonic steroids by a secretory transfer mechanism. Naunyn-Schmiedeberg's Arch Pharmacol 293:R45

Sund RB, Lauterbach F (1978) Intestinal secretion of sulphanilic acid by the isolated mucosa of guinea pig jejunum. Acta Pharmacol Toxicol 43:331–338

Turnheim K, Lauterbach F (1972) Intestinal transport of quaternary ammonium compounds in vivo. Naunyn-Schmiedeberg's Arch Pharmacol 274:R118

Turnheim K, Lauterbach F (1977a) Absorption and secretion of monoquaternary ammonium compounds by the isolated intestinal mucosa. Biochem Pharmacol 26:99–108

Turnheim K, Lauterbach F (1977b) Secretion of monoquaternary ammonium compounds by guinea pig small intestine in vivo. Naunyn-Schmiedeberg's Arch Pharmacol 299:201–205

Turnheim K, Lauterbach F (1980) Interaction between intestinal absorption and secretion of monoquaternary ammonium compounds in guinea pigs – A concept for the absorption kinetics of organic cations. J Pharmacol Exp Ther 212:418–424

Turnheim K, Lauterbach F, Kolassa N (1977) Intestinal transfer of the quaternary ammonium compound N-methyl-scopolamine by two transport mechanisms in series. Biochem Pharmacol 26:763–767

Coupling Stoichiometry and the Energetic Adequacy Question

G. KIMMICH[1]

Introduction: Na^+-Dependent Transport Systems – A Historical Overview

The Early Models

Historically, scientific consideration of the energetics of intestinal Na^+-dependent transport systems has evolved through five distinct conceptual stages (see Table 1). The earliest stage involved recognition of a mechanistic role for Na^+ (Riklis and Quastel 1958, Crane et al. 1961), Characterization of its kinetic effects (Crane et al. 1965, Curran et al. 1967, Goldner et al. 1969) and proposal of transport models which considered flow of Na^+ down a gradient of chemical potential as the means of energy input to the transport system (for a review see Schultz and Curran 1970). These models typically involved membrane components with binding sites for both Na^+ and an organic solute. On the basis of kinetic information, Na^+ was thought to alter either the affinity of the carrier for its substrate (kinetics) (Crane et al. 1965, Curran et al. 1967) or the carrier permeability (V kinetics) (Goldner et al. 1969) or

Table 1. Stages in the evolution of Na^+-dependent transport models

Stage	Dates	Events	Energetic concepts
I	1958–1968	Early model formulation Kinetic effects of Na	Flow of Na^+ down a $\Delta\mu_{Na^+}$ 1:1 Na: sugar stoichiometry
II	1968–1972	"Challenge"	$\Delta\mu_{Na^+}$ is insufficient; second energy input required
III	1972–1980	A role for $\Delta\psi$	Flow of Na^+ down a $\Delta\tilde{\mu}_{Na^+}$
IV	1977–1980	Properties of serosal sugar transfer – enhanced sugar gradients	$\Delta\tilde{\mu}_{Na^+}$ is insufficient for usual models
V	1980–	New methods for stoichiometry measurement – new models and concepts	Actual Na^+: sugar ratio is 2.0 Measurement of theoretical sugar gradient

1 Department of Radiation Biology and Biophysics, University of Rochester, School of Medicine and Dentistry, Rochester, N.Y. 14642, USA

Intestinale Transport
(ed. by M. Gilles-Baillien and R. Gilles)

both (mixed kinetics) (see Fig. 1). In each case, the flow of Na^+ across the cell membrane and down a gradient of chemical potential was believed to drive the flow of organic solute into the cell against a gradient of chemical potential (Fig. 2). At the steady state, it should be mandatory for the epithelial cell to maintain a larger gradient

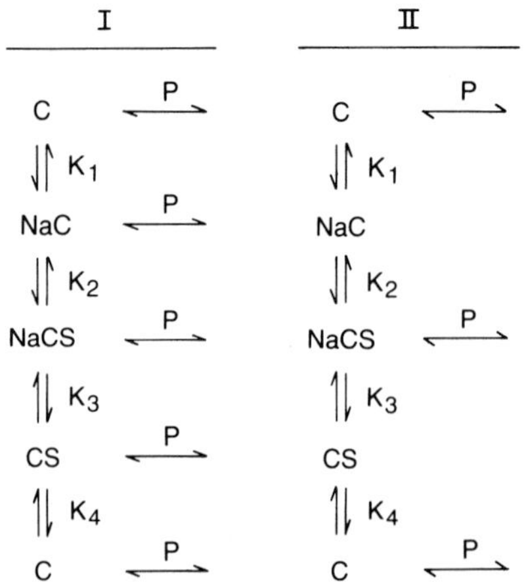

Fig. 1. Two early models for Na^+-dependent transport systems. For *Model I* each form of the carrier is thought to have equal permeability to the plasma membrane. Na^+ can alter the affinity of the carrier for its substrate ($K_2 < K_3$) such that the carrier will have differing affinities at the inner and outer membrane interfaces due to the transmembrane gradient of Na^+ concentration. In *Model II*, Na^+ is proposed to modify the peremeability of the carrier such that the ternary complex has higher permeability than either binary complex. For detailed descriptions of similar models see Curran et al. (1967) and Goldner et al. (1969)

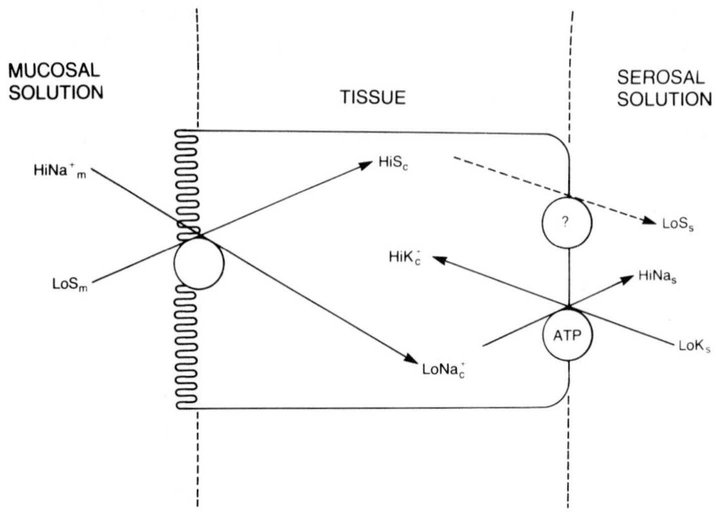

Fig. 2. Schematic representation summarizing early ideas for the role of Na^+-dependent transport systems in mediating trans-epithelial solute transfer. The Na^+-dependent systems are localized in the brush-border and allow the epithelial cell to accumulate solute against a concentration gradient. Energy was thought to be provided by flow of Na^+ down a gradient of chemical potential which is maintained by the serosally localized (Na^+ + K)ATPase. Other transport systems for the accumulated solute were believed to exist in the serosal cell boundary and to allow transfer from the cell to an extracellular compartment accessible to the circulatory system. These basic ideas have been expanded as described. (Kimmich 1981b)

of chemical potential for Na^+ than for the organic solute if this concept is correct. In fact, the steady state ratio of the two chemical potential gradients

$$\left(\frac{\Delta\mu_{\text{solute}}}{\Delta\mu_{Na^+}}\right) \times 100$$

has been considered to be a measure of efficiency for coupled transport systems.

Challenges to the Early Models

In the late 1960's and early 1970's, a number of reports surfaced which challenged the developing dogma of Na^+-dependent transport systems driven strictly by the flow of Na^+ down a trans-membrane gradient of chemical potential. These challenges originated in work with ascites cell amino acid transport systems (Jacquez and Schafer 1969, Schafer and Heinz 1971, Potashner and Johnstone 1971) and were extended to include sugar transport systems in isolated intestinal epithelial cells prepared from chickens *(Gallus gallus)* (Kimmich 1970, review by Kimmich 1973). In every instance, they were based on observations in which under certain circumstances the difference in chemical potential for the organic solute can be greater than the measured difference in chemical potential for Na^+ (see Table 2). For the case with isolated intestinal cells (Kimmich 1970, 1973), concentrative sugar accumulation can actually occur under conditions in which the Na^+ gradient is reversed from normal (i.e., $[Na^+]_i > [Na^+]_o$). The latter observations imply that, at the very least, there must be an energy input provided to Na^+-dependent transport systems which can compensate for the unfavorable driving force represented by experimentally imposed reversed Na^+ gradients.

Table 2. Early examples of solute gradient formation where $\Delta\mu_S > \Delta\mu_{Na^+}$

Solute	$[Na]_o$	$[Na]_i$	$[S]_i$	$[S]_o$	$\frac{[Na]_o}{[Na]_i}$	$\frac{[S]_i}{[S]_o}$	Reference
AIB	40.2	31.7	14.06	2.0	1.27	5.53	Jacquez and Schafer (1969)
AIB	38.06	56.53	22.0	3.51	0.68	6.27	Shafer and Heinz (1971)
Methionine	145	110	6.86	2.0	1.32	3.43	Potashner and Johnstone (1971)
Glycine	145	110	21.0	2.0	1.32	10.5	Potashner and Johnstone (1971)
Galactose	80	80	1.94	1.25	1.0	1.55	Kimmich (1970)
3-O-Methylglucose	80	80	5.62	1.25	1.0	4.5	Kimmich (1970)
Valine	80	80	3.5	1.0	1.0	3.5	Tucker and Kimmich (1973)

The Membrane Potential as an Energy Input

The first clue to the nature of this alternative energy input was again provided in experiments with ascites tumor cells. In 1972, Gibb and Eddy demonstrated that ATP-depleted ascites cells can accumulate methionine against a concentration gradient by a mechanism which depends in part on the magnitude of the membrane potential. These investigators induced changes in the membrane potential by experimentally imposing K^+ gradients and adding valinomycin to create a diffusion potential for K^+. When the imposed Na^+ gradient and induced membrane potential are comparable in magnitude to that maintained by normally energized cells, then the methionine gradients observed for the two situations are also similar (Gibb and Eddy 1972, Reid et al. 1974). In 1974, a similar dependence on membrane potential was described for Na^+-dependent glucose transport in isolated brush border membrane vesicles prepared from rat *(Rattus rattus)* small intestine (Murer and Hopfer 1974). Subsequent work with similar epithelial membrane vesicle preparations from both intestinal and renal tissue of rats or rabbits *(Oryctolagus cuniculus)* provided evidence for the fact that the capability of Na^+-dependent transport systems can be modified when the membrane potential is manipulated with a variety of imposed ion gradients and ion selective ionophores (Hopfer et al. 1975, Sigrist-Nelson et al. 1975, Beck and Sacktor 1975). Either anion or cation gradients can be used to elicit the changes in membrane potential. Similar observations have also been provided from studies with ATP-depleted isolated intestinal epithelial cells as shown in Fig. 3 (Carter-Su and Kimmich 1979, 1980). The latter preparation offers the technical advantage of larger intracellular volumes so that imposed gradients of ions and/or potential persist for a much

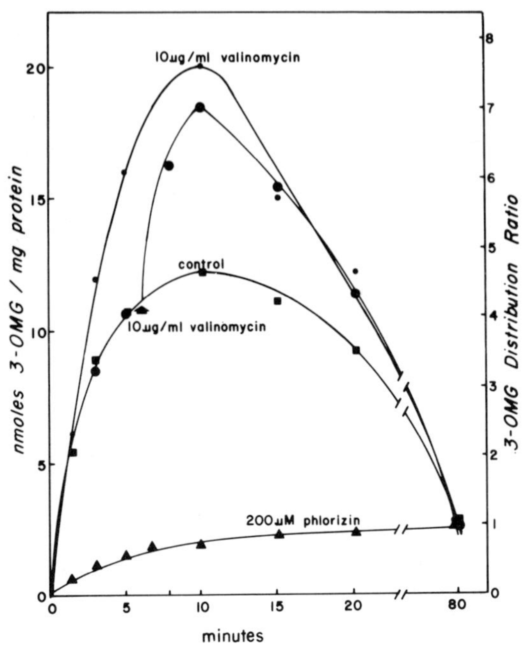

Fig. 3. Accumulation of 3-OMG in ATP-depleted intestinal epithelial cells induced by an imposed membrane potential. The ATP-depleted cells were loaded with K^+ during a pre-incubation interval and suddenly introduced to a medium in which Na^+ replaced K^+. In two cases the diffusion potential for K^+ was enhanced by adding valinomycin either at the outset or part way through the experiment. Note the marked enhancement in Na^+-dependent sugar accumulation in each case. (Kimmich et al. 1977)

longer interval of time before being dissipated by diffusion. This allows the opportunity for easier kinetic studies and for discerning the mechanistic role of chemical versus electrical driving forces.

The general principle established by all of these studies is the fact that electrical driving forces play a key role in the mechanism of Na^+-dependent transport for a number of organic solutes. In general, on the basis of this information, attention shifted from the old idea of Na^+ flow down a gradient of *chemical* potential ($\Delta\mu_{Na^+}$) as the energetic basis of transport to a new paradigm which regards Na^+ flow down an *electro-chemical* potential gradient ($\Delta\tilde{\mu}_{Na^+}$) as the energizing force (see Fig. 4). In every instance where electrical forces were implicated, an interior-negative membrane potential was found to provide an enhancement of Na^+-dependent transport capability for the organic solute.

The new insight for the mechanistic basis of Na^+-dependent transport seemed to fully resolve earlier reservations regarding the adequacy of energy input for these systems. Indeed, when a steady-state gradient of cellular solute accumulation has been achieved, the trans-membrane difference in chemical potential for the organic solute typically was shown to be less than the difference in electrochemical potential

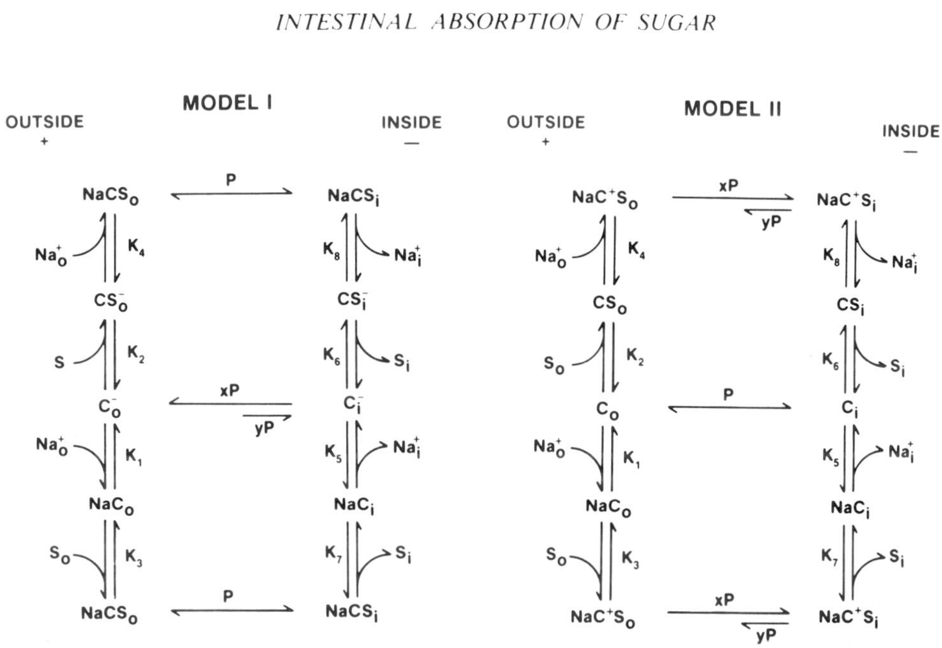

Fig. 4. Two models for Na^+-dependent sugar transport which include a role for the membrane potential. In *Model I* the free carrier is postulated to be anionic and driven to the outer membrane face by an interior negative membrane potential where it can load with Na^+ and sugar to form a neutral ternary complex. In *Model II* the free carrier is neutral but after binding Na^+ and sugar a cationic ternary complex is formed which can be drawn inward by the membrane potential. In addition to the electrical driving force acting on one carrier form in each model, there is a driving force represented by the difference in chemical potential for Na^+ which exists across the membrane. (Kimmich and Carter-Su 1978)

difference for Na^+ in accord with thermodynamic expectation (Armstrong et al. 1973, Kessler and Semenza 1979). This expectation can be written notationally as follows:

$$RT \ln \frac{(S)_i}{(J)_o} < \left(RT \ln \frac{(Na^+)_o}{(Na^+)_i} + F\Delta \right) n \tag{1}$$

where the subscripts i and o designate intracellular and extracellular compartments respectively and n is the coupling stoichiometry between Na^+ and solute flow. Because of the earlier reported stoichiometries of 1.0 or less for sugars (Goldner et al. 1969) and for amino acids (Curran et al. 1967), most analyses of energetic adequacy usually have presumed n = 1.0 as shown for the two models depicted in Fig. 4.

For intestinal tissue, one of the most complete attempts at testing the thermodynamic relationship given in Eq. (1) was reported by Armstrong et al. (1973). They made use of potential sensing and ion selective microelectrodes to obtain a direct brush border membrane of columnar epithelial cells in intact rat small intestine. They concluded that the sugar gradient which these cells maintained could have been established by flow of Na^+ across the membrane down its gradient of electrochemical potential ($\Delta\tilde{\mu}_{Na^+}$) by a mechanism involving 1:1 coupling between Na^+ and sugar molecules. Unfortunately, the epithelial preparation used in this study established only a 3–4-fold gradient of sugar which is much less than intestinal cells are capable of achieving under optimal conditions. Even for a 4-fold sugar gradient, Armstrong et al. (1973) found that the Na^+-dependent carrier would need to operate at approximately 60% efficiency using the usual definition of efficiency

$$\frac{\Delta\mu_{sugar}}{\Delta\tilde{\mu}_{Na^+}} \times 100.$$

More recently, Kessler and Semenza (1979) utilized a preparation of brush border vesicles to study the relationship between the magnitude of imposed electrochemical potential gradients for Na^+ and the $\Delta\mu$ for sugar which such vesicles can establish. They too concluded that the data were compatible with a highly efficient system exhibiting a 1:1 coupling ratio.

Alternative Flux Routes as a Limitation to Gradient Formation

Because of these and similar reports it appeared that the energetic adequacy controversy raised by the challenges to the early models for gradient coupled transport had been resolved. However, all of the work aimed at resolution of this problem suffered from a fundamental flaw in interpretation. In each case, no consideration was given to the fact that sugar fluxes into and out of the epithelial cell are mediated in part by routes other than via the Na^+-dependent carrier system. In general, these alternative flux routes are non-concentrative and represent diffusional or facilitated diffusional fluxes. For an intact epithelium the facilitated diffusional flux routes often relate to the function of transport systems which in vivo allow transfer of the organic solute across the serosal boundary of each epithelial cell. When only uni-directional influxes of solute are considered these non-concentrative pathways can seem inconsequential in magnitude. It is important to recognize, however, that at a steady-state of solute accumulation the gradient forming capability of the Na^+-dependent transport systems is severely diminished by continual loss of solute from the cell by dissipative "leaks" represented by the non-concentrative routes.

For a long time the extent to which "leak" pathways limited gradient formation was overlooked in the consideration of Na^+-dependent transport energetics. This was, in part, due to limited information regarding the characteristics of epithelial flux routes associated with the serosal cell membrane. The next stage of conceptual development depended on a growing body of information about the nature of serosal solute transfer. This work originated in studies with isolated intestinal epithelial cells which are ideally suited for such considerations because all boundaries of the cell are equally exposed to the suspending medium and therefore accessible for solute transfer. In addition, the isolated cell retains the capability for generating a stable transmembrane $\Delta\tilde{\mu}_{Na^+}$ so that the relationship between solute gradient-forming systems and dissipative pathways can be studied.

The separate pathways for sugar influx into isolated intestinal cells have been characterized with the aid of chemical agents which selectively inhibit their function. For instance, when the sugar concentration is 100 μM, about 95% of the total sugar influx is Na^+-dependent and inhibites by phlorizin. This predominant flux pathway therefore, has the properties established for the brush-border localized, Na^+ gradient coupled transport system. Another 3% of the total flux is carrier-mediated but not dependent on Na^+, nor is it concentrative in capability. It is inhibited by a variety of inhibitors which in decreasing order of effectiveness include phloretin, cytochalasin B, flavones, flavanones and theophylline (Kimmich and Randles 1975, 1978, 1979, Randles and Kimmich 1978). This facilitated diffusion flux pathway was originally postulated to be localized in the serosal boundary of the epithelial cell where it could transfer sugar either to or from the circulatory system depending on the direction of the sugar gradient at a particular instant in time. Subsequent work with membrane vesicles prepared from the serosal membrane of intestinal epithelium confirmed the predicted locus (Hopfer et al. 1975, Murer 1976). The remaining 2% of total cellular sugar influx shows no evidence of saturability or competition by sugar analogs and is apparently due to a diffusional pathway (Kimmich and Randles 1979). The separate unidirectional flux pathways are shown schematically in Fig. 5.

Although the diffusional and facilitated diffusional flux routes together account for only 5% of the total unidirectional influx of sugar to an intestinal epithelial cell, they play a much greater role in the overall trans-membrane flows of sugar in a cell maintaining a steady-state concentration gradient. As shown in the lower half of Fig. 5 for cells which can typically establish a 10–15-fold sugar gradient, the passive flux routes will account for 50%–75% of the total cellular efflux of sugar. Stated differently, the Na-dependent sugar transport system operates against a steady-state "leak" flux of sugar which is equal to 2/3 of the active entry rate. This implies that agents which block a passive flux pathway should allow the active system to create a much better sugar gradient than under control conditions. Such gradient enhancement has in fact been observed (Kimmich and Randles 1975, 1978, 1979, Randles and Kimmich 1978). With cytochalasin B, which is the most effective inhibitor of

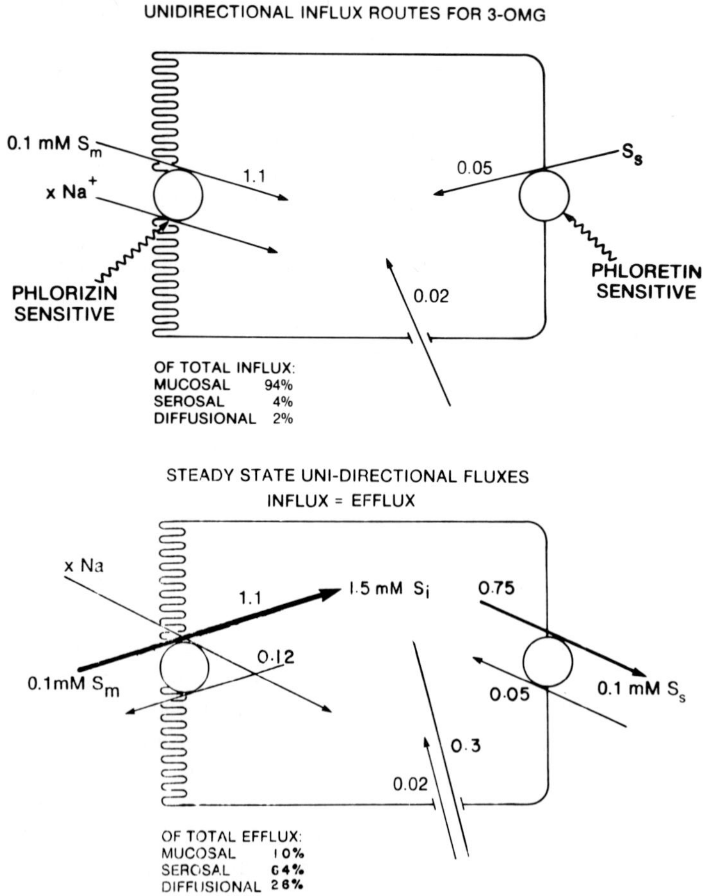

Fig. 5. Schematic representation of flux routes for 3-O-methylglucose which exist in isolated intestinal epithelial cells. *Top diagram* shows values for unidirectional influx of the sugar by three different entry routes. More than 90% of the total flux is catalyzed by the Na^+-dependent carrier. *Lower diagram* values are given for unidirectional fluxes for the same systems operating when the cell has established a 15-fold steady state gradient of sugar. Note that most of the efflux of 3-OMG is catalyzed by the passive flux routes (diffusion and facilitated diffusion). (Kimmich 1981b)

the serosal sugar carrier, sugar gradients as large as 70-fold can be established by the isolated epithelial cells (Kimmich and Randles 1979). More recently, it has been found that certain sugars (such a α-methylglucoside) are not substrates for the serosal carrier (Kimmich and Randles 1981). With these sugars the isolated cells spontaneously establish sugar gradients of nearly 100-fold. Because there is still a very significant sugar efflux via the diffusional pathway under these circumstances, it is clear that the inherent gradient forming capability for the sugar carrier must be even greater than 100-fold.

Sugar gradients of this magnitude are not compatible with an energy input driven strictly by the flow of Na^+ down a gradient of electrochemical potential, if the early ideas regarding 1:1 coupling stoichiometry between Na^+ and sugar are correct. Indeed, using Eq. (1) and reported values for the membrane potential and Na^+ gradient maintained by intact intestinal epithelium, it is easy to calculate that a maximum sugar gradient of approximately 30-fold would be established (Schultz 1977). This maximum would only be achieved for a situation in which all of the sugar flux (influx and efflux) occurs via the Na^+ coupled sugar carrier (Kimmich 1981a,b).

The foregoing discussion implies that one of two alternatives must be true. Either an energy input exists for sugar transport in addition to the flow of Na^+ down a $\Delta\tilde{\mu}_{Na^+}$ *or* the coupling stoichiometry between Na^+ and sugar is greater than 1.0.

New Insights to Na^+-Dependent Transport Energetics

The Na^+: Solute Coupling Stoichiometry

Recently, we have shown that an accurate value for coupling stoichiometry can only be obtained when appropriate precautions are taken to prevent changes in the membrane potential during the period of measurement (Kimmich and Randles 1980). Coupling ratios determined without this precaution are underestimates of the true value. We have reported an experimental approach for obtaining accurate values using ATP-depleted epithelial cells in which the membrane potential was maintained near zero with the aid of valinomycin and relatively high but equal intracellular and extracellular K^+ concentrations. In this situation the potential is "set" near zero due to the high permeability of K^+ ions relative to other ionic species. Under ordinary circumstances when valinomycin is absent, the addition of sugar induces a Na^+ influx with concomitant disturbance of the membrane potential. The inclusion of valinomycin avoids this problem. A representative experiment for the voltage-clamped situation is shown in Fig. 6 in which phlorizin was employed to identify Na^+ and sugar fluxes associated with the sugar carrier. Note that the Na^+:sugar flux ratio on the carrier is 2.0 under these conditions. The same experiment performed with no control of the membrane potential gives an apparent, but inaccurate, coupling ratio near 1.0, similar to the value reported for intact epithelial sheets when changes in the potential were not controlled (Goldner et al. 1969).

Because of the exponential relationship between the theoretical limit to sugar gradient formation and the Na^+:sugar coupling ratio, a ratio of 2.0 implies that the carrier is potentially capable of establishing sugar gradients near 400-fold. However, this limit could only occur if no sugar leak pathways were operative.

Transport Efficiency

An interesting result of the work using inhibitors of facilitated diffusional sugar transport to induce sugar gradient enhancement by the Na^+-dependent carrier is that

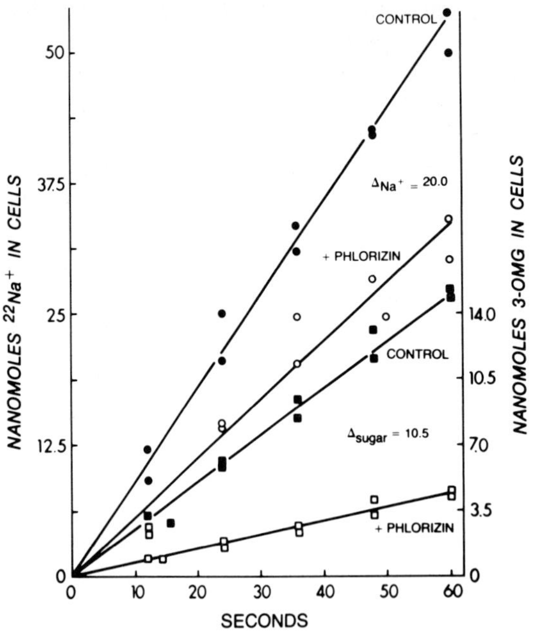

Fig. 6. Determination of Na^+:sugar flux stoichiometry using ATP-depleted intestinal epithelial cells "clamped" at a potential near zero with the aid of valinomycin and elevated K^+ concentration ($K_o^+ = K_I^+ = 30$ mM). Phlorizin was used to identify that part of the total Na^+ and sugar fluxes which occur on the Na^+-dependent sugar carrier. The ratio of phlorizin-sensitive fluxes ($\Delta Na^+/\Delta S$) indicates a coupling ratio of 2.0 (Kimmich and Randles 1980)

it identifies an error in the usual definition of transport efficiency. In reality, the ratio between

$$\frac{\text{Obs. sugar gradient}}{\text{Theor. sugar gradient}} \times 100$$

is a measure of efficiency for the cellular "system" of sugar transport pathways rather than for the Na^+-dependent sugar carrier itself. Indeed, in the absence of "leak" pathways the Na^+-dependent sugar carrier would establish an equilibrium between the difference in chemical potential for sugar and the difference in electrochemical potential for Na^+ which exists across the brush border membrane (i.e., $\Delta\mu_{sugar} = \Delta\tilde{\mu}_{Na^+}$). An experimental measure of the theoretical limit to sugar gradient formation will therefore provide an exact measure of the $\Delta\mu_{Na^+}$ which the cell can maintain (Kimmich 1981a, 1982).

Measuring $\Delta\tilde{\mu}_{Na^+}$ via Unidirectional Sugar Fluxes

Recently we have begun to use the latter fact in an effort to determine the cellular $\Delta\tilde{\mu}_{Na^+}$ by a completely non-invasive procedure. The approach depends on obtaining a measure of the inward and outward apparent rate constants for the Na^+-dependent carrier system. The ratio of these constants is a measure of the equilibrium constant for the transfer of the carrier form which delivers sugar into or out of the cell. Conceptually, the method is analogous to the use of the ratio of forward and reverse rate constants for a chemical reaction as a measure of the equilibrium constant for that reaction. It is not necessary in either case for the reaction to be *at* equilibrium in order to use the ratio of constants to measure the equilibrium. Two factors need to be taken into consideration, however. The first is to recognize that the *apparent* rate constants for a $2Na^+$:1 sugar transport mechanism are expected to be a function of the membrane potential. As we have pointed out previously (Kimmich 1981b), a plausible working model for describing $2Na^+$ transport mechanism is shown in Fig. 7. In this model an anionic carrier is envisioned being driven to the outer face of the membrane by an interior negative membrane potential where it can bind a sugar molecule and two Na^+ ions. The cationic "loaded" carrier is then drawn to the inner membrane face by the membrane potential. The x and y coefficients of the permeability constants shown in the model are therefore functions of the membrane potential. In determining the ratio of "apparent" permeability or rate constants (xP/yP) it is important that the measurement should be done for the membrane potential maintained normally by the cells.

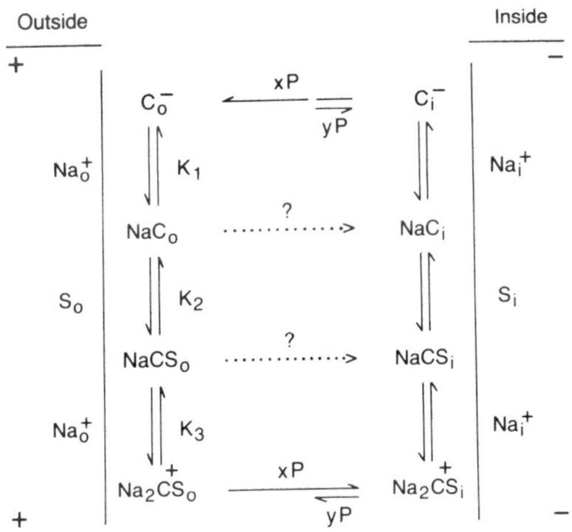

Fig. 7. Schematic representation for a transport model involving a 2.0 coupling stoichiometry and a role for the membrane potential. In this model the potential plays a dual role by driving on anionic free carrier outward where it can bind 2 Na^+ and a sugar molecule to form a cationic quaternary complex which is pulled inward by the potential. It is not clear whether the intermediate carrier forms (NaC_o and $NaCS_o$) have any permeability to the membrane, but the weight of experimental evidence suggests they do not

A second consideration is whether the transport mechanism might have internal "leaks" in the sense that partially loaded carrier forms (such as NaC, CS, or NaCS) might have some permeability to the membrane. The result of such processes would be a partial uncoupling of sugar transport from the full energy available for a mechanism in which *only* the fully loaded (Na_2CS) and unloaded (C) carrier forms have permeability. Again the determination of permeabilites for different carrier forms must be achieved for the same potential expected for cells functioning at a steady state of sugar accumulation.

Permeability of either binary complex (NaC or CS) can largely be ruled out by the absence of any phlorizin sensitivity for Na^+ or sugar fluxes in the absence of the other solute. Transfer of the ternary complex (NaCS) is more difficult to assess. A particularly annoying possibility is that sugar *influx* might be largely catalyzed by the $2Na^+$ (i.e., Na_2CS) route, but in the low Na^+ intracellular milieu one Na^+ might dissociate off and the ternary complex (Na CS) could return to the outer membrane face. The net result of such a process would be to deliver Na^+ but not sugar to the cell interior via the transport mechanism.

We have attempted to detect permeability of the ternary complex by measuring the coupling stoichiometry at low Na^+ concentrations. Under these conditions one might expect that both the ternary and the quaternary carrier complexes should exist and to the extent that the ternary complex has permeability the stoichiometry will decrease from 2.0. In general, however, we have found the stoichiometry to remain at or near 2.0 even at Na^+ concentrations as low as 30 mM (see Table 3).

Table 3. The Na^+:sugar coupling stoichiometry determined at three different Na^+ concentrations

$[Na^+]$	Na influx [a]			Sugar influx [b]			$\frac{\Delta Na}{\Delta S}$
	Con	$+ P_Z$	ΔNa^+	Con	$+ P_Z$	ΔS	
112 mM	67.0	44.5	22.7	17	5.2	11.8	1.9
60 mM	52.4	33.8	18.6	14.6	4.4	10.2	1.8
30 mM	24.2	19.8	4.4	9.9	7.0	2.9	1.6

All experiments were performed using rotenone-treated cells which were equilibrated with 30 mM KCl in the presence of 5 $\mu g\ ml^{-1}$ valinomycin in order to "clamp" the membrane potential at a value near zero

[a] All rates are given in nmoles $min^{-1}\ mg^{-1}$ cell protein

[b] 20 mM 3-O-methylglucose was used in each case

Another approach has been to create a reversed membrane potential (cell interior positive) which would tend to hold the cationic quaternary complex at the outer face of the membrane but not impede inward movement of the neutral ternary complex. The degree of phlorizin sensitivity of sugar flux under these conditions may be a measure of ternary complex mobility but only if the quaternary complex is *totally* excluded by the unfavorable potential. Note that with a reversed polarity the measured phlorizin-sensitive sugar flux is only 8% of that observed under conditions in which the membrane potential has normal polarity (Table 4). This indicates that the ternary complex has very limited (if any) permeability to the plasma membrane of the epithelial cell.

Table 4. Comparison of phlorizin-sensitive sugar fluxes in normally energized cells and cells with a reversed polarity for the membrane potential

	Sugar influx [a]		ΔS
	Con	$+ P_Z$	
Normally energized	0.60	0.055	0.545
Reversed membrane potential [b]	0.056	0.011	0.045

[a] 100 μM α-methylglucoside was used as the test sugar

[b] The membrane potential was reversed by diluting K^+ depleted cells with a medium containing 140 mM K^+ and valinomycin in order to induce a diffusion potential. The Na^+ concentration was 20 mM

Having ruled out appreciable fluxes of carrier forms other than that associated with the quaternary complex, we are in a position to determine the ratio of "apparent" permeability constants for this carrier (i.e., xP/yP) and hence to define the equilibrium constant for the sugar transfer mechanism. The pertinent fluxes of sugar are identified by their sensitivity to phlorizin as summarized in Fig. 8. The difference between rates measured by conventional influx experiments performed with and without phlorizin provides a value for the carrier mediated sugar influx (I). For efflux, the cells are allowed to establish a steady state gradient of sugar. Phlorizin is then added in order to immobilize the carrier and the observed rate of sugar efflux is that part of the total efflux occurring by routes not associated with the sugar carrier (D) (i.e., diffusional). Because total influx = total efflux at the steady state, the difference between sugar influx (I + D) and diffusional efflux (D) provides a value for the amount of sugar efflux which must have been occurring on the carrier (E).

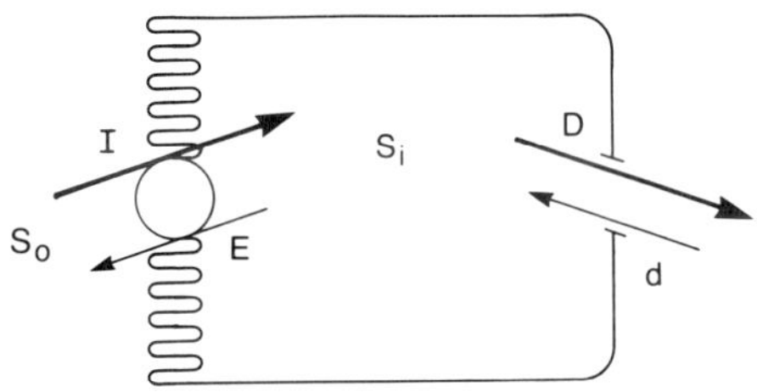

Fig. 8. Summary of the procedure for determining the theoretical sugar gradient which isolated cells are capable of forming. A detailed description is given in Kimmich (1981b)

Measuring the theoretical sugar gradient

Step 1: At a low sugar concentration ($S \ll K_T$), measure the steady state $[S]_i$ and $[S]_o$.

Step 2: At $[S]_o$: Measure unidirectional influxes ± phlorizin
Influx without phlorizin = total influx
Influx with phlorizin = d
The difference = I

Step 3: Allow cells to establish steady state $[S]_i$
a) Add phlorizin – measure efflux = D
b) Total influx = total efflux (D + E), so
Total influx – D = E

Step 4: $\frac{I}{E} \cdot \frac{[S]_i}{[S]_o}$ = theoretical gradient = $\Delta\tilde{\mu}_{Na^+}$

When this procedure is applied to a study of α-methylglucoside transport, the flux ratio data provides an estimate of xP/yP of approximately 300-fold (Kimmich 1981a). Because this ratio, in essence, is a value for the distribution of the quaternary complex at the inner and outer face of the membrane,

$$\frac{(Na_2CS)_i}{(Na_2CS)_o},$$

it will be a measure of the sugar distribution across the membrane *when* sugar concentrations at each interface are low enough so that the quaternary complex is proportional to sugar concentration ($[S] < K_T$).

Thus,

$$\frac{xP}{yP} = \frac{k_1\,(S)_i}{k_2\,(S)_o}$$

where k_1 and k_2 are proportionality constants which relate to the dissociation of Na^+ and sugar from the quarternary complex at the two membrane surfaces. If $k_1 = k_2$, as suggested by the observation that the K_T for sugar transport is not a function of the membrane potential or Na^+ concentration (Carter-Su and Kimmich 1980) then the ratio of apparent rate constants equals the equilibrium sugar distribution *or* the theoretical limit to sugar gradient formation.

Therefore, the analysis outlined above indicates an upper limit to sugar gradient formation of 300-fold. The upper limit could only be achieved when *all* sugar flux in and out of the cell is catalyzed by the sugar carrier (i.e., absence of diffusional flux pathways). The important aspect, however, is that the theoretical limit can be measured when other flux pathways are operative, and that the value obtained is a a precise measure of the $\Delta\tilde{\mu}_{Na^+}$ which the cells maintain. It can be obtained by totally non-invasive techniques in contrast to current procedures which require micro-electrode impalement to measure either component of the $\Delta\tilde{\mu}_{Na^+}$ (i.e., either membrane potential or cellular activity of Na^+). Furthermore, the microelectrode techniques measure a value for a particular cell in which it is presumed that the membrane properties have *not* been altered by the impalement procedure. The sugar flux procedure provides an average value for $\Delta\tilde{\mu}_{Na^+}$ for a cell population which has not run the risk of membrane damage. To the extent that *either* component of $\Delta\tilde{\mu}_{Na^+}$ can be measured by other non-invasive procedures the unmeasured component can be calculated. For instance, we are evaluating an SCN-flux procedure as a means of non-invasively measuring the membrane potential. It this method proves reliable, we will be able to calculate the $\Delta\tilde{\mu}_{Na^+}$ from the following relationship:

$$\Delta\tilde{\mu}_{Na^+} = \Delta\mu_{Na^+} + \Delta\psi \qquad (2)$$

where $\Delta\tilde{\mu}_{Na^+}$ has been measured by the sugar flux ratio procedure outlined here.

Ultimately, any gradient coupled transport system offers the possibility for non-invasive measurement of the trans-membrane difference in chemical or electrochemical potential for the energizing ion as long as a specific inhibitor is available to define carrier-mediated solute fluxes. The gradient coupled transport systems, therefore, represent natural sensors which can be used as detection devices for certain parameters of interest in electrophysiology in lieu of conventional microelectrode techniques.

Acknowledgements. This paper has been sponsored in part by a grant from the Public Health Service No. AM 15365 and also by Contract No. DE-AC02-76EV03490 with The U.S. Department of Energy at the University of Rochester Department of Radiation Biology and Biophysics and has been assigned Report No. UR-3490-2202.

References

Armstrong WMcD, Byrd D, Hamang PM (1973) Energetic adequacy of Na^+ gradients for sugar accumulation in epithelial cells of small intestine. Biochim Biophys Acta 330:237–241

Beck JC, Sacktor B (1975) Energetics of the Na^+-dependent transport of D-glucose in renal brush border membrane. J Biol Chem 250:8674–8680

Carter-Su C, Kimmich GA (1979) Membrane potentials and sugar transport by ATP-depleted intestinal cells: Effects of anion gradients. Am J Physiol 237:C67–C74

Carter-Su C, Kimmich GA (1980) Effect of the membrane potential on Na^+-dependent sugar transport by ATP depleted intestinal cells. Am J Physiol 238:C73–C80

Crane RK, Miller D, Bihler I (1961) The restrictions on possible mechanisms of intestinal active transport of sugars. In: Kleinzeller A, Kotyk A (eds) Membrane transport and metabolism. Academic Press, New York, pp 439–449

Crane RK, Forstner G, Eicholz A (1965) Studies on the mechanism of the intestinal absorption of sugars. An effect of Na^+ on the apparent Michaelis constant for intestinal sugar transport in vitro. Biochim Biophys Acta 109:467–477

Curran PF, Schultz SG, Chez RA, Fuisz RC (1967) Kinetic relations of the Na^+-amino acid interaction at the mucosal border of intestine. J Gen Physiol 50:1261–1267

Gibb LE, Eddy AA (1972) An electrogenic sodium pump as a possible factor leading to the concentration of amino acids by mouse ascites tumor cells with reversed sodium ion concentration gradients. Biochim J 129:979–981

Goldner AM, Schultz SG, Curran PF (1969) Sodium and sugar fluxes across the mucosal border of rabbit ileum. J Gen Physiol 53:362–383

Hopfer U, Sigrist-Nelson K, Murer H (1975) Intestinal sugar transport: Studies with isolated plasma membranes. Ann NY Acad Sci 264:414–427

Jacquez JA, Schafer JA (1969) Na^+ and K^+ electrochemical potentials and the transport of α-amino-isobutyric acid in Ehrlich ascites tumor cells. Biochim Biophys Acta 193:368–383

Kessler M, Semenza G (1979) On the efficiency of energy conversion in sodium-driven D-glucose transport across small intestinal brush border membrane vesicles: An estimation. FEBS Lett 108:205–208

Kimmich GA (1970) Active sugar accumulation by isolated intestinal epithelial cells: A new model for sodium-dependent metabolite transport. Biochemistry 9:3669–3677

Kimmich GA (1973) Coupling between Na^+ and sugar transport in small intestine. Biochim Biophys Acta 300:31–38

Kimmich GA (1981a) The Na^+-dependent sugar carrier as a sensor of the cellular electrochemical Na^+-potential. In: Dinno M (ed) Membrane biophysics: Structure and function in epithelia. Alan Liss, New York, pp 129–142

Kimmich GA (1981b) Intestinal absorption of sugar. In: Johnson L (ed) Physiology of the gastrointestinal tract, vol 2. Raven Press, New York, pp 1035–1061

Kimmich GA (1982) Intestinal transport of sugar – the energetics of epithelial "pump-leak" systems. In: Martonosi A (ed) Membranes and transport. Plenum Press, New York (in press)

Kimmich GA, Carter-Su C (1978) Membrane potentials and the energetics of intestinal Na^+-dependent transport systems. Am J Physiol 235:C73–C81

Kimmich GA, Randles J (1975) A Na^+-independent phloretin sensitive monosaccharide transport system in isolated intestinal epithelial cells. J Membr Biol 23:57–76

Kimmich GA, Randles J (1978) Phloretin-like action of bioflavanoids on sugar accumulation capability of isolated intestinal cells. Membr Biochem 1:221–237

Kimmich GA, Randles J (1979) Energetics of sugar transport by isolated intestinal epithelial cells: Effects of cytochalasin B. Am J Physiol 237:C56–C63

Kimmich GA, Randles J (1980) Evidence for an intestinal Na^+:sugar transport coupling stoichiometry of 2.0. Biochim Biophys Acta 596:439–444

Kimmich GA, Randles J (1981) α-Methylglucoside satisfies only Na^+-dependent transport system of intestinal epithelium. Am J Physiol 241:C227–C232

Kimmich GA, Carter-Su C, Randles J (1977) Energetics of intestinal Na^+-dependent sugar transport by isolated intestinal cells: evidence for a major role for membrane potentials. Am J Physiol 233:E357–E362

Murer H (1976) Differences in amino acid and glucose transport between bruch border and basolateral plasma membrane of intestinal epithelial cells. J Cell Physiol 89:805–810

Murer H, Hopfer U (1974) Demonstration of electrogenic Na^+-dependent D-glucose transport in intestinal brush border membranes. Proc Natl Acad Sci USA 71:424–488

Potashner SJ, Johnstone RM (1971) Cation gradients, ATP and amino acid accumulation in Ehrlich ascites tumor cells. Biochim Biophys Acta 233:91–103

Randles J, Kimmich AG (1978) Effects of phloretin and theophylline on 3-O-methylglucose transport by intestinal epithelial cells. Am J Physiol 234:C64–C72

Reid M, Gibb LE, Eddy AA (1974) Ionophore mediated coupling between ion fluxes and amino acid absorption in mouse ascites tumor cells. Restoration of the physiological gradients of methionine by valinomycin in the absence of adenosine triphosphate. Biochem J 140:383–393

Riklis E, Quastel JH (1958) Effect of cations on sugar absorption by isolated surviving guinea pig intestine. Can J Biochem Physiol 36:347–362

Schafer JA, Heinz E (1971) The effect of reversal of Na^+ and K^+ electrochemical gradients on the active transport of amino acids in Ehrlich ascites tumor cells. Biochim Biophys Acta 249: 15–33

Schultz SG (1977) Sodium-coupled solute transport by small intestine: a status report. Am J Physiol 233:E249–E254

Schultz SG, Curran PF (1970) Coupled transport of sodium and organic solutes. Physiol Rev 50: 637–718

Sigrist-Nelson K, Murer H, Hopfer U (1975) Active adenine transport in isolated brush border membranes. J Biol Chem 250:5674–5680

Tucker AM, Kimmich GA (1973) Characteristics of amino acid accumulation by isolated intestinal epithelial cells. J Membr Biol 12:1–22

Several Compartments Involved in Intestinal Transport

M. GILLES-BAILLIEN [1]

Introduction

Historically intestinal transport processes were first approached by studies performed in loops in vivo and in situ but very rapidly by means of in vitro techniques. Incubation of rings of intestine and of everted sacs has allowed to establish some concepts in the field of intestinal absorption which are still valid nowadays. In after years, a definitive taking-off in intestinal transport studies was observed by the use of Ussing chambers as experimental tool. Indeed investigation carried out with preparations of isolated intestinal epithelium has led to a better knowledge of the part played by the mucosa itself in absorption processes. Progressively greater interest was taken in the secretion processes at work at the level of the isolated mucosa. Indeed from the evaluation of transepithelial fluxes and electrical properties of the intestinal epithelium, useful informations have been provided and will still be provided, allowing to forge the scheme of intestinal function and of its regulation.

In this experimental approach however the intestinal mucosa has to be considered as a "Black Box". Attempting to get insight within this black box several trends of investigation have evolved, the more successful being an approach which could be termed biochemical in so far as transport mechanisms, mostly identified with carrier mechanisms, are studied at the level of brushborder and basolateral membranes or vesicles with biochemical tools, aiming at an elucidation at the molecular level of these transport mechanisms. Recent developments in this research area will be presented in following chapters. Another trend of investigation which has won less attention results from the study of transport mechanisms at the level of isolated cells. In both approaches, however, it is mostly the mature enterocyte which is taken into consideration. Nonetheless it should be envisaged that the different cell types encountered in the intestinal epithelium could act as distinct compartments introducing their own component in the establishment of the overall transepithelial fluxes.

Convincing evidence was already brought up in 1972 by Roggin et al. that villous cells more specifically deal with absorptive processes while crypt cells are responsible for secretory processes. Indeed the unimpaired secretory response of rabbit jejunum to choleratoxin after selective damage to villous epithelium favours such a hypothesis.

1 Laboratory of General and Comparative Biochemistry, University of Liège, 17 Place Delcour, 4020 Liège, Belgium

Intestinal Transport
(ed. by M. Gilles-Baillien and R. Gilles)

Moreover Field (1980) postulates that cyclic AMP would have an antiabsorptive action in villous cells, but in contrast would stimulate secretion in crypt cells, thus suggesting different transport mechanisms at both levels.

This has incited us to review briefly some aspects of the morphological structure of intestinal epithelia.

Morphological Structure of Intestinal Epithelia

The morphological structure of the mammalian intestinal epithelium is very complex and our purpose here is not to study its histology exhaustively but only to point out some of its features which may be relevant when interpreting data concerning transport properties.

Cell Types Encountered

The intestinal epithelium of mammalian species is essentially constituted by villi which are leaf-like, tongue-like or finger-like projections of the mucosa and by crypts. Striking differences exist when comparing villous cells to crypt cells and however villous cells result from the transformation or differentiation of crypt cells. According to Cheng and Leblond (1974) stem cells at the basis of the crypt would differentiate into four main cell types migrating together along the walls of the crypt towards the villus tips where they are released in the lumen. These cells are: (a) absorptive or columnar cells (enterocytes); (b) goblet cells; (c) Paneth cells; (d) entero-endocrine cells. Histologically these four cell types encountered in the villi show important differences when compared to their own precursors present in the crypts.

For instance a columnar cell in the crypt has little microvilli and smooth lateral membranes when compared to a mature columnar cell in the villus. The proportion of the various cells changes along migration from the crypt towards the villus. The base of the crypt is occupied for 50% by Paneth cells which, in contrast, are rare in the villi. It is commonly assumed that columnar cells of the villi are responsible for absorption but would also be responsible for ultimate steps of digestion by the enzymatic equipment of their brush border. Goblet cells in the villi, and in the crypts as well, secrete mucus which coats the entire epithelium. Paneth cells are rich in Zn; they would elaborate lysozyme (Erlandsen et al. 1974) and contain immunoglobulin A (Erlandsen et al. 1976). Enteroendocrine cells are probably of different kinds, some producing serotonin, others secretin and cholecystokinin (see Ham and Cormack 1979). The liberation of cells at the tip of the villi would be attributed according to Altmann (1976) to a progressive decrease in the synthesis by each cell of a glycosylated surface protein (glycocalyx), this protein being responsible for their adhesion to the basement membrane and to each other. More recently it has been assumed that the cell-cell dissociation needed for cell extrusion was facilitated by an increased amount of sialic acid in the membranes of the villus tip cells (Breimer et al. 1981).

Regional Differences

Differences between crypts and villi have already been mentioned. Numerous differences also exist according to the level of intestine which is taken into consideration. Some striking characteristics can be briefly summarized. In duodenum the villi are broader and higher than farther down in the small intestine. At the level of the duodenum mucus-secreting glands, called Brunner's glands, found in the submucosa, are emptying their mucous secretion in ducts connected to the crypts of Lieberkühn. These glands are responsible for a thick coating of the entire mucosa with mucus. However, at the level of jejunum and ileum the epithelium is also covered by a layer of mucus which is secreted by goblet cells present in the crypts as well as on the villi. The number of goblet cells in the villi increases progressively from duodenum down to the end of the ileum and pH of the mucus would also increase regularly. In the colon there is an even thicker layer, of alkaline mucus, which results from a larger number of goblet cells (Weiss and Greep 1977).

At the level of the ileum are found lymphoid nodules which also exist at the anterior level but in much smaller amount. They are termed Peyer's patches and would play an important role in the local antibody system.

The epithelium of the colon differs from the one of the small intestine mainly by the absence of villi and a more important thickness due to deeper crypts. It has no Paneth cells, more goblet cells than in the small intestine, as already mentioned, and enteroendocrine cells of various types.

Species Differences

The different characteristics of the intestinal epithelium just described could apply to a majority of mammals but of course species differences exist which relate, for instance, to the length of the intestine versus the weight of the animal or to a varying proportion of the different portions of the intestine. In the latter case this is often related also to the kind of diet to which the animal is adapted.

But when going from marine invertebrates to higher vertebrates, the epithelium of the digestive tract in the course of evolution has undergone deep alterations.

For the purpose of this paper let us just mention similarities and differences between reptilian and mammalian intestinal epithelia. Macroscopically the two epithelia have rather the same general organization: (a) A small intestine with three portions, duodenum, jejunum and ileum. The thickness of the epithelium decreases from duodenum to ileum. (b) A large intestine preceded by a more or less important caecum and ending in a more or less important rectum according to the species considered. But when looking at the general aspect of the small intestinal mucosa the velvet appearance caused by villi in mammals is replaced by longitudinal lines due to folds of the mucosa in reptiles. Confirmed by microscopic studies, the intestinal epithelium of reptiles has no villi, no crypts and so far only two cell types: enterocytes and goblet cells (Andrew 1959). Personal investigation and numeration of cells in the jejunum mucosa of the active tortoise (*Testudo hermanni hermanni* Gmelin) has given for goblet cells a proportion of 14.2% ± 4.8%. Another important feature of the reptilian intestinal mucosa would be its very slow turn-over. In the tortoise again, which is a hibernating species, it is observed that undifferentiated interstitial cells accumulate at the basis of the epithelium during hibernation, the epithelium being renewed at the awakening from hibernation.

Compartmental Analysis of the Isolated Mucosa in Relation with Transport

Transepithelial fluxes measured in vitro across the isolated intestinal mucosa have been generally interpreted on the basis of a three-compartment model: mucosal saline cellular compartment, serosal saline. We have seen in the previous section how complex the cellular compartmentation can be when considering the various cell types involved. In order to evaluate the part played by each cell type in the establishment of the overall fluxes of ions or molecules, a more complex compartmental analysis has to be performed. As a first approach, the isolated intestinal mucosa of the tortoise has been selected for several reasons. The main one is, of course, because of its simpler morphological structure: two cell types, enterocytes and goblet cells, no crypts, no finger-like processes. Then the intestinal epithelium is very easily cleared of its muscular layers when compared to the mammalian intestine. Finally the isolated intestinal epithelium can sustain long in vitro incubation without losing its integrity (Baillien and Schoffeniels 1961).

The experimental procedure has been described elsewhere (Gilles-Baillien 1976, 1982, Fernandez-Tejero and Gilles-Baillien 1977). Briefly, the isolated epithelium is loaded either by the mucosal surface or by the serosal one with radioactive material during a period of time (preliminary determined) sufficient to reach the steady-state level in all the compartments involved. Then both faces of the epithelium are washed out at 1-min intervals, cold saline being also added at 1-min intervals. The radioactivity present in the washing-out samples is plotted against time. Moreover at the end of the washing-out procedure the radioactivity remaining in the epithelium is determined and summed up with the activity that has been washed during the preceding minute so as to build up a curve representing the evolution of the residual activity present in the mucosa as a function of washing-out time. The washing-out curves as well as the curve representing the residual radioactivity, plotted on semi-log paper, generally appear as complex exponentials which may be solved into simple ones by a peeling procedure. Effluxes, half-renewal times and amounts of radioactivity contained in the different compartments involved in the intestinal epithelium are evaluated. In Fig. 1 have been gathered the different results obtained when loading a preparation of isolated jejunum mucosa with either radioactive Na (115 mEq/l), cycloleucine (20 mM) or inulin (traces), from the mucosal saline or from the serosal one.

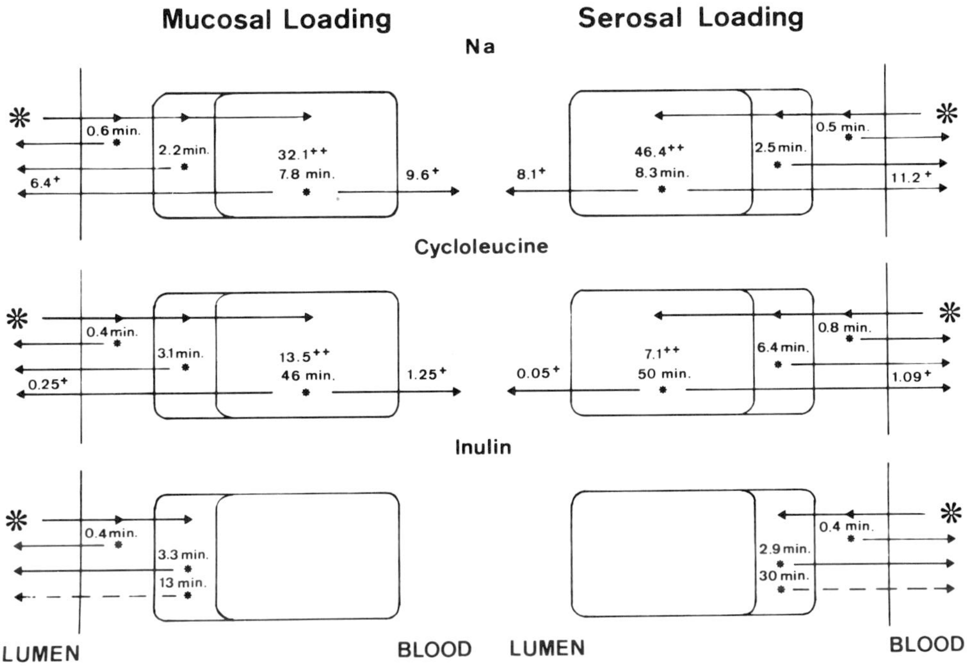

Fig. 1. Mucosal and serosal effluxes, contents and half-renewal times of the different compartments identified in the tortoise isolated jejunum mucosa by compartmental analysis. The epithelium is loaded either from the mucosal side *(left drawings)* or from the serosal one *(right drawings)* with radioactive Na, cycloleucine or inulin. Results are means of 3–6 sets of experiments, mucosal and serosal loading being performed in each set on symmetricl portions of jejunum mucosa. + the unit of this value is μEq cm^{-2} h^{-1} in the case of Na and μmol cm^{-2} h^{-1} in the case of cycloleucine, ++ the unit of this value is μEq kg^{-1} wet weight for Na and μmol kg^{-1} wet weight for cycloleucine

Na Compartments

As far as Na is concerned the mucosal efflux after mucosal loading can be best fitted by a triple exponential which can be decomposed in three simple ones. The half-renewal time of compartment I is less than 1 min and represents extracellular space on the mucosal side with some contamination by the incubation solution. Compartment II with a half-renewal time of 2.2 min does not exchange with the serosal saline; it is therefore located on the mucosal surface and could correspond to an unstirred layer, but more probably to a mucus layer, as will be ascertained in the following section. Compartment III is the only one exchanging with both salines and it would correspond to the cellular compartment.

After serosal loading three compartments can be assumed from the serosal efflux, only one from the mucosal efflux. Compartment I is again extracellular. Compartment II could correspond to intercellular spaces and would then be limited by tight-junctions on the mucosal side and by the basement membrane on the serosal side. To our knowledge little is known about the permeability of the basement membrane.

It is, however, interesting to point that Gupta et al. (1978) have reported that the lateral intercellular spaces are hypertonic to the incubation medium which could be considered as an indication that the basement membrane act as a barrier to Na diffusion. In contrast, tight-junctions are reported to be "leaky" specially to Na in mammalian intestine (see for instance Schultz et al. 1974). But in the present experiments if compartment II really represents intercellular spaces, since this compartment does not exchange with the mucosal saline, it should be assumed that tight-junctions are less "leaky" in the tortoise. A higher electrical resistance reported in the tortoise is also favouring such an assumption (Gilles-Baillien and Verbert 1978). As to compartment III, which is the only one to exchange with both salines, its half-renewal time is very close to the one of compartment III identified after mucosal loading. However, the same experiment have been performed in May when the animals are awakening from hibernation at a time of the year when cell turn-over is important, different half-renewal times are found: 12 min after mucosal loading, 6 min after serosal loading (Gilles-Baillien 1976). Moreover in fully active animals, in the presence of ouabain, data concerning compartment III, obtained after mucosal loading, are seriously modified, while after serosal loading, only slight modifications are recorded. Results are summarized in Fig. 2. After mucosal loading in the presence of ouabain there is an important decrease of both mucosal and serosal effluxes which confirms an old hypothesis postulating the existence of an active Na extrusion mechanism at both borders of the intestinal epithelium (Baillien and Schoffeniels 1961). Moreover, the Na content is decreased and the half-renewal time of the compartment is increased. After serosal loading there is a slight decrease in effluxes and in the Na content, a small increase of the half-renewal time but these modifications are not

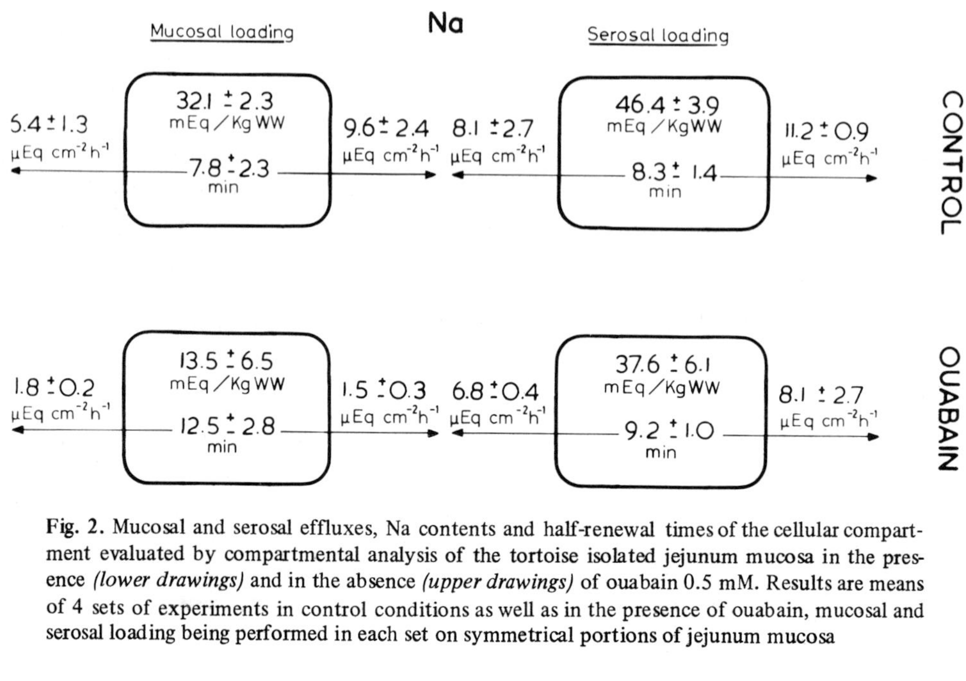

Fig. 2. Mucosal and serosal effluxes, Na contents and half-renewal times of the cellular compartment evaluated by compartmental analysis of the tortoise isolated jejunum mucosa in the presence *(lower drawings)* and in the absence *(upper drawings)* of ouabain 0.5 mM. Results are means of 4 sets of experiments in control conditions as well as in the presence of ouabain, mucosal and serosal loading being performed in each set on symmetrical portions of jejunum mucosa

significant. These different results tend to demonstrate that the cellular compartment reached from the mucosal side is different from the one which is reached from the serosal side. Either two distinct Na pools exist within the main cell type, that is to say in the absorptive cell, or absorptive cells can be loaded only from the lumen while goblet cells would be loaded only from blood side. In this latter case, however, Na would reach enormous concentration levels since 46.4 mEq is the amount of Na which would be present in a kg wet weight where goblet cells represent only 15%, that is a much smaller volume than absorptive cells. Finally another possibility has to be taken into consideration. Indeed it cannot be ruled out that the compartment III loaded from the serosal side would correspond to intercellular spaces in which case the mucosal efflux would reflect the paracellular pathway involved in mammals (Schultz et al. 1974). Indeed a high Na concentration in these intercellular spaces has been reported in the rabbit ileum (Gupta et al. 1978); in the tortoise jejunum a high Na content is also the case for compartment III loaded from blood side. Moreover results obtained in the presence of ouabain do not desagree with such an explanation. But data presented below are nonetheless in favour of a cellular origin for this compartment III.

In the results presented above it should also be mentioned that the serosal efflux obtained after mucosal loading (9.6 μEq cm^{-2} h^{-1}) and the mucosal efflux obtained after serosal loading (8.1 μEq cm^{-2} h^{-1}) should correspond to Na transepithelial influx and outflux measured in "flux chambers" (Table 1). In fact they are almost twice as high. Of course it should be pointed out that the estimation of fluxes in washing-out experiments is probably a little less accurate. Moreover it is not excluded that the technique in itself involving the renewal by fresh well oxygenated saline every minute would stimulate Na fluxes. An alternative explanation could also reside in the fact that brief exposure to air as well as changes of hydrostatic pressure occurring every minute could also result in damaging the preparation. However, the total Na content of the preparation after such experiments do not differ significantly from the one estimated directly (Gilles-Baillien 1972) and is not indicative of a Na leak. Moreover transepithelial fluxes obtained for cycloleucine closely correspond to the ones obtained in washing-out experiments as will be shown below.

In isolated guinea-pig intestinal mucosa a similar compartmental analysis has been performed by Lauterbach (1976). The conclusion of this work was that the same cellular Na pool was reached from lumen side or blood side, this pool having a half-renewal time close to 1.25 min. In contrast Warner (1978), by performing a compartmental analysis of the Na efflux across the rabbit ileal mucosal membrane, assumes

Table 1. Transepithelial influx (J_{in}) and outflux (J_{out}) of Na and cycloleucine across the isolated jejunum mucosa of the tortoise

	J_{in}	J_{out}
Na 111 mEq l^{-1} [a] (μEq cm^{-2} h^{-1})	4.32 ± 1.64 (7)	4.82 ± 2.05 (5)
Cycloleucine 20 mM M. or S. [b] (μmol cm^{-2} h^{-1})	0.915 ± 0.236 (15)	0.026 ± 0.013 (11)

[a] Gilles-Baillien et al. (1979)
[b] Gilles-Baillien et al. (1978)
Results are means ± S.D. Number of experiments between brackets

that the half-renewal time of the intracellular Na pool is 13.5 min, which is more in the range of the values obtained for the tortoise.

Cycloleucine Compartments

When the intestinal mucosa is loaded with radioactive cycloleucine (20 mM) either from lumen side or from blood side, the number of compartments which are identified is the same as for Na (Fig. 1). On the mucosal surface compartment II exchanges with a half-renewal time of 3.1 min and is probably mucus. On the serosal side compartment II, which has been assumed to be intercellular spaces in the case of Na, exchanges cycloleucine with a half-renewal time of 6.4 min, that is to say more slowly than Na.

Compartment III exchanges much more slowly than in the case of Na after mucosal loading (46 min) as well as after serosal loading (50 min). The serosal efflux measured after mucosal loading and the mucosal efflux obtained after serosal loading reasonably correspond to transepithelial influx and outflux as determined with "flux chambers" (Table 1). Now the question is, as for Na, whether compartment III loaded from lumen side is the same as the one loaded from blood side since they have similar half-renewal times. By another experimental approach and using not only cycloleucine but also different amino acids which are susceptible of being metabolized, we came to the conviction that influx and outflux of amino acids go through two distinct cellular pathways. In fact using 6 different amino acids at 4 different concentrations, the following experiments have been performed (Gilles-Baillien 1980). Influx and outflux of radioactivity have been measured on symmetrical portions of jejunum mucosa by sampling during 4 h. Then mucosal saline and serosal saline were taken up and concentration as well as radioactivity of the substrate amino acid were determined with an amino-acid analyser (Technicon) equipped with a flow cell for the detection of radioactivity. Other amino acids appearing in the salines as metabolites of the substrate amino acid were also detected. The fragment of intestinal mucosa exposed during the experiment was blotted on filter paper, weighed, total radioactivity was measured and amino-acid analysis was performed. The dialysis residue was digested with soluene and radioactivity was estimated. From the difference of radioactivity inside and outside the dialysis tubing, the amount of amino acid (or of its metabolites) bound to or incorporated into macromolecules (probably proteins) was calculated. For the 6 amino acids assayed at the concentration 20 mM, Table 2 gives the values obtained for the fluxes of radioactivity as well as the percentages of metabolites that have appeared in the fluxes while Fig. 3 summarizes the informations relevant to metabolization in the intracellular fluid. In Table 2 it is shown that: (a) no metabolites appear in the radioactive fluxes of cycloleucine (which was of course expected since cycloleucine is reported not to be metabolized in the intestinal mucosa); (b) in the case of lysine 30% of the influx is metabolites while the outflux is constituted only by lysine; (c) in the case of glutamate, most of the radioactive influx is no more glutamate. In the reverse direction the flux is so small that the proportion of metabolites could not be estimated; (d) in the case of leucine and alanine the radioactivity appearing in the serosal saline after influx experiment is almost totally attributable to the substrate amino acid while in outflux experiments 50%–70% of the flux is metabloites of the substrate amino acid.

Table 2. Transepithelial fluxes of radioactivity with different labelled amino acids at a concentration 20 mM. Percentages of metabolites appearing in these fluxes. (Gilles-Baillien 1980)

	J_{in} (μmol cm^{-2} h^{-1})	Metabolites %	J_{out} (μmol cm^{-2} h^{-1})	Metabolites %
Cycloleucine	0.784 ± 0.151	0	0.010 ± 0.003	0
Lysine	0.103 ± 0.037	29.8	0.085 ± 0.065	0
Glutamate	0.075 ± 0.047	79.6	0.002 ± 0.001	?
Leucine	0.625 ± 0.248	0	0.080 ± 0.032	66.5
Alanine	1.301 ± 0.521	4.9	0.263 ± 0.222	52.4
Serine	0.424 ± 0.166	0	0.031 ± 0.023	0

Results are means ± S.D. 4 experiments in each case

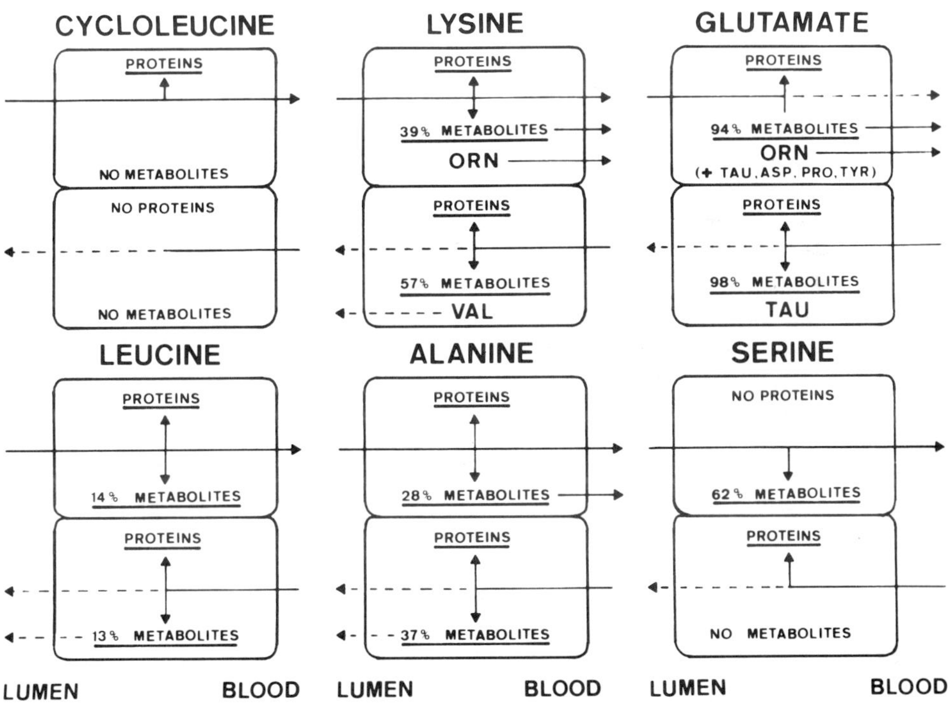

Fig. 3. Schematical representation of the metabolization and incorporation into proteins that occur in influx and in outflux experiments of different labelled amino acids (20 mM) performed with tortoise isolated jejunum mucosa

In Fig. 3 the following facts are reported:
(a) in the case of cycloleucine no metabolites are appearing in the cells whether cycloleucine comes from lumen side or blood side, but in influx experiments it is shown that cycloleucine is incorporated into macromolecules: (b) in the case of lysine, important metabolization as well as incorporation occur in both flux directions but while ornithine is the only amino acid identified among metabolites in the influx experiments, valine is the only one in outflux experiments; (c) with glutamate, metabolization takes place in a very important way wherever it comes from. Incorporation is weak but exists in both conditions. But in influx experiments, it is essentially ornithine which is produced, while in the opposite direction taurine is the main metabolite that appears, no ornithine at all being detected; (d) in the case of leucine and alanine, incorporation and metabolization are effective in both directions but the metabolites formed are not amino acids; (e) when serine is used as substrate amino acid again striking differences exist between influx and outflux experiments. An important metabolization and no incorporation are recorded in the cells when the epithelium is supplied from lumen side. But when serine crosses the epithelium coming from blood side, no metabolization occurs but serine is incorporated into macromolecules.

Initially the purpose of this study was not to get involved in the different pathways followed by amino acids within intestinal cells but to determine the importance of metabolization and the error ensuing from assimilation of amino-acid fluxes to radioactive fluxes as commonly found in literature. But the obtained results are furthermore bringing support to the assumption that influx and outflux of amino acids cross two distinct pools of cellular origin, since metabolization and (or) macromolecule synthesis occur(s) in both directions. Now it could be speculated that these two cellular pools correspond respectively to enterocytes and goblet cells since only these two cell types would be encountered in the tortoise intestinal mucosa. An argument favouring the view that it is essentially the goblet cells that are supplemented from blood side can be found in the results obtained with serine. Indeed the values of the amino-acid pool fed from blood side and of the amount of amino acid incorporated into macromolecules are the highest for serine when compared to the five other amino acids assayed (see Gilles-Baillien 1980). On the other hand it has been recently reported that serine is one of the major amino-acid constituent of the native glycoprotein purified from pig small intestinal mucus (Mantle and Allen 1981). Mucus synthesis being the main function ascribed to goblet cells, it seems reasonable therefore to assume that serine is taken up by goblet cells from blood side and is there further incorporated into mucus. Thus it can be tentatively concluded that the transepithelial influx of Na as well as of amino acids would cross the enterocytes while the transepithelial outflux would go through goblet cells.

Mucus Coating: a Compartment by Itself?

The mucosa of the whole digestive tract is known to be covered by a mucus coating. From histologists its existence has been recognized at all levels. Besides a mechanical function the possible roles of the mucus are only starting to draw attention. At the level of the pig gastric mucosa a potential role in mucosal protection has already been ascertained (see, for instance, Williams and Turnberg 1980). In the rat colon, the mucus layer would be responsible for pH-stable microclimate despite pH variations in the lumen (Rechkemmer et al. 1979) and would hence favour the resorption of short-chain fatty acids (Engelhardt and Rechkemmer, this vol.). In some cases, attempts have been made in order to evaluate the thickness of the mucus layer. In the rat, this thickness would be about 151 μm in the proximal colon (Sakata and Engelhardt 1981) and 166 μm on the fundic part of the gastric mucosa (Bickel and Kauffman 1981). The possible interference of the mucus layer in transport processes has not yet deserved much attention.

In the previous section it has been shown that on the mucosal surface of the tortoise intestinal epithelium there is a compartment exchanging Na, cycloleucine or inulin rather slowly (see Fig. 1). It has already been reported that this compartment tends to disappear when the mucosal surface is blotted on filter paper after the loading period and before the washing-out procedure (Fernandez-Tejero and Gilles-Baillien 1977).

In order to ascertain that this compartment corresponds to mucus, an investigation on isolated mucus has been undertaken. Indeed after 1 h incubation of the isolated intestinal mucosa under intense oxygenation, an important mucus layer develops which can be collected by gentle scraping of the mucosa (Gilles-Baillien 1981).

Composition of the Mucus

Some components of this preparation of isolated mucus have been determined. No DNA was found in this preparation, meaning that no cells or cell debris are present. The amount of protein and sugar has been determined (Gilles-Baillien 1981). But our major concern was to examine the possibility that the mucus layer could represent a compartment with specific inorganic ion composition. In Table 3 are compared pH and Na, K and Cl concentrations of the isolated mucus and of the saline which is used for incubation and collection of the mucus. First it appears that the mucus would be able to maintain a pH different from that of the saline, a conclusion which was reached also for the mucus of the rat colon (Rechkemmer et al. 1979). But the

Table 3. pH and Na, K and Cl concentrations in isolated mucus and in the saline used for collection of the mucus (Gilles-Baillien 1981)

	Mucus	Saline
pH	7.63 ± 0.29 (10)	7.0
Na	110.9 ± 2.1 (5)	114
K	5.4 ± 0.7 (5)	2.55
Cl	170.2 ± 6.1 (5)	114.8

Na, K and Cl concentrations, expressed in μEq ml^{-1}, are means ± S.D. Number of determinations between brackets

fact that this pH is slightly alkaline is at variance with microelectrode studies of the microclimate existing on the surface of the rat jejunum (Lucas et al. 1975) which would be acid. As far as Na, K and Cl concentrations are concerned it is shown that K and Cl concentrations are higher in the isolated mucus than in the saline.

The mucus compartment is therefore a compartment with its own chemical composition, with regard to inorganic ions. The attempt has been made to isolate rat mucus in the same way but the preparation is always contaminated with cells. In contrast, biochemical studies of the glycoprotein contained in the mucus of the digestive tract of several mammals are available: pig gastric mucus (Starkey et al. 1974), human gastric mucus (Pearson et al. 1980), pig colonic mucus (Marshall and Allen 1978), pig small intestinal mucus (Mantle and Allen 1981), human small intestinal mucus (Forstner et al. 1979). From these studies it appears that the mucus glycoprotein is different from species to species and, for a same species, according to the level where it is produced. This glycoprotein would be bound non-covalently with a protein which could contribute to the rheological properties of the mucus (List et al. 1978).

Our main interest being to evaluate the possible interference of the mucus layer in transport processes, tortoise intestinal mucus has not been further studied on biochemical grounds, but the preparation of mucus has been used to study its exchange rates towards Na, cycloleucine and inulin as has been done for the whole intestinal mucosa.

Compartmental Analysis of the Mucus

The mucus isolated as describe above, can be loaded with radioactive material and then submitted to a washing-out procedure by using a slightly modified millipore filtering device (Gilles-Baillien 1981). By summing up the radioactivity remaining in the mucus sample at the end of the washing-out procedure with the radioactivity that has come out of the sample during the preceding minute, an exponential can be built up which represents the residual radioactivity of the mucus as a function of time of washing out. This exponential in all the cases so far investigated is a triple exponential which can be peeled and decomposed in three simple exponentials allowing to estimate the radioactive content as well as the half-renewal time of each of the three compartments. Table 4 summarizes the data obtained when mucus samples are loaded with either ^{22}Na (114 mEq l^{-1}) ^{14}C-cycloleucine (20 mM) or ^{3}H-inulin (traces). In each condition, three compartments are identified. The slowly exchanging compartment (III) contains very small amounts of radioactivity in all three conditions.

In the case of Na, most of the radioactivity is found in compartment I which exchanges very rapidly (0.7 min); when this radioactivity is expressed in terms of concentration per ml mucus, the obtained value (98.9 μEq ml^{-1} mucus) is very close to the one of the saline used for the loading period and washing-out (114 μEq ml^{-1}). This result agrees with the estimation of the Na concentration obtained by flame photometry (Table 3).

As far as cycloleucine is concerned, compartment I exchanging very rapidly (0.9 min) reaches the same concentration as in the saline (19.4 μmol ml^{-1} mucus versus

Table 4. Na, cycloleucine and inulin amounts and half-renewal times of the three compartments identified in washing-out experiments performed with tortoise isolated mucus (Gilles-Baillien 1981)

Compartments	I	II	III
Na contents (μEq ml^{-1} mucus)	98.9 t/2 0.7 min	1.9 t/2 1.9 min	0.7 t/2 17.5 min
Cycloleucine contents (μmol ml^{-1} mucus)	19.4 t/2 0.9 min	31.9 t/2 2.9 min	0.2 t/2 27.2 min
Inulin contents (μl ml^{-1} mucus)	tr. t/2 0.6 min	932.8 t/2 1.5 min	50.0 t/2 23 min

Only mean values are reported for sake of clarity

20 μmol ml^{-1} solution). But compartment II exchanging with a half renewal time of 2.9 min has a higher concentration (31.9 μmol ml^{-1} mucus). Therefore the total concentration of the mucus is approximately 52 μmol ml^{-1} mucus, that is to say much more concentrated than the saline. As to inulin, most of the radioactivity is in compartment II which exchanges with a half-renewal time of 1.5 min. When these results are compared with those where the whole epithelium is loaded (Fig. 1), compartment II identified in mucus corresponds to compartment II identified after mucosal loading of the epithelium. This is specially true in the cases of Na and cycloleucine. As far as inulin is concerned the half-renewal times of the compartments II are different (3.3 min in mucosal loading of the whole epithelium, 1.5 min in the mucus). This could possibly be explained in terms of binding or affinity. Indeed one could speculate that compartment I represents the water of mucus where Na and cycloleucine move freely, obeying the laws of simple diffusion, while compartment II would correspond to pool(s) where inulin and cycloleucine move less freely with some interaction with the mucus constituents. In the case of inulin there is no accumulation within this pool but in the case of cycloleucin, the fact that it is accumulated within this pool suggest the presence of binding sites. In the case of inulin the interaction with the mucus would be different after the mucus has been isolated.

Possible Role of Mucus in Transport Processes

The exact relevance of the mucus layer in intestinal transport processes is far from being established. Its role is certainly more important than just to maintain an unstirred water layer (UWL) on the mucosal surface of the epithelium. Indeed the existence of UWL has been invoked and its effect on kinetic parameters of active transport processes has been studied (see for instance Dietschy et al. 1971, Winne 1976, 1977, Thomson 1979, Thomson and Dietschy 1980). But considering the fact that the mucus layer could on the one hand, maintain a pH zone different from the pH existing in the lumen and on the other hand, be responsible for a zone of accumulation favouring transport of certain substances across the brush border, its role in transport processes is undeniable.

Its possible interference in methodology used to study transport processes should also be taken into consideration. For instance, it is classical to measure uptake of sugars and amino acids across the brush border by differential loading between the sugar or the amino acid, and inulin during very short periods (45–60 s). If in these experiments a mucus layer exists, such short periods would not allow the mucus layer to be fully loaded and moreover the mucus itself could represent the zone of uptake, at least for certain substances. The use of a preincubation period could also favour the development of a thick layer of mucus.

It is therefore very important at this stage to evaluate the thickness of the mucus layer in in vitro experiments as well as in the in vivo situation. If the layer is about 150 μm as is the case for rat colonic and gastric mucus (Sakata and Engelhardt 1981, Bickel and Kauffman 1981), when compared to the height of an enterocyte (25 μm), this mucus layer would probably have an important part to play in overall transport processes. Also it should be established if this mucus coating is a wide-mesh net or if it acts as a continuous barrier in the small intestine. As far as gastric mucus is concerned it seems already well demonstrated that it acts as a barrier to hydrogen ion diffusion (Williams and Turnberg 1980, Pfeiffer 1981). From our own experiments the importance of the mucus coating, evaluated from the amount of Na, cycloleucine or inulin present in compartment II when loading the epithelium from lumen side (Fig. 1), is very variable from one experiment to the other; oxygenation, stirring of solutions during the experiment, time required for dissection may be parameters which should be carefully monitored. Occasionally it has been mentioned that the output of mucus could be enhanced by osmotic shock or under certain pathological conditions (for a review see Forstner 1978). Regulation or control of the thickness of the mucus layer could also be involved in the regulation of the overall transport processes. In this respect regulation of intestinal goblet cell secretion has already deserved some attention. Neutra et al. (1982) have recently studied the effect of several potential secretagogues on the goblet cell secretion. Information concerning the neurohormonal control of secretion processes are also available (see Turnberg, this vol). But further investigations and more informations are required before the role of mucus in transport processes and possibly in their regulation can be fully assessed.

Conclusions and Prospects

When striking the balance of research performed in the field of intestinal transport, one realizes that most effort has been, and is still, directed towards the understanding of the mechanisms involved in intestinal "absorption". From studies performed with loops in vivo until sophisticated investigations achieved with brushborder and basolateral membranes, interest has been essentially focused on mature enterocytes. However, in the recent years a current of investigations has emerged, concerning secretion processes and their regulation. It is indeed well established that amongst the different cell-types involved in the complex morphological structure of the mammalian intestinal mucosa, certain cells are absorptive, others are secreting. A simplified view consists of considering that villous cells would be mainly devoted to absorption while crypt cells would deal with secretion (see Field 1980), a view which may reveal very productive in the coming years because techniques for isolation of these two categories of cells are available (Harrison and Webster 1969, Weiser 1973, Kimmich 1975, Kremski et al. 1977, Towler et al. 1978, Lawson et al. 1982). In fact the transepithelial fluxes of any substance across the intestinal mucosa are the complex resultant of fluxes crossing different compartments (cellular or paracellular) arranged in series or in parallel. To approach such a compartmental analysis, tortoise intestinal mucosa is a choice material because of its simple morphological structure (only two cell types, enterocytes and goblet cells, and no villi, no crypts) and because it stands long in vitro experiments without losing its integrity.

The tortoise intestinal mucosa has been loaded from lumen side or from blood side with either ^{22}Na, ^{14}C-cycloleucine or ^{3}H-inulin and then submitted to washing-out procedure. The results of such experiments allow to postulate that influx and outflux of Na and of cycloleucine go through two distinct cellular pools namely the enterocyte in the lumen-blood direction and the goblet cell in the opposite one. This is confirmed for Na by the study of ouabain effect. For cycloleucine and five other amino acids (lys, gly, leu, ala, ser) a study of metabolization and of incorporation into macromolecules occurring when the amino acid crosses the epithelium in one direction or in the other also firmly supports the assumption that transepithelial influx crosses nearly exclusively the enterocyte while the outflux goes through the goblet cell.

Moreover compartmental studies give evidence of the presence on the blood side and on the lumen side of two compartments retarding fluxes, which are arranged in series with respect to the two parallel cellular compartments. On the blood side it would correspond to the intercellular spaces which are delimited by the basement membrane, about which little is known concerning its permeability properties (tight-junctions would be much less permeable in tortoises than in mammals). On the lumen side it is identified with mucus. Indeed a clean preparation of isolated mucus has similar half-renewal times towards Na and cycloleucine. Moreover it is shown that cycloleucine accumulates in the mucus layer at a higher concentration than in the incubation medium. Such an accumulation, if also valid for other amino acids or other substances, would confer to the mucus layer a very important part in intestinal transport, by maintaining a zone of high concentration close to the brush border, hence favouring entrance within the absorptive cells. Another point of interest is the fact that the mucus layer would maintain a slightly alkaline pH, different from the pH in the lumen, which could also promote the movement of certain molecules.

More information are necessary on the transport properties of the different cell types encountered in the intestinal epithelium, and also concerning the possible roles or interferences of the mucus coating as well as of the basement membrane, before an integrate picture of intestinal transport across the sole epithelium can be produced.

Acknowledgments. The experimental work reported in this paper has been financially supported by grant No. 2.4544.76 from the Fonds de la Recherche Fondamentale Collective.

References

Altmann GG (1976) Factors involved in the differentiation of the epithelial cells in the adult rat small intestine. In: Cairnie AB, Lala PK, Osmond DG (eds) Stem cells of renewing cell populations. Academic Pres, New York, p 51

Andrew W (1959) Textbook of comparative histology. University Press, New York Oxford

Baillien M, Schoffeniels E (1961) Origine des potentiels bioélectriques de l'épithelium intestinal de la tortue grecque. Biochim Biophys Acta 53:537–548

Bickel M, Kauffman GL Jr (1981) Gastric gel mucus thickness. Effect of distention, 16,16-dimethyl prostaglandin E_2 and carbenoxolone. Gastroenterology 80:770–775

Breimer ME, Hansson GC, Karlsson KA, Leffler H (1981) Glycosphingolipids and the differentiation of intestinal epithelium. Exp Cell Res 135:1–13

Cheng H, Leblond CP (1974) Origin, differentiation and renewal of the four main epithelial cell types in the mouse small intestine. I. Columnar cells. Am J Anat 141:461–479

Dietschy JM, Sallee VL, Wilson FA (1971) Unstirred water layers and absorption across the intestinal mucosa. Gastroenterology 61:932–934

Erlandsen SL, Parsons JA, Taylor TD (1974) Ultrastructural immunocytochemical localization of lysozyme in the Paneth cells of man. J Histochem Cytochem 22:401–413

Erlandsen SL, Rodning CB, Montero C, Parsons JA, Lewis CA, Wilson ID (1976) Immunocytochemical identification and localization of Immunoglobulin A within Paneth cells of the rat small intestine. J Histochem Cytochem 24:1085–1092

Fernandez-Tejero C, Gilles-Baillien M (1977) Sodium cycloleucine and inulin compartments in the intestinal mucosa of the tortoise. J Physiol (Lond) 266:35P–36P

Field M (1980) Regulation of small intestinal ion transport by cyclic nucleotides and calcium. In: Field M, Fordtran JS, Schultz SG (eds) Secretory diarrhea. Waverly Press, Baltimore, p 21

Forstner JF (1978) Intestinal mucus in health and disease. Digestion 17:234–263

Forstner JF, Jabbal I, Qureshi R, Kells DIC, Forstner GG (1979) The role of disulphide bonds in human intestinal mucin. Biochem J 181:725–732

Gilles-Baillien M (1972) Inexchangeable fraction of the cationic content in the intestinal epithelium of the tortoise *Testudo hermanni hermanni* Gmelin and its modification during hibernation. Arch Int Physiol Biochim 80:789–797

Gilles-Baillien M (1976) Na^+ compartmentation in the jejunal mucosa of the tortoise. In: Robinson JWL (ed) Intestinal ion transport. MTP Press, Lancaster, p 75

Gilles-Baillien M (1980) Transepithelial fluxes of amino acids and metabolism in the tortoise intestinal mucosa. Arch Int Physiol Biochim 88:15–23

Gilles-Baillien M (1981) Na, cycloleucine and inulin compartments in tortoise intestinal mucus. Possible role of the mucus in intestinal absorption processes. Mol Physiol 1:265–272

Gilles-Baillien M (1982) Kinetic aspects of Na transport across the intestinal mucosa. In: Oosawa F (ed) Dynamic aspects of biopolyelectrolytes and biomembranes. Kodansha, p 375

Gilles-Baillien M, Verbert M (1978) Seasonal changes in the electrical parameters of the small intestine, colon and bladder mucosa of land tortoises (*Testudo hermanni hermanni* Gmelin). Experientia 34:1174–1175

Gilles-Baillien M, Aguilar J, Fernandez-Tejero C (1978) Análisis de un modela de transporte en membranas biologicas: transporte de cicloleucina. Rev Esp Fisiol 34:25–32

Gilles-Baillien M, Havaux JC, Chapelle S (1979) Seasonal variation in Na^+ transport and (Na^+-K^+) ATPase activity in tortoise intestinal epithelium. J Comp Physiol 129:355–360

Gupta BL, Hall TA, Naftalin RJ (1978) Microprobe measurement of Na, K and Cl concentration profiles in epithelial cells and intercellular spaces of rabbit ileum. Nature 272:70–73

Ham AW, Cormack DH (1979) Histology, 8th ed. Lippincott, Philadelphia Toronto

Harrison DD, Webster HL (1969) The preparation of isolated intestinal crypt cells. Exp Cell Res 55:257–260

Kimmich GA (1975) Preparation and characterization of isolated intestinal cells and their use in studying intestinal transport. In: Korn E (ed) Methods in membrane biology, vol IV. Plenum, New York, p 51

Kremski VC, Varani L, Desaive C, Miller P, Nicolini C (1977) Crypt cell isolation in the small intestine of the mouse. J Histochem Cytochem 25:554–559

Lauterbach FO (1976) Ion fluxes in isolated guinea-pig intestinal mucosa. In: Robinson JWL (ed) Intestinal ion transport. MTP Press, Lancaster, p 51

Lawson AJ, Smit RA, Jeffers NA, Osborne JW (1982) Isolation of rat intestinal crypt cells. Cell Tissue Kinet 15:69–80

List SJ, Findlay BP, Forstner GG, Forstner JF (1978) Enhancement of the viscosity of mucin by serum albumin. Biochem J 175:565–571

Lucas ML, Schneider W, Haberich FJ, Blair JA (1975) Direct measurement by pH-microelectrode of the pH microclimate in rat proximal jejunum. Proc R Soc Lond B 192:39–48

Mantle M, Allen A (1981) Isolation and characterization of the native glycoprotein from pig small intestinal mucus. Biochem J 195:267–276

Marshall T, Allen A (1978) The isolation and characterization of the high-molecular-weight glycoprotein from pig colonic mucus. Biochem J 173:569–578

Neutra MR, O'Malley LJ, Specian RD (1982) Regulation of intestinal goblet cell secretion. II. A survey of potential secretagogues. Am J Physiol 242:G380–G387

Pearson JP, Allen A, Venables CW (1980) Gastric mucus: Isolation and polymeric structure of the undegraded glycoprotein: Its breakdown by pepsin. Gastroenterology 78:709–715

Pfeiffer CJ (1981) Experimental analysis of hydrogen ion diffusion in gastrointestinal mucus glycoprotein. Am J Physiol 240:G176–G182

Rechkemmer G, Wahl M, Kuschinsky W, Engelhardt W v (1979) pH-microclimate at the surface of the intestine in guinea-pig and rat. Pfluegers Arch 382:R31

Roggin GM, Banwell JG, Yardley JH, Hendrix TH (1972) Unimpaired response of rabbit jejunum to choleratoxin after selective damage to villous epithelium. Gastroenterology 63:981–989

Sakata T, Engelhardt W v (1981) Luminal mucin in the large intestine of mice, rats and guinea pigs. Cell Tissue Res 219:629–635

Schultz SG, Frizzell RA, Nellans HN (1974) Ion transport by mammalian small intestine. Ann Rev Physiol 36:51–91

Starkey BJ, Snary D, Allen A (1974) Characterization of gastric mucoproteins isolated by equilibrium density-gradient centrifugation in Caesium chloride. Biochem J 141:633–639

Thomson ABR (1979) Kinetic constants for intestinal transport of four monosaccharides determined under conditions of variable effective resistance of the unstirred water layer. J Membr Biol 50:141–163

Thomson ABR, Dietschy JM (1980) Experimental demonstration of the effect of the unstirred water layer on the kinetic constants of the membrane transport of D-glucose in rabbit jejunum. J Membr Biol 54:221–229

Towler CM, Pugh-Humphreys GP, Porteous JW (1978) Characterization of columnar absorptive epithelial cells isolated from rat jejunum. J Cell Sci 29:53–75

Warner RR (1978) Compartmental analysis of Na efflux across ileum mucosal membrane. Pfluegers Arch 378:155–160

Weiser MM (1973) Intestinal epithelial cell suface membrane glycoprotein synthesis. I. An indicator of cellular differentiation. J Biol Chem 248:2536–2541

Weiss L, Greep RO (1977) Histology, 4th ed.McGraw-Hill, New York St Louis

Williams SE, Turnberg LA (1980) Retardation of acid diffusion by pig gastric mucus: a potential role in mucosal protection. Gastroenterology 79:299–304

Winne D (1976) Unstirred layer thickness in perfused rat jejunum in vivo. Experientia 32:1273–1279

Winne D (1977) Correction of the apparent Michaelis constant biased by an unstirred layer if a passive transport component is present. Biochim Biophys Acta 464:118–126

Part 2
Brush Border and Basolateral Membranes

Mechanisms of Sodium Transport Across Brush Border and Basolateral Membranes

E.M. WRIGHT, R.D. GUNTHER, J.D. KAUNITZ, B.R. STEVENS, V. HARMS, H.J. ROSS and R.E. SCHELL [1]

Introduction

Sodium is absorbed across the intestinal epithelium by a number of processes including primary active, secondary active and passive transport processes. To quantitate the contributions of each process we have elected to study transport using membrane vesicles isolated from the brush border and basolateral surfaces of mature enterocytes. Vesicles offer the potential to measure transport kinetics under well defined experimental conditions that avoid the complications encountered in the intact epithelium due to paracellular shunts, unstirred layers, and cellular metabolism. In this review we present our studies bearing on the role of the Na/K pump, coupled transport processes, and diffusion in the absorption of sodium across the jejunum. We conclude that diffusion and cotransport with organic solutes are the major transport systems in brush borders, while the pump accounts for net Na transport out of the epithelium into the blood.

Methods

Plasma membrane vesicles were prepared from the epithelial cells of the rabbit and rat jejunum as described previously. Brush borders were isolated by the Ca^{2+}-precipitation procedure (Stevens et al. 1982a) and these were enriched 19–26-fold. Basolateral membranes were obtained by differential and density gradient centrifugation procedures (Mircheff and Wright 1976, Wright et al. 1980, Mircheff et al. 1980) and in these studies were enriched 13–25-fold.

Transport into the membrane vesicles was measured using radioactive tracers and a rapid filtration technique (see Stevens et al. 1982a, Kaunitz et al. 1982). The properties of basolateral Na/K pumps were deduced from measurements of Na/K-ATPase activity, (^{3}H-)-ouabain binding and ^{32}P incorporation (Harms and Wright 1980). All experiments were carried out at 22°C, except for ^{32}P incorporations which were carried out at 0°C.

1 Department of Physiology, University of California, Medical Center, Los Angeles, CA 90024, USA

Intestinal Transport
(ed. by M. Gilles-Baillien and R. Gilles)

Results

Na Transport Across Brush Borders

Figure 1 shows the time course of ^{22}Na (100 mM NaCl) uptake into rabbit brush border vesicles. The $t_{1/2}$ was 3–4 min and, as shown in the insert, uptake was linear for the first 2–30 s. This linear uptake corresponded to an initial rate (J_{Na}) of 550 pmol mg^{-1} s^{-1}. The 0 time intercept on the ordinate, 1,500 pmol mg^{-1}, is interpreted as binding to the external surface of the vesicles: additional experiments (Gunther and Wright 1983) have demonstrated that the intercept is independent of J_{Na} at each Na concentration.

J_{Na} was measured as a function of Na activity and the results plotted (Fig. 2) as J_{Na} vs. J_{Na}/a_{Na}. The plot can be resolved into a two component curve: one saturable uptake with a J_{max} of 41 pmol mg^{-1} s^{-1} and a K_t of 5 mM; and another diffusional flux with an apparent permeability coefficient (P'_{Na}) of 1.9 nl mg^{-1} s^{-1}. Therefore, J_{Na} is described by:

$$\begin{aligned} J_{Na} &= [(J_{max} \times a_{Na})/(K_t + a_{Na})] + (P'_{Na} \times a_{Na}) \\ &= [(41 \text{ pmol mg}^{-1} \text{ s}^{-1} \times a_{Na} \text{ mM})/(5 \text{ mM} + a_{Na} \text{ mM})] \\ &\quad + (1.9 \text{ nl mg}^{-1} \text{ s}^{-1} \times a_{Na} \text{ mM}) \text{ pmol mg}^{-1} \text{ s}^{-1}. \end{aligned}$$

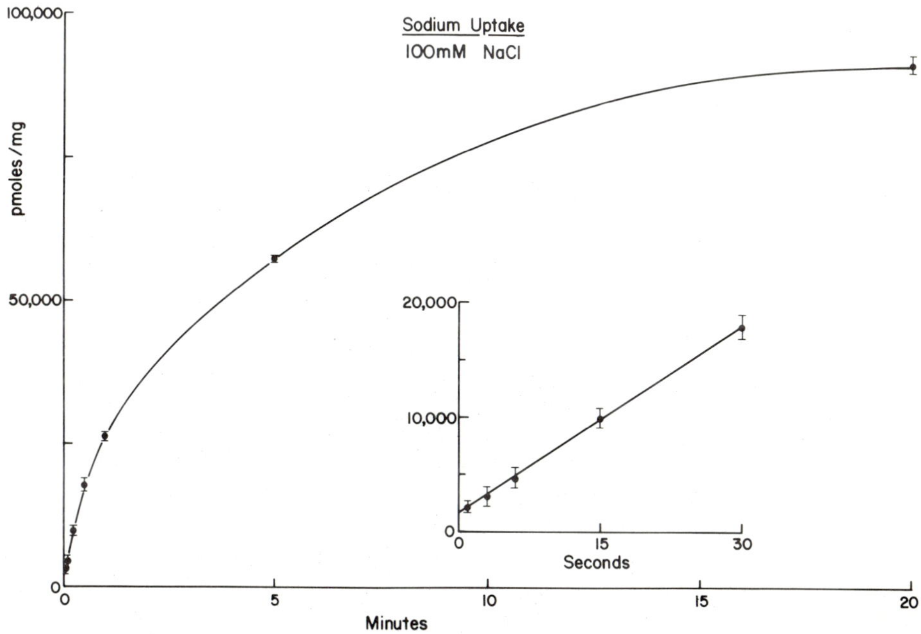

Fig. 1. Time course of sodium uptake into jejunal brush border membrane vesicles. The intravesicular solution contained 50 mM HEPES/Tris (pH 7.5). The extravesicular solution contained 100 mM NaCl in 50 mM HEPES/Tris (pH 7.5). Osmolarity was maintained at 350 mosm l^{-1} inside and out with D-mannitol. Data points are the mean of five samples ± standard error of the mean. The linear regression line ($r = 0.965$) shown in the insert has a slope of 550 pmol/(mg s^{-1}) and a Y-intercept of 1,500 pmol mg^{-1}. The sodium uptake at 5 h was 58,700 pmol mg^{-1} ± 1,050 S.E. (Gunther and Wright 1983)

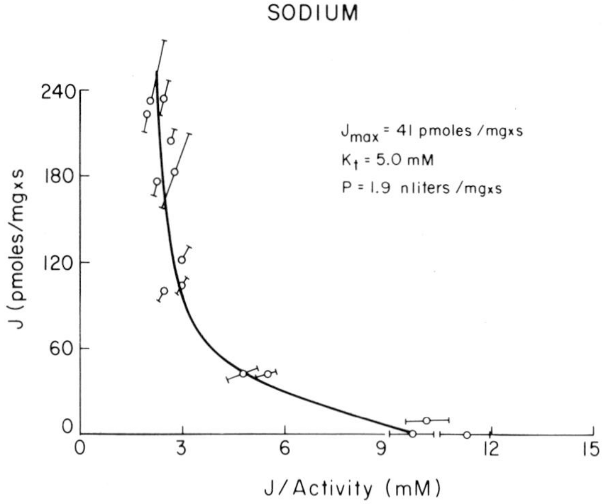

Fig. 2. Woolf-Augustinsson-Hofstee plot of sodium kinetics where the initial rate of Na influx *(J)* is plotted against J/sodium activity. The intravesicular solution contained 50 mM Tris/HEPES (pH 7.5). The extravesicular solution contained NaCl at concentrations ranging from 10 mM to 150 mM and 50 mM Tris/HEPES (pH 7.5). Osmolarity was maintained at 350 mOsm l^{-1} inside and out with D-mannitol. Initial flux rates were determined by linear regression through uptakes between 2 and 30 s at each concentration (see Fig. 1, insert). The *points* shown are data points ± standard deviation. Computer-generated courses were fitted to the data points, and the best-fitting curve was selected visually. The result is a two component curve consisting of a saturable mechanism (J_{max} = 41 pmol mg^{-1} s^{-1} and K_t = 5 mM) and diffusion with an apparent permeability coefficient (P'_{Na}) of 1.9 nl mg^{-1} s^{-1}. (Gunther and Wright 1983)

Diffusion. In six separate experiments the average P_{Na} was 4.6 ± 1.0 nl mg^{-1} s^{-1}, and so at physiological Na concentrations (150 mM) uptake by diffusion amounts to 700 pmol mg^{-1} s^{-1}. The selectivity of this diffusional pathway for sodium across brush borders has been characterised using bi-ionic diffusion potentials. These were measured by an optical technique with the voltage sensitive fluorescent dye 3,3′-dipropylthiadicarbocyanide iodide [diS-C_3-(5)].

The approach is illustrated in Fig. 3. In the presence of impermeant ions (choline and gluconate, which are used to maintain a constant ionic strength) K gradients generate changes in fluorescence that are virtually identical to those (Nernst) in the presence of the K ionophore valinomycin. This confirms that $P_K \gg P_{choline}, P_{gluconate}$. In the presence of permeable anions, e.g., NO_3, the magnitude of the fluorescent change was less than expected for a K Nernst potential. In fact, replacing 100 mM potassium gluconate with 100 mM KNO_3 reversed the polarity of the response, i.e., $P_{NO_3} > P_K$. Subsequent addition of valinomycin restored the signal to that expected for the K equilibrium potential, but in this case the fluorescence decayed towards baseline due to the presence of the permeable anion. We have carried out experiments with Na and Cl loaded vesicles with similar results.

After calibration of the fluorescence signal in volts (using ionophores and ion gradients; see Wright et al. 1981), it is then possible to extract relative ion permeability coefficients from these experiments using the constant field equation. The results are summarized in Table 1.

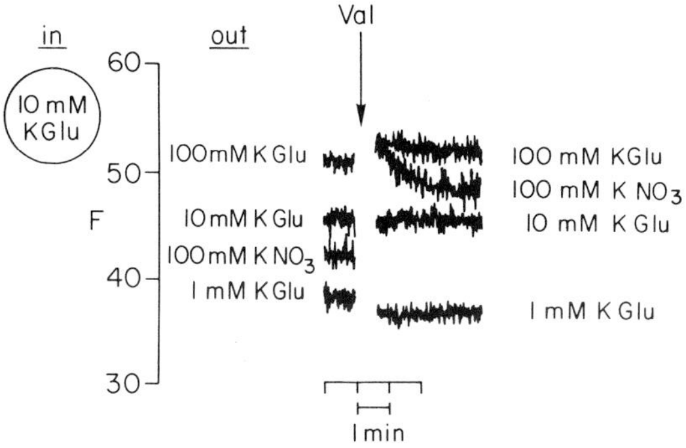

Fig. 3. Measurement of bi-ionic diffusion potentials across intestinal brush borders using the fluorescent dye diS-C_3-(5). Vesicles were preloaded with 10 mM K gluconate and 90 mM choline gluconate, and then added to a cuvette containing 1.5 μM diS-C_3-(5) and: (1) 10 mM K gluconate and 90 mM choline gluconate; (2) 100 mM K gluconate; (3) 1 mM K gluconate and 99 mM choline gluconate; and (4) 100 mM KNO_3. The fluorescence signal was obtained by activating at 620 nm and recording the emissions at 669 nm. Valinomycin (25 μM) was added at the point indicated by the *arrow*. For additional details see Wright et al. (1981) and Schell et al. (1983)

Table 1. Relative ion permeabilities (Gunther, Schell and Wright, unpubl. data)

Ion	P_i/P_{Na}	Ion	P_i/P_{Cl}
Ise^-	0.07	$NH_4{}^+$	6.79
Glu^-	0.09	Li^+	2.24
F^-	0.04	Cs^+	1.51
Cl^-	1.46	Rb^+	1.44
NO_3^-	1.49	K^+	1.39
Br^-	2.30	Na^+	0.58
I^-	39.7	$Chol^+$	0.06

Permeability ratios were estimated from bi-ionic diffusion potentials using the constant field equations as described in Fig. 3. Diffusion potentials were measured using the fluorescent dye DiS-C_3-(5). Vesicles were loaded with either 10 mM Na gluconate and 90 mM choline gluconate or 10 mM choline chloride and 90 mM choline gluconate. Vesicles were incubated in 100 mM salt solutions to produce the ion gradients across the membrane

The alkali metal cation selectivity sequence is similar, but not identical to the Eisenman sequence I (see Diamond and Wright 1969); this implies that cation permeation is controlled by negatively charged sites with a low field strength. However, the high relative permeabilities of Li and NH_4 indicate that polarization phenomena play a role in regulating cation permeation. The halide permeability sequence corresponds to the Eisenman sequence I, and the whole anion sequence agrees qualitatively and quantitatively with that predicted from the selectivity isotherms constructed by Wright and Diamond (1977).

Na/H Exchange. There is a substantial body of evidence to suggest that the saturable component of Na uptake (Fig. 2) is mediated by Na/H exchange (Gunther and Wright 1982). This includes: (1) J_{Na} is stimulated by a pH gradient, i.e., decreasing the intravesicular pH increased J_{Na} and actually produced a transient overshoot in the intravesicular content. A pH gradient of 2 units increased the J_{max} by a factor of 2–3 with no change in K_t. (2) J_{Na} is inhibited by both harmaline and amiloride, drugs which are known to inhibit Na/H exchange in other systems. (3) The specificity of Na uptake, as judged by competition studies (Li, NH_4 > K, Rb, Cs), also agrees with that observed for Na/H exchange in renal brush borders. Li uptake into jejunal brush border vesicles is also stimulated by pH gradients and is blocked by amiloride, while K and Rb are both transported by saturable system with J_{max}'s very close to that for Na (Gunther and Wright 1983). (4) Others have shown that Na gradients stimulate H transport across renal and intestinal brush borders (Murer et al. 1976, Burnham et al. 1982).

Na-Glucose Cotransport. It is generally accepted that the concentrative glucose transport across intestinal brush borders involves coupling with an inward Na flux down its electrochemical potential gradient (see Schultz and Curran 1970). Much work with intact tissues and brush border vesicles has been done to elucidate the kinetics of this sugar carrier (see Schultz and Curran 1970, Kimmich 1981, Murer and Kinne 1980). Recently, we have reported both direct and indirect experiments to establish the coupling stoichiometry using vesicles (Kaunitz et al. 1982), and we are currently exploring the kinetics of glucose transport.

Our indirect experiments involved measuring the intitial rate of glucose uptake (J_{gluc}) as a function of external sodium concentration, and subjecting the results to a Hill analysis. Figure 4 shows the results of one experiment, where it can be observed that we obtained a sigmoid relationship betwen J_{gluc} and (Na). The analysis yielded a Hill coefficient of 2, suggesting that under these experimental conditions there are at least 2 Na ions transported along with each molecule of glucose.

Direct experiments designed to measure the coupling coefficient require the determination of 4 initial rates of uptake: (1) J_{Na} in the absence of glucose; (2) J_{gluc} in the absence of Na; (3) J_{Na} in the presence of glucose; (4) J_{gluc} in the presence of Na. In four separate experiments at Na and glucose concentrations of 30 and 1 mM the mean ratio of the glucose-dependent J_{Na} to the Na-dependent J_{gluc} was 3.2 ± 0.7.

Thus both sets of experiments are consistent with the conclusion that two or more Na ions are coupled to the transport of each glucose molecule across jejunal brush borders.

More recent experiments indicate that glucose transport is even more complex than originally anticipated. Our kinetic studies indicate that there are two separate Na-dependent glucose carriers. At a Na concentration of 100 mM there is one with a high capacity (J_{max} 165 pmol mg^{-1} s^{-1}) and a low affinity (K_t 1.1 mM), and another with a lower J_{max} (65 pmol mg^{-1} s^{-1}) and a much higher affinity (K_t 0.03 mM).

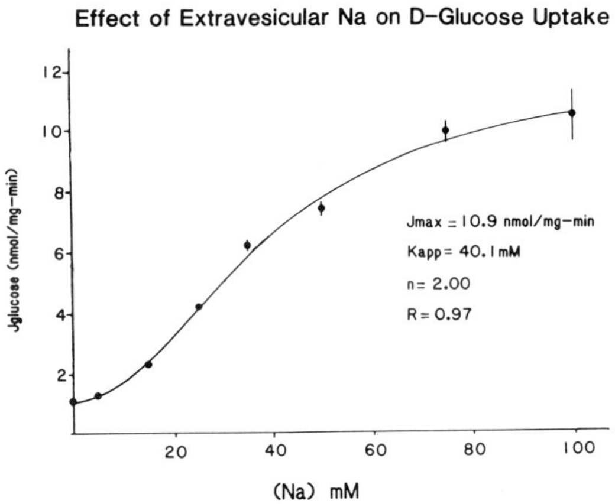

Fig. 4. Effect of extravesicular Na on the initial rate of D-glucose uptake. Vesicles were pre-equilibrated in 300 mM D-mannitol/50 mM HEPES/Tris (pH 7.5/100 mM KCl). Transport buffers consisted of 0.4 mM ^{14}C (U) D-glucose/2 μM valinomycin/7 μM FCCP/10 μM amiloride/100 mM KCl with varying concentrations of sodium. The line was calculated by a best-fit nonlinear regression of the data points to the Hill equation after subtraction of sodium-independent glucose uptake (J^{ind}_{gluc}).

$$J^{dep}_{gluc} = \frac{(J^{dep}_{gluc\,(max)} \times (Na)^n)}{K_{app} + (Na)^n}$$

where $J^{dep}_{gluc\,(max)}$ is the maximum J^{dep}_{gluc} at infinite sodium concentration, K_{app} is a constant and n is the Hill coefficient. (Data from Kaunitz et al. 1982)

While it is premature to discuss the coupling coefficients for these systems, we can estimate the maximum rate of Na uptake through the glucose-Na cotransport system, $(J'_{max} \times n') + (J''_{max} \times n'')$ where J_{max} and n are the glucose J_{max} s and coupling coefficients. This amounts to between 230 and 690 pmol mg^{-1} s^{-1} depending on the value of n.

Na-Amino Acid Cotransport. Sodium-dependent amino acid transport also occurs in intestinal brush borders (see Schultz and Curran 1970, Murer and Kinne 1980). Our studies of the specificity and kinetics of amino acid transport across jejunal brush borders (Stevens et al. 1982a) has led us to the conclusion that there are three separate Na-amino acid carriers with overlapping specificities. Although all three handle neutral amino acids, we were unable to detect significant Na-dependent transport of either β- or dibasic amino acids (see also Schell et al. 1983).

One of the three carriers handles a wide variety of neutral amino acids, including phenylalanine, one is fairly specific for phenylalanine, and the third transports proline, hydroxy-proline, and MeAIB. To estimate the contributions of these carriers to Na transport across brush borders we have estimated the coupling coefficients from indirect experiments. The Hill coefficient for phenylalanine was 1, and this, together with the phenylalanine J_{max}, suggests that the maximal Na flux through this coupled transporter is 625 pmol mg^{-1} s^{-1} (Table 2). In the case of proline the Hill coefficients

Table 2. Kinetic parameters for Na-cotransport systems

	J_{max} (pmol mg^{-1} s^{-1})	Hill coefficient n	J^{Na}_{max} (pmol mg^{-1} s^{-1})
Glucose	240	1–1.9	240–480
Proline	170	1–2.2	170–510
Phenylalanine	625	1	625

Kinetic parameters for Na-dependent solute transport across jejunal brush borders are from Kaunitz et al. (1982), Stevens et al. (1982a) and unpublished observations. Hill coefficients (n) were estimated as described in Fig. 4 at substrate concentrations ranging from tracer to the K_t s. The maximum rates of Na transport via each coupled system (J^{Na}_{max}) was estimated from the substrate J_{max}s and the coupling coefficients (1–3) estimated from Hill coefficients

ranged from 1–2.2, which implies that there are 1–3 Na ions transported for each proline molecule. Therefore, the Na flux through the proline pathway is in the range of 170–510 pmol mg^{-1} s^{-1} (Table 2). Consequently, the maximal rate of Na transport across the brush border membrane via amino acid carriers (1,135 pmol mg^{-1} s^{-1}) is about 2–3 times greater than that through the glucose carrier (Table 2). In this calculation we assume that the J_{max} for phenylalanine represents the sum of the J_{max} s for the specific carrier for phenylalanine and the common neutral amino acid carrier.

Rates of Na-dependent transport of sulphate, lactate, succinate and chloride across rabbit jejunal brush borders are negligible relative to those for sugars and amino acids (Stevens et al. 1982b, Gunther and Wright 1983).

Na Transport Across Basolateral Membranes

In both the rabbit and rat small intestine the Na/K-pumps (Na/K-ATPase) are virtually restricted to the basolateral membrane (Stirling 1972, Mircheff and Wright 1976). Our first approach to the question of Na transport across basolateral membranes was to study the properties of the Na/K-ATPase (Harms and Wright 1980). Using purified basolateral membrane vesicles we obtained estimates of the density and turnover number of pumps by measuring Na/K-ATPase activity, (^{3}H)-ouabain binding, and ^{32}P phosphorylation. Our results are summarized in Table 3. Ouabain binding gave a pump density of 1.5×10^6 sites $cell^{-1}$ and a turnover number of 10 s^{-1}. Enzyme activity and ^{32}P incorporation on the other hand gave a density of 0.15×10^6 sites $cell^{-1}$ and a turnover of 140 s^{-1}. These turnover numbers are close to those reported for other cells and tissues at 22°C (Harms and Wright 1980).

It should be noted that the K_m's obtained for ouabain binding (2.9×10^{-5} and 1.5×10^{-5} M) are indistinguishable from the ouabain inhibitor constant for the enzyme ($K_i = 3 \times 10^{-5}$ M).

Our phosphorylation studies also demonstrate that the intestinal Na/K-ATPase resembles the enzyme in other cells and tissues. Figure 5 contains the electrophoretic pattern of basolateral membrane exposed to gamma ^{32}P-labelled ATP in the presence of (1) Na, (2) K, and (3) Na + K. Phosphorylation in the presence of Na led to incor-

Table 3. Properties of the basolateral (Na + K-ATPase)

Ouabain binding
k_{+1} 1.3 × 10^3 M^{-1} s^{-1}
k_{-1} 3.6 × 10^{-2} s^{-1}
k_{-1}/k_{+1} 2.95 × 10^{-5} M
K_m 1.5 × 10^{-5} M
B_{max} 175 pmol mg^{-1} protein
Pump density: 1.5 × 10^6 $cell^{-1}$
Turnover number: 10 s^{-1}
Phosphorylation
^{32}P incorporation: 10 ± 1 pmol mg^{-1} protein
Pump density: 0.15 × 10^6 $cell^{-1}$
Turnover number: 140 s^{-1}

Data are from Harms and Wright (1980) obtained on rat basolateral membranes enriched 13-fold. The specific activity of the Na/K-ATPase was 6 μmol/(mg^{-1} protein h^{-1}) and the ouabain inhibitor constant was 3 × 10^{-5} M. All experiments were performed at 22°C, except for ^{32}P incorporation which was carried out at 0°C

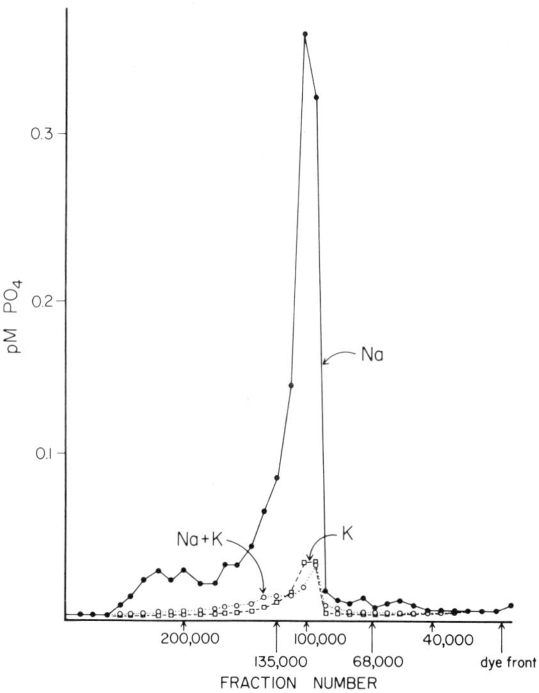

Fig. 5. Electrophoretic pattern of basal lateral membranes purified 25-fold by digitonin density perturbation, electrophoresed, and stained with Coomassie Blue. Shown is the pattern of ^{32}P incorporation in the presence of 50 mM NaCl (Na) or 50 mM NaCl, 10 mM KCl (Na-K) or 10 mM KCl, 40 mM choline chloride (K) and in all cases 0.5 μM ^{32}P ATP, 12 μM $MgCl_2$, and 5 mM Tris/HEPES (pH 7.4). Molecular weights of standard proteins are indicated at the bottom. Note that the peak of phosphate incorporation is at 100,000 M.W. 100μg of membrane protein were run on slab gels of 4.5% acrylamide using Fairbanks buffers. ^{32}P was assayed by liquid scintillation counting of 3 mM slices. (After Harms and Wright 1980)

poration of ^{32}P into a 100,000 M.W. protein. The amount of incorporation was markedly reduced by the addition of K, the absence of Na, or treatment with hydroxylamine. Thus the phosphorylated intermediate of intestinal Na/K-ATPase is an acyl phosphate with a molecular weight of 100,000, i.e., the alpha subunit of the dimeric enzyme.

If we assume that as in other leaky epithelia 3 Na ions are exchanged for 2 K ions during each cycle of the pump (see Zeuthen and Wright 1981, Saito and Wright 1982), we may estimate the rate of Na pumping across the intestinal basolateral membrane from the pump density and turnover number (Table 3). At 22°C about 45×10^6 ions are transported out of each cell per second. At this rate the pump could deplete the intracellular Na in 1–2 min, and this agrees with more direct measurements of pump rates in the amphibian gall bladder and choroid plexus (Graf and Giebisch 1979, Spring and Hopte 1979, Zeuthen and Wright 1981).

So far, we have no direct information about the Na permeability of basolateral membranes. In other leaky epithelia the membrane with the Na/K pumps generally has an ionic conductance 4–10 times greater than the membrane without pumps, but the high conductance is due to a high K permeability (see for example Zeuthen and Wright 1981). This suggests that the diffusional flux of Na across intestinal basolateral membranes is lower than that across brush border membranes. Likewise we have no experiments regarding passive Na carriers in basolateral membranes, even though there is reason to believe that a NaCl symport system exists in these membranes. However, we do have clear evidence that the basolateral membranes do not contain Na-glucose cotransporters, and that the major pathway for amino acid transport is the classical Na-independent "L" system (Wright et al. 1980, Mircheff et al. 1980). There is Na-amino acid cotransport, but the J_{max} is less than 5% of that for the "L" transporter.

Conclusions

On the basis of our studies of jejunal brush border and basolateral membrane vesicles, we can estimate the contributions of each transport system to sodium absorption across the intact epithelium (Fig. 6). The major contribution to Na influx across the brush border membrane (3 μmol cm^{-2} h^{-1}) are from diffusion (1 μmol cm^{-2} h^{-1}) and Na-cotransport with sugars and amino acids (2 μmol cm^{-1} h^{-1}) It should be noted that our estimate of the diffusional Na flux across rabbit brush borders is within an order of magnitude of that reported for the intact rabbit ileum (Goldner et al. 1969).

Our calculations suggest that the basolateral Na/K pumps are adequate to maintain the intracellular Na in the face of this influx across the brush border by pumping Na out of the cell into the blood across the basolateral membrane. Thus the Na/K pumps are able to account for an active Na flux across the epithelium of 4–5 μmol cm^{-2} h^{-1}. The observed rate of active Na transport across the jejunum is about 8 μmol cm^{-2} h^{-1} (Barry et al. 1965). The agreement between the observed and predicted rates of transport is quite encouraging given that: (1) we have based our estimates on vesicle experiments carried out at 22°C. Activation energies for transport are in the range of

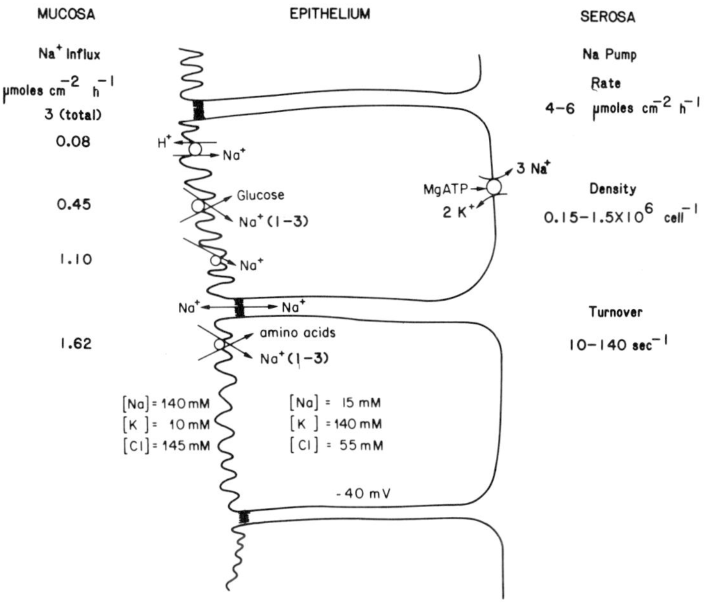

Fig. 6. A summary of Na fluxes across brush border and basolateral membranes of the intestinal epithelium. All fluxes from our vesicle studies at 22°C are expressed in units of μmol cm^{-2} h^{-1}, when the area refers to the serosal surface of the intestine. The conversion factor for rates in vesicles (mol mg^{-1} s^{-1}) to rates in the intact tissue (mol cm^{-2} h^{-1}) requires: (1) the enrichment factor for the vesicles (20 times brush borders and 13 times basolaterals); (2) the protein per cell (4×10^{-7} mg $cell^{-1}$); and the number of cells cm^{-2} (1.6×10^7 cells cm^{-1}). All fluxes are the values calculated from the J_{max}'s and the upper limits of the coupling coefficients

15–30 kcal mol^{-1}, and (2) there is room for error in estimating transport rates for the intact tissue from parameters measured in membrane vesicles. Nevertheless, vesicle transport data taken in concert with that obtained from the intact epithelium with tracer and electrophysiological techniques offer unique clues about intestinal transport mechanisms. Questions yet to be tackled include elucidation of the passive Na transport processes in the basolateral membrane, and the mechanisms of intestinal secretion.

Acknowledgements. This work was supported by grants from the USPHS (AM 19567 and AM 29737).

References

Barry RJC, Smyth DH, Wright EM (1965) Short circuit current and solute transfer by rat jejunum. J Physiol (Lond) 181:410–431

Burnham C, Munzesheimer C, Rabon E, Sachs G (1982) Ion pathways in renal brush border membranes. Biochim Biophys Acta 685:260–272

Diamond JM, Wright EM (1969) Biological membranes: The physical basis of ion and nonelectrolyte selectivity. Ann Rev Physiol 31:581–646

Goldner AM, Schultz SG, Curran PF (1969) Sodium and sugar fluxes across the brush border of rabbit ileum. J Gen Physiol 53:362–383

Graf J, Giebisch G (1979) Intracellular sodium activity and sodium transport in *Necturus* epithelium. J Membr Biol 47:327–355

Gunther RD, Wright EM (1983) Na^+, Li^+ and Cl^- transport by brush border membranes from rabbit jejunum. J Membr Biol 74 (in press)

Harms V, Wright EM (1980) Some characteristics of Na/K ATPase from rat intestinal basal lateral membranes. J Menbr Biol 53:119–128

Kaunitz JD, Gunther R, Wright EM (1982) Involvement of multiple sodium ion in intestinal D-glucose transport. Proc Natl Acad Sci USA 79:2315–2318

Kimmich GA (1981) Intestinal absorption of sugar. In: Johnson LR (ed) Physiology of the gastrointestinal tract. Raven Press, New York, p 1035

Mircheff AK, Wright EM (1976) Analytical isolation of plasma membranes of intestinal epithelial cells. Identification of Na, K-ATPase rich membranes and the distribution of enzyme activities. J. Membr Biol 28:309–333

Mircheff AK, Os CH van, Wright EM (1980) Pathways for alanine transport in intestinal basal lateral membrane vesicles. J Membr Biol 52:83–92

Murer H, Kinne R (1980) The use of isolated membrane vesicles to study epithelial transport processes. J Membr Biol 55:81–95

Murer H, Hopfer U, Kinne R (1976) Sodium/proton antiport in brush border membrane vesicles isolated from rat small intestine and kidney. Biochem J 154:597–604

Saito Y, Wright EM (1982) Kinetics of the sodium pump in the frog choroid plexus. J Physiol (Lond) 328:229–243

Schell RE, Stevens BR, Wright EM (1982) Kinetics of Na-dependent solute transport by rabbit renal and jenunal brush border vesicles using a fluorescent dye. J Physiol (Lond) 335:307–318

Schultz SG, Curran PF (1970) Coupled transport of sodium and organic solutes. Physiol Rev 50: 637–718

Spring KR, Hope A (1979) Dimensions of cells and lateral intercellular spaces in living *Necturus* gall bladder. Fed Proc 38:128–133

Stevens BR, Ross HJ, Wright EM (1982a) Multiple transport pathways for neutral amino acids in rabbit jejunal brush border vesicles. J Membr Biol 66:213–225

Stevens BR, Wright SH, Hirayama B, Ross HJ, Gunther RD, Nord E, Kippen I, Harms V, Wright EM (1982b) Liquid nitrogen preservation of organic and inorganic solute transport in renal and intestinal vesicles. Membr Biochem 4:271–282

Stirling CE (1972) Radiographic localization of sodium transport sites in rabbit intestine. J Cell Biol 53:704–714

Wright EM, Diamond JM (1977) Anion selectivity in biological systems. Physiol Rev 57:109–156

Wright EM, Os CH van, Mircheff AK (1980) Sugar uptake by basolateral membrane vesicles. Biochem Biophys Acta 597:112–124

Wright SH, Krasne S, Kippen I, Wright EM (1981) Na^+-dependent transport of tricarboxylic acid cycle intermediates by renal brush border membranes. Effects on fluorescence of a potential-sensitive cyanine dye. Biochim Biophys Acta 640:767–778

Zeuthen T, Wright EM (1981) Epithelial potassium transport: Tracer and electrophysiological studies in choroid plexus. J Membr Biol 60:105–128

Transport of Inorganic Anions Across the Small Intestinal Brush Border Membrane

H. MURER [1], J. BIBER [1], V. SCALERA [1], G. CASSANO [1], B. STIEGER [1], G. DANISI [2], B. HILDMANN [3], G. BURCKHARDT [3], and H. LÜCKE [4]

Introduction

The small intestine absorbs various inorganic anions (e.g., chloride, phosphate, sulfate) from the lumen and delivers them to the blood. Similarly to most solutes, this absorption can occur against their concentration gradients and is sodium-dependent. Active transport is always transcellular and involves translocation across the brush border and the basolateral membrane.

This review is restricted to the transcellular transport of anions. For a more complete discussion of transepithelial (transcellular and paracellular) anion transport in small intestine we refer to recent reviews on related topics (e.g., Binder 1981, Field 1981, Schultz 1979, 1981a,b).

For the driving forces of anion transport across the luminal and contraluminal border of the epithelial cell the following considerations have to be taken into account. Across the membranes of an epithelial cell an inside negative electrical potential difference exists. Therefore, passive anion distribution across the membranes leads to anion concentrations lower in the cell than in the surrounding medium, e.g., for a monovalent anion a tenfold smaller intracellular concentration is expected at an electrical potential difference of −60 mV. Whenever intracellular anion concentrations exceed the passive equilibrium distribution, energy-requiring, active transport across the plasma membranes has to occur. If for a reabsorbed anion the intracellular concentration is higher than predicted from equilibrium distribution, the active step can be expected to be located in the luminal cell border (Fig. 1). Exit of an anion from the cell at the contraluminal border might proceed passively, driven by its electrochemical potential difference.

The energy for active transport, i.e., transport against an electrochemical potential difference may be provided by direct coupling to a chemical reaction ("primary active transport"). An example for a primary active transport system is the $(Na^+ + K^+)$-ATPase

1 Department of Physiology, University of Zürich, Rämistrasse 69, CH-8028 Zürich, Switzerland
2 Department of Pharmacology, University of Geneva, 20, rue de l'Ecole-de-Médecine, CH-1211 Genève 4, Switzerland
3 Max-Planck-Institut für Biophysik, Kennedyallee 70, D-6000 Frankfurt-M. 70, Fed. Rep. of Germany
4 Division of Gastroenterology, Department of Medicine, University of Göttingen, D-3400 Göttingen, Fed. Rep. of Germany

Intestinal Transport
(ed. by M. Gilles-Baillien and R. Gilles)

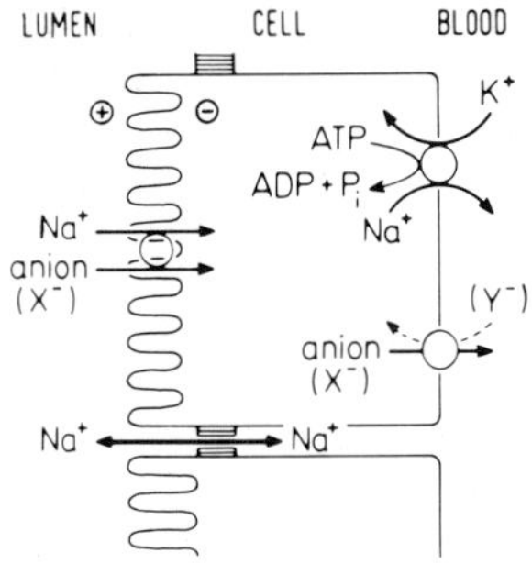

Fig. 1. Schematic representation of the polar distribution of transport systems possibly involved in the transepithelial absorption of anions

which is located in the antiluminal membrane (e.g., Murer and Kinne 1980). This enzyme maintains a high intracellular potassium concentration and a low intracellular sodium concentration. Active transport may also be achieved by flux coupling to another solute, e.g., sodium, which is transported down its electrochemical potential difference. The electrochemical potential difference of sodium is maintained by the ($Na^+ + K^+$)-ATPase which thereby indirectly energizes sodium-coupled transport processes ("secondary active transport"). Active anion transport mechanisms seem to fall into the category of secondary active, sodium-coupled transport (Murer and Burckhardt 1982).

Sodium-Coupled Anion Absorption

Chloride

Small intestinal epithelia, similarly to renal proximal tubular epithelia, are characterized as leaky, i.e., they have a high permeability to small ions and water (Frömter et al. 1973, Schultz 1979, 1981a,b). Thus due to the small contribution of transcellular chloride movement to overall transepithelial chloride movement it was difficult to identify the cellular absorptive mechanisms.

In rabbit ileum the luminal chloride entry mechanism and the intracellular chloride accumulation depends on the presence of sodium (for reviews see: Frizzell et al. 1979, Schultz 1979, 1981a,b). These findings suggested the existence of a sodium/chloride cotransport mechanism in the luminal membrane. This mechanism was apparently inhibited by furosemide and by increased intracellular c-AMP levels (for reviews see: Frizzell et al. 1979, Duffey et al. 1979, Field 1981). An alternative mechanism for electroneutral sodium chloride absorption is the coupled action of an Na^+/H^+ and a Cl^-/OH^- (HCO_3^-) exchanger in the luminal membrane of enterocytes which has been postulated for the ileum by Turnberg et al. (1970a,b). Coupling of these two exchange mechanisms could be possible via pH changes in the microclimate of the transport systems. The source of intracellular H^+ and HCO_3^- is carbonic acid formed by mediated hydration of CO_2 (carbonic anhydrase). The clinical observation of an impaired anion exchange mechanism (Bieberdorf et al. 1972) and the experimental observation of inhibition of sodium chloride absorption in intact epithelia by acetazolamide (inhibition of carbonic anhydrase) are indications that these exchange mechanisms are indeed operating in intact small intestine and are parts of electroneutral sodium-chloride cotransport (Nellans et al. 1975).

As assumed recently by Murer and Burckhardt (1982) and by Warnock and Eveloff (1982), similar membrane mechanisms are involved in the renal handling of chloride. However, at least in the rat renal proximal tubule the passive flux of chloride is predominant (Frömter et al. 1973).

Liedtke and Hopfer (1982a) recently denied, based on kinetic experiments with brush border membrane vesicles isolated from rat small intestine, the existence of a NaCl cotransport. On the other hand, the existence of a Na^+/H^+ exchanger as well as of a Cl^-/OH^- exchanger was documented in rat and rabbit small intestinal and proximal tubular brush border membrane vesicles by pH-meter, tracer, and spectroscopic techniques (Kinsella and Aronson 1981, Liedtke and Hopfer 1977, 1982a,b, Murer et al. 1976, 1980a, Reenstra et al. 1981, Sachs et al. 1982, Warnock and Yee 1981, Warnock and Eveloff 1982, Cassano et al., unpublished data).

In agreement with the inhibition of transepithelial chloride movements in small intestine by furosemide and stilbenes, Liedtke and Hopfer (1982b) found also an inhibition by these drugs of the tracer chloride transport via the anion exchanger in brush border membrane vesicles. Thus, vesicle studies using tracers have revealed the existence of Na^+/H^+ and Cl^-/OH^- exchange mechanisms but failed to demonstrate their combined action, i.e., sodium-stimulated electroneutral chloride flux.

In our laboratory we recently reexamined Na^+/H^+ exchange and Cl^-/OH^- exchange in rat small intestinal brush border membrane vesicles by using an acridine orange fluorescence quenching technique. Similar studies have employed this technique to follow the rate of Na^+/H^+ exchange in renal brush border membrane vesicles (Reenstra et al. 1981, Warnock and Eveloff 1982). From the experiment presented in Fig. 2, it can be seen that efflux of sodium is paralleled by an acidification of the intravesicular medium (Fig. 2a), as indicated by the transient fluorescence quenching. This quenching of fluorescence is partially due to indirect coupling between conductive Na^+-entry and conductive H^+-efflux (Cassano et al., unpublished data). However, the signal observed under conditions of short circuited membrane potential – potassium at equal concentrations at both membrane sites in the presence of valinomycin – must be related to the Na^+/H^+ exchange mechanism. Surprisingly an inside directed chloride gradient is unable to provoke a significant fluorescence signal (Fig. 2b). If the rates for Na^+/H^+ and Cl^-/OH^- exchangers would be similar, we would have expected a similar acidification in the Cl^-/OH^- experiments (Fig. 2b) as in the Na^+/H^+ exchanger experiments (Fig. 2a). Therefore, we have to conclude that the Cl^-/OH^- exchange system identified with tracer studies (Liedtke and Hopfer 1982b), has a much smaller rate than the Na^+/H^+ exchanger. This conclusion finds further experimental support by a direct comparison of the acidification of the intravesicular volume in the presence of outwardly directed Na^+-gluconate- or Na^+-chloride-gradients, respectively (Fig. 2c). If Na^+/H^+ exchanger and Cl^-/OH^- exchanger would operate in the same membrane and with comparable rates, the signal in the Na^+-chloride condition should be considerably lower than the signal in the presence of Na^+-gluconate. This is not observed. However, the faster reequilibration in the presence of chloride of the transmembrane pH-gradient created by the Na^+/H^+ exchanger

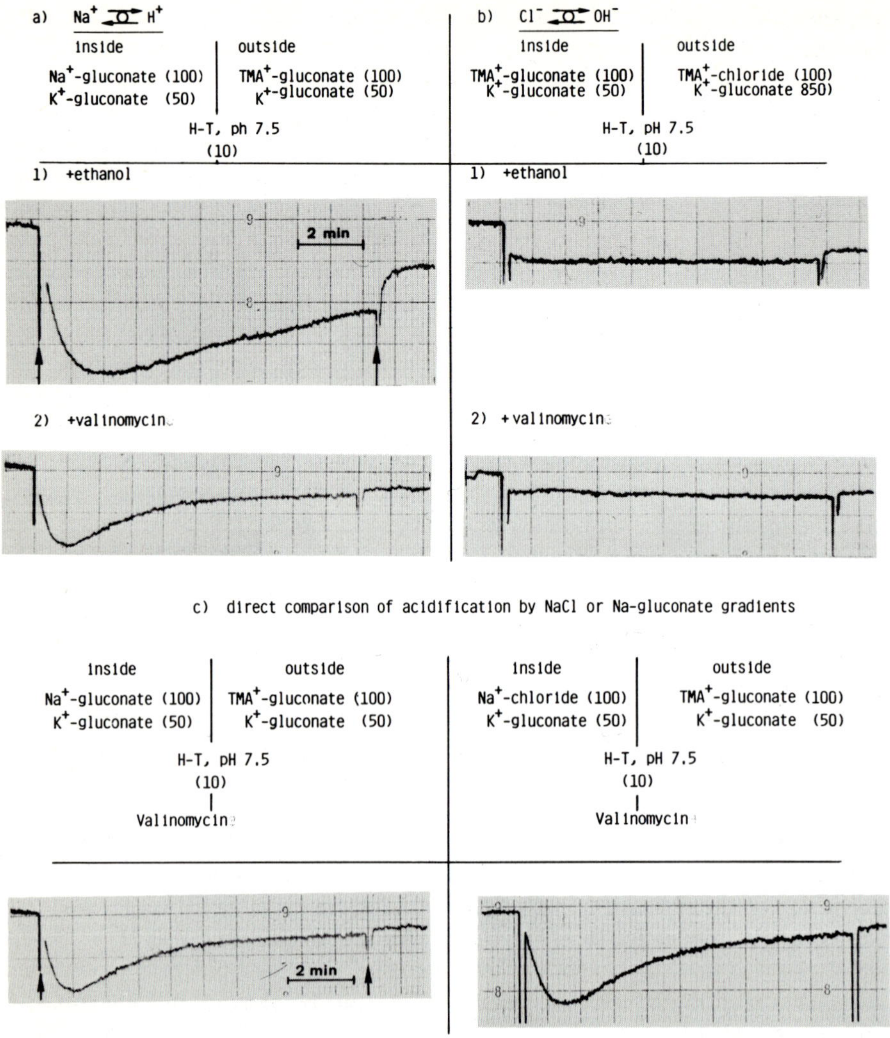

Fig. 2. Measurement of Na^+/H^+ exchange and Cl^-/OH^- exchange in rat small intestinal brush border membrane vesicles by the acridine orange fluorescence quenching technique. The methods used for this experiment are similar to that described by Reenstra et al. (1981). Important experimental conditions are indicated in the figure. The details of this study will be published later. (Cassano and Murer, unpublished data)

(Fig. 2c) is an indication for the existence of the Cl^-/OH^- exchanger in this membrane. This latter conclusion is in agreement with the tracer data, showing a stimulation of chloride flux by transmembrane pH-gradient (Liedtke and Hopfer 1982b). Unfortunately, the tracer experiments on Cl^-/OH^- exchange were not performed with the same membranes in the presence and absence of sodium (Liedtke and Hopfer 1982b, Murer et al. 1980a). An inhibitory effect on Cl^-/OH^- exchange via a Na^+/H^+ exchanger-dependent breakdown of the present pH-gradient would have been an elegant experimental proof for the coexistence of the two exchangers in the same membrane.

Little is known about the exit step at the contraluminal cell membrane. As intracellular chloride activities are above equilibrium distribution, the movement of chloride out of the cell could follow its electrochemical potential difference. Stilbene derivatives, possible inhibitors of anion exchangers in different biological membranes, did not significantly influence chloride efflux from basolateral membrane vesicles isolated from renal and small intestinal epithelium (Grinstein et al. 1980). As these authors also found no effect of stilbenes on chloride efflux from brush border membranes the significance of this finding is unclear. A stilbene sensitivity of chloride fluxes mediated by anion exchange has been demonstrated in studies with rat small intestinal brush border membrane vesicles (Liedtke and Hopfer 1982b). Langridge-Smith and Field (1981) provided convincing evidence for a stilbene sensitive anion exchanger in the basal-lateral membrane of intact rabbit ileum. This mechanism also accepted chloride as a substrate.

Phosphate

Inorganic phosphate transport against an electrochemical potential difference in the small intestine depends on the presence of sodium (for reviews see: Bikle et al. 1981, Murer et al. 1980c, Murer and Hildmann 1981). This has been demonstrated in studies with everted loops and in experiments on the unidirectional influx of phosphate across the brush border membrane into the epithelial cell (Harrison and Harrison 1963, Danisi and Straub 1980). The observed sodium-dependence led to the speculation that transepithelial transport of phosphate is energized by the sodium gradient across the brush border membrane and can therefore be considered as a secondary active transport process. A similar mechanism was documented for the renal proximal tubule (Murer et al. 1980c, Dennis et al. 1979, Murer and Burckhardt 1982).

Conclusive evidence for a sodium-phosphate cotransport system was first presented by Hoffmann et al. (1976) for rat renal brush border membranes and by Berner et al. (1976) for rat intestinal brush border membranes. Arsenate inhibited competitively phosphate transport. Meanwhile sodium-phosphate cotransport mechanisms have also been identified in other brush border membranes such as chicken, rabbit, and mouse intestinal and renal preparations (Cheng and Sacktor 1981, Danisi et al. 1982, Fuchs and Peterlik 1980, Matsumoto et al. 1980, Tenenhouse and Scriver 1978).

In both, intact intestinal epithelia and isolated intestinal membrane vesicles, a decrease in pH increased phosphate transport (Fig. 3; Berner et al. 1976, Danisi et al. 1982, Walling 1978). At first this was taken as an evidence for the preferential transport of monovalent phosphate (Berner et al. 1976). However, altering of the pH of the incubation medium may alter the transport system itself. Assuming that the transport system accepts preferentially one ionic species of phosphate, an influence of a trans-membrane pH difference on sodium-dependent phosphate transport is expected. With monovalent phosphate being the preferred substrate, an increase in intravesicular pH should increase phosphate transport rate. In experiments with isolated rabbit small intestinal brush border membranes we were unable to detect a significant alteration in the transport rate by variation of the intravesicular pH (Danisi et al. 1982).

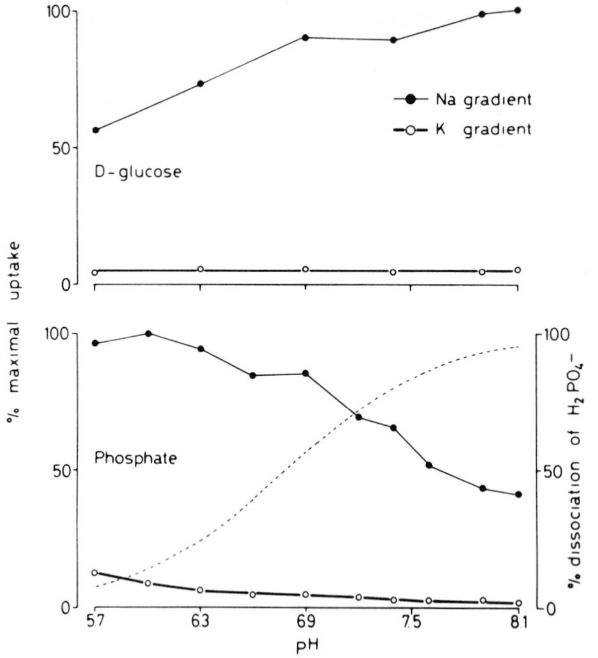

Fig. 3. pH dependence of inorganic phosphate transport in brush border membrane vesicles isolated from rabbi duodenum. Membranes were prepared in 300 mmol l^{-1} mannitol, 20 mmol l^{-1} HEPES adjusted with Tris to pH 7.4. Incubation was performed in 300 mmol l^{-1} mannitol, 100 mmol l^{-1} NaCl, 0.1 mmol l^{-1} $KH_2\,^{32}PO_4$, 20 mmol l^{-1} MES-Tris or HEPES-Tris to give the final pH as indicated on the figure. For D-glucose uptake measurement, substrate concentration was also 0.1 mmol l^{-1}. 15 s uptake values are given in the figure

The experiments with rabbit small intestinal brush border membrane vesicles (Danisi et al. 1982) suggested that the pH-dependence of phosphate transport in small intestine is provoked by a pH sensitivity of the transporter and does not reflect preferential transport of monovalent phosphate. Decreasing pH from 8 to 6, i.e., changing primary/secondary phosphate concentration ratio by a factor of 100 led only to an about twofold increase in sodium-dependent phosphate transport (Fig. 3). As phosphate transport was measured at total phosphate concentrations far below K_m, this finding can only be explained by a transport of both monovalent and divalent phosphate.

Only limited knowledge is available on the exit step at the contraluminal cell border. A study with basolateral membrane vesicles isolated from rat kidney cortex (Hoffmann et al. 1976) and preliminary results with basolateral membrane vesicles from rat small intestine (Murer et al., unpublished data) indicate that the exit step at the contraluminal cell surface proceeds via sodium-independent pathways. Experiments with isolated basolateral membrane vesicles from rat small intestine and dog renal proximal tubule did not provide convincing evidence for a specific transport system (Grinstein et al. 1980).

Sulfate

Inorganic sulfate was shown to be reabsorbed in intact ileal epithelia against its concentration gradient. This transport was sodium-dependent (Anast et al. 1965, Turnberg et al. 1970a). More recently, unidirectional fluxes of sulfate across and into rabbit ileal epithelium were measured under short circuit conditions (Smith et al. 1981). These experiments clearly suggested the existence of a sodium-coupled sulfate influx across the brush border membrane. Active proximal tubular sulfate transport was also sodium-dependent (Ullrich et al. 1980a, Murer and Burckhardt 1982).

Studies with brush border membrane vesicles isolated from rat ileum and rat and rabbit proximal tubule revealed a sodium-dependent sulfate transport system (Fig. 4 for rabbit ileum). Sodium-sulfate cotransport was electroneutral (Lücke et al. 1979, 1981, Ullrich et al. 1980a,b, Schneider et al. 1980) in agreement with electrophysiological data obtained with rabbit ileal epithelium (Smith et al. 1981) and proximal tubule in vivo (Samarzija et al. 1981). The sodium-sulfate cotransport mechanism was shared by thiosulfate, but not by phosphate (Lücke et al. 1979, 1981, Ullrich et al. 1980a,b). A recent study on sulfate influx across the rabbit ileal brush border membrane suggested Na^+-H^+-SO_4^{2-} cotransport at low sulfate concentrations but Na^+-Na^+-SO_4^{2-} cotransport at high sulfate concentrations (Langridge-Smith et al. 1982).

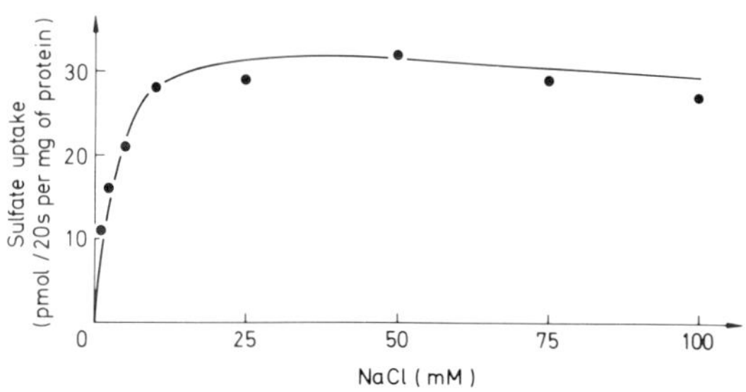

Fig. 4. Influence of Na^+-concentration on sulfate uptake into Na^+-preequilibrated vesicles. Membranes were prepared from rat ileum in 100 mmol l^{-1} mannitol, 20 mmol l^{-1} HEPES adjusted with Tris to pH 7.4 and preincubated with different NaCl and/or KCl containing solutions for equilibration with different sodium concentrations. Incubation was carried out in the same sodium and/or potassium containing solutions containing in addition 0.75 mmol l^{-1} $Na_2{}^{35}SO_4$. Total concentration of (KCl + NaCl) was always 100 mmol l^{-1}

Langridge-Smith and Field (1981) provided evidence for a stilbene-sensitive anion exchange mechanism accepting the sulfate anion in the basolateral membranes of intact rabbit ileal epithelia. Similar findings were obtained in isolated rabbit proximal tubules (Brazy and Dennis 1981). Stilbene derivatives inhibited sulfate flux also in basolateral membrane vesicles isolated from rat small intestine and dog kidney cortex (Grinstein et al. 1980).

Regulatory Mechanisms in Anion Absorption at the Level of the Brush Border Membrane?

Chloride

Experimental evidence suggests that c-AMP exerts its antiabsorptive effect on NaCl-reabsorption in the villus cells and stimulates secretion in the crypt cells. The c-AMP effect might involve protein phosphorylation and be mediated by calcium probably via calmodulin (Field 1981, Frizzell et al. 1979). The mechanisms involved in chloride secretion are similar to those of chloride absorption. Secretion is sodium-dependent and ouabain-sensitive, i.e., there is evidence that the secretion of chloride is a secondary active process. The sodium-coupled entry mechanism has to be located in the basolateral membrane (for review see: Field 1981).

As well-defined membrane vesicles are not available from the crypt region of the small intestinal epithelium, studies with vesicles on regulatory mechanisms can only contribute to an understanding of the inhibited NaCl entry across the brush border membrane of the villus cells. As discussed above, studies with vesicles were unable to document sodium-coupled chloride influx. However, evidence for Na^+/H^+ exchange and Cl^-/OH^- exchange – possible elements of "coupled" NaCl entry in the intact tissue – was obtained. Studies with vesicles isolated from small intestinal segments perfused either with control-solutions, dibutyryl c-AMP, or choleratoxin containing solutions did not show reduced sodium uptake rates but significantly increased sodium-dependent D-glucose uptake (Fig. 5; Murer et al. 1980b). This finding indicates

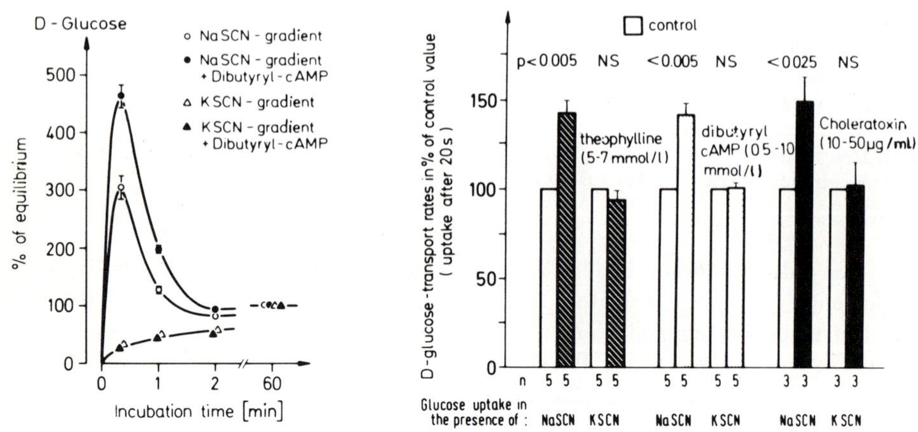

Fig. 5. Influence of in vivo perfusion with different compounds on transport of D-glucose by brush border vesicles. Rat jejunum was perfused in vivo for 2.5 h with a Krebs-Henseleit phosphate buffer containing the different drugs as indicated in the figure. Secretion of chloride and fluid and stimulation of 3-O-methyl-D-glucose was observed in the in vivo perfusion. After in vivo perfusion brush border membrane vesicles were isolated and analyzed for sodium gradient-dependent D-glucose transport (for details see: Murer et al. 1980b). Significant stimulation of sodium-dependent D-glucose uptake was observed under sodium gradient conditions in membrane vesicles isolated from intestinal segments perfused with agents known to increase cellular c-AMP levels

that the membrane retains after isolation altered membrane functions. Unfortunately, chloride uptake was not examined in these vesicles. As Na^+/H^+ exchange and Cl^-/OH^- exchange could represent partial activities of an electroneutral NaCl entry mechanism and might operate coupled under physiological conditions, it is feasable, that a c-AMP-dependent reduction in Cl^-/OH^- exchange would also reduce "coupled" electroneutral NaCl absorption. This would represent a situation similar to that observed for furosemide. Liedtke and Hopfer (1982b) have shown an inhibition of NaCl absorption in the intact rat small intestine by this drug and inhibition of Cl^-/OH^- exchange in vesicles.

Because many effects of c-AMP are suggested to be mediated through regulation of protein kinase activity, it seems likely that phosphorylation and dephosphorylation reactions of single membrane proteins could lead to an altered brush border membrane function (transport properties). Lucid and Cox (1972) reported enhanced incorporation of ^{32}P into the brush border membrane under the influence of choleratoxin by an in vivo method, and De Jonge (1981) and Schlatz et al. (1978, 1979) reported enhanced ^{32}P incorporation into a single polypeptide when isolated brush border membranes were phosphorylated by γ-^{32}P-ATP in the presence of c-AMP and c-GMP. As some choleratoxin or dibutyryl-c-AMP-induced changes of the brush border membrane permeability are retained at the level of the isolated membrane (Fig. 5; Murer et al. 1980b), it seems reasonable to look for an altered membrane phosphorylation after the small intestine has been perfused with (^{32}P) orthophosphate and choleratoxin or dibutyryl-c-AMP. As shown in Fig. 6, the brush border membrane phosphoprotein pattern was not significantly changed under the influence of choleratoxin or dibutyryl-c-AMP. Sometimes, a rather general slight dephosphorylation was observed after perfusion with choleratoxin or dibutyryl-c-AMP.

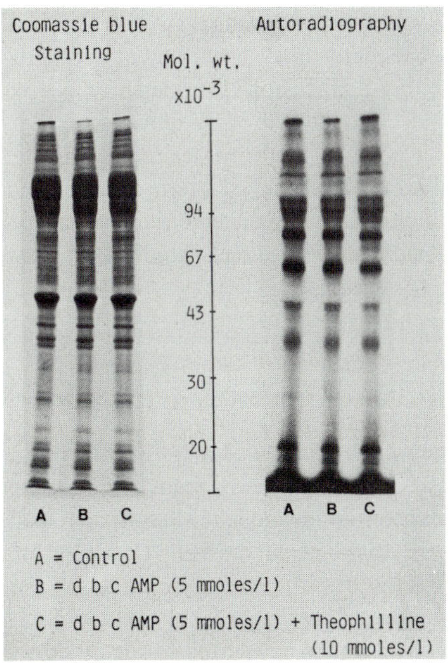

Fig. 6. In vivo phosphorylation of rat jejunum brush border membrane proteins. Rat intestinal segments were perfused in the absence *(A)* or presence *(B, C)* of dibutyryl-c-AMP with ^{32}P-orthophosphate containing solutions. Perfusion was performed as in the experiments presented in Fig. 5. Brush border membranes were isolated from the perfused segments and analyzed by autoradiography for incorporation of ^{32}P into membrane proteins separated on polyacrylamide gels

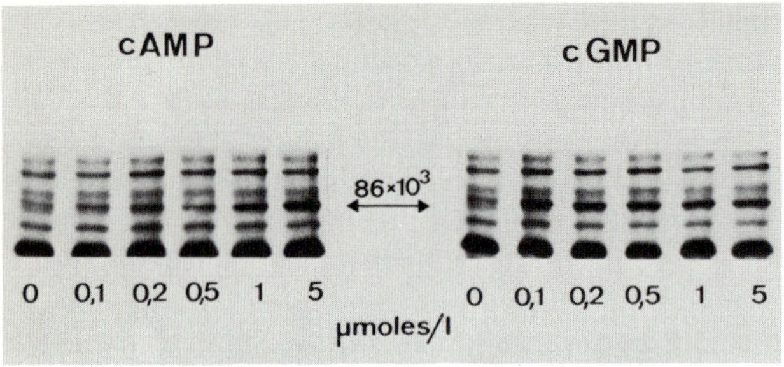

Fig. 7. Effect of the cyclic nucleotides c-AMP and c-GMP on brush border membrane phosphorylation. Autoradiograms of brush border membranes phosphorylated by γ^{32}P-ATP (20 μmol l^{-1}) in the presence of 0.1% (w/v) saponin and in the presence of the indicated concentrations (in μmol l^{-1}) of the cyclic nucleotides for 20 s at room temperature are shown. Only the molecular weight region (~ 86,000 M.W.) with cyclic nucleotides dependent changes is presented

The effect of c-AMP on the brush border membrane phosphorylation was also analyzed by phosphorylating freshly isolated (without prior perfusion) brush border membrane vesicles by γ-^{32}P-ATP in the presence of 0.1% Saponin (Fig. 7). Cyclic-AMP clearly stimulated the phosphorylation of a 86,000 M.W. protein at concentration higher than 0.5 μmol l^{-1}, but no other c-AMP-dependent phosphorylation could be observed. Because this 86,000 M.W. protein was reported to be identical with a cyclic GMP-dependent protein kinase, the effect of c-GMP was also tested. In agreement with the data reported by De Jonge (1976, 1981), c-GMP was much more effective with respect to the phosphorylation of the 86,000 M.W. protein. Phosphorylation of this protein was clearly stimulated at 0.1 μmol l^{-1} c-GMP already, but the phosphorylation of the other proteins again did not change.

In conclusion, no evidence for a c-AMP-dependent brush border membrane protein phosphorylation could be obtained by either a perfusion of rat jejunum in vivo with (^{32}P)-orthophosphate in the presence of choleratoxin or by dibutyryl-c-AMP or by an in vitro phosphorylation of freshly isolated brush border membranes by γ-^{32}P-ATP. Therefore, it remains questionable whether the secretion of water and electrolytes induced by intoxication with choleratoxin is mediated directly via membrane protein kinase activity which would be activated by the elevated intracellular level of c-AMP.

Phosphate

Small intestinal phosphate transport is influenced by various control mechanisms. Regulatory phenomena are mediated by 1.25 $(OH)_2$ Vit D_3. All maneuvres leading to an increased serum concentration of this secosteroid hormone lead to an increased transepithelial transport rate (for review see: Bikle et al. 1981). The effect of 1.25 $(OH)_2$ Vit D_3 is expressed at the level of the brush border membrane and can be analyzed in isolated brush border membrane vesicles (Fig. 8; for review see: Bikle et al. 1981, Murer and Hildmann 1981). The experiment presented in Fig. 8 shows

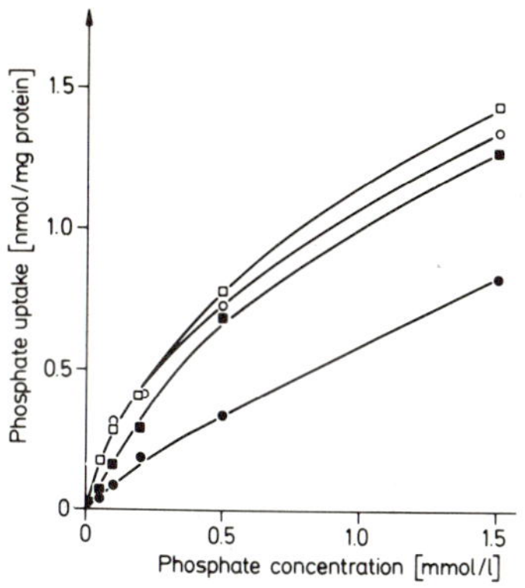

In vivo treatments: ■ control

● EHDP

□ + 1.25 $(OH)_2$ Vit. D_3

○ EHDP + 1.25 $(OH)_2$ Vit. D_3

Fig. 8. Influence of the 1.25 $(OH)_2$ Vit D status of rabbits on the sodium gradient dependent uptake of inorganic phosphate at various medium phosphate concentrations. The experimental details are presented elsewhere (Hildmann et al. 1982). Four rabbits of similar weight (range of 300 g from lightest to heaviest) were selected for each experiment. To induce 1.25-D_3 deficiency, two of them were injected s.c. with 40 mg kg^{-1} EHDP dissolved in sterile water at a concentration of 125 mg ml^{-1} in the morning of three consecutive days. 8–10 h before sacrifice, one EHDP-trated rabbit and a sham-treated rabbit received an i.v. injection of 600 ng kg^{-1} 1.25-D_3. The duodena (60 cm distal of the pyloric bulb) were immediately removed and bathed in icecold Ringer's solution. 40 μl of membrane suspension in 300 mmol l^{-1} mannitol, 20 mmol l^{-1} HEPES/Tris, pH 7.4 were added to 40 μl of incubation media containing in addition 200 mmol l^{-1} sodium chloride and different concentrations of KH_2PO_4 leading to the final concentrations of P_i indicated in the figure. Samples of 50 μl were taken after 10 s. Each experiment was performed with triplicates of incubation media. The values represent averages of four experiments. *EHDP* disodium ethane-1-hydroxy-1-1-diphosphonate; this drug is known to lower plasma levels of 1.25-dihydroxy-vitamin D_3. Brush border membranes were isolated by the divalent cation precipitation method

that the sodium-dependent saturable component of phosphate uptake into isolated rabbit duodenal brush border vesicles is decreased by maneuvres leading to a decreased 1.25 $(OH)_2$ Vit D_3 level in the animal prior to sacrifice. On the other hand, increased levels lead to an increased transport (Hildmann et al. 1982). With regard to the cellular mechanisms of 1.25 $(OH)_2$ Vit D_3 leading to altered brush border membrane properties it is unclear at the moment whether the action is mediated by de novo protein synthesis or occurs via changes in the lipid composition of the brush border membrane (for review see: Bikle et al. 1981, Isselbacher 1981, Murer and Hildmann 1981, Peterlik et al. 1981).

Concluding Remarks

Figure 9 summarizes the experimental evidence for membrane mechanisms involved in anion transport in small intestine. There is considerable knowledge on the proper-

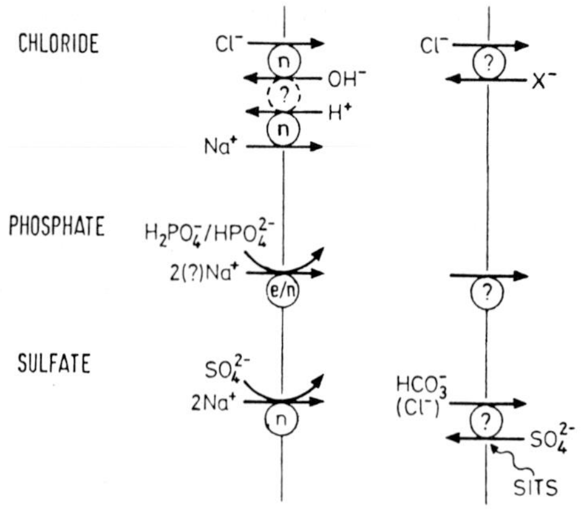

Fig. 9. Schematic representation of membrane transport mechanisms involved in the intestinal absorption of anions. (For extensive discussions see Murer and Burckhardt 1982)

ties of the anion transport mechanisms in the luminal membrane. Inspection of the figure may indicate where future efforts can complete our knowledge on the mechanisms involved in transepithelial anion transport. Progress will only be possible when different in vivo and in vitro methods are applied. Comparison of the findings with respect to driving forces, specificity, cellular and segmental localization will allow to draw a more conclusive picture.

Acknowledgment. The work discussed in this review was supported by the Schweizerischer Nationalfonds.

References

Anast C, Kennedy R, Volk G, Adamson L (1965) In vitro studies of sulfate transport by the small intestine of the rat, rabbit and hamster. J Lab Clin Med 65:903–911

Berner WR, Kinne R, Murer H (1976) Phosphate transport into brush border membrane vesicles isolated from small intestine. Biochem J 160:467–474

Bieberdorf FA, Gorden P, Fordtran JS (1972) Pathogenesis of congenital alkalosis with diarrhea. Implications for the physiology of normal ileal electrolyte absorption and secretion. J Clin Invest 51:1958–1968

Bikle DD, Morrissey RL, Zolock DT, Rasmussen H (1981) The intestinal response to vitamin D. Rev Physiol Biochem Pharmacol 89:63–142

Binder HJ (1981) Colonic secretion. In: Johnson LR (ed) Physiology of the gastrointestinal tract. Raven Press, New York, pp 1003–1020

Brazy PC, Dennis VW (1981) Sulfate transport in rabbit proximal convoluted tubules: presence of anion exchange. Am J Physiol 241:F300–F307

Cheng L, Sacktor B (1981) Sodium gradient-dependent phosphate transport in renal brush border membrane vesicles. J Biol Chem 256:1556–1564

Danisi G, Straub RW (1980) Unidirectional influx of phosphate across the mucosal membrane of rabbit small intestine. Pfluegers Arch 385:117–122

Danisi G, Murer H, Straub R (1982) Phosphate transport in rabbit duodenal membrane vesicles. Experientia (in press)

Dennis VW, Stead WW, Myers JL (1979) Renal handling of phosphate and calcium. Ann Rev Physiol 41:257–271

Duffey ME, Thompson SM, Frizzell RA, Schultz SG (1979) Intracellular chloride activities and active chloride absorption in the intestinal epithelium of the winter flounder. J Membr Biol 50:331–341

Field M (1981) Secretion of electrolytes and water by mammalial small intestine. In: Johnson LR (ed) Physiology of the gastrointestinal tract. Raven Press, New York, pp 963–981

Frizzell RA, Field M, Schultz SG (1979) Sodium-coupled transport by epithelial tissues. Am J Physiol 236:F1–F8

Frömter E, Rumrich G, Ullrich KJ (1973) Phenomenological description of Na^+, Cl^- and HCO_3^- absorption. Pfluegers Arch 343:189–220

Fuchs R, Peterlik M (1980) Vitamin D induced phosphate transport in intestinal brush border membrane vesicles. Biochem Biophys Res Commun 93:87–92

Grinstein S, Turner RJ, Silverman M, Rothstein A (1980) Inorganic anion transport in kidney and intestinal brush border and basolateral membranes. Am J Physiol 238:F452–F460

Harrison HE, Harrison HC (1963) Sodium, potassium, and intestinal transport of glucose, L-tyrosine, phosphate, and calcium. Am J Physiol 205:107–111

Hildmann B, Storelli C, Danisi G, Murer H (1982) Regulation of Na^+-P_i cotransport by 1.25 dihydroxyvitamin D_3 in rabbit duodenal brush border membrane. Am J Physiol 242:G533–G544

Hoffmann N, Thees M, Kinne R (1976) Phosphate transport by isolated brush border membrane vesicles. Pfluegers Arch 362:147–156

Isselbacher KJ (1981) Introduction: calcium and phosphate transport by the intestinal cell. In: Bronner F, Peterlik M (eds) Calcium and phosphate transport across biomembranes. Academic Press, New York, pp 129–133

Jonge HR de (1976) Cyclic nucleotide dependent phosphorylation of intestinal epithelium proteins. Nature 262:590–592

Jonge HR de (1981) Cyclic GMP-dependent protein kinase in intestinal brush borders. Adv Cyclic Nucleotide Res 14:315–333

Kinsella JL, Aronson PS (1981) Interaction of NH_4^+ and Li^+ with the renal microvillus membrane Na^+-H^+ exchanger. Am J Physiol 241:C220–C226

Langridge-Smith JL, Field M (1981) Sulfate transport in rabbit ileum: characterization of the serosal border anion exchange process. J Membr Biol 63:207–214

Langridge-Smith JL, Sellin JE, Field M (1982) Sulfate influx across the rabbit ileal brush border membrane: sodium and proton dependence, and substrate specificity. J Membr Biol (in press)

Liedtke C, Hopfer U (1977) Anion transport in brush border membranes isolated from rat small intestine. Biochem Biophys Res Commun 76:579–585

Liedtke C, Hopfer U (1982a) Mechanism of Cl^- translocation across the small intestinal brush border membrane. I. Absence of NaCl cotransport. Am J Physiol 242:G263–G271

Liedtke C, Hopfer U (1982b) Mechanism of Cl^- translocation across the small intestinal brush border membrane. II. Demonstration of Cl^-/OH^- exchange and Cl^- conductance. Am J Physiol 242:G272–G280

Lucid SW, Cox AC (1972) The effect of cholera toxin on the phosphorylation of protein in epithelial cells and their brush borders. Biochem Biophys Res Commun 49:1183–1186

Lücke H, Stange G, Murer H (1979) Sulfate-ion/sodium-ion cotransport by brush border membrane vesicles from rat kidney cortex. Biochem J 182:223–229

Lücke H, Stange G, Murer H (1981) Sulfate sodium cotransport by brush border membrane vesicles isolated from rat ileum. Gastroenterology 80:22–30

Matsumoto J, Fontaine O, Rasmussen HT (1980) Effect of 1.25 dihydroxy-vitamin D_3 on phosphate transport into chick intestinal brush border membrane vesicles. Biochim Biophys Acta 599:13–23

Murer H, Burckhardt G (1982) Membrane transport of anions across epithelia of mammalian small intestine and kidney proximal tubule. In: Reviews of physiology, biochemistry and pharmacology, Springer, Berlin Heidelberg New York, pp 1–51

Murer H, Hildmann B (1981) Transcellular transport of calcium and inorganic phosphate in the small intestinal epithelium. Am J Physiol 240:G409–G416

Murer H, Kinne R (1980) The use of isolated membrane vesicles to study epithelial transport processes. J Membr Biol 55:81–95

Murer H, Hopfer U, Kinne R (1976) Sodium/proton antiport in brush border membrane vesicles isolated from rat small intestine and kidney. Biochem J 154:597–604

Murer H, Kinne-Saffran E, Beauwens R, Kinne R (1980a) Proton fluxes in isolated renal and intestinal brush border membranes: In: Schultz I, Sachs G, Forte JG, Ullrich KJ (eds) Hydrogen ion transport in epithelia. Elsevier/North Holland, Amsterdam, pp 267–285

Murer H, Lücke H, Kinne R (1980b) Isolated brush border vesicles as a tool to study disturbances in intestinal solute transport. In: Field M, Fordtran JS, Schultz SG (eds) Secretory diarrhea. American Physiological Society, Bethesda, pp 31–42

Murer H, Stern H, Burckhardt G, Storelli C, Kinne R (1980c) Sodium-dependent transport of inorganic phosphate across the renal brush border membrane. In: Massry SG, Jahn H (eds) Phosphate and minerals in health and disease. Plenum Publishing Corp, New York, pp 11–23

Nellans HN, Frizzell RA, Schultz SG (1975) Effect of acetazolamide on sodium and chloride transport by in vitro rabbit ileum. Am J Physiol 228:1808–1814

Peterlik M, Fuchs R, Sing Cross H (1981) Phosphate transport in the intestine: cellular pathways and hormonal regulation. In: Bronner F, Peterlik M (eds) Calcium and phosphate transport across biomembranes. Academic Press, New York, pp 173–179

Reenstra WW, Warnock DG, Yee VJ, Forte JG (1981) Proton gradients in renal cortex brush-border membrane vesicles. Demonstration of a rheogenic proton flux with acridine orange. J Biol Chem 256:11663–11666

Sachs G, Faller LD, Rabon E (1982) Proton/hydroxyl transport in stomach and intestine. J Membr Biol 64:123–135

Samarzija I, Molnar V, Frömter E (1981) The stoichiometry of Na^+ coupled anion absorption across the brush border membrane of rat renal proximal tubule. In: Takacs L (ed) Advances in physiological sciences, vol II. Kidney and body fluids. Pergamon Press, New York, pp 419–423

Schlatz LJ, Kimberg DV, Cattieu KA (1978) Cyclic nucleotide dependent phosphorylation of rat intestinal microvillus and basal-lateral membrane proteins by an endogenous protein kinase. Gastroenterology 75:838–846

Schlatz LJ, Kimberg DV, Cattieu KA (1979) Phosphorylation of specific rat intestinal microvillus and basal-lateral membrane proteins by cyclic nucleotides. Gastroenterology 76:293–299

Schneider EG, Durham JC, Sacktor B (1980) The sodium-dependent transport of inorganic sulfate by rabbit renal brush border membranes. Fed Proc 39:1711

Schultz SG (1979) Transport across small intestine. In: Giebisch G, Tosteson DC, Ussing HH (eds) Membrane transport in biology. Springer, Berlin Heidelberg New York, pp 749–777

Schultz SG (1981a) Salt and water absorption by mammalian small intestine. In: Johnson LR (ed) Physiology of the gastrointestinal tract. Raven Press, New York, pp 983–990

Schultz SG (1981b) Ion transport by mammalian large intestine. In: Johnson LR (ed) Physiology of the gastrointestinal tract. Raven Press, New York, pp 991–1002

Smith PL, Orella SA, Field M (1981) Active sulfate absorption in rabbit ileum: dependence on sodium and chloride and effects of agents that alter chloride transport. J Membr Biol 63: 199–206

Tenenhouse HS, Scriver CR (1978) The defect in transcellular transport of phosphate in the nephron is located in brush border membranes in X-linked hypophosphataemia. Can J Biochem 56:640–646

Turnberg LA, Bieberdorf FA, Morawski SG, Fordtran JS (1970a) Interrelationships of chloride, bicarbonate, sodium, and hydrogen transport in the human ileum. J Clin Invest 49:557–567

Turnberg LA, Fordtran JS, Carter NW, Rector FC (1970b) Mechanism of bicarbonate absorption and its relation to sodium transport in the human jejunum. J Clin Invest 49:557–567

Ullrich KJ, Rumrich G, Klöss S (1980a) Active sulfate reabsorption in the proximal convolution of the rat kidney: specificity, Na^+ and HCO_3^- dependence. Pfluegers Arch 383:159–163

Ullrich KJ, Rumrich G, Klöss S (1980b) Bidirectional active transport of thiosulfate in the proximal convolution of the rat kidney. Pfluegers Arch 387:127–132

Walling MW (1978) Intestinal inorganic phosphate transport. In: Massry SG, Ritz E, Rapado A (eds) Homeostasis of phosphate and other minerals. Plenum Press, New York, pp 131–148

Warnock DG, Eveloff J (1982) NaCl entry mechanisms in the luminal membrane of the renal tubule. Am J Physiol 242:F561–F574

Warnock DG, Yee VJ (1981) Chloride uptake by brush border membrane vesicles isolated from rabbit renal cortex. Coupling to proton gradients and K^+ diffusion potentials. J Clin Invest 67:103–115

Mechanisms of Sugar Transport Across the Intestinal Brush Border Membrane

E. BROT-LAROCHE and F. ALVARADO [1]

Introduction

This article deals with certain aspects of intestinal sugar transport, notably the coupling mechanism in sugar and Na^+ cotransport across the brush border membrane (BBM [2]). Although there are various possible mechanisms for monosaccharide transport across the BBM, we will focus on what we call the main transport system, which is specific for D-glucose and D-galactose, occurs uphill [3] and is both sodium dependent and phlorizin sensitive (review: Crane 1968). In the absence of a better name, we refer to this as the D-glucose or sugar BBM carrier or, simply, as System 1. In addition, the BBM possesses at least one other hexose transport system, specific for D-fructose, and perhaps a third for the transport of both D-glucose and D-galactose, but different from System 1. The physiological significance of this latter system, however, remains obscure and more work is needed to characterize it clearly (see Robinson et al., this vol.).

The Glucose and Galactose Malabsorption Syndrome

Certain human mutants cannot absorb either D-glucose or D-galactose (see Schneider et al. 1966). Given regular milk, whose lactose is split by brush border lactase into glucose and galactose, these babies develop an osmotic diarrhea that can be fatal.

1 Centre de Recherches sur la Nutrition, Centre National de la Recherche Scientifique, 92190 Meudon, France

2 Abbreviations used: BBM, brush border membrane; BLM, basolateral membrane; ECS, extracellular space; S, substrate, organic solute

3 The term uphill transport means that the substrate is transported against its own electrochemical gradient. Because, by definition, this type of transport requires an energy input, it is common to use the trivial name, active transport, when uphill transport is meant. If the energy input comes directly from an exergonic chemical reaction, then we speak of primary uphill transport (true active transport). If the energy derives from the dissipation of the "osmotic" gradient of another substrate or ion, then we speak of secondary uphill transport. This is believed to be the mechanism underlying sugar transport across the BBM. In contrast, equilibrating or downhill transport systems usually operate in the direction of the substrate's electrochemical gradient: the term facilitated diffusion may be applied here (for further details on nomenclature, see Alvarado 1970, Crane 1977, Schultz 1980)

Intestinal Transport
(ed. by M. Gilles-Baillien and R. Gilles)

However, they thrive if fed an artificial milk containing D-fructose as the only sugar. In our view, this is the best evidence thus far available to prove that: (1) D-glucose and D-galactose share a common transport mechanism that is genetically determined and probably involves a single protein or carrier; and (2) D-fructose can be taken up by a genetically different transport system. At the same time, these observations cast doubt on the existence of a second system for glucose and galactose transport, at least in the human intestine.

Interestingly, D-xylose, similarly to D-fructose, does not provoke diarrhea in babies with the glucose and galactose malabsorption syndrome (Schneider et al. 1966), thus suggesting that there may be an additional, pentose-preferring intestinal transport pathway whose significance remains to be established.

The Polarity of the Enterocyte

In vertebrates, the small intestine, the proximal renal tubule, the choroid plexus and, in certain species, the large intestine and the gall bladder, are all able to transport sugars uphill. We assume that the BBM sugar carrier in all vertebrates derives from a common ancestor (Alvarado and Monreal 1967) and the question may be posed whether the same is true for the renal BBM system. We know that the kidney is more complex than the intestine since it has at least two main aldose transport systems (review: Silverman 1976). One, the M system, seems to be specific for D-mannose and is absent in the intestine. The other, the G system, is specific for D-glucose and can be equated tentatively with System 1 in the enterocyte. In contrast with the various epithelia mentioned above, all other vertebrate cells appear to transport sugars downhill. The prototype of such systems is found in the red blood cell, and the one in the basolateral membrane (see below) is very similar if not identical to it. For convenience, we group these systems under the term plasma membrane, red cell-type, downhill sugar transport systems to emphasize the point that, most probably, the BBM is not a typical plasma membrane. As a result of cell differentiation, the BBM seems to have developed a sugar carrier quite unique to it.

Epithelia are functionally distinct because of their capacity for transcellular uphill transport. This vectorial transport capacity derives from the functional linking of two different transport systems, located at opposite sides of a polar cell and operating in series. The cellular basis of the intestinal system is a monolayer of cylindrical cells, the enterocytes, joined together by apical intercellular junctions separating the plasma membrane into two morphologically and functionally distinct domains (Fig. 1). The apical or luminal region, with its characteristic microvilli, is the BBM. Below the junctions, there is the basolateral membrane (BLM) region, facing intercellular spaces that separate the enterocytes.

The transmural transfer of solutes, therefore, can follow either of two routes. (1) The cellular or transcellular route involves the sequential crossing of the BBM and the BLM; if a "valve" is located in at least one of these membranes, transcellular uphill transport can result (the "valve" concept is discussed further below). (2) The paracellular route involves solute transfer across the junctions, bypassing the enterocytes.

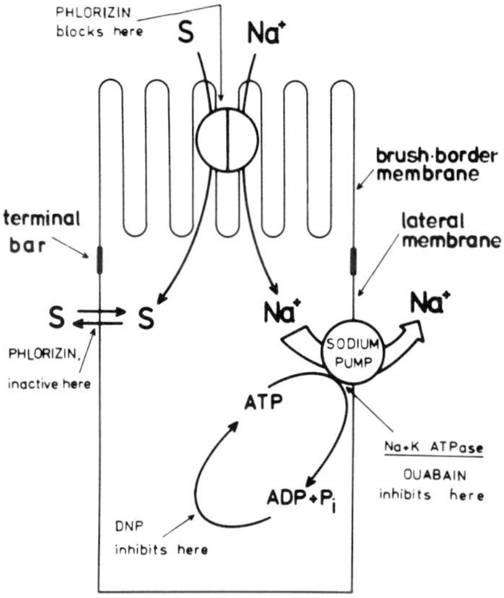

Fig. 1. Current concept of the enterocyte as a transcellular transport machine. The brush border membrane is the site of sugar *(S)* and Na^+ cotransport where phlorizin inhibits by competing with the sugars for the D-glucose binding site in a two-site carrier. The energy for this secondary uphill transport of sugars derives from the Na^+ transmembrane electrochemical gradient to which sugar influx is coupled. The Na^+ gradient is maintained by the sodium pump, located in the lateral membrane (BLM). This is a primary uphill transport mechanism since it gets its energy directly from the hydrolysis of cellular ATP. Sugar uphill transport, therefore, can be inhibited indirectly, either by inhibiting the sodium pump (the Na + K ATPase) with ouabain; or by inhibiting the synthesis of ATP with dinitrophenol (DNP) or similar agents. Efflux of intracellular accumulated sugars into the intercellular spaces involves a separate, downhill transport system *(left)* located in the lateral membrane: phlorizin is not active here

Since the junctions behave as truly passive molecular sieves, this route operates exclusively as a leak. Depending on their relative leakiness, epithelia have been classified as either leaky or tight (Frömter and Diamond 1972). The small intestine is the prototype of the leaky epithelia since its leaky tight junctions permit passage of certain electrolytes and nonelectrolytes, including the monosaccharides (see Robinson and Antonioli 1980). However, the question whether sugars can leak is a quantitative one. Particularly in short-term experiments (time scale: minutes), any passive leaks across the luminal surface of the intestine are quantitatively unimportant when compared with the carrier-mediated fluxes taking place across the BBM. As mentioned above, in the glucose and galactose malabsorption syndrome luminal hexoses are not taken up to any significant extent: because of this, they can exert an osmotic effect that is the primary cause of the diarrhea characterizing this disease. Identical mechanism can be adduced to explain the well-known "mannitol diuresis" (Schneider et al. 1966).

The Brush Border Membrane is the Site of the Uphill Step in Sugar Transport in the Small Intestine

Work with intact tissue (McDougal et al. 1960), confirmed with radioautography (Kinter and Wilson 1965) and, more recently, with isolated membrane vesicles (review: Murer and Kinne 1980) has revealed that the uphill step in intestinal sugar transport occurs at the BBM. Sugars first accumulate within the enterocytes, then move downhill into the underlying tissue. As schematized in Fig. 1, the BBM uphill step involves a sodium-coupled, secondary active transport mechanism to be discussed in detail below. In contrast, sugar movements across the BLM are thought to occur downhill.

To explain cell function, decomposition into building blocks and, eventually, reconstitution, is a fundamental approach in modern biology. The pioneering work of Kaback and colleagues (review: Kaback 1976) with isolated bacterial membrane vesicles strongly influenced the intestinal and renal transport field when similar methods began to be applied here in the early 1970's. Today, it has become routine to experiment with isolated BBM and BLM vesicles, both from intestinal and from kidney cortex tissue. But the great successes achieved by using isolated membrane vesicles should not obscure the fact that the intestine and the kidney are first of all epithelia exhibiting a very peculiar geometry. A look at Fig. 1 should show that the whole is superior to its parts, and any knowledge obtained with isolated membrane vesicles would be largely futile if not used to explain the functioning of the entire machine, the enterocyte in its epithelial niche. Vesicle work should therefore be run in parallel with intact cell or intact-epithelium work. Trying to reconcile the often conflicting results obtained with either experimental tool seems a constructive approach that we propose to develop in this article.

Identical molecular mechanisms undoubtedly underlie the transport observed in intact tissue and in vesicle preparations, but their respective macroscopic behavior needs not be identical, for many reasons. For instance, under physiological conditions (notably, absence of metabolic inhibitors), intact tissue preparations are energized, i.e., they exhibit a membrane potential that is tightly regulated. In this respect, bacterial membrane vesicles are similar since they can be energized if an appropriate electron donor such as D-lactate is provided. In other words, these membranes possess the metabolic machinery necessary to establish long-lasting transmembrane potentials and electrochemical ion gradients, similar to those characterizing the mother cell. This is not the case with BBM vesicles which behave as truly passive "bags". Electrical and ionic gradients can be imposed on them at will, but they tend to dissipate rapidly in an impossible-to-control, difficult-to-assess manner. We shall see that to perform kinetic studies with BBM vesicles, exceedingly short incubation times (e.g., 1 or 2 s) are often needed, requiring special equipment but, nevertheless, leaving largely unanswered the question of to what extent the initial gradients have changed during the experiment. But before describing the virtues and possible pitfalls of research with isolated membrane vesicles, a phenomenological description of sugar transport across the BBM seems necessary. It is against this background, derived largely from kinetic experiments with intact tissue preparations, that work with vesicles needs to be compared.

Mechanism of Sugar Transport in the Small Intestine: Summary from Studies with Intact Tissue Preparations

One technique was at the base of the revolution in intestinal transport research that took place in the early 1960's: the so-called tissue accumulation method (see Crane 1960, 1968). In effect, this technique opened the way for the kinetic analysis of transport, to be developed in the following.

Phlorizin Competes with the Sugars for the D-Glucose Binding Site in the Brush Border Membrane Carrier

Phlorizin (I) is a classical inhibitor of intestinal and renal sugar transport (review: Crane 1960). Since it possesses a β-D-glucopyranoside moiety, the hypothesis seemed reasonable that competition for D-glucose-binding sites plays a key role in phlorizin inhibition. Kinetic proof substantiating this idea was first provided by Alvarado and Crane (1962), completed shortly thereafter with the necessary proof (Alvarado 1967) that, as demanded by theory, the sugar transport rate tends to zero as the concentration of the (fully competitive) inhibitor increases. These observations revealed two facts of considerable relevance to the mechanism of sugar transport in the intestine and, indirectly, in the renal tubule.

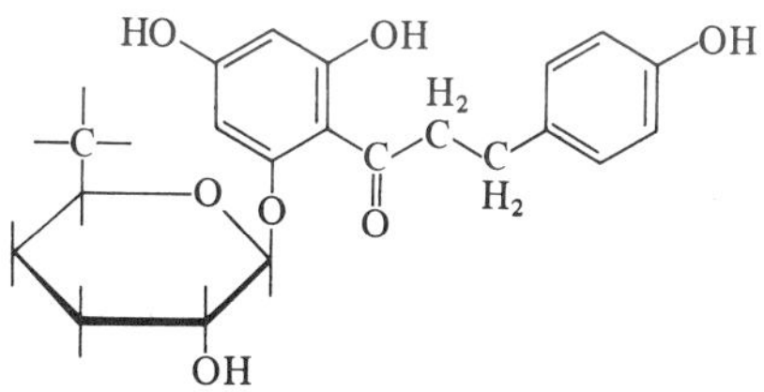

(I)

(1) Sugar transport as measured by net uptake into rings of intact tissue, in vitro, exhibits apparently perfect Michaelis-Menten kinetics. In effect, it is not simply that transport rates when plotted according to appropriate linear transformations give straight lines. More meaningfully, the results are highly consistent internally insofar as identical K_i values are obtained, either in experiments where (S) is varied at constant (I) or in the reciprocal situation where (I) is the variable quantity. No less striking is the fact that essentially identical phlorizin K_i values were calculated by Stirling (1967) who used an entirely different technique. (2) As predicted, phlorizin behaves as a fully competitive inhibitor of intestinal sugar transport.

These observations paved the way for the use of phlorizin as a high-affinity probe for the study of the BBM carrier from both intestinal and renal tissue although, to the best of our knowledge, a kinetic demonstration, equivalent to that summarized above for the intestine, has never been provided for the renal system. Studies on phlorizin binding to both intestinal and renal BBM vesicles seem to support the notion that this glucoside competes for the glucose-binding site in the BBM carrier. Such evidence, however, must be considered with caution. Demonstration of direct competition between inhibitor and substrate for exactly the same site on an enzyme or carrier is not, as generally believed, an easy matter. As an example, harmaline exhibits apparently typical kinetics of fully competitive inhibition with sucrose in intestinal brush border sucrase. But, in all likelihood, harmaline does not bind to the substrate site in this enzyme (Mahmood and Alvarado 1977). It follows that the demonstration of fully competitive inhibition kinetics is not in itself sufficient to prove that phlorizin interacts directly with the BBM glucose-binding site. Nonetheless, this assumption seems very likely because, as mentioned, phlorizin possesses a β-D-glucopyranoside moiety that can be expected to direct the inhibitor towards the appropriate receptor in the carrier. In fact, Alvarado and Crane (1964) have suggested that phlorizin may be a high-affinity substrate for System 1.

The Brush Border Region as a Black Box: Operational Definition of the Intestinal Sugar Transport "Carrier"

The fact that net tissue uptake by intact tissue preparations conforms to Michaelis-Menten kinetics seems now to be established beyond reasonable doubt. Sceptics can peruse the carefully designed, mathematically sophisticated work of Robinson and associates, dealing directly with this question (see Robinson et al., this vol.).

For unexplained reasons, however, the existence of this type of kinetics in intact tissue experiments seems not to have aroused the interest of researchers in this field. To the best of our knowledge, we are the only ones to have concluded that, as far as intestinal transport is concerned, Michaelis kinetics are not simply a curiosity, but reveal something very fundamental about the transport mechanism itself.

In cellular and mechanistic terms, the kinetics observed can be rationalized as follows. Extracellular (mucosal) substrate, S_o, interacts with a membrane receptor, C_m, to give a Michaelis-type complex, S-C_m. This stereospecific recognition step would be fast and freely reversible, thereby meeting the equilibrium condition necessary to explain the kinetics. By mechanisms unknown, a substrate translocation step follows whereby S_i, the chemically unchanged, osmotically active substrate, appears on the inner side of the membrane. This translocation would be rate-limiting and essentially irreversible, the sum of the two steps just mentioned yielding the overall reaction

$$S_o + C_m \rightleftharpoons S\text{-}C_m \rightarrow C_m + S_i \tag{1}$$

which, needless to say, fulfills the basic assumptions in Michaelis kinetics. The problem is to explain why C_m interacts readily with S_o but fails to do so with S_i, but this will be dealt with later. At this point we wish simply to explain why, operationally, intact tissue preparations exhibit Michaelis kinetics. If S_i crosses the BLM (see Fig. 1), then it will be trapped within the subepithelial tissue layers acting as a sink (McDougal et al. 1960). Since unspecific leaks back into the medium are negligibly small when short incubation times are used, it can be assumed that net substrate accumulation within the tissues is a measure of the unidirectional influx across the BBM, i.e., the initial transport rate.

Following Crane (1968, 1977), it is generally thought that both the recognition and the translocation steps implicate a single agency or "carrier", located in the BBM. But the brush border is a complex subcellular organelle formed by three anatomically distinct elements: the membrane or "coat"; a fibrillar skeleton or "core"; and a free space separating the two (see Crane 1966). Although the work of Kinter and Wilson (1965) showed that radioactive substrates bind first to the brush border region, the resolution of the autoradiographic technique employed is not sufficient to ascertain whether this initial step involves exclusively the brush border coat. Does intestinal transport involve the sequence brush border coat → core → cytosol? Seen this way, the brush border region is a black box and the "BBM carrier" is defined operationally by reaction sequence (1), the complexity of which remains to be established. The recognition step may be attributed in principle to a specific substrate-binding site, presumably located in the outer face of the brush border coat. Kinetically, it is defined by a Michaelis constant, K_T, which can be used as an index of "affinity" because, if true Michaelis kinetics hold, then K_T can be equated to a dissociation constant and the assumption be made that affinity = $1/K_T$.

The translocation reaction might be more or less complex, but it should include a rate-limiting step that operationally behaves as "irreversible". The overall step will be defined by a maximal flux rate, V_{max} (or J_{max}). We also call V_{max} the capacity factor to distinguish it from K_T, the affinity factor. We shall see that it is very useful to dissociate these two factors, not only kinetically, but also mechanistically.

Substrate Specificity of the Main Sugar Brush Border Membrane Carrier

By 1960, it had become clear that, together with the renal tubule, the small intestine is unique in being capable of transporting sugars against their own concentration gradient. This uphill transport was demonstrated beyond doubt at the end of the 1950's by means of the everted sac technique, which showed that sugars appearing in the serosal compartment are free (osmotically active) and chemically unchanged. This technique allowed Crane and Wilson to classify the monosaccharides into two distinct groups: those which are and those which are not, actively transported (Crane 1960). On the basis of this, Crane and Wilson defined the "minimal structural features" necessary for a sugar to interact with the BBM carrier, illustrated here with the well-known scheme:

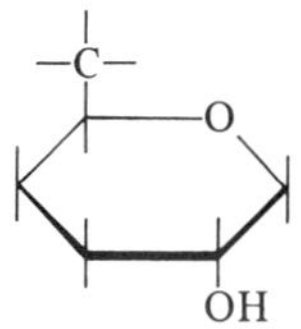

(II)

In our studies on the mechanism of action of phlorizin, we found that the attachment of bulky phenol groups at C-1 does not prevent interaction of certain phenyl-β-D-glycopyranosides with the BBM carrier. Some glucosides such as arbutin (III) are actively transported by hamster intestine with transport characteristics identical to those of D-glucose. Others, like salicin (IV) do interact with the carrier as revealed by their competitive inhibition on sugar transport, but are not taken up to any demonstrable extent (Alvarado and Crane 1964). Such observations helped to develop the concept that transport involves the two phenomenologically distinct steps, recognition and translocation. But an interesting, thus far unexplained example of species differences may be mentioned here. Whereas salicin, as mentioned, is not taken up (translocated?) by hamster intestine, the chicken intestine appears to transport salicin uphill, similar to arbutin and the free sugars (Alvarado and Monreal 1967). This observation suggests that the recognition and the translocation steps may exhibit different substrate specificities.

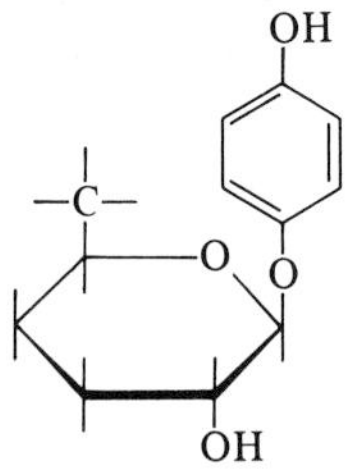

(III)

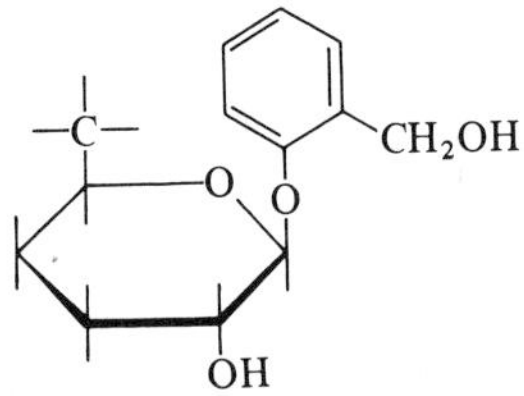

(IV)

The Affinity Scale

In the course of these studies we also noted that the pentose analog, phenyl-β-D-xylopyranoside (V) is a competitive inhibitor of the BBM carrier. This observation matched well with that made at about the same time by Salomon et al. (1961), showing that D-xylose is transported in the small intestine by a facilitated diffusion mechanism, i.e., downhill: these workers called it thermal transport to emphasize their thinking that xylose uptake does not involve the D-glucose "active transport" mechanism.

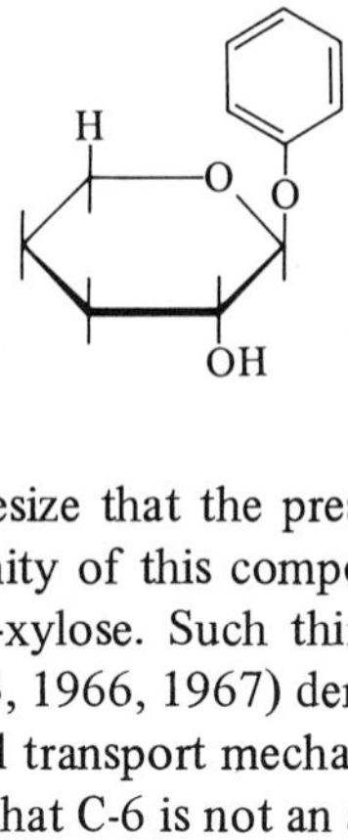

(V)

It seemed logical to hypothesize that the presence of a phenyl group in the xyloside (V) increases the net affinity of this compound for the D-glucose carrier when compared with that of free D-xylose. Such thinking led inevitably to new experiments whereby Alvarado (1963, 1966, 1967) demonstrated that D-xylose is indeed a substrate of the D-glucose uphill transport mechanism[4].

This work clearly indicated that C-6 is not an absolute requirement for interaction with the D-glucose carrier, although its absence entails a drastic fall in the apparent affinity of the carrier for D-xylose, the pentose most closely resembling D-glucose. In contrast, D-lyxose, the pentose structurally related to D-mannose, behaves in

4 It is worth noting that D-xylose, although a low-affinity substrate, is clearly transported through System 1 (Alvarado 1966). In contrast, the phenylxyloside, which exhibits greater affinity for the carrier, seems not to be transported through this pathway (Alvarado and Crane 1964). This would constitute, therefore, still another example where the recognition and the translocation steps seem to exhibit different specificities

practice as inert. In view of these observations, and consistent with the idea that carriers, like enzymes, exhibit a specificity that is never absolute, Alvarado (1966) proposed the elimination of the classical distinction between actively and non-actively transported sugars. Rather, he proposed an affinity scale in which all sugars would belong to a single series (see Table 1). Obviously, sugars located very low on the affinity scale would approach the ideal behavior of a true "non-substrate". Experimentally, such sugars would be of use as markers of the extracellular space (ECS), the space accessible to an analog distributing itself exclusively by diffusion.

Table 1. The affinity scale. Sugars are listed in decreasing affinity order. Names in parenthesis indicate analogs believed to bind to the carrier but where the evidence for translocation requires further study. This very incomplete and preliminary list reflects the experience of the authors with hamster intestine. Much additional work will be needed to complete it with quantitative data that would take into account differences between animal species. Further details in the text

(Phlorizin)
•
•
D-glucose
6-deoxy-D-glucose, arbutin
β-methyl-D-glucopyranoside
D-galactose
α-methyl-D-glucopyranoside
1,5-anhydro-D-glucitol
(Salicin)
3-O-methyl-D-glucose
(L-fucose)
D-fucose
(β-phenyl-D-xylopyranoside)
D-xylose
•
•
D-mannose, α-methyl-D-mannopyranoside
L-glucose
D-lyxose, L-rhamnose, 2-deoxy-D-glucose
D-mannitol, D-sorbitol

Correcting for the ECS is a delicate question, particularly when dealing with borderline cases where the difference between "weak transport" and simple diffusion may prove to be difficult. In our laboratory, D-mannitol is used routinely for this purpose since it behaves, in practice, as inert (Alvarado 1966). Consequently, in Table 1 we place it at the bottom of the affinity scale which is logical since, although structurally very close to D-glucose, it cannot adopt the pyranose structure that seems to be an important requisite for interaction with the System 1 carrier. Use of mannitol

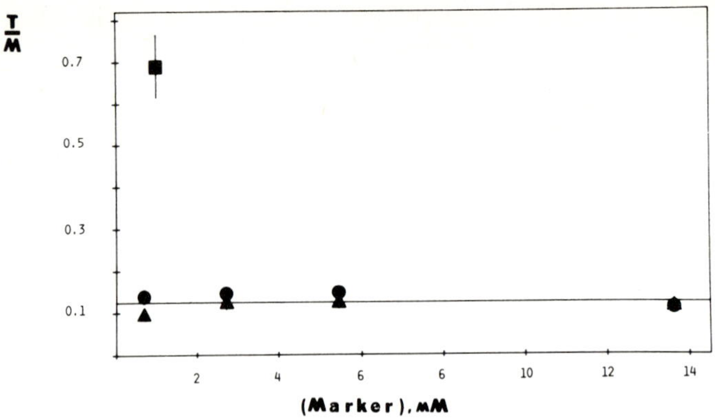

Fig. 2. D-mannitol and L-glucose as extracellular space markers with hamster small intestine. The tissue accumulation method was used under standard conditions (Alvarado and Lherminier 1982), involving 2-min incubations at 37°C in 5 ml of oxygen-saturated Krebs-Henseleit phosphate buffer with 145 mM Na^+ and the indicated concentrations of the markers: [^{14}C]-D-mannitol (▲) and [^{3}H]-L-glucose (●). The results are expressed as tissue/medium concentration ratios (T/M) which are equivalent to a relative uptake or clearance (see Shapiro and Heinz 1980) and have the units ml g^{-1} fresh tissue: compare with the data in Fig. 5. By definition, T/M ratios greater than 1 signify uphill transport. The symbols are larger than the E.E.M.; n = 16 determinations per point. The entire series proved to be homogeneous according to a one-way analysis of variance. *Horizontal line* indicates the average T/M value (0.124 ± 0.014; n = 128). (■) shows a separate test concerning the uptake of 1 mM β-methylglucoside (n = 8) in a sodium-free medium composed of 145 mM LiCl + 10 mM Tris/HCl, pH 7.2

as an ECS marker is illustrated in Fig. 2. Here, the uptake data demonstrate diffusion kinetics because the T/M ratios are constant and independent of the marker concentration. But, according to this criterion, the same can be said of L-glucose (Fig. 2), a point whose significance will become apparent below. In Table 1, L-glucose marks the frontier separating substrates from non-substrates, although it is understood that this distinction is an operational one. For instance, we have placed D-mannose above L-glucose although its being a substrate is open to question (Alvarado 1966).

Mechanism of Sugar and Na^+ Cotransport

We have mentioned that sugar uptake across the BBM involves a secondary active transport mechanism where the energy for the uphill sugar influx comes from the dissipation of an inward-directed Na^+ electrochemical gradient. The evidence for this model, the so-called sodium gradient hypothesis, has been reviewed extensively and will not be repeated here (see Crane 1968, 1977, Schultz and Curran 1970, Alvarado 1976, Kimmich 1981). Sugar and Na^+ cotransport is a special case within the large family of organic solute and Na^+ (S/Na^+) cotransport systems which has given origin to numerous kinetic models. In our laboratory, we have developed a general cotransport model which: (1) explains in formally identical terms the cotransport with Na^+ of both sugars and amino acids in vertebrates; and (2) opens the way to explaining within the same theoretical framework the kinetics of cotransport either with intact tissue or with vesicle preparations. A detailed account of the theoretical basis and equations describing our model has appeared recently (Alvarado and Lherminier 1982). For brevity, we will give here only the essential concepts, with emphasis on those points where our model diverges from the classical ones (cf. Schultz and Curran 1970). In the last part of this chapter, experiments with BBM vesicles will be described, summarizing initial attempts at testing with this type of preparation the mechanistic implications of our general model.

Definitions

Stoichiometry. All S/Na^+ cotransport models are based on the premise that a carrier with specific binding sites for each S and Na^+ exists. Of course, the cotransport concept implies that interactions are reciprocal, so that Na^+ is also a substrate (cosubstrate). As illustrated in Fig. 3, we assume a stoichiometry S/Na^+ = 1/1, which is the one that fits best our results[5].

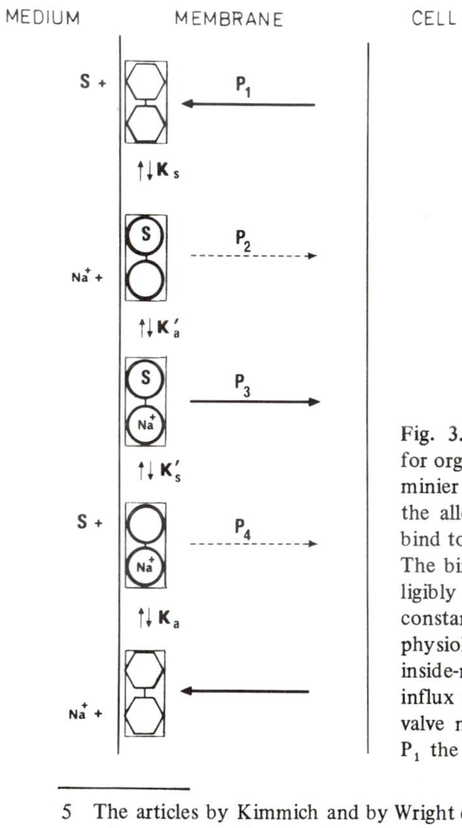

Fig. 3. Schematic representation of the obligatory model for organic solute and Na^+ cotransport (Alvarado and Lherminier 1982). The change of hexagons into circles illustrates the allosteric transition taking place when either S or Na^+ bind to their respective sites in a two-site carrier *(rectangles)*. The binary complexes are not mobile, P_2 and P_4 being negligibly small. Only the ternary complex (with permeability constant P_3) can translocate: obligatory kinetics. Under physiological, intact-tissue conditions which include an inside-negative membrane potential, the obligatorily coupled influx of S and Na^+ occurs irreversibly and constitutes the valve mentioned in the text. P_3 is therefore the opening, P_1 the closing, of the rectified carrier or gate

5 The articles by Kimmich and by Wright et al. (this vol.) show that the stoichiometry question remains controversial today. More work is clearly needed to solve this problem

With Intact Tissue Preparations, the Brush Border Membrane Carrier is Highly Asymmetrical and in Practice Functions Irreversibly. Most cotransport models envisage a classical, reversible carrier system, meaning that substrate influx and efflux utilize the same pathway. In contrast, we consider that, with intact tissue preparations under physiological conditions (involving, notably, an inside-negative membrane potential), the BBM carrier functions essentially irreversibly and is not involved in substrate efflux. Therefore, in our model, we consider only influx reactions (Fig. 3). With intact tissue preparations, evidence for the irreversiblity of sugar and amino acid influx has been reviewed by Alvarado and Lherminier (1982). Additional, preliminary evidence on how work with vesicles can be construed to explain and support this postulate will be discussed later.

The Cotransport of Na^+ with Either Sugars or Amino Acids Involves Formally Identical, Random Ordered, Mixed-Type Activation Mechanisms. Schultz and Curran (1970) considered the possibility of a general model for cotransport in the small intestine. They rejected it, however, because they felt that the kinetics of sugar and amino acid transport in rabbit intestine were entirely different. But, as explained in detail by Alvarado and Mahmood (1974) and by Alvarado and Lherminier (1982), the purported differences in the kinetic behavior of sugar and amino acid transport systems in rabbit ileum cannot be construed as indicating the existence of any real difference in mechanism. First, and concerning the compulsory model advocated for amino acid transport, Alvarado and Mahmood (1974) showed that it is an unwarranted oversimplification of the more general, non-compulsory one which, we demonstrated, can fit very well all the amino acid transport data, irrespective of the animal species used. An ordered mechanism is theoretically conceivable, but it would require additional restrictions. Our position is that, as long as evidence for the existence of such restrictions remains unavailable, we should adopt the most general model, the noncompulsory one (Fig. 3).

Secondly, the rabbit ileum results (Schultz and Curran 1970) only superficially suggest different mechanisms. According to widely-accepted terminology, sugar transport would be a case of apparently pure V-type or capacity-type activation kinetics where K_T is constant but V_{max} increases with $[Na^+]_m$. In contrast, the amino acid transport system would be an example of K-type or affinity-type activation kinetics where V_{max} remains constant but K_T decreases (the affinity increases) as $[Na^+]_m$ increases. But we wish to empahsize that these two types of kinetics are by no means mutually exclusive and attempts at classifying animal species according to this criterion are doomed to failure. It is much more useful to consider that the two distinct kinetic effects coexist in all species; i.e., mixed-type kinetics is the rule, a fact of parammount importance in what concerns mechanism, as we shall see.

The Sodium Ion is an Obligatory Activator in Intestinal Organic Solute and Na^+ Cotransport Systems. Here we arrive at the most critical point in our article, for the concept of Na^+ as an obligatory activator was developed by Alvarado (1979) on theoretical grounds and, at least superficially, goes against most of the experimental evidence available at present (our argument is that this evidence is biased by a key artifact, as discussed in the next section).

By obligatory activator we mean that there is no reaction (translocation) when the activator is absent. Since the fluxes of each S and Na^+ are linked, this means that only the ternary complex translocates: P_3 is positive but P_2 and P_4 are nil. Seen in this light, cotransport would be an all-or-none phenomenon, the carrier switching between two possible conformations, open and closed. The P_1 parameter in Fig. 3, therefore, represents the closing of the gate after S and Na^+ are released simultaneously to the inner side of the membrane: the carrier then returns to the ground state.

Moreover, if the ternary complex S-C-Na^+ carries with it a net positive charge, then the gate opening would allow the coupled influx of the two cosubstrates to occur irreversibly because Na^+ would be moving towards its position of electrical equilibrium (remember, we are discussing the intact tissue situation, characterized by an inside-negative membrane potential).

Finally, the proposed model implies that the carrier itself is electrically neutral, a question we deal with in greater detail later.

The Na^+ Apical Reservoir

According to our obligatory model, sugar and amino acid transport should be nil when Na^+ is absent from the bulk of the incubation medium. As mentioned, however, with intact tissue preparations the experimental evidence seems to negate this postulate. In effect, it is well known that most intestinal cotransport systems exhibit identical V_{max} in the presence and in the absence of Na^+. In the nomenclature of Fig. 3, this means that $P_2 = P_3 = P_4$ whereas the obligatory model demands that $P_2 = P_4 = 0 \ll P_3$. One experiment illustrating that P_2 indeed seems to be greater than zero was given above (Fig. 2). When incubated in a sodium-free, lithium-substituted medium, β-methylglucoside uptake was much greater than the extracellular space, although certainly much lower than when Na^+ was present (T/M about 2.5 in this particular experiment, not illustrated). One explanation of this result could be that Li^+ can substitute for Na^+, at least partially. Another explanation, first proposed by Alvarado (1976), is that Li^+ is inert and the net sugar uptake observed in the presence of this ion is due to the existence of a self-perpetuating layer of Na^+ in the microenvironment of the outer face of the BBM. This layer we call the apical Na^+ reservoir. Alvarado and Lherminier (1982) have shown how all available data on the kinetics of intestinal S/Na^+ cotransport can be explained with the obligatory model, provided that the Na^+ reservoir is taken into consideration. Robinson and van Melle appear to have confirmed our prediction and, furthermore, have been able to calculate that Na^+ exhibits an activity of about 4.8 mM in the apical reservoir of guinea pig intestine (see Robinson et al., this vol.). But more work is obviously required to confirm the existence of the Na^+ reservoir whose actual size (activity × volume) may be negligibly small.

Experiments with Isolated Brush Border Membrane Vesicles

If intact tissues exhibit a self-perpetuating apical Na^+ reservoir, to demonstrate conclusively that Na^+ is indeed an obligatory activator of S/Na^+ cotransport is virtually impossible with such preparations. Consequently, we untertook experiments with isolated BBM vesicles which permit controlling the composition of the media in contact with both sides of the membrane. In the experiments that follow, D-glucose transport was studied by using conventional filtration techniques (references in Murer and Kinne 1980) and BBM vesicles obtained from the frozen jejunum of either guinea pigs or 15-day-old chickens. Experimental details are given in the legends to the figures.

Operational Definition of Vesicular D-Glucose Transport Through System 1

The first problem encountered was to establish the limit between "transport" and "diffusion". This distinction is essential but is technically difficult considering that our aim is to establish whether or not System 1 is operative in the absence of Na^+, i.e., under conditions where transport is either zero, or very close to it. With vesicles, it is customary to define sugar transport as the difference in uptake between the D- and L-glucose isomers, the latter being assumed not to be transported at all. To make things more difficult, the vesicles are leaky, meaning that L-glucose is in fact taken up rather fast and eventually occupies the same space as D-glucose. We assume that this "leakiness" results from damage to the membrane during its isolation. As already discussed, the in situ BBM is not leaky.

In principle, use of L-glucose as a non-transported (diffusional?) marker would seem to be at odds with the claim (see Crane 1968, Bihler 1969) that this sugar is a substrate for the BBM sugar carrier. We have seen, however, that L-glucose is not transported by intact tissue preparations of either hamster or guinea pig intestine: together with D-mannitol and sorbitol, it seems to behave as an appropriate extracellular marker. Our experience indicates that the same applies to isolated BBM vesicles and, therefore, we operationally define D-glucose transport as that uptake remaining after the L-glucose uptake, under identical conditions, has been substracted.

We have defined System 1 as both sodium dependent and phlorizin sensitive. The sodium dependence of our BBM vesicle preparations was verified by the classical "overshoot" test (not illustrated). The experiment in Fig. 4 shows that, in accord with the results found with intact tissue preparations, phlorizin behaves as a fully competitive inhibitor of D-glucose uptake when Na^+ is present, the data yielding kinetic constants of the appropriate order of magnitude for each $K_{i\ (phlorizin)}$ and $K_{T\ (glucose)}$. Additional evidence that, in the presence of Na^+, our BBM vesicle preparations behave according to the rules is provided by the observation (not illustrated) that the initial rate of D-glucose transport is not affected by phloretin, an inhibitor believed to be specific for the BLM carrier but inert towards the BBM System 1 (see Kimmich 1981).

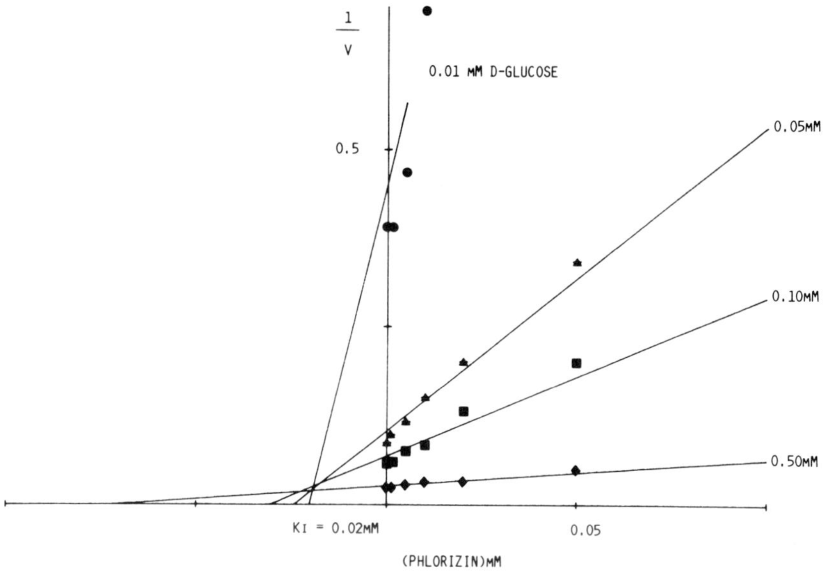

Fig. 4. Effect of phlorizin on the D-glucose uptake by guinea pig brush-border membrane vesicles in the presence of sodium. Conditions: 1.6-s incubations at room temperature in 20 mM HEPES/Tris buffer, pH 7.4, with 100 mM NaCl, 200 mM D-mannitol, [^{14}C]-D-glucose (10–500 μM) and phlorizin (1–200 μM). The data, corrected for the L-glucose uptake, correspond to a single experiment (n = 6) and are plotted according to Dixon (Dixon and Webb 1964). Uptake units: v = pMol mg^{-1} protein s^{-1}. *Straight lines* were calculated by the method of least squares but are illustrated only in the range from 0–50 μM phlorizin

Experiments in Sodium-Free Media

A preliminary survey revealed that, as expected, D-glucose transport falls drastically when Na^+ is absent from the incubation mixtures. However, there seemed to be a significant amount of residual D-glucose uptake, both in the presence and in the absence of other alkali-metal ions. The following orders of relative apparent activation of D-glucose uptake were found: $Na^+ \gg Li^+ > K^+ = Rb^+ = Cs^+$ = sorbitol for guinea pig intestine: and $Na^+ \gg Li^+ = K^+ = Rb^+ = Cs^+$ = sorbitol for the chicken. The 1-min incubations used in this initial experiment can be criticized (see below). Nonetheless, the results indicate quite clearly that Li^+ behaves as an activator of D-glucose transport in guinea pig BBM vesicles, and the question is whether such apparent activation involves System 1. According to the operational definition given above, phlorizin should be useful in establishing whether the lithium-activated D-glucose uptake involves this system or not. In effect, such studies should not only supply a qualitative answer to the question but, if the answer is yes, they also will give each of the key kinetic constants, K_i (for phlorizin), K_T and V_{max} (for glucose), which should be of great use in establishing whether Li^+ behaves mainly as an affinity-type activator, a capacity-type activator, or both.

D-Glucose Transport is Nil in the Absence of Alkali-Metal Ions. Figure 5 illustrates the results with guinea pig BBM vesicles in the absence of any alkali-metal ions. The uptakes of each D- and L-glucose are indistinguishable. Furthermore, the relative uptake of each of these sugars is a horizontal line at concentrations ranging from 25 to 150 mM, both in the absence and in the presence of phlorizin. We conclude that the initial influx of the two isomers involves simple diffusion and System 1 is not operative under these conditions, a conclusion that agrees fully with the observation by Hopfer and Groseclose (1980) that D-glucose uptake by rabbit BBM vesicles is zero in the absence of Na^+. In their experiments, Hopfer and Gloseclose used K^+ to substitute for Na^+ whereas in the experiment in Fig. 5 we used sorbitol. Nevertheless, and in accord with the results already mentioned, there is reason to expect that K^+, Rb^+ and Cs^+ should give results similar to those of sorbitol: complete lack of activation of System 1.

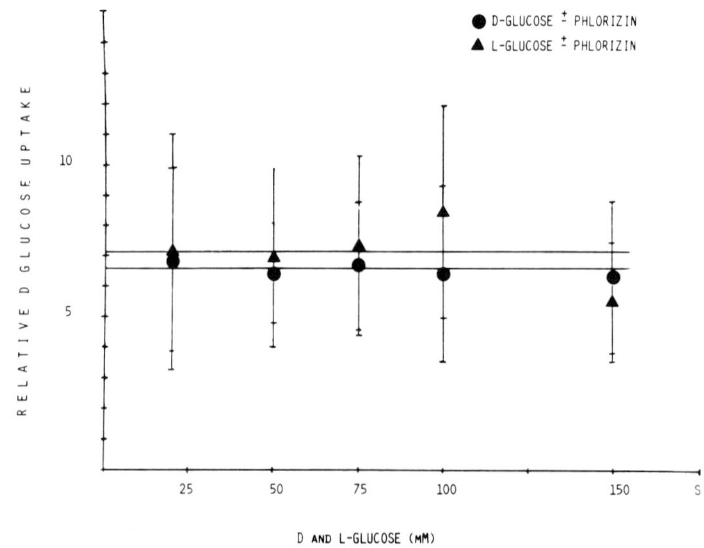

Fig. 5. Lack of D-glucose uptake by brush-border membrane vesicles from guinea pig jejunum in the absence of alkali-metal ions. Conditions: 10-s incubations in 10 mM HEPES/n-butylamine buffer, pH 7.4, with the indicated concentrations of each D- and L-glucose with or without phlorizin (100–700 μM). Osmolarity was maintained constant with sorbitol. The results are plotted as relative uptakes where the units are nL mg^{-1} protein s^{-1} (see legend of Fig. 2). The entire body of data (corresponding to either D- or L-glucose, with or without phlorizin) was found to be homogeneous according to a one-way analysis of variance

D-Glucose Uptake in Lithium Media Does Not Involve System 1. Uptake experiments in lithium media gave the unexpected result that phlorizin behaves under these conditions as a fully noncompetitive inhibitor. When plotted according to Dixon (Dixon and Webb 1964), the lithium data give a series of straight lines intersecting at a common point on the x-axis (Fig. 6). This is the classical proof of fully noncompetitive

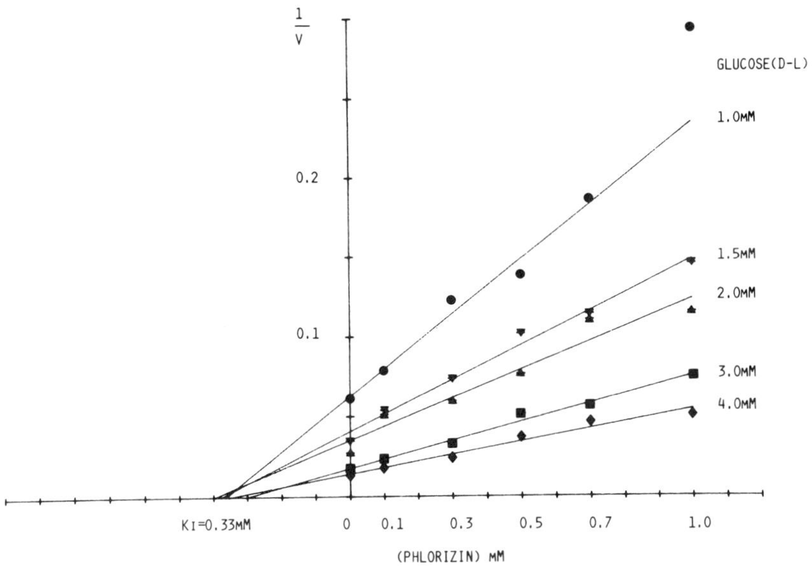

Fig. 6. Effect of phlorizin on the D-glucose uptake by guinea pig brush-border membrane vesicles in the presence of lithium. Experiment similar to that in Fig. 4, except for the use of LiCl instead of NaCl and 15 s incubation times. The buffer consisted of 10 mM HEPES and 7 mM n-butyl-amine, adjusted to pH 7.4 with maleic acid (about 2 mM)

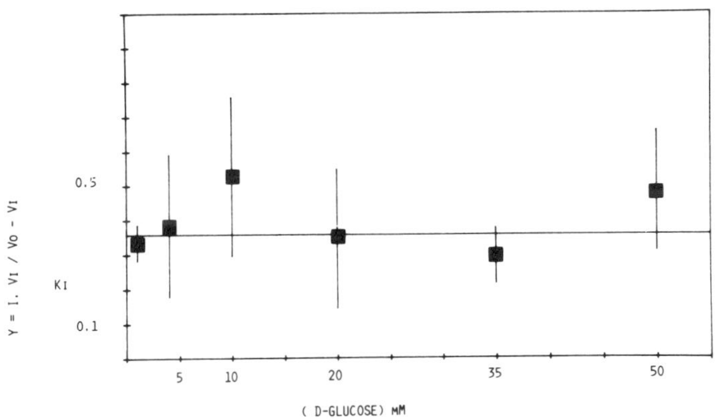

Fig. 7. Effect of phlorizin on the D-glucose uptake by guinea pig brush-border membrane vesicles in the presence of lithium. Experiment similar to that in Fig. 6, except that the D-glucose concentration range was enlarged as shown. The data are plotted according to Hunter and Downs (see the text for further details

inhibition, and an identical result was obtained when other tests were applied to these or separate data. For instance, Fig. 7 illustrates another experiment where substrate concentrations as high as 50 mM were used. Plotted according to Hunter and Downs (see Dixon and Webb 1964), these data give a horizontal line, again indicating noncompetitive inhibition. Moreover, all of these tests yielded essentially identical values of $K_{i\ (phlorizin)}$ = 0.3 mM. The apparent K_T of glucose in these experiments was about 50 mM. However, there is not much certainty in this estimate and K_T might be much greater, meaning that the carrier has very little affinity for D-glucose in the presence of Li^+. We conclude that D-glucose uptake in lithium media does not involve System 1.

But what is the mechanism of D-glucose transport in lithium media? Because of its low affinity and absence of sodium dependence, an analogy with the BLM carrier seems conceivable. However, our BBM preparations are not inhibited by phloretin, indicating that contamination by BLM material is negligible. The existence of an apparently specific activation by Li^+, together with a noncompetitive inhibition effect by phlorizin suggest that a different, as yet unidentified sugar transport system might be involved. But more work is clearly needed to answer this question.

The Question of the Net Charge of the Ternary S-Carrier-Na^+ Complex

Our obligatory model for S/Na^+ cotransport proposes that when there is an inside-negative membrane potential (the situation prevailing in physiological, intact-tissue preparations), the coupled influx of S and Na^+ occurs irreversibly because Na^+ is moving towards its position of electrical equilibrium. This implies that the carrier is neutral or, more precisely, the net charge of the carrier, assumed to be a protein, does not enter the picture. In other words, because Na^+ activation is highly specific, we assume that electrostatic attraction to negative charges in the carrier protein, although possibly contributing to it, is not the key factor in Na^+ binding.

In contrast, other workers have proposed that the free carrier bears a negative charge, so that the ternary complex would be neutral. Perusal of the literature shows that there is one main argument supporting the concept that the carrier is negatively charged. Experiments with both renal and intestinal BBM vesicles indicate that the binding of phlorizin, a purportedly non-transported sugar analog, depends critically on the existence of an inside-negative membrane potential (Aronson 1978, Toggenburger et al. 1982). This observation is taken as evidence that the negatively charged carrier (or part of it) is repulsed towards the outer face of the BBM, thereby becoming available for phlorizin binding. As discussed in the next section, however, there is reason to believe that phlorizin is indeed translocated across the BBM, which would obviate the necessity of postulating that the carrier bears any net negative charge.

Is Phlorizin a Substrate for the Brush Border Membrane Sugar Carrier?

The idea seems to have gained hold that phlorizin, although binding to the sugar site in System 1, is not translocated. Due to its exceedingly high affinity and apparent lack of mobility, it is often assumed that it acts mainly by blocking the carrier. But this concept suggests a rigidity in phlorizin inhibition that is entirely unwarranted. Phlorizin inhibition is freely reversible, in accord with the fact that it is a fully competitive inhibitor. Phlorizin should therefore be thought of as constantly binding and debinding: no matter how little its relative affinity for the carrier is, free sugar molecules should always have a statistical chance to bind and be transported; i.e., the transport machinery is not blocked.

We wish also to point out that the frequently-made distinction between a substrate and a fully competitive inhibitor or non-transported ligand is kinetically unsound. By definition, a substrate always behaves as a fully competitive inhibitor of another. As mentioned, the kinetic evidence available cannot show whether phlorizin is translocated or not. Nevertheless, the reasonable suggestion has been made that phlorizin is a high-affinity substrate (Alvarado and Crane 1964): evidence supporting this idea has been provided by others (Lyon 1967, Hoshi and Komatsu 1970, Heath and Aurbach 1973).

An analysis of the various arguments adduced against the notion that phlorizin is a substrate (see Silverman 1976, Crane 1977, Aronson 1978, Toggenburger et al. 1982) indicates quite clearly that the weight of the evidence derives from the work of Stirling. By using a quantitative light microscope radioautographic technique, Stirling (1967) performed a brilliant analysis showing that: (1) phlorizin binding to the intestinal BBM follows a Langmuir adsorption isotherm with a half-saturation constant of about 13 μM, entirely consistent with phlorizin K_i values calculated from kinetic analyses. (2) Binding to the BBM is apparently not followed by any transfer of phlorizin to the cytosol. Before analyzing more closely the situation, however, let us note that although this work shows the brush border to act as a barrier to phlorizin, this barrier refers to the entire brush border black box, not necessarily only to the external face or "coat", as it seems to have been interpreted by many researchers. In Stirling's words, "the accumulation of phlorizin was limited to the brush border *band* of the epithelium" (emphasis ours).

Alvarado and Crane (1964) argued that the apparent lack of phlorizin uptake by the intestinal and renal epithelia might be explained in terms of its exceedingly high affinity for the sugar carrier. In effect, carrier-mediated efflux could become saturated at low intracellular phlorizin concentrations, thereby stopping any further net influx. Stirling (1967) rejected this interpretation by arguing that phlorizin uptake is also negligible when very low concentrations are used, i.e., when the carrier may be expected not to be saturated. But there are ways in which this argument can be refuted. One is that the affinity of phlorizin might be greater for the efflux than for the influx reaction. Another is that phlorizin may bind to additional sites located in the brush border core. Earlier, we argued that net sugar transport may involve two steps in series, binding to the core being a step preceding substrate release to the cytosol. Several arguments can be adduced in support for the hypothesis that the barrier to net phlorizin transfer into the cytosol may occur at the core, not at the coat, of the brush border black box.

First, BBM vesicles often contain core material which no one has thus far taken the trouble to quantitate. For instance, Aronson (1978) specifically mentions without further comment that his vesicle preparations "consisted predominantly of individual microvilli with intact core structures". The question may be posed whether differences in absolute phlorizin "binding" quoted by this and other investigators reflect differences in the amount of core material contained in the various BBM vesicle preparations.

Secondly, phlorizin has a definite tendency to bind to proteins, both specifically and unspecifically. Work with isolated BBM vesicles has revealed the existence of at least three different types of phlorizin-binding receptors (see Bode et al. 1972) and sequestering of phlorizin inside membrane vesicles has already been suggested (e.g., Chesney et al. 1974). The elegant in vivo experiments of Silverman (1974) seem worthy of mention here. Dog kidneys perfused with a radioactive phlorizin pulse retained 90%–95% of the label, which could be washed off by a second pulse of either cold phlorizin or of D-glucose. This was taken as evidence that the retained phlorizin was bound specifically to the glucose transport receptor (carrier). In reality, however, such long-lasting binding of phlorizin contrasts sharply with the ready reversibility of phlorizin inhibition, mentioned above. Indeed, this inhibition can be reversed fully by washing briefly with substrate-free buffer, which leads us to the conclusion that the phlorizin retained by the kidney in Silverman's experiments probably does not involve the coat receptor. There are obviously additional sites for specific, more or less rapidly reversible phlorizin binding, and the core material is in our view a likely candidate for furnishing such sites.

Thirdly, both Aronson (1978) and Toggenburger et al. (1982) find that, although phlorizin binding depends on the existence of an inside-negative membrane potential difference ($\Delta\Psi$), phlorizin debinding is not accelarated when the polarity of $\Delta\Psi$ is inverted. This fact can be explained readily if phlorizin is not bound to the external face of the membrane, but to intravesicular material after having been transported. In effect, there is no need to expect that phlorizin release from a core-bound pool should be affected by $\Delta\Psi$. In our view, the dependence of phlorizin uptake on $\Delta\Psi$ constitutes the best evidence thus far available that phlorizin is indeed a substrate of the BBM carrier. This conclusion carries with it the additional one that there is no need to postulate that the carrier bears a negative charge.

Concluding Remarks

One of the key premises of our organic solute and Na^+ cotransport model is that activation is obligatory. Full occupancy of both the S and the Na^+ sites constitutes the signal that triggers the opening of the transport gate. When there is a transmembrane electrical potential of the appropriate sign, and since Na^+ carries a net positive chage, the obligatorily linked influx of the two cosubstrates should occur irreversibly as Na^+ will be moving towards its position of electrical equilibrium. An analogy with the nerve seems possible here. In the nerve, acetylcholine is the signal that triggers the opening of the Na^+ channel. In the intestinal BBM, sugars and amino acids have a similar role, except that they are also cosubstrates and move across the membrane together with the Na^+.

The difference between our model and classical ones based on reversible, mobile carrier kinetics, is that we see the carrier behaving as a valve and the coupled S plus

Na^+ influx as rectified. With isolated BBM vesicles facing a Na^+ electrochemical gradient, this situation is reflected by the well-known overshoot phenomenon which, however, is short-lived due to the rapid dissipation of the driving force. With intact tissue preparations, on the contrary, the transmembrane electrical potential is maintained within certain limits and constant-rate, unidirectional substrate influx can occur for relatively long periods. Our point is that, although S/Na^+ contransport causes membrane depolarization, this depolarization is only partial and the remaining membrane potential is sufficient to keep the influx of S at its maximal theoretical rate. This is why, with intact tissue preparations, Michaelis-Menten kinetics are observed which, furthermore, seem to be independent of the membrane potential. In effect, we define the maximal substrate transport rate as $V_s^{max} = P_3\ [C_t]$ (where $[C_t]$ is the total carrier concentration, the sum of the four carrier forms in Fig. 3). We further assume that P_3 is a function of the membrane electrical potential: $P_3 = f\Delta\Psi$. However, it seems that P_3 may exhibit its limiting, upper value (i.e., behave in practice as constant) at "physiological" $\Delta\Psi$ values ranging from, say, -80 mV to about -20 mV. Consequently, the system in situ gives Michaelis kinetics and V_s^{max} is constant, thus behaving as if it were independent of the membrane potential.

In closing, we would like to emphasize that the driving force in S/Na^+ cotransport across the intestinal BBM is the Na^+ electrical gradient: the Na^+ chemical gradient plays little role, if any, in mechanism. This conclusion is fully in accord with the observations of Carter-Su and Kimmich (1980) with isolated chicken enterocytes.

Acknowledgements. We are indebted to Mrs. A. Candido and Mrs. D. Gerbaud-Ballue for skilful technical help. We also thank Miss I. Coquelet for typing the manuscript. This work was supported in part by Research Contracts with the D.G.R.S.T. (MB-80.7.0194), the I.N.S.E.R.M. (82.7001) and the C.N.R.S (ATP International 4812).

References

Alvarado F (1973) Difusión facilitada, primera etapa en el transporte activo de azúcares por el intestino. VII Giornate Biochim Latin, S Margherita Ligure (Genoa, Italy), p 87

Alvarado F (1966) D-xylose active transport in the hamster small intestine. Biochim Biophys Acta 112:292–306

Alvarado F (1967) Hypothesis for the interaction of phlorizin and phloretin with membrane carriers for sugars. Biochim Biophys Acta 135:483–495

Alvarado F (1970) La membrane celular como mosaico de functiones. Bol R Soc Esp Hist Nat (Biol) 68:33–68

Alvarado F (1976) Sodium-driven transport. A re-evaluation of the sodium-gradient hypothesis. In: Robinson JWL (ed) Intestinal ion transport. MTP, Lancaster, England, pp 117–152

Alvarado F (1979) Reversible binding and irreversible translocation: two distinct stages in sodium and solute cotransport in the small intestine. J Physiol (Lond) 292:77P–78P

Alvarado F, Crane RK (1962) Phlorizin as a competitive inhibitor of the active transport of sugars by hamster small intestine, in vitro. Biochim Biophys Acta 56:170–172

Alvarado F, Crane RK (1964) Studies on the mechanism of intestinal absorption of sugars. VII. Phenylglycoside transport and its possible relationship to phlorizin inhibition of the active transport of sugars by the small intestine. Biochim Biophys Acta 93:116–135

Alvarado F, Lherminier M (1982) Phenylalanine transport in guinea pig jejunum. A general mechanism for organic solute and sodium cotransport. J Physiol (Paris) 78:131–145

Alvarado F, Mahmood A (1974) Cotransport of organic solutes and sodium ion in the small intestine. A general model. Amino acid transport. Biochemistry 13:2882–2890

Alvarado F, Monreal J (1967) Na^+ dependent active transport of phenylglucosides in the chicken small intestine. Comp Biochem Physiol 20:471–488

Aronson PS (1978) Energy-dependence of phlorizin binding to isolated renal microvillus membranes. J Membr Biol 42:81–98

Bihler I (1969) Intestinal sugar transport: Ionic activation and chemical specificity. Biochim Biophys Acta 183:169–181

Bode F, Baumann K, Diedrich DF (1972) Inhibition of [^{3}H] phlorizin binding to isolated kidney brush border membranes by phlorizin-like compounds. Biochim Biophys Acta 290:134–149

Carter-Su C, Kimmich GA (1980) Effect of membrane potential on Na^+-dependent sugar transport by ATP-depleted intestinal cells. Am J Physiol 238:C73–C80

Chesney R, Sacktor B, Kleinzeller A (1974) The binding of phlorizin to the isolated luminal membrane of the renal proximal tubule. Biochim Biophys Acta 332:263–277

Crane RK (1960) Intestinal absorption of sugars. Physiol Rev 40:789–825

Crane RK (1966) Structural and functional organization of an epithelial cell brush border. In: Warren KB (ed) Intracellular transport. Academic Press, New York, pp 71–102

Crane RK (1968) Absorption of sugars. In: Code CF (ed) Handbook of physiology, sect 6: Alimentary canal, vol 3: Intestinal absorption. Am Physiol Soc, Washington, pp 1323–1351

Crane RK (1977) The gradient hypothesis and other models of carrier-mediated active transport. Rev Physiol Biochem Pharmacol 78:99–159

Dixon M, Webb EC (1964) Enzymes, 2nd edn. Academic Press, New York

Frömter E, Diamond JM (1972) Route of passive ion permeation in epithelia. Nature 235:9–13

Heath DA, Aurbach GD (1973) Uptake of ^{125}I-phlorizin by tubules isolated from the renal cortex of the rat. J Biol Chem 248:1577–1581

Hopfer U, Groseclose R (1980) The mechanism of Na^+-dependent D-glucose transport. J Biol Chem 255:4453–4462

Hoshi T, Komatsu Y (1970) Effects of anoxia and metabolic inhibitors on the sugar-evoked potential and demonstration of sugar-outflow potential in toad intestine. Tohoku J Exp Med 100:47–59

Kaback HR (1976) Molecular biology and energetics of membrane transport. J Cell Physiol 89: 575–593

Kimmich GA (1981) Intestinal absorption of sugar. In: Johnson LR (ed) Physiology of the gastrointestinal tract. Raven Press, New York, pp 1035–1061

Kinter WB, Wilson TH (1965) Autoradiographic study of sugar and amino acid absorption by everted sacs of hamster intestine. J Cell Biol 25:19–39

Lyon I (1967) Studies on transmural potentials in vitro in relation to intestinal absorption. IV. Phlorizin-sugar interactions in rat gut. Biochim Biophys Acta 135:496–506

Mahmood A, Alvarado F (1977) Harmaline interactions with sodium-binding sites in intestinal brush border sucrase. Biochim Biophys Acta 483:367–374

McDougal DB, Little KD, Crane RK (1960) Studies on the mechanism of intestinal absorption of sugars. IV. Localization of galactose concentration within the intestinal wall during active transport in vitro. Biochim Biophys Acta 45:483–489

Murer H, Kinne R (1980) The use of isolated membrane vesicles to study epithelial transport processes. J Membr Biol 55:81–95

Robinson JWL, Antonioli JA (1980) Is paracellular movement of importance in the intestinal absorption of organic solutes? Gastroenterol Clin Biol 4:78–86

Salomon LL, Allums JA, Smith DE (1961) Possible carrier mechanism for the intestinal transport of D-xylose. Biochem Biophys Res Commun 4:123–126

Schneider AJ, Kinter WB, Stirling CE (1966) Glucose-galactose malabsorption. N Engl J Med 274:305–312

Schultz SG (1980) Ion-coupled transport across biological membranes. In: Andreoli TE, Hoffman JF, Fanestil DD (eds) Membrane physiology. Plenum Press, New York, pp 273–286

Schultz SG, Curran PF (1970) Coupled transport of sodium and organic solutes. Physiol Rev 50: 637–718

Shapiro MP, Heinz E (1980) Na^+-linked cotransport of glycine in vesicles of Ehrlich cells. Biochem Biophys Acta 600:898–911

Silverman M (1974) The in vivo localization of high-affinity phlorizin receptors to the brush border surface of the proximal tubule in dog kidney. Biochim Biophys Acta 339:92–102

Silverman M (1976) Glucose transport in the kidney. Biochim Biophys Acta 457:303–351

Stirling CE (1967) High-resolution radioautography of phlorizin-^3H in rings of hamster intestine. J Cell Biol 35:605–618

Toggenburger G, Kessler M, Semenza G (1982) Phlorizin as a probe of the small-intestinal Na^+, D-glucose cotransporter. A model. Biochim Biophys Acta 688:557–571

Mechanism of Active Calcium Transport in Basolateral Plasma Membranes of Rat Small Intestinal Epithelium

C.H. VAN OS and W.E.J.M. GHIJSEN [1]

Introduction

In addition to its role in bone formation, calcium ions play a key role in the regulation of many cellular processes. As a consequence the ionized Ca^{2+} concentrations of both intracellular and extracellular fluids have to be precisely regulated.

Maintenance of a constant blood calcium concentration is primarily the function of three regulatory hormones: parathyroid hormone, calcitonin and 1,25-dihydroxycholecalciferol $1,25(OH)_2D_3$, the most biologically active metabolite of vitamin D_3 (Nordin 1976).

The seco-steroid hormone, $1,25(OH)_2D_3$, is the major stimulus for increased calcium absorption by the intestine (DeLuca and Schnoes 1976, Avioli and Birge 1978). Ca-absorption proceeds via two pathways: a transcellular route and a paracellular route. Absorption via the paracellular route is passive and not regulated by $1,25(OH)_2D_3$ (Nellans and Kimberg 1978). The transcellular pathway for Ca-absorption can be dissected into at least three processes: uptake of Ca^{2+} across the brush border membrane, movement of Ca^{2+} through the cytosol and transport of Ca^{2+} out of the cell by the basolateral plasmamembrane. A schematic representation of transcellular Ca^{2+} transport is given in Fig. 1.

Ca^{2+} influx across the brush border is down the electrochemical gradient for Ca^{2+} and the brush border membrane permeability for Ca^{2+} is controlled by $1,25(OH)_2D_3$. A good review on the effect of $1,25(OH)_2D_3$ on brush border permeability has recently been published (Rasmussen et al. 1981). Ca^{2+} in the cytosol is probably buffered by a vit.D-dependent calcium-binding protein (CaBP) (Wasserman 1981). The handling of Ca^{2+} within the cell has also been reviewed recently by Bikle et al. (1981).

The extrusion of Ca^{2+} is against the electrochemical gradient and requires energy. In general two types of Ca^{2+} transport systems can perform Ca^{2+} extrusion across mammalian plasma membranes: Ca-Mg-ATPase with a high affinity site for Ca^{2+} ions and Na^+/Ca^{2+} exchange in which the electrochemical gradient for Na^+ ions provides the energy for Ca-extrusion (Carafoli 1981). The difficulty of isolating pure basolateral membrane preparations of small intestinal cells is reflected in the low number of papers dealing with transport mechanisms in this particular membrane. However,

1 Department of Physiology, University of Nijmegen, P.O. Box 9101, Nijmegen, The Netherlands

Intestinal Transport
(ed. by M. Gilles-Baillien and R. Gilles)

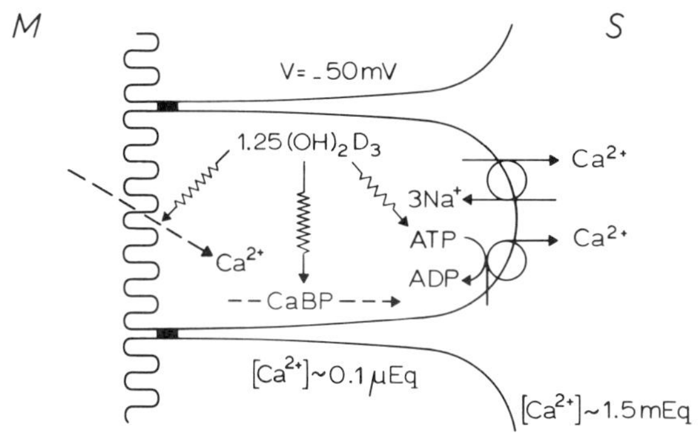

Fig. 1. Model for transcellular Ca-transport across rat intestinal epithelium (for details see text)

recent work in three different groups indicate that intestinal basolateral membranes contain ATP-dependent Ca^{2+} transport as well as Na^+/Ca^{2+} exchange systems (Nellans and Popovitch 1981, Hildmann et al. 1982, Ghijsen et al. 1982). It has also been shown recently that ATP-dependent Ca^{2+} transport in basolateral membranes of rat duodenum is controlled by the vitamin-D status of the animal (Ghijsen and van Os 1982). The subject of this chapter will be the recent progress made with isolated basolateral membranes of rat small intestine.

Localization of Ca-ATPase and ATP-Dependent Ca^{2+} Transport

The first report of Ca-ATPase activity in intestinal epithelium was made by Martin et al. (1969) and the activity was reported in brush border membranes. Haussler et al. (1970) found in brush borders closely correlated responses between Ca-ATPase and alkaline phosphatase to vitamin-D and to various inhibitors and they suggested that both activites are different manifestations of the same enzyme. Subsequently, Mircheff and Wright (1976) reported Ca-ATPase and alkaline phosphatase in rat duodenal basolaterals. Kinetic analysis of Ca-induced ATP-hydrolysis demonstrated the presence of both high and low affinity sites for Ca^{2+} in brush borders and in basolaterals of duodenal cells (Ghijsen and van Os 1979). A detailed examination of the substrate specificity of the Ca-induced activity plus the effects of inhibitors as theophylline and chlorpromazine, led to the conclusion that a specific high affinity Ca-ATPase, the enzymatic expression of a Ca-pump, is exclusively located in basolateral plasmamembranes (Ghijsen et al. 1980). The alkaline phosphatase activity in the brush border and in the basolateral membrane has both high and low affinity sites for Ca^{2+}; it hydrolyses ATP but has no role in active Ca-translocation (Ghijsen et al. 1982). Apparently, the susceptibility of alkaline phosphatase to Ca^{2+} results from Zn^{2+} depletion from the active center due to EDTA used during membrane isolation or to EGTA which is used as a Ca^{2+} buffer in the Ca-ATPase assay. These conclusions could be confirmed by a phosphorylation study in which the phosphorylated intermediates of alkaline phosphatase and of Ca-ATPase were visualized on SDS-acrylamide gels (de Jonge et al. 1981). The enzymatic studies referred to have been done with leaky membrane vesicles in which the demonstration of ATP-driven Ca-accumulation was not possible.

Hildmann et al. (1979) were the first to demonstrate ATP-dependent Ca^{2+} transport in partially resealed basolateral membrane vesicles, which was later confirmed by others (Nellans and Popovitch 1981, Ghijsen et al. 1982). The basolateral membrane preparations used in these studies were still heavily contaminated with smooth endoplasmic reticulum (SER), and it is of importance to know whether SER fragments contribute significantly to ATP-dependent Ca-uptake in this membrane preparation. By giving the results of two different approaches we like to demonstrate that it is highly unlikely that SER fragments also contain an ATP-dependent Ca-uptake system as present in basolateral membranes. The first approach was to further purify basolateral membranes by free-flow electrophoresis or zonal electrophoresis on density gradients. Both methods separate the SER marker NADPH-cytc-reductase from Na-K-ATPase activity (Mircheff et al. 1979, van Os et al. 1980). A typical experiment is shown in Fig. 2.

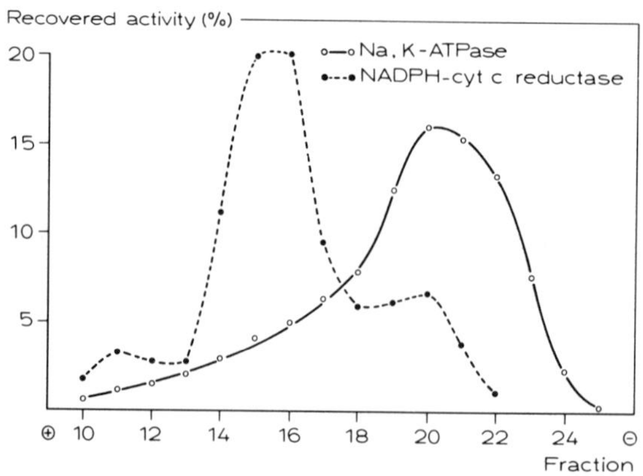

Fig. 2. Distribution of marker enzymes of smooth endoplasmic reticulum (NADPH-cytc-red) and of basolateral membrane (Na-K-ATPase) from rat duodenum, after zonal electrophoresis on density gradients

The specific activity of Na-K-ATPase in the combined fractions 20–22 increased by a factor of 1.57 [21.6 ± 1.7 to 33.9 ± 2.1 μmol P_i h^{-1} mg^{-1} protein (n = 7)]. This membrane fraction also showed an 1.50 increase in specific activity of Ca-ATPase [1.4 ± 0.3 to 2.1 ± 0.3 μmol P_i h^{-1} mg^{-1} protein (n = 7)]. The co-purification of Ca-ATPase along with the Na-K-ATPase presents good evidence that Ca-ATPase activity originates from plasma membranes.

Another approach was to study the effect of digitonin and of oxalate on ATP-dependent Ca^{2+} uptake. Digitonin action depends on a specific interaction with cholesterol-rich plasma membranes (Amar-Costesec et al. 1974). In Fig. 3 it is shown that 0.05% digitonin inhibits ATP-dependent Ca^{2+} uptake by 75%, apparently by destroying the membrane barrier function of cholesterol-rich plasma membranes. Control studies revealed that Ca-ATPase activity is not influenced by this concentration of digitonin. Oxalate is used in Ca^{2+} transport studies in sarcoplasmic reticulum to enhance Ca-uptake by precipitating accumulated Ca^{2+} and thereby maintain a low concentration of free Ca^{2+} in the vesicle (Weber 1966). The potency of oxalate to stimulate Ca-uptake is related with the membrane permeability to oxalate. Since plasma membranes are less permeable than microsomal membranes for oxalate, Ca-uptake into SER vesicles should be stimulated more than Ca-uptake into BLM vesicles when oxalate is present (Moore et al. 1974). In Fig. 3 is shown that 15 mM oxalate has no significant effect on the initial rate of ATP-dependent uptake. We also found no effect of oxalate after 10 min when the Ca-uptake had reached a plateau value. These results support the conclusion that ATP-dependent Ca-uptake is associated with plasma membranes.

Another interesting observation is that the activity of the ATP-dependent Ca-uptake system was highest in the duodenum and decreased sharply towards the ileum. The distribution along the small intestinal tract is shown in Fig. 4. A similar distribution has been reported for net calcium fluxes across intact intestinal epithelium (Walling 1977, Mircheff et al. 1977) and for vitamin D-dependent calcium-binding protein

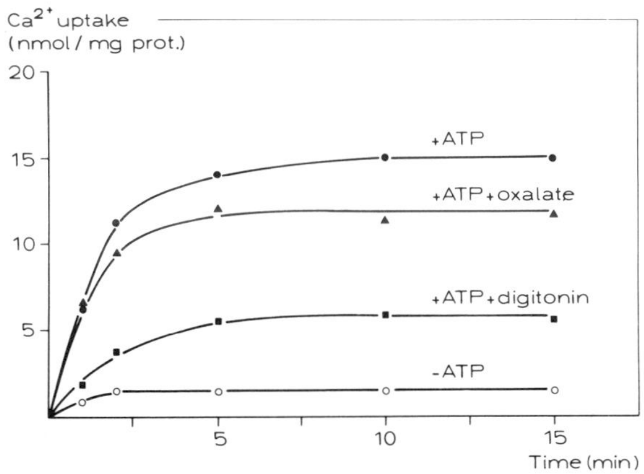

Fig. 3. Effects of digitonin and oxalate on ATP-dependent Ca-uptake in basolateral membranes. Ca-uptake is measured at 1 μM free Ca^{2+} as described in detail by Ghijsen et al. (1982)

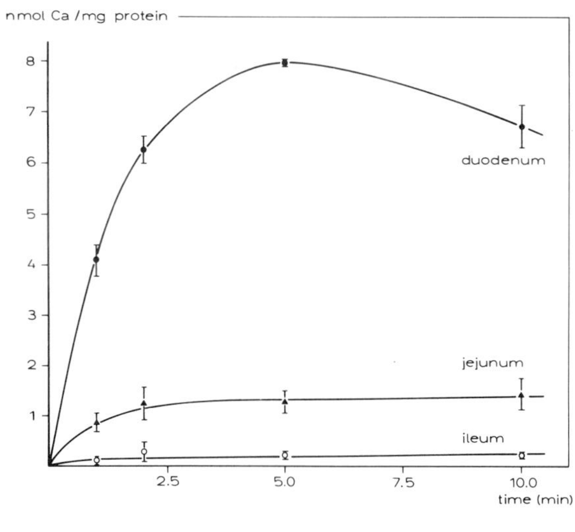

Fig. 4. Distribution of ATP-dependent Ca-transport along the small intestinal tract. Basolateral membranes are isolated from 10 cm pieces of proximal, mid and terminal small intestine (Ghijsen et al. 1982). ATP-dependent Ca-uptake has been corrected for ATP-independent uptake

(Thomasset et al. 1981). This close correlation between ATP-dependent Ca-uptake in basolateral vesicles and net Ca-fluxes across intact epithelium strongly suggests a primary role for Ca-ATPase in intestinal Ca-absorption.

Effects of Calmodulin and Calmodulin Antagonists

Much attention has been given recently to the role of the calcium regulatory protein, calmodulin, in activating Ca-ATPase from erythrocytes and heart plasmalemma (Sarkadi 1980, Carafoli 1981). Nellans and Popovitch (1981) reported that calmodulin increases both the maximal transport rate and the affinity for Ca^{2+} of the ATP-dependent Ca^{2+} transport in intestinal basolateral membranes. We have studied the effect of calmodulin on Ca-ATPase activity, on formation of the phosphorylated intermediate and on the Ca^{2+} transport system. The effect on Ca-ATPase and on phosphorylation are given in Fig. 5.

These studies were done with the leaky membrane preparation which means that the intestinal cells have been disrupted in hypotonic EDTA solutions and that 0.5 mM EDTA is present in all the isolation buffer solutions. The calmodulin content of these membranes is still very high (10–20 μg mg^{-1} protein; de Jonge et al. 1981), but despite this, addition of 0.5 μM bovine brain calmodulin stimulates Ca-ATPase activity by 60% while no effect is found on the level of phosphorylation from [γ^{32}P]-ATP.

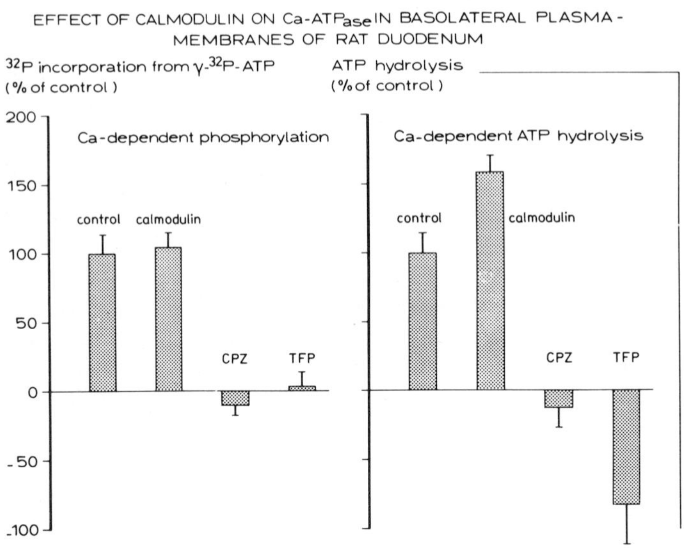

Fig. 5. Effect of calmodulin and phenothiazines on Ca-ATPase and on formation of phosphorylated intermediate of Ca-ATPase. *CPZ* chlorpromazine (10^{-4} M); *TFP* trifluoperazine (10^{-4} M). Phosphorylated intermediates are identified on SDS-polyacrylamide gels as previously published (de Jonge et al. 1981)

This result suggests an effect of calmodulin on the turnover rate of the Ca-pump. With the same membrane preparation the phenothiazines, chlorpromazine and trifluoperazine (10^{-4} M) totally inhibit Ca-dependent phosphorylation and Ca-ATPase activity in the presence or absence of calmodulin. With trifluoperazine also background Mg-ATPase is inhibited. It is known that at 0.1 mM, the action of phenothiazines is not very specific since disruption of the membrane lipid environment of enzymes may occur (Hinds et al. 1981).

The story is different when effects of calmodulin and its antagonists are studied on ATP-dependent Ca-uptake in resealed basolateral membrane vesicles. These membranes have not been exposed to EDTA and EGTA during homogenization and isolation. With freshly prepared membranes, no significant effects of calmodulin antagonists can be found on ATP-dependent Ca^{2+} transport, even with concentrations up to 0.1 mM (Table 1).

Table 1. Effects of calmodulin and phenothiazines on ATP-dependent Ca-uptake in basolateral membranes

	Untreated membranes		Membranes washed with 5 mM EGTA	
	– calmodulin	+ calmodulin	– calmodulin	+ calmodulin
Control	5.3 ± 0.1 (16)	5.2 ± 0.4 (4)	4.2 ± 0.3 (4)	5.1 ± 0.4 (4)
TFP	4.9 ± 0.3 (4)	n.d.	2.9 ± 0.7 (3)	3.9 ± 0.7 (3)
PF	5.2 ± 0.4 (3)	n.d.	4.4 ± 0.1 (3)	4.6 ± 0.3 (3)

Concentrations of phenothiazines, TFP (trifluoperazine) and PF (penfluridol), are 10^{-4} M. Concentration of bovine brain calmodulin is 0.5 μM. Free Ca-concentration is 1 μM. n.d. means not determined. Mean values are given with the number of observations in parentheses

When these membranes are washed with 5 mM EGTA the transport rate at 1 μM free Ca^{2+} decreases by 20%. With calmodulin, the initial transport rate is restored and phenothiazines block the calmodulin effect as shown in Table 1. The calmodulin-independent activity is not inhibited by phenothiazines. An obvious conclusion from these results is that in the intact in vivo system calmodulin is associated with Ca-ATPase but that it is not susceptible to calmodulin antagonists. Moreover, the effects shown in Table 1 are much smaller than those reported for Ca-ATPase in erythrocytes (Hinds et al. 1981) and in heart plasmalemma (Caroni and Carafoli 1981). Apparently, the osmotic shock in the presence of EDTA loosens the association between calmodulin and Ca-ATPase, whereafter phenothiazines can antagonize calmodulin (Fig. 5).

Kinetics and Stoichiometry of the Intestinal Ca-Pump

The kinetic parameters of Ca-ATPase activity and of ATP-dependent Ca-uptake observed in freshly prepared basolateral membranes are shown in Fig. 6.

The affinity for Ca^{2+} is equal in both systems ($K_m \sim 0.15\ \mu M\ Ca^{2+}$). K_m values in Fig. 6 have been derived from Lineweaver-Burk plots. Ca-ATPase activity has been

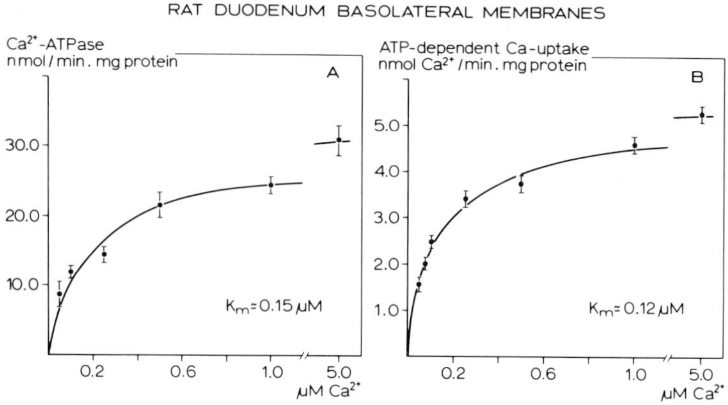

Fig. 6. Kinetics of Ca-ATPase and ATP-dependent Ca-uptake in basolateral membranes of rat duodenum

measured in the presence of 2.5 mM theophylline to suppress Ca-dependent ATP-hydrolysis by alkaline phosphatase. This latter activity exceeds Ca-ATPase activity fourfold (Ghijsen et al. 1982). Nellans and Popovitch (1981) reported an affinity for Ca^{2+} one order of magnitude higher ($K_m \sim 0.03\ \mu M\ Ca^{2+}$) but this difference is probably due to using different association constants in the Ca-EGTA buffer system. The maximal velocity of both activities in Fig. 6 differ by a factor of 5. In order to compare ATP-dependent Ca^{2+} transport rates with Ca-ATPase activities we need information on the orientation of the vesicles. Only vesicles resealed for Ca^{2+} and with their ATP-binding sites on the outside (inside-out orientation) contribute to ATP-dependent Ca-uptake. On the other hand all leaky plus inside-out vesicles contribute to Ca-ATPase activity. So far, we were unable to solve satisfactorily the problems met in orientation studies with basolateral membrane vesicles. This was primarily due to complicating side-effects of detergent treatment on membrane enzymes studied. For example, digitonin inhibited Na-K-ATPase as resported previously by Mircheff et al. (1979). Triton X-100 treatment appeared the most reliable tool to unmask latency in Na-K-ATPase activity (Ghijsen et al. 1982). SDS-treatment stimulated Na-K-ATPase activity by 500%, more than twice the effect of Triton. From the stimulation of Na-K-ATPase by Triton we calculate that the percentage of vesicles sealed for ATP varies between 50% and 70% among different preparations incubated for 30 min at 25°C. Incubations at 37°C lead to lower degree of sealing (40%; Ghijsen et al. 1982). The channel forming peptide, alamethacin, has successfully been used in Na-K-ATPase latency studies with heart sarcolemmal vesicles (Jones et al. 1980). We have used 0.5 mg alamethacin per mg membrane protein in Ca-ATPase assays to make resealed vesicles leaky for ATP. From the increase in Ca-ATPase activity it can be concluded that approximately 30% of the vesicles have an outside-out orientation (ATP binding site inside the vesicles). Alamethacin also increased K^+ stimulated p-nitrophenylphosphatase activity to the same extend as Ca-ATPase. Since the PNPP binding site is most likely on the extracellular side of the plasma membrane (Bers et al. 1980), this alamethacin effect suggests that approximately 30% of the vesicles are oriented inside-out.

In summary, it appears that one third of the vesicles are leaky, one third is inside-out and one third is outside or rightside-out.

Correcting the V_{max} values for Ca-ATPase and ATP-dependent Ca^{2+} transport according to this information, then a ratio is found of 0.5 Ca^{2+} ions transported per mol ATP hydrolyzed. Previously we calculated a ratio of 1 Ca^{2+} ion per mol ATP, but that calculation was based on sideness studies done at 37°C which induces more leakiness (Ghijsen et al. 1982). The free energy of the electrochemical Ca^{2+} gradient across intestinal cells (see Fig. 1) can be calculated and comes out to be $\Delta\tilde{\mu}$ Ca^{2+} = 7.3 kcal mol^{-1} (Ghijsen et al. 1980). Since hydrolysis of 1 mol ATP in intact cells can generate between 7.2 and 11.9 kcal mol^{-1} (Slater 1979) a stoichiometry of 1 Ca^{2+} per mol ATP is theoretically possible.

A potentially large error can be made in estimating the V_{max} value of Ca-ATPase activity. In the assay, theophylline is used to suppress alkaline phosphatase activity. If theophylline does not inhibit totally, but for example 80%, that would lead to an overestimation by a factor of 2 in the Ca-ATPase activity. However, without too much emphasis on the actual value of the Ca-ATP ratio, these results support the conclusion that Ca-ATPase activity and ATP-dependent Ca-uptake are expressions of a Ca-pump similar as the one described in plasma membranes of a variety of tissues (Ghijsen et al. 1982).

Influence of Vitamin-D Status on ATP-Dependent Ca^{2+} Transport

In cooperation with Walling, Mircheff and Wright we have reported previously on effects of $1{,}25(OH)_2D_3$ on alkaline phosphatase and Ca-ATPase activities in basolateral membranes of rat duodenum (Mircheff et al. 1977). However, in that early report no distinction was made between high and low affinity Ca-ATPase. Therefore we decided to study the effect of $1{,}25(OH)_2D_3$ on ATP-dependent Ca^{2+} transport and on high affinity Ca-ATPase activity. Male rats were raised from weaning under vitamin D-deficient conditions for 6–8 weeks exactly as done previously by Walling (1977) and Mircheff et al. (1977). Plasma calcium levels verified the vit.D-deficient (–D) and $1{,}25(OH)_2D_3$-repleted (+D) state of the animals. Repleted rats received 160 ng $1\alpha25(OH)_2D_3$ (courtesy of Hoffmann-La Roche, Basle) intraperitoneally 48 and 24 h before sacrifice. After repletion, plasma Ca-levels had increased from 1.55 ± 0.05 (–D) to 2.61 ± 0.04 mM (+D) (n = 62).

Figure 7 shows that in duodenal basolateral vesicles from repleted rats, the initial rate of ATP-dependent Ca-uptake has doubled with respect to the deficient controls. Kinetic analysis of the effect of $1{,}25(OH)_2D_3$ revealed only the V_{max} of the transport system has increased but not the affinity for Ca^{2+} ions (Ghijsen and van Os 1982). This result suggests that $1{,}25(OH)_2D_3$ either increases the number of pump sites or enhances the turnover rate of the pump. It has been reported that calmodulin levels in duodenal cells are not influenced by the vitamin-D status of rats (Thomasset et al. 1981). Therefore, it is highly unlikely that calmodulin plays a role in $1{,}25(OH)_2D_3$-dependent regulation of the Ca-ATPase.

We have done several control studies with basolateral membranes from deficient and repleted animals. Firstly, the D-glucose space of the two vesicle populations have

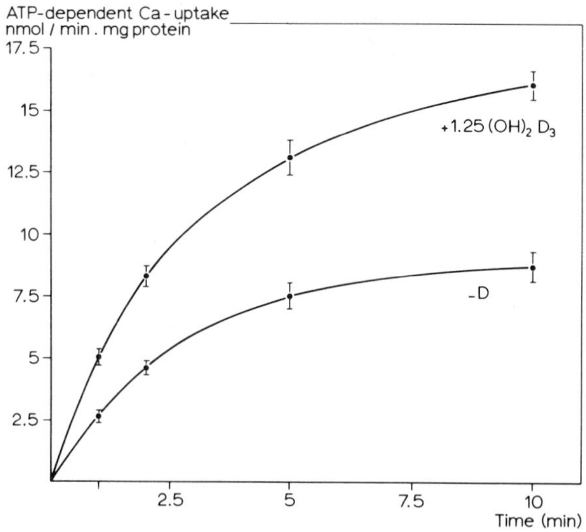

Fig. 7. Effect of repletion with 1,25(OH)$_2$D$_3$ on the ATP-dependent Ca-transport capacity of basolateral membrane vesicles (– D), vitamin-D-deficient control, + 1,25 (OH)$_2$D$_3$, 48 h after repletion with the hormone)

been measured and amount to 2.9 ± 0.2 (– D) and 2.6 ± 0.3 μl mg^{-1} protein (+ D) (n = 4). This means that it is unlikely that differences in Ca^{2+} transport capacities are due to different intravesicular volumes. Secondly, Ca-ATPase activity in the leaky vesicle preparation increased to the same extent as ATP-dependent Ca-uptake after repletion with 1,25(OH)$_2$D$_3$. No change in Na-K-ATPase activity was detected between + D and – D basolateral preparations. Thirdly, the difference in ATP-dependent Ca-uptake capacity is not due to aspecific Ca^{2+} binding to the membranes. In control Ca-uptake experiments in the absence of ATP the following uptake values were found after 1 min incubation at 1 μM free Ca^{2+}: 1.20 ± 0.08 (– D) and 1.23 ± 0.07 nmol Ca^{2+} mg^{-1} protein (+ D), hence no significantly different binding of Ca^{2+} in both preparations. The last control consisted in measuring ATP-dependent Ca-uptake in basolateral membrane vesicles from rat kidney cortex of deficient and repleted animals. The ATP-dependent Ca-uptake in these kidney cortex basolaterals was not significantly different in – D and + D preparations (details of ATP-dependent Ca^{2+} transport in kidney basolateral membranes will be published elsewhere). Therefore, 1,25(OH)$_2$D$_3$ stimulates the small intestinal Ca-pump but not the one in kidney proximal tubules.

Na^+/Ca^{2+} Exchange and Intestinal Ca-Absorption

The extrusion of Ca^{2+} by way of a Na^+/Ca^{2+} exchange mechanism is a possibility suggested by some authors who report a dependency of Ca-absorption on Na^+ ions (Martin and DeLuca 1969, Holdsworth et al. 1975). However, a Na^+ dependency has also been questioned by others (Wasserman and Taylor 1963, Behar and Kerstein 1976). Moreover, inconsistant responses to ouabain have been reported (Adams and Norman 1970, Birge et al. 1974). In isolated basolateral membranes the presence of a Na^+/Ca^{2+} exchanger has been claimed by Hildmann et al. (1982). These authors showed that ATP-dependent Ca-accumulation was inhibited by Na^+ in the uptake medium, apparently by inducing a Ca^{2+} leak for intravesicular Ca^{2+} via the Na^+/Ca^{2+} exchanger. When we repeated these experiments with duodenal basolateral membranes, we obtained somewhat different results. As shown in Fig. 8, the presence of 70 mM Na^+ in the uptake medium reduces ATP-dependent Ca-uptake by 50%, almost independent of the time of incubation. When ouabain is present, the effect of Na^+ is significantly reduced and only 20% of ATP-dependent Ca-uptake is inhibited. This effect of ouabain is only observed when membranes were preincubated with ouabain for 45 min or longer, which indicates that oubain exerts its effect inside the vesicles. The ouabain effect cannot be due to inhibition of ATP-hydrolysis by Na-K-ATPase, since the effect is also observed at 1 and 2 min incubation points when the ATP supply is certainly not exhausted. The ATP concentration in these experiments was 6 mM. The inhibiting effect of Na^+ and ouabain could be explained as follows: with ATP and Na^+ present on the outside and K^+ on the inside of the vesicles, the Na-pump accumulates Na^+ inside the vesicles. This high Na^+ inside could displace Ca^{2+} from intravesicular binding sites. Ouabain, by inhibiting the Na-pump, prevents Na^+ accumulation and thereby Ca^{2+} displacement. A critical experiment for this Na^+ accumulation

Table 2. Effects of ouabain and monensin on Na^+ inhibition of ATP-dependent Ca^{2+} uptake in basolateral membranes from rat duodenum. The effects are expressed as percentages of the control value for ATP-dependent Ca^{2+}-uptake (4.6 ± 0.4 nmol Ca^{2+} min^{-1} mg^{-1} protein)

	Control	+ Monensin (2 μM)
Control	100%	98.9 ± 6.5% (4)
+ 100 mM Na^+	62.4 ± 5.1% (5)	74.6 ± 4.3% (9)
+ 100 mM Na^+ + ouabain (2 mM)	75.7 ± 4.3% (7)	79.7 ± 2.9% (6)

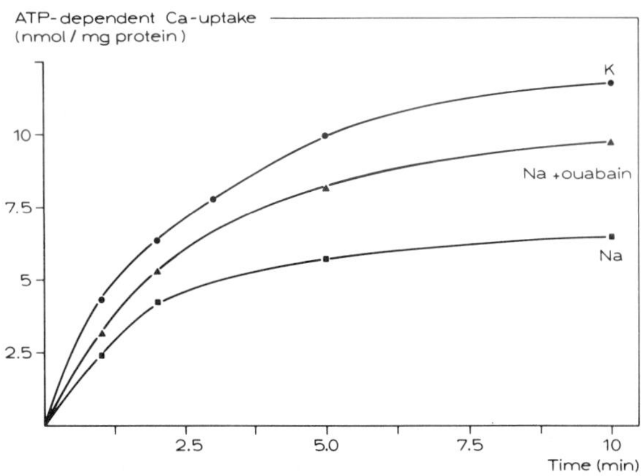

Fig. 8. Effects of Na^+ ions and Na^+ ions plus ouabain on ATP-dependent Ca-uptake in basolateral membranes from rat duodenum. Control uptake is measured in presence of 150 mM K^+, Na^+ effects are determined in the presence of 75 mM K^+ and 75 mM Na^+. Ouabain concentration is 2 mM

theory would be to make the vesicles leaky for Na^+ by means of a Na^+ ionophore as for example monensin. The results of such experiments are shown in Table 2. Table 2 shows clearly that the effect of Na^+ on ATP-dependent Ca^{2+} accumulation is reduced by monensin to the same extent as preincubation with ouabain. From the results of Fig. 8 and Table 2 it is obvious that no firm conclusions can be drawn from these indirect experiments with regard to the presence and capacity of a Na^+/Ca^{2+} exchange in basolateral membranes.

The effect of ouabain shown in Fig. 8 is at variance with the results of Hildmann et al. (1982), who found no effect of ouabain. Two explanations may explain that difference: firstly, it cannot be retrieved whether Hildmann et al. (1982) preincubated membranes with ouabain, because without preincubation we do not observe the ouabain effect; secondly, Hildmann et al. (1982) used duodenum and jejunum as a source for basolateral membranes while we used duodenum only. In view of the steep fall in Ca-pump capacity when cells are taken from more distal parts of the small intestine (see Fig. 4), one may anticipate that Na^+/Ca^{2+} exchange becomes relatively more important further down the small intestinal tract. Therefore, Na^+ effects on ATP-dependent Ca-uptake may differ with the location in the intestine.

We have also studied the influence of the vitamin-D status on Na^+ inhibition of ATP-dependent Ca-uptake in basolateral membranes from duodenum. These results are given in Fig. 9.

In vesicles from deficient as well as from repleted animals, 70 mM Na^+ inhibits 50% of ATP-dependent Ca^{2+} transport, while ouabain reduces the Na^+ inhibition to 15%. Ascribing the 15% inhibition to the action of a Na^+/Ca^{2+} exchanger, which is not proven beyond doubt, then obviously Na^+/Ca^{2+} exchange in duodenum is not influenced by the vitamin-D status of the animal.

In order to be able to assess the relative importance of Na^+/Ca^{2+} exchange in Ca-absorption it is imperative to design experiments in which it can be shown that a Na^+ gradient across the vesicle membrane, $[Na]_i > [Na]_o$, accumulate Ca^{2+} ions in the absence of ATP. These experiments are currently in progress.

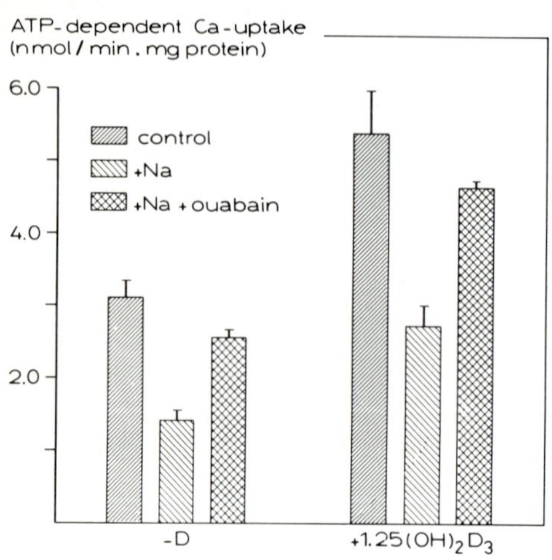

Fig. 9. Effect of vitamin-D status on Na^+ inhibition of ATP-dependent Ca-uptake in basolateral membranes from rat duodenum. Conditions and concentrations of Na^+ and ouabain as in Fig. 8

Summary

A high affinity Ca-ATPase activity, the enzymatic expression of a Ca-pump is exclusively located in the basolateral plasma membrane of rat enterocytes. The contamination with smooth endoplasmatic reticulum does not contribute significantly to this Ca-ATPase activity. With partially resealed vesicles of isolated basolateral membranes it can be demonstrated that ATP induces Ca^{2+} accumulation. The activity of the Ca-pump is highest in duodenum and decreases sharply towards the ileum. Oxalate is not able to further enhance Ca^{2+} accumulation while treatment of the vesicles with digitonin inhibits the ATP-dependent Ca^{2+} uptake by destroying the membrane barrier function. Both observations support the idea that the Ca-pump is of plasmalemmal origin. Calmodulin and calmodulin antagonists have little or no effect on freshly prepared membranes but effects are inducible after washing the membranes with EDTA or EGTA.

The affinity for Ca^{2+} of Ca-ATPase and ATP-dependent Ca-transport is equal in both systems ($K_m \sim 0.15\ \mu M\ Ca^{2+}$). Maximal velocities of both systems differ by a factor of 5. The stoichiometry of the Ca-pump is most likely between 0.5 and 1 Mol Ca^{2+} transported per Mol ATP hydrolyzed.

Effects of Na^+ ions on ATP-dependent Ca-transport suggest the presence of Na^+/Ca^{2+} exchange in parallel to the Ca-pump. However, the inhibitory effect of Na^+ is sensitive to ouabain and monensin. These observations suggest that Na-K-ATPase accumulates Na^+ inside the vesicles and that high Na^+ competes with Ca^{2+} for Ca^{2+} binding sites.

Basolateral membranes of duodenum isolated from vitamin D_3-deficient animals have a decreased capacity for Ca^{2+} transport. This is reflected in a 50% smaller Ca-ATPase activity and ATP-dependent Ca-transport rate. Repletion with $1{,}25(OH)_2D_3$ restores the Ca-transport capacity of these membranes to normal. From the effects of Na^+ on ATP-dependent Ca^{2+} transport it can be concluded that the vit-D status of the animal has no influence on the capacity of the Na^+/Ca^{2+} exchanger.

Acknowledgement. We are indebted to Mr. M.D. de Jong for excellent technical assistance. This study was in part supported by the Netherlands Organization for the Advancement of Basic Research (ZWO) via the foundation for Medical Research (FUNGO). $1\alpha{,}25(OH)_2D_3$ was generously provided by Hoffmann-La Roche, Basle.

References

Adams TH, Norman AW (1970) Studies on the mechanism of action of calciferol. J Biol Chem 245:4421–4431

Amar-Costesec A, Wibo M, Thines-Sempoux D, Beaufray H, Berthet J (1974) Analytical study of microsomes and isolated subcellular membranes from rat liver, IV. J Cell Biol 62:717–745

Avioli LV, Birge SJ (1978) Mechanisms of calcium absorption: a reappraisal. In: Andreoli TE, Hoffman JF, Fanestil DD (eds) Physiology of membrane disorders. Plenum Medical Book Company, New York London, pp 919–940

Behar J, Kerstein MD (1976) Intestinal calcium absorption: difference in transport between duodenum and ileum. Am J Physiol 230:1255–1260

Bers DM, Philipson KD, Nishimoto AY (1980) Sodium-calcium exchange and sideness of isolated cardiac sarcolemmal vesicles. Biochim Biophys Acta 601:358–371

Bikle DD, Zolock DT, Morrisey RL (1981) Action of vitamin D on intestinal calcium transport. Ann NY Acad Sci 372:481–500

Birge SI, Switzer SC, Leonard DR (1974) Influence of sodium and parathyroid hormone on calcium release from intestinal mucosal cells. J Clin Inv 54:702–709

Carafoli E (1981) Ca^{2+} pumping systems in the plasma membrane. In: Bronner F, Peterlik M (eds) Calcium and phosphate transport across biomembranes. Academic Press, New York, pp 9–14

Caroni P, Carafoli E (1981) The Ca^{2+} pumping ATPase of heart sarcolemma. Characterization, calmodulin dependence and partial purification. J Biol Chem 256:3263–3270

DeLuca HF, Shnoes HK (1976) Metabolism and mechanism of action of vitamin D. Ann Rev Biochem 45:631–666

Ghijsen WEJM, Os CH van (1979) Ca-ATPase in brush border and basolateral membranes of rat duodenum with high affinity sites for Ca^{2+} ions. Nature 279:802–803

Ghijsen WEJM, Os CH van (1982) 1α,25-dihydroxy-vitamin D_3 regulates ATP-dependent calcium transport in basolateral plasma membranes of rat enterocytes. Biochim Biophys Acta 689: 170–172

Ghijsen WEJM, Jong MD de, Os CH van (1980) Dissociation between Ca-ATPase and alkaline phosphatase activites in plasma membranes of rat duodenum. Biochim Biophys Acta 599: 538–551

Ghijsen WEJM, Jong MD de, Os CH van (1982) ATP-dependent calcium transport and its correlation with Ca-ATPase activity in basolateral plasma membranes of rat duodenum. Biochim Biophys Acta 689:327–336

Haussler MR, Nagode LA, Rasmussen H (1970) Induction of intestinal brush border alkaline phosphatase by vitamin D and identity with Ca-ATPase. Nature 228:1199–1201

Hildmann B, Schmidt A, Murer H (1979) Ca^{2+} transport in basal-lateral plasma membranes isolated from rat small intestinal epithelial cells. Pfluegers Arch 382:R23

Hildmann B, Schmidt A, Murer H (1982) Ca^{2+} transport across basal-lateral plasma membranes from rat small intestinal epithelial cells. J Membr Biol 65:55–62

Hinds TR, Raess BU, Vincenzi FF (1981) Plasma membrane Ca^{2+} transport: Antagonism by several potential inhibitors. J Membr Biol 58:57–65

Holdsworth ES, Jordan JE, Keenan E (1975) Effects of cholecalciferol on the translocation of calcium by non-everted chick ileum in vitro. Biochem J 152:181–190

Jones LR, Maddock SW, Besch HR Jr (1980) Unmasking effect of alamethacin on the (Na^+ + K^+)-ATPase, β-adrenergic receptor-coupled adenylate cyclase, and cAMP-dependent protein kinase activities of cardiac sarcolemmal vesicles. J Biol Chem 255:9971–9980

Jonge HR de, Ghijsen WEJM, Os CH van (1981) Phosphorylated intermediates of Ca-ATPase and alkaline phosphatase in plasma membranes from rat duodenal epithelium. Biochim Biophys Acta 647:140–149

Martin DL, DeLuca HF (1969) Influence of sodium on calcium transport by rat small intestine. Am J Physiol 216:1351–1359

Martin DL, Melancon MJ, DeLuca HF (1969) Vitamin D stimulated, calcium-dependent adenosine triphosphatase from brush borders of rat small intestine. Biochem Biophys Res Commun 35:819–823

Mircheff AK, Wright EM (1976) Analytical isolation of plasma membranes of intestinal epithelial cells: Identification of Na-K-ATPase rich membranes and the distribution of enzyme activities. J Membr Biol 28:309–333

Mircheff AK, Walling MW, Os Ch van, Wright EM (1977) Distribution of alkaline phosphatase and Ca-ATPase in intestinal epithelial cell plasma membranes: Differential response to 1,25 $(OH)_2 D_3$. In: Norman AK, Schaefer K, Coburn JW, DeLuca HF, Fraser D, Grigoleit HG (eds) Vitamin D, biochemical, chemical and clinical aspects related to calcium metabolism. de Gruyter, Berlin New York, pp 281–284

Mircheff AK, Sachs G, Hanna SD, Labiner CS, Rabon E, Douglas AP, Walling MW, Wright EM (1979) Highly purified basal lateral plasma membranes from rat duodenum. Physical criteria for purity. J Membr Biol 50:343–363

Moore L, Fitzpatrick DF, Chen TS, Landon EJ (1974) Calcium pump activity of the renal plasmamembrane and renal microsomes. Biochim Biophys Acta 345:405-418

Nellans HN, Kimbert DV (1978) Cellular and paracellular calcium transport in rat ileum. Effects of dietary calcium. Am J Physiol 235:E716–E727

Nellans HN, Popovitch TE (1981) Calmodulin regulated, ATP driven calcium transport by basolateral membranes of rat small intestine. J Biol Chem 256:9932–9936

Nordin BEC (1976) Calcium, phosphate and magnesium metabolism. Churchill Livingstone, Edinburgh London New York

Os CH van, Jonge HR de, Jong MD de, Ghijsen WEJM, Walters JALI (1980) Separation of basolateral plasma membranes from smooth endoplasmic reticulum of the rat enterocyte by zonal electrophoresis on density gradients. Biochim Biophys Acta 600:730–738

Rasmussen H, Fontaine O, Matsumoto T (1981) Liponomic regulation of calcium transport by 1,25$(OH)_2 D_3$. Ann NY Acad Sci 372:518–528

Sarkadi B (1980) Active calcium transport in human red cells. Biochim Biophys Acta 604:159–190

Slater EC (1979) Measurement and importance of phosphorylation potentials: Calculation of free energy of hydrolysis in cells. In: Colowick SP, Kaplan NO (eds) Methods in enzymology, vol 55, part F, pp 235–245

Thomasset M, Molla A, Parkes O, Demaille JG (1981) Intestinal calmodulin and calcium-binding protein differ in their distribution and in the effect of vitamin D steroids on their concentration. FEBS Lett 127:13–16

Walling MW (1977) Intestinal Ca and phosphate transport: differential responses to vitamin D_3 metabolites. Am J Physiol 233:E488–E494

Wasserman RH (1981) Intestinal absorption of calcium and phosphorus. Fed Proc 40:68–72

Wasserman RH, Tqylor AN (1963) Nonessentiality of sodium ions for intestinal calcium transport. Proc Soc Exp Biol Med 114:479–482

Weber A (1966) Energized calcium transport and releasing factors. In: Sanadi DR (ed) Current topics in bioenergetics. Academic Press, New York, pp 203–254

The Small Intestinal Na^+, D-Glucose Cotransporter: a Likely Model

G. SEMENZA [1]

The co-transport mechanism of the coupled movement of D-glucose and Na^+ across the brush border membrane is now universally accepted (for a review, see Crane 1977).

The transport agency shows a stable asymmetry with respect to the plane of the membrane, as demonstrated by its asymmetric inactivation (or reactivation) by little-permeant or impermeant reagents. Comparison was made between little permeant vs. very permeant reagents and between sealed, right-side out membrane vesicles vs. deoxycholate-extracted membrane fragments (Klip et al. 1979a,b, 1980a,b). The Na^+, D-glucose cotransporter shows a functional asymmetry also (in addition, of course, to that which can be imposed onto it by asymmetric distributions of substrate(s) and/or by a $\Delta\psi \neq 0$). In fact, in right-side out vesicles: (1) trans-inhibitions and trans-stimulations by substrates are much larger in influx than in efflux experiments; (2) D-glucose influx responds to $\Delta\psi$ much more than its efflux; (3) the apparent K_m for efflux is more than one order of magnitude larger than the apparent K_m for influx; (4) influx and efflux rates may differ by one order of magnitude when measured at equivalent, but mirrored, conditions (Kessler and Semenza 1983).

The observed asymmetric insertion of the cotransporter in the membrane obviously agrees with current ideas on the biosynthesis and vectorial insertion of intrinsic membrane proteins. It rules out, also, freely "diffusing" or "tumbling over" carrier models for its mode of operation, whereas it makes more likely a "gated channel" (Semenza 1982; see Crane and Dorando 1979, 1980, 1983), i.e., a "channel with multiple conformational states", which is a limiting case of the general channel mechanisms (Läuger 1980).

Small intestinal Na^+, D-glucose cotransporter shows the functional characteristics of a "mobile carrier": in particular, it shows counterflow. In addition, optimal binding of phlorizin to the cotransporter in vesicles requires, in addition to Na^+, a $\Delta\psi \leqslant 0$ (negative inside the vesicles) which indicates that under the action of the electric field a portion of the channel must move to allow phlorizin (which is uncharged, Toggenburger et al. 1978) to bind.

It is possible to obtain some information on the nature of this "gate". As mentioned above, intravesicular D-glucose (and also Na^+, but not both at the same time) exert a strong transinhibition on the Na^+-dependent out → in (influx) of D-glucose.

1 Laboratorium für Biochemie der Eidgenössischen Technischen Hochschule Zürich, Universitätsstrasse 16, CH-8092 Zürich, Switzerland

Intestinal Transport
(ed. by M. Gilles-Baillien and R. Gilles)

This shows that the in → out translocation probabilities of the binary complexes of the translocator are small as compared to that of the free translocator. The translocation probabilities of the "gate" will be affected in different ways, depending on whether it is electrically neutral in the unloaded form (z = 0) (and thus has a positive change of 1 when bound to Na^+ or to Na^+-glucose) or whether it carries a negative charge of 1 (z = – 1) (and thus is uncharged when bound to Na^+ or to Na^+-glucose). As it will be shown elsewhere (Kessler and Semenza 1983), a $\Delta\psi \ll 0$ (negative inside the vesicles) reduces the trans inhibition if z = – 1, it enhances it if z = 0. In actual fact, it reduces it, which rules out the model with an "unacharged" gate (z = 0) and is perfectly compatible with a gate carrying a negative charge of 1 (z = – 1). Furthermore, if the tertiary complex with Na^+, phlorizin is electrically neutral, the velocity of release of phlorizin from the vesicles should not be affected by a $\Delta\psi$ across the membrane. This is the case (Fig. 1).

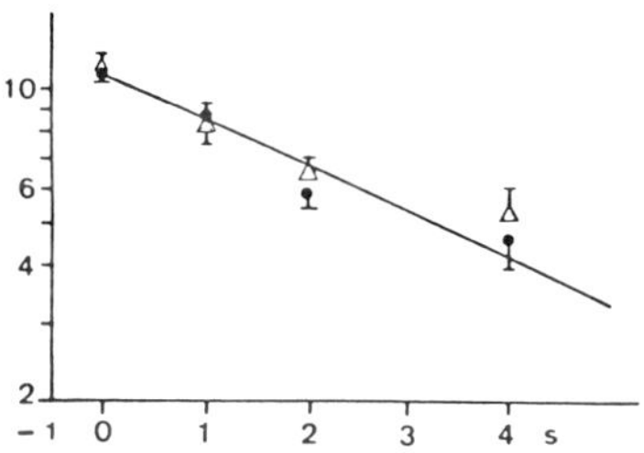

Fig. 1. Time course of the release of bound phlorizin in the presence of Na^+ and of either a membrane potential, negative inside the vesicles *(triangles)* or of a negligible membrane potential *(black dots)*. The scale at *left* indicates the pmol of phlorizin retained per mg of membrane protein. *Bars* indicate the S.E. (Toggenburger et al. 1982)

It seems thus very likely that the "gate" is composed of, or contains, a negatively charged group – presumably a COO^--group, because DCCD inhibits irreversibly both Na-dependent phlorizin binding and glucose transport (Weber and Semenza, unpublished data). Finally, the much higher K_m for D-glucose$_{out}$ at pH 5.4 as compared to K_m at neutral pH, also would agree with a carboxylate playing a functional role in the co-transporter (Toggenburger et al. 1978).

As Na^+ much prefers O over other potential ligands, it is very tempting to suggest that the Na^+ binding site is identical with or encompasses the "gate-COO^-".

A partial kinetic analysis only was possible in this complicated and rather tricky system (Kessler and Semenza 1983). It could rule out, however, Iso Ping Pong Bi Bi mechanisms, and render linkely a Random Bi Bi with preferred binding sequence Na^+_{out} first. The kinetic model proposed is compatible with various features of the models proposed by Crane and Dorando (1979, 1983) and by Hopfer and Groseclose (1980).

A mechanistic model, which combines all known kinetic characteristics with the results from chemical modifications, has been proposed recently (Kessler and Semenza 1983). The part of the model describing phlorizin binding (and thus presumably describing the first steps in the interaction of the cotransporter with substrates) is reported below as Fig. 2 (Toggenburger et al. 1982).

The rationale for suggesting form III as the carrier form binding phlorizin optimally is the following: (1) in membrane fragments the presence of Na^+ is mandatory for

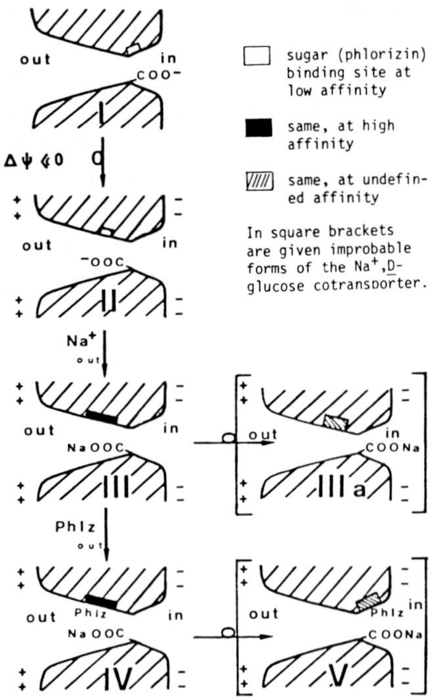

Fig. 2. Proposed minimum model for phlorizin binding to the Na^+, D-glucose cotransporter of the small-intestinal brush border membrane. The cotransporter is suggested to be a gated channel (or pore), the mobile gate consisting of, or encompassing, a COO^- group. (Toggenburger et al. 1982)

phlorizin binding (Klip et al. 1979a,c) which makes form I an unlikely candidate as a good phlorizin binder; (2) that the presence of $\Delta\psi$ (negative inside) alone (i.e., in the absence of Na^+) does not lead to optimal phlorizin binding, which makes form II an unlikely candidate as a good phlorizin binder; (3) that external Na^+ alone with small or no $\Delta\psi$ is not conductive to optimal phlorizin binding, which indicates again that form I, even in the presence of an out → in Na^+ gradient hardly binds phlorizin. In actual fact, the K_d-values for phlorizin binding at only moderately negative $\Delta\psi$ are fairly large and cannot be determined reliably at very low $\Delta\psi$-values; (4) that the hypothetical transition III → IIIa (gate-COONa snapping towards the inside) is very unlikely. In fact, if form IIIa existed, internal Na^+, even in the absence of external Na^+, should favour phlorizin binding from the outside. This is not the case, however, internal Na^+ inhibiting, rather than stimulating, phlorizin binding. Moreover, form IIIa would be a part of an Iso Ping Pong Bi Bi mechanism and/or of a mechanism transporting Na^+ in the absence of sugars, both of which are unlikely.

If the accessibility (or translocation) of the binding sites for Na^+ and for sugar is such that both these sites have to "look" to the same side of the membrane, it is reasonable to expect, also, that the inwardly oriented substrate binding site may exhibit high-affinity phlorizin-binding if the (equally inwardly oriented and presumably neighbouring) Na^+-binding site is occupied. That is, form I could exhibit Na^+-dependent (and probably $\Delta\psi$-independent) high affinity phlorizin binding if both phlorizin and Na^+ reach it from the "in" side. Unfortunately it has been impossible to date to obtain inside-out brush border vesicles and thus to carry out the pertinent experiment. However, deoxycholate-disrupted membranes, in which Na^+ and phlorizin have access to both sides and the $\Delta\psi$ is, of course, equal to zero (and in which, by analogy form I, both Na^+ and substrate binding sites are likely to have an inwardly orientation), do exhibit Na^+-independent phlorizin binding, the K_d being 9.5 μM (Klip et al. 1979c). In comparison, brush border membrane vesicles pre-equilibrated in Na^+ and with $\Delta\psi \approx 0$ show very poor binding for phlorizin added to the "out" side only.

Thus, the model in Fig. 2 accommodates satisfactorily a number of observations, both old and new; to the best of our knowledge, it does not disagree with any. It seems very likely to us, therefore, that it has fair chances of being essentially correct.

Acknowledgements. The author's work was partially supported by the SNSF, Berne. Thanks are also due to Drs. M. Kessler, A Klip, S. Grinstein, G. Toggenburger and to Mrs. B. O'Neill.

References

Crane RK (1977) Rev Physiol Biochem Pharmacol 78:101–159
Crane RK, Dorando F (1979) In: Quagliariello E et al (eds) Functional and molecular aspects of biomembrane transport. Elsevier, Amsterdam, pp 271–278
Crane RK, Dorando F (1980) Ann NY Acad Sci 339:46–52
Crane RK, Dorando F (1983) In: Martonosi A (ed) Membranes and transport, vol 2. Plenum, New York, pp 153–160
Hopfer U, Groseclose R (1980) J Biol Chem 255:4453–4462
Kessler M, Semenza G (1983) J Membr Biol (in press)
Klip A, Grinstein S, Semenza G (1979a) Biochim Biophys Acta 558:233–245
Klip A, Grinstein S, Semenza G (1979b) FEBS Lett 99:91–96
Klip A. Grinstein S, Semenza G (1979c) J Membr Biol 51:47–73
Klip A, Grinstein S, Semenza G (1980a) Ann NY Acad Sci 358:374–377
Klip A, Grinstein S, Biber J, Semenza G (1980b) Biochim Biophys Acta 589:100–114
Läuger P (1980) J Membr Biol 57:163–178
Semenza G (1982) In: Martonosi A (ed) Membranes and transport, vol 2. Plenum Press, New York, pp 175–181
Toggenburger G, Kessler M, Rothstein A, Semenza G, Tannenbaum C (1978) J Membr Biol 40: 269–290
Toggenburger G, Kessler M, Semenza G (1982) Biochim Biophys Acta 688:557–571

Protein-Lipid Interactions and Lipid Dynamics in Rat Enterocyte Plasma Membranes

Th.A. BRASITUS [1]

Introduction

The luminal (brush border and microvillus) and contraluminal (basolateral) plasma membranes of the rat enterocyte, the predominant cell type lining the small intestine, are highly differentiated to perform a variety of digestive and transport functions. These antipodal membranes have been previously demonstrated to differ from each other in ultrastructure (Oda 1976), electrophysiologic properties (Rose and Schultz 1971, Okada et al. 1977), enzyme and transport activities (Douglas et al. 1972, Lewis et al. 1975, Murer et al. 1974), protein components (Fujita et al. 1973), and lipid composition (Forstner et al. 1968, Douglas et al. 1972, Kawai et al. 1974, Lewis et al. 1975, Brasitus and Schachter 1980a). While protein-lipid interaction has previously been studied in a number of other plasma membranes (Fox 1975, Lee 1975), until recently, these interactions had not been examined in the plasma membranes of the rat enterocyte. The functional interaction between protein and lipids of these membranes, using steady state fluorescence polarization (SSFP) and differential scanning calorimetry (DSC), have now been studied and the results serve as the basis for this review.

Lipid-Protein Interactions in the Microvillus Membranes

Schachter and Shinitzky (1977) first demonstrated that rat microvillus membranes possessed a low lipid fluidity [2], as determined by SSFP studies, utilizing the lipid

1 Department of Medicine, Columbia University, College of Physicians and Surgeons, New York, NY 10032, USA

2 The term "lipid fluidity" as applied to model bilayers and natural membranes is used to express the relative motional freedom of the lipid molecules or substituents thereof. It bears emphasis, however, that this term is broad and includes different types of motion, e.g., rotational or lateral diffusion of a molecule in an array, movements of substituent groups of a molecule, and flow of molecules under a pressure gradient in accord with a fluidity which is 1/viscosity of the molecular array (Blank 1976, Lakowicz et al. 1979a,b). In this review "lipid fluidity" was assessed by the steady state fluorescence polarization of lipid soluble probes. The anisotropy parameters so obtained are probe dependent and reflect the overall motional freedom of these molecules without distinguishing the specific mechanisms affecting its motion such as viscous drag of the environment, anisotropic rotations and hindred motions due to structural factors (Lackowicz et al. 1979a,b, Chen et al. 1977, Kinosita et al. 1977, Brasitus and Schachter 1980a)

Intestinal Transport
(ed. by M. Gilles-Baillien and R. Gilles)

soluble fluorophor 1,6-diphenyl-1-3-5-hexatriene (DPH). The degree of fluorescence polarization of DPH, as assessed by the anisotropy parameter[3] $[({}^{r}o/r) - 1]^{-1}$ was shown to be the highest yet reported in a mammalian membrane. Both membrane proteins and lipids appeared to contribute to this low fluidity. The fluidity of this membrane was shown to decrease with the distance of the intestinal segment from the pylorus, i.e., the ileal segment had the lowest fluidity.

These authors also demonstrated that this membrane's lipid exhibited a characteristic thermotropic phase transition[4] at 26 ± 2°C. Membrane protein, while influencing fluidity, was not shown to influence this transition temperature. Since the mean transition temperature was below 37°C, its physiological significance was unclear at that time. Further studies, however, by Brasitus et al. (1980), using SSFP and DSC, revealed that this membrane exhibited a broad reversible lipid thermotropic transition in the range of 23°–39°C with an enthalpy of approximately 0.1 cal g^{-1} (Fig. 1). Total lipid extracts of this membrane, both hydrated and unhydrated, yielded similar transitions but with enthalpies that exceeded the values of the intact membrane (Fig. 2). SSFP studies appeared to detect only the lower critical temperature of the transition, possibly because they were only performed up to 40°C (Fig. 3). Cholesterol and proteins present in this membrane were shown to influence the enthalpy observed in the intact membrane. These studies demonstrated that the microvillus

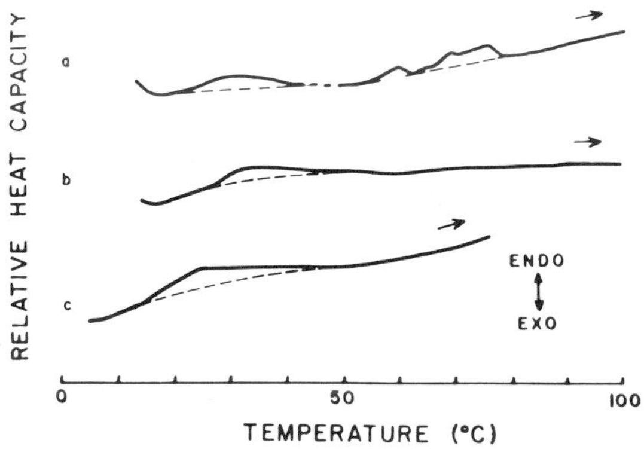

Fig. 1. Differential scanning calorimetry heating curves of *a* intact microvillus membranes, *b* the same sample following heating to 100°C, and *c* hydrated lipids of microvillus membranes. The sample contained 4.5 mg of lipid in *a* and *b* and 3 mg of lipid in *c*. Samples were heated at 5°C min^{-1}. Sensitivity settings for *a* were 0.1 mcal s^{-1} at temperatures *below the dashed curve* (i.e., in the region of the lipid transition) and 1.0 mcal s^{-1} at the higher temperatures (i.e., in the region of the denaturation endotherms). Sensitivity settings for *b* and *c* respectively were 0.1 and 0.2 mcal s^{-1}. *Endo* and *exo* represent the directions of endothermic and exothermic transitions. (Brasitus et al. 1980)

3 $[({}^{r}o/r) - 1]^{-1}$, i.e., the anisotropy parameter was calculated using a value of o = 0.362 for DPH as previously described (Schachter and Shinitzky 1977, Shinitzky and Barrenholz 1974)

4 The term "thermotropic phase transition" is used to denote a thermally induced change in the physical state of the membrane lipids. The changes may involve order-disorder transitions of the liquid-crystalline to gel type, lateral phase separations, or other mechanisms (Lee 1975, Brasitus et al. 1979)

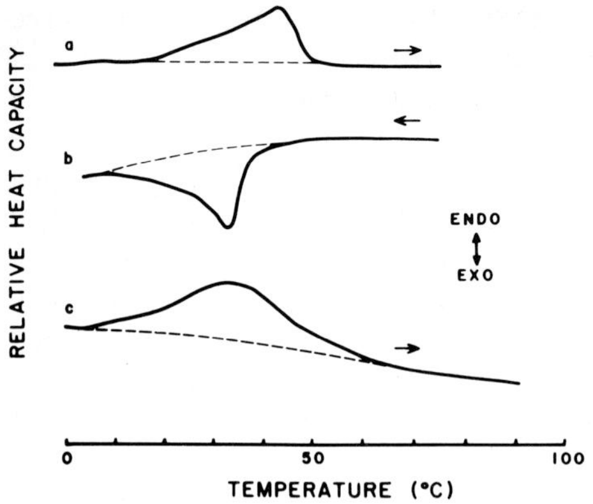

Fig. 2. Differential scanning calorimetry heating curve *(a)* and cooling curve *(b)* of dried microvillus membrane lipids and heating curve *(c)* or dried basolateral membrane lipids. The samples contained 0.84 mg of lipid in *a* and *b* and 0.70 mg of lipid in *c*. All samples were heated or cooled at 5°C min^{-1} at a sensitivity of 0.2 mcal s^{-1}. *Endo* and *exo* represent the directions of endothermic and exothermic transitions. (Brasitus et al. 1980)

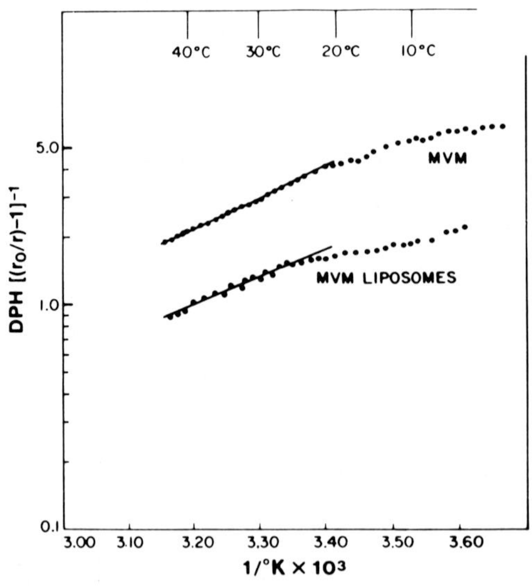

Fig. 3. Arrhenius plot of the anisotropy parameter of diphenylhexatriene in a sample of isolated microvillus membranes *(MVM, upper curve)* and in liposomes prepared from a lipid extract of these membranes *(lower curve)*. (Brasitus et al. 1980)

membrane functioned in vivo near or at the upper critical temperature of the phase transition. While the physiological importance of this arrangement is unknown, previous studies have shown that the lateral compressibility of membrane lipids is influenced when the temperature is lowered through the upper portion of a lipid transition and two phases exist (Linden et al. 1973, Lee 1975). This increased compressibility presumably might facilitate the insertion of substances into the membrane and enhance membrane assembly and/or transport.

In *Escherichia coli* and *Mycoplasma* membranes, which possess well-defined lipid thermotropic transitions, it had previously been shown that certain protein-dependent membrane activities, in particular, transmembrane transport mechanisms are influenced by temperature-dependent changes in the physical state of the membrane lipid (Overath et al. 1970, Wilson et al. 1970, Rottem et al. 1973). Brasitus et al. (1979) and Brasitus and Schachter (1980b), therefore, conducted experiments on the ability of the lipid fluidity to influence enzyme activities in the rat microvillus membrane. Under V_{max} conditions, enzyme activities were determined between 10° and 40°C and Arrhenius plots constructed for each enzyme. This membrane's enzyme activities could be categorized into two groups. Enzymes of the first group: sucrase, maltase, lactase, trehalase, leucine aminopeptidase and gamma-glutamyl transpeptidase were shown to exhibit a single slope in their Arrhenius plots from 10°–40°C and appeared not to be functionally influenced by the effects of the lipid thermotropic transition. These enzymes were operationally defined as extrinsic activities. Enzymes of the second group, however, including guanylate cyclase, calcium-dependent adenosine triphosphatase, p-nitrophenyl phosphatase and D-glucose transport across membrane vesicles, all demonstrated a break in the slope of their Arrhenius plot, i.e., a change in energy of activation, in the range of 25°–30°C, corresponding to the region of the lower critical temperature of the lipid transition. This group was operationally defined as intrinsic activities.

To further confirm that the lipid influenced these intrinsic activities, p-nitrophenyl phosphatase was delipidated. This resulted in loss in the discontinuity of the slope in the Arrhenius plot. Subsequent relipidation of this enzyme, using extracted microvillus membrane lipid, restored the breakpoint to its original temperature, i.e., 25°–29°C (Brasitus et al. 1979). Relipidation, using seven other lipids with varying thermotropic transition temperatures, however, also uniformly restored the breakpoint in the range of 25°–29°C. Delipidation of a number of enzymes including *Azotobacter* nitrogenase (Ceuterick et al. 1978), rabbit kidney sodium-potassium dependent adenosine triphosphatase (Na^+ K^+ ATPase) (Kimelberg and Papahadjopoulos 1974) and calcium-dependent adenosine triphosphatase of sarcoplasmic reticulum (Anzai et al. 1978) have also resulted in breakpoints which did not correspond completely to the phase transition temperature of the lipid used. The exact explanation for this behavior is unknown. Some authors (Thorneley et al. 1975, Anzai et al. 1978, Gomez-Fernandez et al. 1979) have emphasized the importance of temperature-induced conformational changes or alterations in aggregation of proteins, processes which presumably require a lipid environment. Others (Ceuterick et al. 1978) have stressed the role of the annular lipids and the strengths of their interaction with the protein as determinants of the break temperature. These studies would suggest that in the intact membrane, the bulk and annular lipids appear to be of similar composition, in as much as different activities such as glucose transport and divalent cation ATPases show a similar breakpoint in their Arrhenius plots and this breakpoint corresponds to the thermotropic transition of the lipid. After delipidation, the solubilized enzyme retains a portion of its annular lipid. The discontinuity, however, is lost, because it apparently requires a critical mass of lipid. Upon reconstitution with the various lipids, the discontinuity is restored, with the temperature of the breakpoint determined in each instance by the endogenous annular lipid (Brasitus et al. 1979).

Cholesterol Biosynthesis and Modulation of Membrane Cholesterol and Lipid Dynamics in Rat Ileal Microvillus Membranes

Previous studies (Schachter et al. 1976, Schachter and Shinitzky 1977), using SSFP, have demonstrated that the lipid fluidity of rat microvillus membranes was least in the distal (ileal) segment of intestine and increased in the proximal segments. Corresponding to these studies, it has also been shown that both the rate of incorporation of precursors into cholesterol and the specific activity of 3-hydroxy-3-methylglutaryl coenzyme A (HMG-CoA) reductase, the rate limiting enzyme of cholesterol synthesis, were greatest in ileal segments compared to their proximal counterparts (Dietschy and Siperstein 1965, Merchant and Heller 1977). Sinensky (1977, 1978) has also described a Chinese hamster ovary cell mutant which is defective in the regulation of cholesterol synthesis by exogeneous cholesterol. An increase in cholesterol of the ambient medium, failed to reduce cholesterol biosynthesis by this mutant and resulted in an increase in cholesterol content in its plasma membrane. The parental cell line, however, under identical conditions, decreased its biosynthetic rate and maintained normal levels of cholesterol in its membrane. The author suggested that "the role of cholesterol-synthesizing enzymes of the mammalian fibroblast was to regulate membrane fluidity" (Sinensky 1978).

Brasitus and Schachter (1982), therefore, investigated the role of cholesterol biosynthesis in rat ileal enterocytes, in vivo, in regulating plasma membrane fluidity and cholesterol composition. Using procedures described by Andersen and Dietschy (1977), increased synthesis was induced by biliary ligation or by feeding cholestryramine, whereas decreased synthesis was elicited by feeding sodium taurocholate or fasting the animals. The results obtained by these studies (Brasitus and Schachter 1982) demonstrated that cholesterol biosynthesis in the rat enterocyte appeared to modulate the cholesterol content of the microvillus membrane as well as its lipid fluidity. Similar effects on the basolateral membrane of these cells could not be demonstrated.

Additionally, Na^+-dependent flux of D-glucose into ileal microvillus membrane vesicles was shown to increase under conditions where this membrane's fluidity was increased, suggesting that this transport mechanism normally operates below its full capacity. The effect was apparently not due to a general increase in membrane permeability since the D-glucose flux independent of sodium was not effected.

Lipid-Protein Interactions in the Basolateral Membranes

Rat enterocyte basolateral membrane lipid was also shown to undergo a broad reversible thermotropic phase transition between 27° and 40°C, as assessed by DSC studies (Brasitus et al. 1980) (Figs. 4 and 2c). The membrane enthalpy values were shown to be low (0.15 cal g^{-1}) and were increased in both the hydrated and unhydrated extracted lipid from this membrane. Protein and cholesterol present in the basolateral membrane appeared to influence these enthalpies. Arrhenius plots of the anisotropy parameter of DPH in these membranes and their liposomes revealed breaks in their

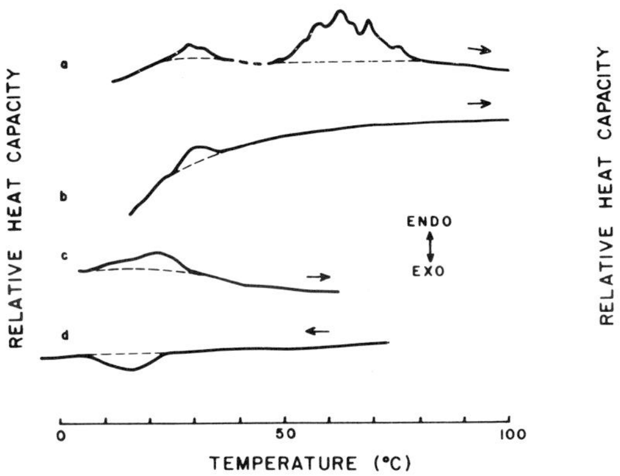

Fig. 4. Differential scanning calorimetry heating curves of *a* intact basolateral membranes, *b* the same sample following heating to 100°C and *c* hydrated lipids of basolateral membranes. In *d* is shown the cooling curve of the hydrated lipids. The sample in *a* and *b* contained 1.3 mg of lipid and that in *c* and *d* contained 2.3 mg of lipid. Samples were heated or cooled at 5°C min^{-1}. Sensitivity settings for *a* were 0.1 mcal s^{-1} *below the dashed curve* (i.e., in the region of the lipid transition) and 0.2 mcal s^{-1} *above the dashed curve* (i.e., for the denaturation endotherm). The sensitivity setting for *b* was 0.1 mcal s^{-1} and for *c* and *d* it was 0.2 mcal s^{-1}. *Endo* and *exo* represent the directions of endothermic and exothermic transitions. (Brasitus et al. 1980)

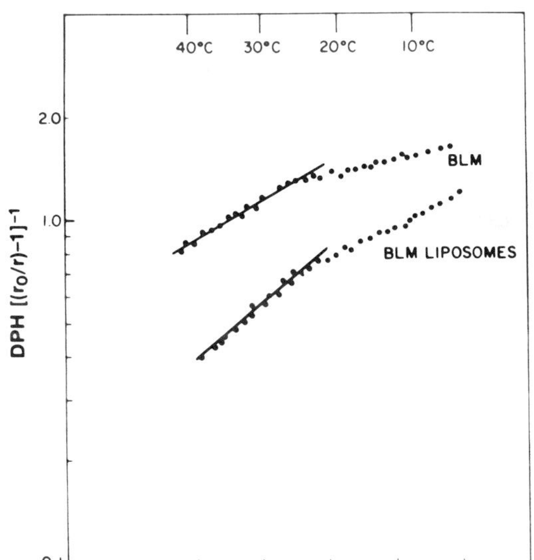

Fig. 5. Arrhenius plots of the anisotropy parameter of diphenylhexatriene in a sample of basolateral membranes *(BLM)* and in liposomes prepared from a lipid extract of these membranes *(lower curve)*. (Brasitus et al. 1980)

slopes at 26° ± 1.5°C and 25° ± 1.7°C, respectively (Fig. 5). These breakpoint temperature again corresponded closely to the lower critical temperature of the phase transition detected by DSC (Brasitus et al. 1980).

Arrhenius plots of this membrane's enzymes provided evidence for two groups of intrinsic activities (Brasitus and Schachter 1980a). 5′nucleotides and basal and stimulated adenylate cyclase (Naf and prostaglandin E_1) showed breaks in their plots at 28°–30°C, corresponding to the midpoint of the transition: Na^+K^+ ATPase, ouabain-sensitive potassium dependent p-nitrophenyl phosphatase and magnesium-dependent adenosine triphosphatase showed breaks at 20°–22°C, i.e., approximately 5°–7°C below the lower critical temperature of the transition. These latter enzymes appeared to be influenced by strong lipid-protein interactions, which lowered the breakpoint temperature of the Arrhenius plot in relationship to the bulk temperature, either by triggering temperature induced changes in protein structure or by providing a micro-environment of increased lipid fluidity as emphasized by previous authors (Thorneley et al. 1975, Anzai et al. 1978, Gomez-Fernandez et al. 1979).

Comparison of the Lipid Fluidity of Microvillus and Basolateral Membranes

Comparison of the anisotropy parameter values of basolateral membranes and their liposomes (Fig. 5) to microvillus membranes and their liposomes, using DPH or other fluorescent probes (Table 1), in each case indicated that lipid molecules in basolateral membranes experience less restraint to motional freedom, and therefore, basolateral membranes had greater "lipid fluidity" than microvillus membranes (Brasitus and Schachter 1980a). Subsequent studies (Gray et al. 1981), using electron spin resonance techniques, have confirmed these findings.

Table 1. Fluorescence polarization and excited-state lifetime studies

Probe	Membrane type	Anisotropy parameter [a] $[(r_0/r) - 1]^{-1}$	P	Mean fluorescence anisotropy, r	Excited-state lifetime [b] τ (ns)	App. rotational relaxation time [c] ρ (ns)
Diphenylhexatriene	Microvillus	3.69 ± 0.14 (16)	<0.001	0.285	11 (3)	133
	Basolateral	1.53 ± 0.10 (11)		0.219	11 (3)	51
Retinol	Microvillus	3.00 ± 0.35 (5)	<0.01	0.275	8 (1)	72
	Basolateral	1.52 ± 0.10 (5)		0.221	8 (1)	37
12-Anthroylstearate	Microvillus	0.62 ± 0.04 (9)	<0.01	0.109	14 (2)	26
	Basolateral	0.40 ± 0.04 (10)		0.081	16 (2)	19
2-Antrhoylstearate	Microvillus	0.89 ± 0.04 (5)	<0.02	0.134	12 (2)	32
	Basolateral	0.61 ± 0.02 (5)		0.108	12 (2)	22
Dansylphosphatidyl-ethanolamine	Microvillus	1.02 ± 0.09 (5)	<0.025	0.162	14 (2)	43
	Basolateral	0.80 ± 0.01 (9)		0.142	14 (2)	34

[a] Values are means ± SE; values in parentheses are the numbers of preparations examined. P values for the difference between microvillus and basolateral membranes are indicated

[b] Values are means; values in parentheses are the numbers of preparations examined

[c] Calculation discussed under Experimental Procedures: Brasitus and Schachter (1980a)

Table 2. Composition of intestinal cell plasma membranes

	MVM	BLM
Protein/lipid [a]	2.02	0.60
Cholesterol/phospholipid [b]	0.87	0.63
Sphingomyelin/lecithin [b]	0.62	0.43

[a] MG/MG
[b] mol/mol

Differences in lipid dynamics of the antipodal membranes appeared to be secondary, at least in part, to differences in their composition. As can be seen in Table 2, the microvillus membrane has higher ratios of protein/lipid (w/w), cholesterol/phospholipid (mol/mol) and sphingomyelin/lecithin (mol/mol) than the basolateral membranes. An increase in the molar ratios of cholesterol/phospholipid and sphingomyelin/lecithin have previously been correlated with a decrease in lipid fluidity in model bilayers and natural membranes (Chapman and Penkett 1966, Hubbell and McConnell 1971, Shinitzky and Inbar 1976) as has the greater ratio (w/w) of protein/lipid (Schachter and Shinitzky 1977, Brasitus et al. 1980). Further studies on these membranes should hopefully clarify the mechanism responsible for maintaining these differences as well as elucidate the functional reasons for such differences.

References

Andersen JM, Dietschy JM (1977) Regulation of sterol synthesis in 16 tissues of rat. J Biol Chem 252:3646–3651

Anzai K, Kirino Y, Shimizu H (1978) Temperature-induced changes in the Ca^{2+}-dependent ATPase activity and in the state of the ATPase protein of sarcoplasma reticulum membrane. J Biochem (Tokyo) 84:815–821

Blank M (1976) In: Bolis L, Hoffman JF, Leaf A (eds) Membranes and diseases. Raven Press, New York, pp 81–88

Brasitus TA, Schachter D (1980a) Lipid dynamics and lipid-protein interactions in rat enterocyte basolateral and microvillus membranes. Biochemistry 19:2763–2769

Brasitus TA, Schachter D (1980b) Membrane lipids can modulate guanylate cyclase activity of rat intestinal microvillus membranes. Biochim Biophys Acta 630:152–156

Brasitus TA, Schachter D (1982) Cholesterol biosynthesis and modulation of membrane cholesterol and lipid dynamics in rat intestinal microvillus membranes. Biochemistry 21:2241–2246

Brasitus TA, Schachter D, Mamouneas TG (1979) Functional interactions of lipids and proteins in rat intestinal microvillus membranes. Biochemistry 18:4136–4144

Brasitus TA, Tall AR, Schachter D (1980) Thermotropic transitions in rat intestinal plasma membranes studied by differential scanning calorimetry and fluorescence polarization. Biochemistry 19:1256–1261

Ceuterick F, Peeters J, Hermans K et al. (1978) Effect of high pressure, detergents and phospholipase on the break in the Arrhenius plot of *Azotobacter* nitrogenase. Eur J Biochem 87:401–407

Chapman D, Penkett SA (1966) Nuclear magnetic resonance spectroscopic studies of the interaction of phospholipids with cholesterol. Nature (Lond) 211:1304–1305

Chen LA, Dale RE, Roth S, Brand L (1977) Nanosecond time-dependent fluorescence depolarization of diphenyl hexatriene in dimyristoyl lecithin vesicles and the determination of microviscosity. J Biol Chem 252:2163–2169

Dietschy JM, Siperstein MD (1965) Cholesterol synthesis by the gastrontestinal tract: Localization and mechanism of control. J Clin Invest 44:1311–1327

Douglas AP, Kerley R, Isselbacher KJ (1972) Preparation and characterization of lateral and basal plasma membranes of the rat intestinal epithelial cell. Biochem J 128:1329–1338

Forstner GG, Tanaka K, Isselbacher KJ (1968) Lipid composition of the isolated rat intestinal microvillus membrane. Biochem J 109:51–59

Fox CF (1975) In: Fox CF (ed) Biochemistry of cell walls and membranes, vol 2. University Park Press, Baltimore, pp 197–306

Fujita M, Kawai K, Asano S, Nakao M (1973) Protein components of two different regions of an intestinal epithelial cell membrane. Biochim Biophys Acta 307:141–151

Gomez-Fernandez JC, Goni FM, Bach D, Restall C, Chapman D (1979) Protein-lipid interactions. FEBS Lett 98:224–228

Gray JP, Henderson GI, Dunn GD et al. (1981) Membrane fluidity and lipid composition of normal rat enterocytes: Differentiation between brush border (BB) and basolateral membranes (BLM). Gastroenterology 80:1162A

Hubbell WL, McConnell HM (1971) Molecular motion in spin labeled phospholipids and membranes. J Am Chem Soc 93:314–326

Kawai K, Fujita M, Nakao M (1974) Lipid composition of different regions of intestinal epithelial cell membranes of mouse. Biochim Biophys Acta 369:222–233

Kimelberg HK, Papahadjopoulos D (1974) Effects of phospholipid acyl chain fluidity, phase transitions, and cholesterol on (Na^{+} + K^{+})-stimulated adenosine triphosphatase. J Biol Chem 249:1071–1080

Kinosita K, Kanato S, Ikegami A (1977) A theory of fluorescence polarization decay in membranes. Biophys J 20:289–305

Lakowicz JR, Prendergast FG, Hogen D (1979a) Differential polarized phase fluorometric investigations of diphenyl hexatriene in lipid bilayers: Quantitation of hindred depolaring rotations. Biochemistry 18:508–519

Lakowicz JR, Prendergast FG, Hogan D (1979b) Fluorescence anisotropy measurements under oxygen quenching conditions as a method to quantify the depolaring rotations of fluorophores: Application to dephenyl hexatriene in isotropic solvents and in lipid bilayers. Biochemistry 18:520–527

Lee AG (1975) Functional properties of biological membranes: A physical-chemical approach. Prog Biophys Mol Biol 29:3–56

Lewis BA, Gray GM, Coleman R et al (1975) Differences in the enzymatic, polypeptide, glycopeptide, glycolipid and phospholipid compositions of plasma membranes from the two surfaces of intestinal epithelial cells. Biochem Soc Trans 3:752–753

Linden CD, Wright KL, McConnell HM et al (1973) Lateral phase separations in membrane lipids and the mechanism of sugar transport in *Escherichia coli.* Proc Natl Acad Sci USA 70:2271–2275

Merchant JL, Heller RA (1977) 3-Hydroxy-3-methylglutaryl coenzyme A reductase in isolated villous and crypt cells of the rat ileum. J Lipid Res 18:722–733

Murer H, Hopfer U, Kinne-Saffran E, Kinne R (1974) Glucose transport in isolated brush-border and lateral basal plasma-membrane vesicles from intestinal epithelial cells. Biochim Biophys Acta 345:170–179

Oda T (1976) In: Yamada E, Mizuhira U, Jurosumi K, Nagano T (eds) Recent progress in electron microscopy of cells and tissues. University Park Press, Baltimore, pp 13–17

Okada Y, Irimajiri A, Inouye A (1977) Electrical properties and active solute transport in rat small intestine. J Membr Biol 31:221–232

Overath P, Schairer HU, Stoffel W (1970) Correlation of in vivo and in vitro phase transitions of membrane lipids in *Escherichia coli.* Proc Natl Acad Sci USA 67:606–612

Rose RC, Schultz SG (1971) Studies on the electrical potential across rabbit ileum. J Gen Physiol 57:639–663

Rottem S, Cirillo VP, Dekruyff B et al (1973) Cholesterol in mycoplasma membranes. Biochim Biophys Acta 323:505–519

Schachter D, Shinitzky M (1977) Fluorescence polarization studies of rat intestinal microvillus membranes. Clin Invest 59:536–548

Schachter D, Coagan U, Shinitzky M (1976) Interaction of retinal and intestinal microvillus membranes studied by fluorescence polarization. Biochim Biophys Acta 448:620–624

Shinitzky M, Barrenholz Y (1974) Dynamics of the hydrocarbon layer in liposomes of lecithin and sphingomyelin containing diacetylphosphate. J Biol Chem 249:2652–2657

Shinitzky M, Inbar M (1976) Microviscosity parameters and protein mobility in biological membranes. Biochim Biophys Acta 433:133–149

Sinensky M (1977) Isolation of a mammalian cell mutant resistant to 25-hydroxyl cholesterol. Biochem Biophys Res Commun 78:863–867

Sinensky M (1978) Defective regulation of cholesterol biosynthesis and plasma membrane fluidity in a Chinese hamster ovary cell mutant. Proc Natl Acad Sci USA 75:1247–1249

Thorneley RNF, Eady RR, Yates MF (1975) Nitrogenases of *Klebsiella pneumoniae* and *Azobacter chroococcum*. Biochem Biophys Acta 403:269–284

Wilson G, Rose SP, Fox CF (1970) The effect of membrane lipids unsaturation on glycoside transport. Biochem Biophys Res Commun 38:617–623

Part 3
Regulation of Intestinal Transport

Role of Cell Sodium in Regulation of Transepithelial Sodium Transport

K. TURNHEIM [1]

Introduction

How Do Epithelial Cells Transfer Greatly Varying Amounts of Ions Through Their Interior and Stay Alive?

One of the basic properties of living cells is that they maintain an internal electrolyte composition that is optimal for their metabolism and specific functions. For epithelial cells, whose specific function is active electrolyte transport, the problem arises how to preserve the internal ionic milieu and at the same time generate widely varying transcellular ion fluxes. Since transcellular transport involves movements across the two limiting membranes of epithelial cells, large changes in transcellular transport would be expected to profoundly affect the internal electrolyte composition and volume of the epithelial cells and hence threaten their survival, if the pumps and leaks in the membranes were invariant and independent. However, the internal ionic milieu may be kept within relatively narrow limits despite varying rates of transcellular transport, if the rates of movement across the apical or luminal membrane and the contraluminal or basolateral membrane are somehow coupled. In other words, when the rate of ion movement across one membrane changes, the internal composition will not be markedly altered, if similar changes in ion movement take place at the other membrane. Hence, maintenance of a relatively constant intracellular composition of electrolytes requires a mechanism which rapidly coordinates the pumps and leaks at the apical and basolateral cell membranes.

In the case of active transepithelial Na transport, intracellular Na has been discussed as a possible mediator of the interaction between the two limiting membranes. In a number of Na-transporting epithelia intracellular Na was reported to exert a negative feedback effect on the Na conductances of the apical membrane: According to this concept increased intracellular Na decreases the apical Na conductance and vice versa. Indeed, maneuvers that presumably increase intracellular Na decrease apical Na entry and increase apical membrane electrical resistance (MacRobbie and Ussing 1961, Essig and Leaf 1963, Erlij and Smith 1973, Hviid Larsen 1973, Leblanc and Morel 1975, Lewis et al. 1976, Helman et al. 1979, Narvarte and Finn 1980).

1 Pharmakologisches Institut, Universität Wien, Währinger Str. 13a, A-1090 Vienna, Austria

Intestinal Transport
(ed. by M. Gilles-Baillien and R. Gilles)

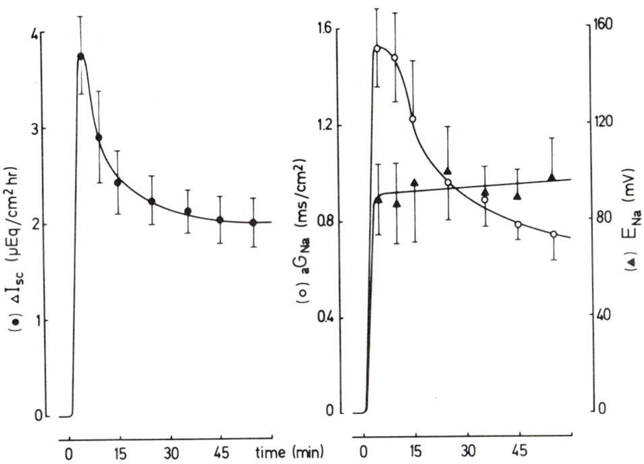

Fig. 1. Effects of suddenly exposing Na-depleted epithelia of rabbit descending colon to a solution containing 140 mM Na on amiloride-sensitive short-circuit current, ΔI_{sc} *(left)*, the conductance of the pathway for active transepithelial Na transport, $_aG_{Na}$, and the equivalent electromotive force for active transepithelial Na transport, E_{Na} *(right)*. (Turnheim et al. 1978, modified)

An example of the operation of the negative feedback between intracellular Na and apical Na conductance is illustrated in Fig. 1. Isolated epithelia of rabbit descending colon, mounted in Ussing-type chambers under short-circuit conditions, were depleted of Na by incubation for at least 1 h in a Ringer solution in which Na had been replaced by choline. When these tissues were suddenly exposed to Na by replacing the Na-free solution with regular Ringer containing 140 mM Na, there was a rapid surge in amiloride-sensitive short-circuit current, ΔI_{sc}, which represents Na current at least across the apical membrane (Turnheim et al. 1978). After a maximum at approximately 5 min after exposure to Na, ΔI_{sc} declined in a quasi-exponential fashion to reach a new steady state after 45–50 min. The transient in ΔI_{sc} was accompanied by similar changes in the amiloride-sensitive conductance of the tissue, $_aG_{Na}$, which stands for the conductance of the active Na transport pathway. Since the electrical resistance of the apical membrane is approximately 10 times higher than that of the basolateral membrane (Schultz et al. 1977), changes in $_aG_{Na}$ reflect, for the most part, changes in the Na conductance of the apical membrane. From the similar time courses of the changes in ΔI_{sc} and $_aG_{Na}$ it is clear that E_{Na}, the equivalent electromotive force of active transepithelial Na transport, remains essentially constant throughout the current transient, since E_{Na} is simply $\Delta I_{sc}/_aG_{Na}$.

These findings indicate that apical Na conductance is maximal when the tissues are totally depleted of Na. As intracellular Na increases, apical Na conductance decreases. The slow time course of this process argues against the notion that extracellular Na causes the decreaase in apical membrane conductance. Further increases in intracellular Na, produced either hy blocking basolateral exit or by opening artificial pores in the apical membrane, abolish the conductance of the physiological, i.e., amiloride-inhibitable Na pathway in the apical membrane (Turnheim et al. 1978);

Although the molecular mechanism of the variations in amiloride-sensitive apical Na conductance is not resolved, an apical membrane protein containing a superficially located sulfhydryl group seems to be involved, since titration of these sulfhydryl groups by macromolecular organic mercurials freezes the conductance of the apical Na entry pathway in a submaximal position and prevents its modulation by amiloride or variations in intracellular Na. Since sulfhydryl groups are known to control the conformation of proteins, it is conceivable that variations in apical membrane Na conductance are a consequence of conformational changes of a specific apical membrane protein (Luger and Turnheim 1981).

In short, in recent years a wealth of evidence has accumulated in favor of the notion, that intracellular Na plays an important role in the autoregulatory or homocellular mechanisms (i.e., mechanisms that reside in the epithelial cells themselves) that protect the intracellular ionic composition and control active transepithelial Na transport. But this evidence is largely indirect, intracellular Na activity, apical Na conductance or permeability, and transcellular Na transport have not been determined simultaneously. It is the aim of the present study to attempt a correlation of these parameters.

The Technical Approach: Current-Voltage Relations of the Apical Na Entry Step

The methodological problems to measure intracellular Na activities by use of ion-selective microelectrodes are still formidable especially in small cells, the difficulties arising from unsatisfactory Na selectivity of glass or liquid ion-exchanger electrodes, the size of the tip diameter and its electrical resistance, and the necessity to measure the electrical membrane potential difference with conventional microelectrodes, preferably in the same cells as impaled with the ion-selective microelectrodes (Civan 1978). In the present study an alternative approach was taken, using the highly selective apical Na entry mechanism itself as an electrode to measure intracellular Na activity. The notion that the apical barrier of Na-transporting epithelia has the properties of a Na-electrode was first proposed by Koefoed-Johnsen and Ussing (1958) and has been substantiated by a host of experimental evidence since then (Fuchs et al. 1977, Schultz et al. 1977, MacKnight et al. 1980, Schultz 1981). Thus, Na enters the epithelial cells via a Na-selective mechanism in the apical membrane, driven by the electrochemical potential gradient across this barrier. Subsequently Na is extruded across the basolateral cell membrane against a steep electrochemical gradient by a process consuming energy. Hence this basolateral Na transport mechanism, which is closely associated with an ouabain-sensitive Na- and K-activated ATPase, is also referred to as the Na pump. In their double-membrane model Koefoed-Johnsen and Ussing (1958) suggested that Na is transported across the basolateral membrane by an electrically neutral exchange for K, which therefore accumulates in the cell interior. A constant intracellular level of K is maintained despite ongoing transcellular Na transport because K leaks back to the inner or serosal side of the epithelium through a K-selective conductive pathway. Accordingly, the electrical potential difference across the apical membrane ψ^{mc}, is a Na diffusion potential and that across the basolateral membrane ψ^{cs}, a K diffusion potential. This now classical model for Na-transporting cells has withstood the test of time with only slight modifications. These modifications concern the mechanism of apical Na entry which is not a simple diffusion process but which was repeatedly shown to be saturable (Biber and Curran 1970, Frizzell and Turnheim 1978, MacKnight et al. 1980), the basolateral Na-K exchange which appears not to be electrically neutral (Schultz 1977, MacKnight et al. 1980), and the involvement of an extracellular shunt pathway which determines the polarity of the intracellular electrical potential with respect to the extracellular solution (Schultz 1972).

It has been shown experimentally (Fuchs et al. 1977, Palmer et al. 1980, Thompson et al. 1982) that apical Na entry is indeed downhill and results from electrodiffusion which closely conforms to the Goldman-Hidgkin-Katz constant field flux equation for a single permeant ion (Goldman 1843; Hodgkin and Katz 1949),

$$I_{Na}^{m} = \frac{P_{Na}^{m} F^2 \psi^{mc}}{RT} \left[\frac{(Na)_m - (Na)_c \exp^{-F\psi^{mc}/RT}}{1 - \exp^{-F\psi^{mc}/RT}} \right] \qquad (1)$$

where I_{Na}^{m} is the rate of apical Na entry, P_{Na}^{m} the Na permeability of the apical membrane, and ψ^{mc} the transapical electrical potential difference; $(Na)_m$ and $(Na)_c$ represent the Na activity in the luminal solution and the intracellular space, respectively. F, R, and T have their conventional meanings. Since the constant field flux equation expresses ionic current as a function of the ionic permeability of the membrane in question, the ion activities on the two sides of the membrane, and the transmembrane electrical potential difference, estimates of both P_{Na}^{m} and $(Na)_c$ can be derived from the relations of Na current and voltage across the apical cell membrane.

Transepithelial current-voltage (I–V) relations can be obtained by generating a staircase of increasing rectangular current pulses of alternating polarity across isolated epithelia mounted in Ussing-type chambers. In order to record the I–V relations of the apical membrane rather than those of the entire tissue it is necessary to monitor ψ^{mc} by use of conventional microelectrodes. Alternatively, the apical membrane may be voltage-clamped directly, if the resistance in series to the apical membrane is reduced to negligible values. Under these conditions a pulse of transepithelial current will produce a voltage step only at the apical membrane so that the transepithelial electrical potential difference, ψ^{ms}, will be very nearly equal to the electrical potential difference across the apical membrane, ψ^{mc}. As pointed out above, the basolateral membrane behaves like a K-electrode, therefore increasing the K concentration of the solution bathing the inner or serosal surface of the epithelium may be expected to depolarize the basolateral barrier and markedly decrease its electrical resistance. Evidence that this is indeed the case has been presented for frog and toad skin (Rawlins et al. 1970, Fuchs et al. 1977) and amphibian and mammalian urinary bladder (Higgins and Frömter 1974, Lewis et al. 1978, Palmer et al. 1980). Direct experimental justification for this non-invasive technique to evaluate the electrical properties of the apical membrane was also provided for rabbit descending colon, which will be used in the present investigation: Thompson et al. (1982) have shown using intracellular microelectrodes that in the presence of a high K solution on the serosal side of the epithelium the electrical potential difference across the basolateral membrane is essentially zero so that $\psi^{mc} \approx \psi^{ms}$ at all values of transepithelial currents and that the resistance of the basolateral membrane is markedly reduced. Further, these authors have shown an excellent agreement between the I–V relations of the apical membrane determined with the serosal high K technique or with the technique using an intracellular microelectrode to measure ψ^{mc} in nondepolarized tissues. In short, under conditions of serosal high K the electrical properties of the epithelium are determined primarily by those of the apical membrane.

The Tissue: Rabbit Descending Colon

The epithelium of rabbit descending colon has similar properties as other so-called tight Na-transporting epithelia such as amphibian skin, amphibian and mammalian urinary bladder and mammalian distal nephron: Under steady-state conditions ψ^{ms} and short circuit current, I_{sc}, are totally attributable to active transpithelial Na transport in the direction of absorption (Frizzell et al. 1976). Basolateral addition of aldosterone stimulates (Frizzell and Schultz 1978), but luminal addition of amiloride blocks (Frizzell and Turnheim 1978, Turnheim et al. 1981) active Na transport. Further positive effectors of active Na transport in rabbit descending colon are, from the luminal side of the epithelium, certain anions such as sulfate and isethionate (Turnheim et al. 1977), the polyene antibiotic amphotericine B (Frizzell and Turnheim 1978), and sulfhydryl reagents when Na absorption is initially low (Luger and Turnheim 1981); negative effectors of active Na transport include, from the luminal side, sulfhydryl reagents when Na absorption is initially high (Luger and Turnheim 1981) and from the serosal side, ouabain, sulfhydryl reagents, and omission of K.

In addition to active Na transport rabbit descending colon actively absorbs Cl by an electrically neutral mechanism, probably via a Cl-HCO_3 exchange located in the apical membrane (Frizzell et al. 1976). Stimuli which increase tissue cyclic AMP levels cause rheogenic (or electrogenic) Cl secretion (Frizzell et al. 1980, Grasl and Turnheim 1982). The cells which are responsible for Cl secretion appear not to be involved in active Na absorption.

The present report will focus only on factors residing within the epithelial cells that control active transepithelial Na transport. External regulatory mechanisms that may influence intestinal Na transport such as hormones, neurotransmitters, etc., will not be dealt with. These external factors are discussed, at least in part, in other chapters of this volume.

Methods

Briefly, segments of descending colon were obtained from white rabbits (2–3 kg) which had been killed by intravenous pentobarbital. The colon was opened along its mesenteric border and rinsed free of luminal contents. The outer muscle layers were stripped off by blunt dissection, and the resulting "partial mucosal strip" preparations were mounted vertically between the two halfs of Ussing-type chambers, the exposed surface being 1.27 cm^2. Both chamber halfs were perfused in a recirculating fashion by 10 ml of electrolyte solutions maintained at 37°C. The composition of the standard electrolyte solution bathing the luminal surface of the epithelium was (in mM) 140 Na, 124 Cl, 21 HCO_3, 5.4 K, 2.4 HPO_4, 0.6 H_2PO_4, 1.2 Mg, and 1.2 Ca. Solutions with lower Na concentrations were prepared by isosmolar replacement of Na by choline. The serosal solution contained 140 K, 25 Cl, 46.7 SO_4, 21 HCO_3, 2.4 HPO_4, 0.6 H_2PO_4, 1.2 Mg, 1.2 Ca, and 57.1 mannitol. All electrolyte solutions contained additionally 10 mM glucose and were gassed with a mixture of O_2 and CO_2 (95 to 5%), resulting in a pH of 7.4.

The tissues were short-circuited using an automatic voltage clamp which was controlled by a small-core computer. I–V relations were obtained by voltage-clamping the tissues in stepwise increments of 20 mV over the range of 0 to + and – 200 mV and recording the corresponding currents. Each rectangular current pulse had a duration of 100 ms with a 400 ms interval between pulses at which ψ^{ms} was zero. The clamping current, I^{ms}, and voltage, ψ^{ms}, were sampled 16 ms after the onset of each pulse; the resulting I–V relations may therefore be termed "instantaneous" or "near-instantaneous". A detailed description of the computer-driven voltage clamp and pulse generator has been reported previously by Thompson et al. (1982).

Usually several tissues from a single animal were mounted, and I–V relations were obtained with 140, 70, 35, 17.5, or 8.75 mM Na in the luminal solution.

Results and Discussion

Regulation of Apical Na Entry

An example of the near-instantaneous transepithelial I–V relations of rabbit colonic epithelium in which the basolateral barrier was functionally removed by a serosal high K solution is illustrated in the upper half of Fig. 2. A rigorous discussion of the distinction between "near-instantaneous" and "steady-state" I–V relations is given by Schultz et al. (1981). The I–V relations appear to be linear over a wide range of voltages, the slope of the regression representing the tissue conductance.

The relations between current and voltage across the entire tissue were determined both in the absence and presence of 0.1 mM amiloride in the luminal solution. Clearly, amiloride decreases tissue conductance. The difference in currents in the absence and presence of 0.1 mM aminoloride, ΔI^{ms}, at a given transepithelial voltage represents the Na current across the apical membrane, I^{m}_{Na}, since amiloride in the concentration used completely blocks active transepithelial Na transport and apical Na entry in this tissue (Frizzell and Turnheim 1978, Turnheim et al. 1981). Plotting I^{m}_{Na} versus ψ^{mc}, which under the conditions of the present experiments may be set equal to ψ^{ms} as discussed above, yields the I–V relation of the amiloride-sensitive Na entry pathway of the apical membrane, which is clearly non-linear (lower half of Fig. 2). The intersect of this curvilinear relation with the abscissa is by definition the reversal potential, E^{m}_{Na}, at which the Na current across the apical membrane is zero. On the other hand, I^{m}_{Na} at $\psi^{mc} = 0$ is the spontaneous rate of active Na transport under short-circuit conditions.

The I–V relations of the apical Na entry step were evaluated by fitting the parameters P^{m}_{Na} and $(Na)_c$ of the Goldman-Hodgkin-Katz constant field flux equation for a single permeant ion [see Eq. (1)] to the experimental values using non-linear regression analysis. In most cases, the non-linear least-square approximation was performed only over the range of voltages over which the data conformed closely to the constant field flux equation. This range was usually – 120 to + 100 mV. The values of $(Na)_c$ obtained in this manner most likely are good estimates of the steady-state intracellular Na activity, since the short duration and alternating polarity of the pulses render changes in ionic composition of the cell interior unlikely.

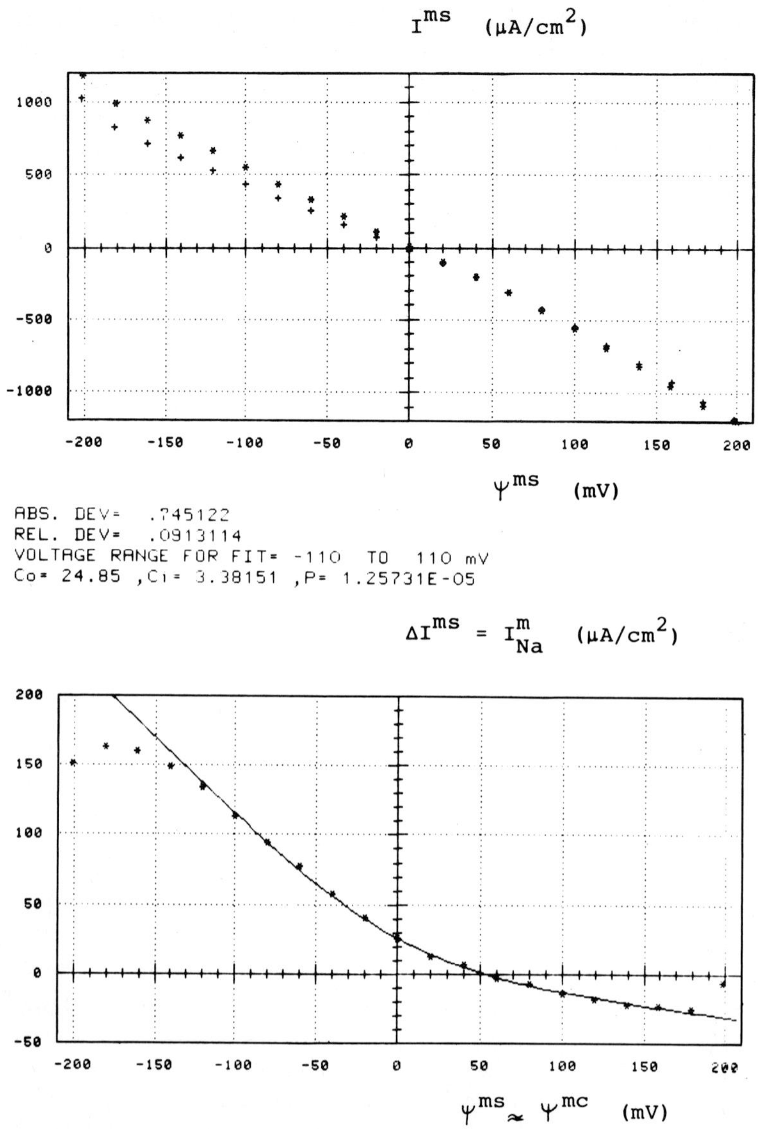

Fig. 2. Example of the current-voltage relations of rabbit colonic epithelium in the presence of serosal high K. *Upper half:* Relations between the transepithelial current, I^{ms}, and the transepithelial electrical potential difference, ψ^{ms}, in the absence (*) and presence of 0.1 mM amiloride (+). *Lower half:* Relations between the amiloride-inhibitiable transepithelial current, ΔI^{ms}, which is identical to the Na current across the apical membrane, I^{m}_{Na}, and ψ^{ms}, which is approximately equal to ψ^{mc}. The curve was calculated by non-linear regression analysis within the indicated voltage range, underlying Eq. (1). The most probable values for the intracellular Na acticity and apical Na permeability obtained in this manner are denoted by C_i (mM) and P (cm s^{-1}), respectively. C_o (mM) is the Na activity in the luminal solution. *ABS. DEV* and *RED. DEV,* absolute and relative deviation between the experimental data points and the calculated curve

$(Na)_c$ at a luminal Na activity, $(Na)_m$, of 99.4 mM was approximately 12 mM. The Na activity coefficient of the solutions bathing the luminal surface of the epithelium was taken to be 0.71 (Robinson and Stokes 1959). Reducing $(Na)_m$ resulted in a decrease in $(Na)_c$, the relation resembling saturation kinetics (Fig. 3). Hence a maximum $(Na)_c$ of 14.4 mM may be extrapolated; $(Na)_m$ at which $(NA)_c$ is half-maximal was approximately 20 mM.

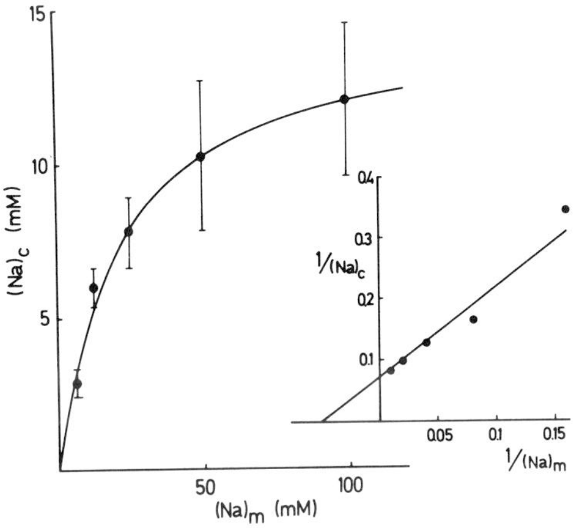

Fig. 3. Dependence of the intracellular Na activity, $(Na)_c$, on the Na activity in the luminal solution, $(Na)_m$. *Inset:* Double reciprocal plot of the relation between $(Na)_c$ and $(Na)_m$

A similar dependence of intracellular Na on the luminal Na concentration was observed in toad urinary bladder (Frazier et al. 1962), *Necturus* proximal nephron (Spring and Giebisch 1977, Kimura and Spring 1979), and *Necturus* gallbladder (Garzia-Diaz and Armstrong 1980), using chemical and radioisotope techniques or Na-selective microlelectrodes.

P^m_{Na} at $(Na)_m$ = 99.4 mM averaged 2.7×10^{-6} cm s^{-1}. From I–V relations of apical Na entry P^m_{Na} of frog skin was estimated to be 15×10^{-6} cm s^{-1} at $(Na)_m$ = 28 mM (Fuchs et al. 1977), 5×10^{-6} cm s^{-1} at $(Na)_m$ = 14 mM in toad urinary bladder (Palmer et al. 1980), and 8×10^{-6} cm s^{-1} at $(Na)_m$ = 34 mM in *Necturus* urinary bladder (Thomas, Suzuki, Thompson, and Schultz, unpublished data).

A reduction in $(Na)_m$ was accompanied by a hyperbolic increase in P^m_{Na} (Fig. 4). The relation between $(Na)_m$ and P^m_{Na} can be linearized by plotting $1/P^m_{Na}$ versus $(Na)_m$. In this manner a maximum P^m_{Na}, i.e., P^m_{Na} at $(Na)_m$ = 0, of approximately 13×10^{-6} cm s^{-1} may be obtained. $(Na)_m$ at which P^m_{Na} is half-maximal was 24 mM. In other words, there is very good agreement between the values of $(Na)_m$ at which the changes in $(Na)_c$ and P^m_{Na} were half-maximal, suggesting that $(Na)_c$ may be involved in the regulation of P^m_{Na}.

A similar relation between P^m_{Na} and $(Na)_m$ was found in frog skin (Fuchs et al. 1977), *Necturus* proximal nephron (Spring and Giebisch 1977), and toad and *Necturus* urinary bladder (Palmer et al. 1982, Thomas, Suzuki, Thompson, and Schultz, unpublished data).

Figure 5 is a plot of P^m_{Na} versus the corresponding values of $(Na)_c$ of the individual experiments. For the sake of clarity in this illustration the experiments were divided

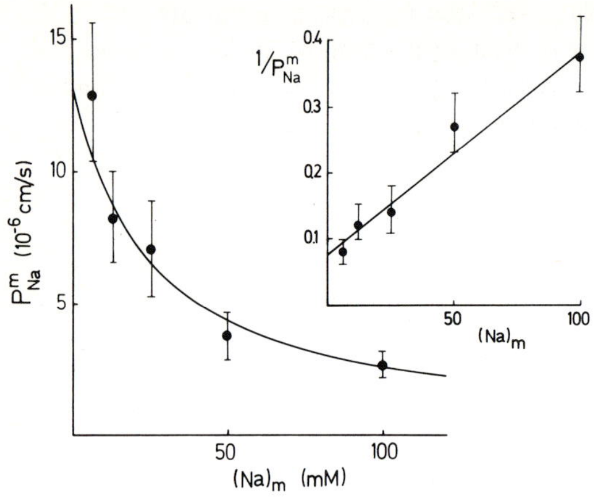

Fig. 4. Dependence of the apical Na permeability, P^m_{Na}, on $(Na)_m$. *Inset:* Plot of the reciprocal of P^m_{Na} versus $(Na)_m$

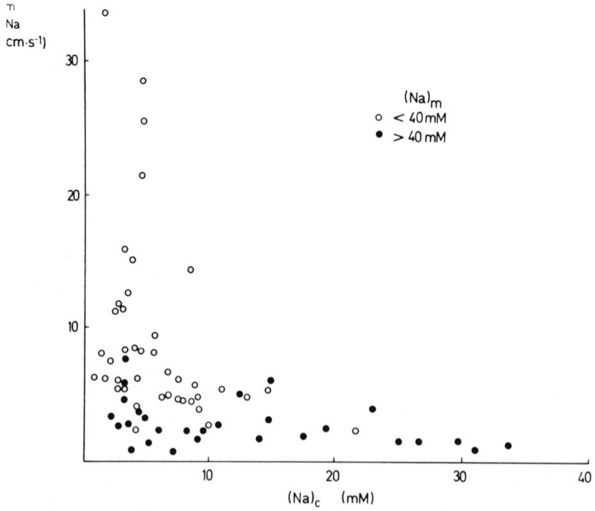

Fig. 5. Relation between P^m_{Na} and $(Na)_c$ of the individual experiments

into only two groups, a group of tissues exposed to high luminal Na and a group exposed to low luminal Na. Obviously, high intracellular Na activities were always associated with low apical Na permeabilities, but a low intracellular Na activities both high and low permeabilities were observed. Hence this finding is consistent with the notion that high $(Na)_c$ decreases P^m_{Na}, whereas at low values of $(Na)_c$ other factors may additionally control P^m_{Na}. But it should be stressed that the value of $(Na)_c$ derived from the I–V relations of the apical Na entry step represents the Na activity in the socalled "intracellular Na transport pool", i.e., the intracellular Na that is in the process of transcellular transport. Hence it cannot be excluded that the level of Na in an intracellular compartment which does not take part directly in transcellular Na transport differs from the level of Na in the transport pool. In fact, it was suggested earlier that the transcellular pathway responsible for active Na transport runs parallel to another intracellular Na compartment and that this cytoplasmic Na compartment exerts the negative feedback on apical Na entry (Turnheim et al. 1978). However, there is very little, if any, hard experimental evidence at present supporting the notion that there indeed exists a Na transport pool which is distinct from the cytoplasmic Na compartment.

When the apical membrane is short-circuited, Na entry may be described by

$$ {}_{o}I^{m}_{Na} = {}_{o}G^{m}_{Na}\, E^{m}_{Na}, \tag{2}$$

where G^{m}_{Na} is the conductance of the apical Na entry pathway, and E^{m}_{Na} is the electromotive force for apical Na entry which is given by the Nernst equation since the mechanism of apical Na entry was shown to be simple electrodiffusion; the subscript o denotes the condition $\psi^{mc} = 0$. ${}_{o}G^{m}_{Na}$ was determined from the slope of the I–V relation of the apical Na entry step at $\psi^{mc} = 0$ and must therefore be termed slope conductance in the sense of Finkelstein and Mauro (1977).

The relation between ${}_{o}I^{m}_{Na}$ and ${}_{o}G^{m}_{Na}$ at $(Na)_m = 99.4$ mM is given in Fig. 6. Clearly, ${}_{o}I^{m}_{Na}$ and ${}_{o}G^{m}_{Na}$ were linearly correlated. It is therefore concluded that the rate of apical Na entry is a function of the conductance of the amiloride-sensitive Na entry pathway, whereas E^{m}_{Na}, which was approximately 70 mV, remains essentially constant.

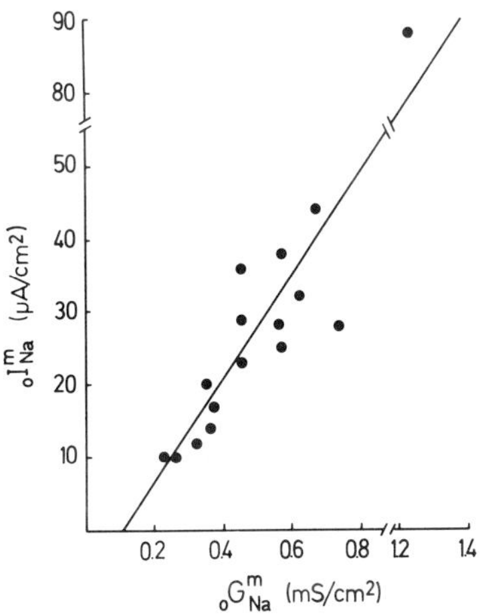

Fig. 6. Relation between the rate of apical Na entry, ${}_{o}I^{m}_{Na}$, and the Na conductance of the apical membrane, ${}_{o}G^{m}_{Na}$, under short-circuit conditions

Regulation of Basolateral Na Extrusion

Under steady-state conditions of transcellular Na transport the Na current across the two limiting membranes of epithelial cells must be identical. It is therefore not only possible to relate $(Na)_m$ and apical Na entry, which can be described by simple Michaelis-Menten kinetics (data not shown, for analogous results see Frizzell and Turnheim 1978), but we can also correlate $(Na)_c$ and basolateral Na extrusion.

Basolateral Na efflux as a function of $(Na)_c$ is given in Fig. 7. Clearly, this relation does not conform to simple saturation kinetics but rather suggests the involvement of

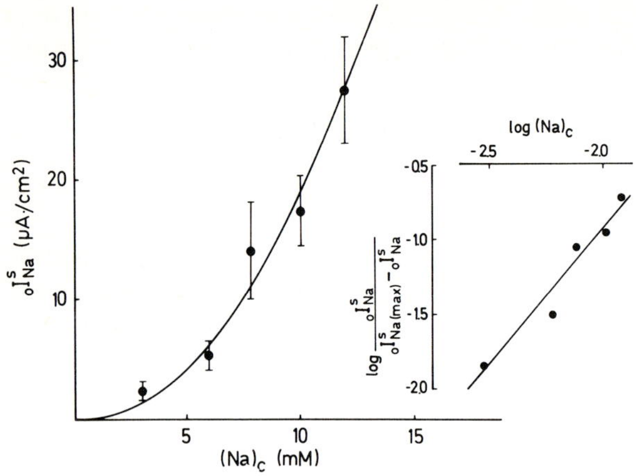

Fig. 7. Relation between the rate of basolateral Na extrusion, $_{o}I^{s}_{Na}$, and $(Na)_c$ under short-circuit conditions. The curve was calculated using the Hill plot given in the *inset* and assuming a maximal $_{o}I^{s}_{Na}$ of 170 $\mu A\ cm^{-2}$

a multisite transport system. The stoichiometry and half-saturation constant of the interaction of Na with the basolateral Na extrusion mechanism (the Na pump) can be derived from a Hill plot of the relation between $(Na)_c$ and the Na flux across the basolateral membrane, assuming a maximal transport rate of the Na pump of 170 μA cm^{-1} (Turnheim et al. 1977, Frrizzell and Turnheim 1978). In this manner it can be calculated that 2–3 Na ions interact with a single pump unit. Multisite Na pumps have also been demonstrated in frog skin (Nagel 1980), urinary bladder (Lewis et al. 1978, Eaton 1981) and many other tissues (see Levitt 1980). The advantage of a multisite pump mechanism is a steeper relation between transport rate and substrate concentration as compared to a single site mechanism. Hence small changes in $(Na)_c$ will cause large changes in basolateral pump rate.

$(Na)_c$ at which basolateral Na extrusion is half-maximal was found to be approximately 24 mM, which is in excellent agreement with the value of 26 mM, which was reported to be the $(Na)_c$ at which the activity of the (Na-K) ATPase of renal tubuli is half-maximal (Jørgensen 1980). Comparing these values with the intracellular Na activities determined in rabbit colon (see Fig. 3) it is clear that the Na pump operates far from saturation. Both this fact and the sigmoidal relation between $(Na)_c$ and the rate of basolateral Na extrusion serve to maintain $(Na)_c$ at fairly constant levels despite wide variations in transcellular Na transport.

Although the present and previous studies (Lewis et al. 1978, Graf and Giebisch 1979, Kimura and Spring 1979, Eaton 1981) provide evidence that an increase in intracellular Na stimulates Na transport, $(Na)_c$ may not be the only factor that determines the transport rate of the Na pump. The relation of active Na transport and $(Na)_c$ illustrated in Fig. 7 was obtained by varying $(Na)_m$. A somewhat different picture emerges when one considers the dependence of active Na transport on $(Na)_c$ at a fixed value of $(Na)_m$. Figure 8 is a plot of the rate of Na extrusion across the basolateral membrane versus the corresponding $(Na)_c$ of the individual experiments at $(Na)_m$ = 99.4 mM. Clearly, there was no correlation between the rate of basolateral Na extrusion and $(Na)_c$. This finding certainly cannot be readily reconciled with the expectation that the transport rate of the Na pump should, in analogy to an enzyme,

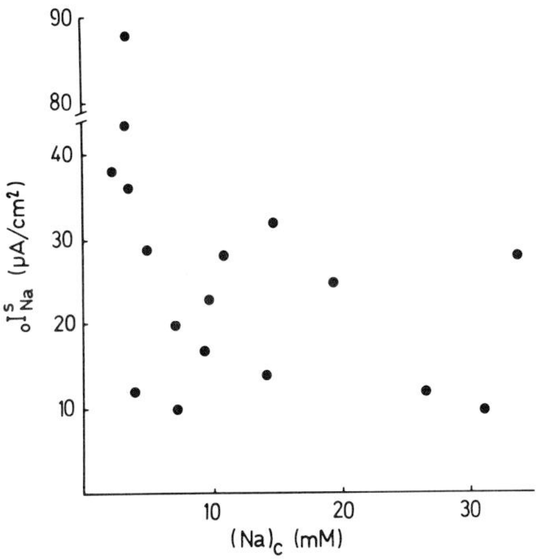

Fig. 8. Relation between $_oI^s_{Na}$ and $(Na)_c$ of the individual experiments at $(Na)_m$ = 99.4 mM

be dependent on the concentration of substrate, in this case intracellular Na. A lack of correlation between $(Na)_c$ and active Na transport was also observed by Wills and Lewis (1980) in rabbit urinary bladder and by Thomas, Suzuki, Thompson, and Schultz (unpublished data) in *Necturus* urinary bladder.

If $(Na)_c$ is not the sole determinant of the pump rate of the basolateral Na extrusion mechanism, as is suggested by the results illustrated in Fig. 8, what other signals could regulate the function of the Na pump? It is unlikely that ψ^{cs}, the electrical potential difference across the basolateral membrane, plays a significant role in determining the transport rate of the Na pump under open-circuit conditions, because ψ^{cs} is essentially independent of the rate of transcellular Na transport (Lewis et al. 1976, Lewis and Wills 1979). Another candidate that may regulate the Na pump is intracellular Ca, which may also modulate both apical Na conductance and basolateral K conductance, as recently discussed by Schultz (1981). Possibly Ca dos not or not only change the kinetic properties of the Na pump, rather Ca may primarily recruit additional pump units. Alternatively, activation of previously resting Na pumps could also be a consequence of transport-induced cell swelling. The concept of recruitment of additional Na pumps is attractive, because this mechanism implies that an increase in overall basolateral Na extrusion may come about although intracellular Na remains unchanged. Clearly, many laborious studies will have to be done to examine all these possibilities.

Conclusions

Although the present experiments may raise more questions than they answer, it is clear that in rabbit colonic epithelia, depolarized by a serosal high K solution, intracellular Na increases in a manner resembling saturation kinetics when luminal Na concentration is increased. The apical Na permeability, on the other hand, decreases in a hyperpolic fashion as the luminal Na concentration is increased. High levels of intracellular Na are always associated with low values of apical Na permeability, but at low levels of intracellular Na other regulatory factors may additionally be operative. At a constant luminal Na concentration the rate of active Na transport is a linear function of apical Na conductance, whereas the driving force for entry remains essentially constant.

Whereas apical Na entry conforms to the kinetics of a single-site process, basolateral Na extrusion exhibits the properties of a multisite transport mechanism which is far from saturation. Both this system and the negative feedback of high intracellular Na on apical Na permeability serve to maintain intracellular Na levels at fairly constant values despite large changes in luminal Na concentration or the rate of transcellular Na transport. Although an increase in intracellular Na results, on the average, in an increased activity of the Na pump, it is also apparent that the transport rate of the pump can vary markedly at a constant value of intracellular Na. Thus, intracellular Na appears to be neither the sole regulator of apical Na permeability nor of the transport function of the basolateral Na pump.

Acknowledgements. Parts of this study were performed in collaboration with S.G. Schultz and S.M. Thompson. K. Turnheim was the recipient of a Max Kade research fellowship.

References

Biber TUL, Curran PF (1970) Direct measurement of uptake of sodium at the outer surface of the frog skin. J Gen Physiol 56:83–99

Civan MM (1978) Intracellular activities of sodium and potassium. Am J Physiol 234:F261–F269

Eaton DC (1981) Intracellular sodium activity and sodium transport in rabbit urinary bladder. J Physiol (Lond) 316:527–544

Erlij D, Smith MW (1973) Sodium uptake by frog skin and its modification by inhibitors of transepithelial sodium transport. J Physiol (Lond) 228:221–239

Essig A, Leaf A (1963) The role of potassium in active transport of sodium by the toad bladder. J Gen Physiol 46:505–515

Finkelstein A, Mauro A (1977) Physical principles and formalism of electrical excitability. In: Geiger S (ed) Handbook of physiology, vol I/1, sect I: Nervous system. Am Physiol Soc, Bethesda MD, p 161

Frazier HS, Dempsey EF, Leaf A (1962) Movement of sodium across the mucosal surface of the isolated toad bladder and its modification by vasopressin. J Gen Physiol 45:529–543

Frizzell RA, Schultz SG (1978) Effect of aldosterone on ion transport by rabbit colon in vitro. J Membr Biol 39:1–26

Frizzell RA, Turnheim K (1978) Ion transport by rabbit colon. II. Unidirectional sodium influx and the effects of amphotericin B and amiloride. J Membr Biol 40:193–211

Frizzell RA, Koch MJ, Schultz SG (1976) Ion transport by rabbit colon. I. Active and passive components. J Membr Biol 27:297–316

Frizzell RA, Heintze K, Stewart CP (1980) Mechanism of intestinal chloride secretion. In: Field M, Fordtran JS, Schultz SG (eds) Secretory diarrhea. Clin Physiol Ser Am Physiol Soc, Bethesda, MD, p 11

Fuchs W, Hviid Larsen E, Lindemann B (1977) Current-voltage curve of sodium channels and concentration dependence of sodium permeability in frog skin. J Physiol (Lond) 267:137–166

Garzia-Diaz JF, Armstrong W McD (1980) Intracellular Na activity and Na transport in *Necturus* gallbladder. Fed Proc 39:1080

Goldman D (1943) Potential, impedance and rectification in membranes. J Gen Physiol 27:37–60

Graf J, Giebisch G (1979) Intracellular sodium activity and sodium transport in *Necturus* gallbladder epithelium. J Membr Biol 47:327–355

Grasl M, Turnheim K (1982) Adenosine-induced chloride secretion in rabbit colon. Naunyn-Schmiedeberg's Arch Pharmacol Suppl. 319:R58

Helman SI, Nagel W, Fisher RS (1979) Ouabain and active transepithelial sodium transport in frog skin. Studies with microelectrodes. J Gen Physiol 74:105–127

Higgins FT Jr, Frömter E (1974) Cell membrane potentials in amphibian urinary bladder. Physiologist 17:A247

Hodgkin AL, Katz B (1949) The effect of sodium ions on the electrical activity aof the giant axon of the squid. J Physiol (Lond) 108:37–77

Hviid Larsen E (1973) Effect of amiloride, cyanide and ouabain on the active transport pathway in toad skin. In: Ussing HH, Thron NA (eds) Transport mechanisms in epithelia. Munksgaard, Copenhagen, p 131

Jørgensen PL (1980) Sodium and potassium ion pump in kidney tubules. Physiol Rev 60:864–917

Kimura G, Spring KR (1979) Luminal Na^+ entry into *Necturus* proximal tubule cells. Am J Physiol 236:F295–F301

Koefoed-Johnsen V, Ussing HH (1958) The nature of the frog skin potential. Acta Physiol Scand 42:298–308

Leblanc G, Morel F (1975) Na and K movements across the membranes of frog skin associated with transient current changes. Pflueger's Arch 358:159–177

Levitt DG (1980) The mechanism of the sodium pump. Biochim Biophys Acta 604:321–345

Lewis SA, Wills NK (1979) Intracellular ion activities and their relationship to membrane properties of tight epithelia. Fed Proc 38:2739–2742

Lewis SA, Eaton DC, Diamond JM (1976) The mechanism of Na^+ transport by rabbit urinary bladder. J Membr Biol 28:41–70

Lewis SA, Wills NK, Eaton DC (1978) Basolateral membrane potential of a tight epithelium: ionic diffusion and electrogenic pumps. J Membr Biol 41:117–148

Luger A, Turnheim K (1981) Modification of cation permeability of rabbit descending colon by sulfphydryl reagents. J Physiol (Lond) 317:49–66

MacKnight ACD, DiBona DR, Leaf A (1980) Sodium transport across toad urinary bladder: a model tight epithelium. Physiol Rev 60:615–715

MacRobbie EAC, Ussing HH (1961) Osmotic behaviour of the epithelial cells of frog skin. Acta Physiol Scand 53:348–365

Nagel W (1980) Rheogenic sodium transport in a tight epithelium, the amphibian skin. J Physiol (Lond) 302:281–295

Narvarte J, Finn AL (1980) Microelectrode studies in toad urinary bladder epithelium. Effects of Na concentration changes in the mucosal solution on equivalent electromotive forces. J Gen Physiol 75:323–344

Palmer LG, Edelman IS, Lindemann B (1980) Current-voltage analysis of apical sodium transport in toad urinary bladder: Effects of inhibitors of transport and metabolism. J Membr Biol 57: 59–71

Palmer LG, Li JHY, Lindemann B, Edelman IS (1982) Aldosterone control of the density of the sodium channels in the toad urinary bladder. J Membr Biol 64:91–102

Rawlins F, Mateu L, Fragachan F, Whittembury G (1970) Isolated toad skin epithelium: transport characteristics. Pflueger's Arch 316:64–80

Robinson RA, Stokes RH (1959) Electrolyte solutions. Academic Press, London New York; Butterworth, London

Schultz SG (1972) Electrical potential differences and electromotive forces in epithelial tissues. J Gen Physiol 59:794–798

Schultz SG (1977) Sodium-coupled solute transport by small intestine: a status report. Am J Physiol 233:E249–E254

Schultz SG (1981) Homocellular regulatory mechanism in sodium-transporting epithelia: avoidance of extinction by "flush-through". Am J Physiol 241:F579–F590

Schultz SG, Frizzell RA, Nellans HN (1977) Active sodium transport and the electrophysiology of rabbit colon. J Membr Biol 33:351–384

Schultz SG, Thompson SM, Suzuki Y (1981) Equivalent electrical circuit models and the study of Na transport across epithelia. Nonsteady-state current-voltage relations. Fed Proc 40: 2443–2449

Spring KR, Giebisch G (1977) Kinetics of Na^+ transport in *Necturus* proximal tubule. J Gen Physiol 70:307–328

Thompson SM, Suzuki Y, Schultz SG (1982) The electrophysiology of rabbit descending colon. I. Instantaneous transepithelial current-voltage relations and the current-voltage relations of the Na-entry mechanism. J Membr Biol 66:41–54

Turnheim K, Frizzell RA, Schultz SG (1977) Effects of anions on amiloride-sensitive, active sodium transport across rabbit colon, in vitro. Evidence for "trans-inhibition" of the Na entry mechanism. J Membr Biol 37:63–84

Turnheim K, Frizzell RA, Schultz SG (1978) Interaction between cell sodium and the amiloride-sensitive sodium entry step in rabbit colon. J Membr Biol 39:233–256

Turnheim K, Luger A, Grasl M (1981) Kinetic analysis of the amiloride-sodium entry site interaction in rabbit colon. Mol Pharmacol 20:543–550

Wills NK, Lewis SA (1980) Intracellular Na^+ activity as a function of Na^+ transporte rate across a tight epithelium. Biophys J 30:181–186

Calcium Regulation of Intestinal Na and Cl Transport in Rabbit Ileum

D.W. POWELL and C.C. FAN[1]

Introduction

In the past decade studies have demonstrated that stimulus-secretion coupling of intestinal water and electrolyte transport is mediated by either cyclic nucleotides and/or calcium (see Rao and Field, this vol.). This is not surprising in view of the ubiquitous role these intracellular messengers play in the coupling of stimuli to cellular action in other tissues. Drugs, hormones, and bacterial toxins that increase intracellular levels of either cyclic AMP, cyclic GMP, or ionized Ca inhibit intestinal water and electrolyte absorption and may reverse absorption to secretion (Binder 1979). Investigations of electrolyte transport both in vivo and in vitro either in the basal or stimulated state have led to hypotheses of the mechanisms whereby the gut epithelium effects the transfer of Na, Cl, and water from lumen to blood and from blood to lumen. This article will review the two major hypotheses of the mechanisms of intestinal electrolyte secretion by rabbit *(Oryctolagus caniculus)* ileum and experiments in our laboratory designed to test these hypotheses.

Models of Na and Cl Transport

Field's Model

One theory of rabbit ileal electrolyte secretion, articulated best by Field (1980) and growing out of studies in the laboratories of Frizzell, Field, and Schultz (1979), separates the absorptive from the secretory mechanisms both conceptually and anatomically. As shown in Fig. 1, Field proposed that cAMP inhibits a brush border coupled NaCl influx process in the villus and stimulates electrogenic Cl secretion in the crypt epithelium. The active Cl secretion comes about because a NaCl influx process on the basolateral membrane of the crypt cells increases intracellular Cl activity above electrochemical equilibrium (Frizzell and Duffey 1980), and then cyclic AMP, cyclic GMP, or an increase in cytosolic free calcium levels increases the apical membrane permeability to Cl allowing it to escape from the crypt cell by moving

1 Department of Medicine, University of North Carolina, Chapel Hill, NC 27514, USA

Intestinal Transport
(ed. by M. Gilles-Baillien and R. Gilles)

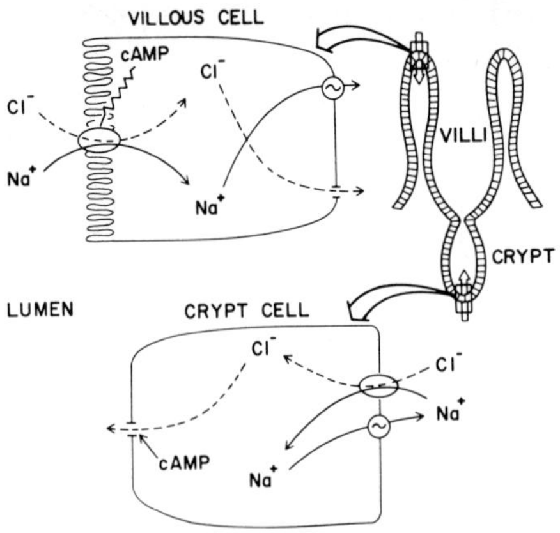

Fig. 1. Proposed scheme for cAMP-mediated electrolyte secretion by rabbit ileum. This hypothesis separates the antiabsorptive (villous cell) from the secretory (crypt cell) effects of the intracellular messengers. Since cyclic GMP and Ca-ionophores have effects on transport similar to cyclic AMP, this hypothesis might apply to all cyclic nucleotide- and Ca-stimulated secretion. (Field 1980. Reproduced by permission of the American Physiological Society)

down the thermodynamic gradient. Thus this hypothesis involves both antiabsorptive and secretory mechanisms in different parts of the mucosa.

The necessity of proposing an antiabsorptive process grew out of observed changes in unidirectional Na and Cl fluxes measured after cyclic nucleotide or Ca stimulated secretion by rabbit ileum mounted in the Ussing chamber. As shown in Fig. 2, the changes in net Na and Cl transport from absorption to zero transport or even to secretion comes about as much from decreases in the mucosal to serosal fluxes (J_{ms}) of Na and Cl as it does from increases in the serosal to mucosal fluxes (J_{sm}) of these ions. These changes occur with both cAMP (Field 1980), cGMP (Guandalini et al. 1982a,b) mediated secretion as well as that caused by calcium (Bolton and Field 1977).

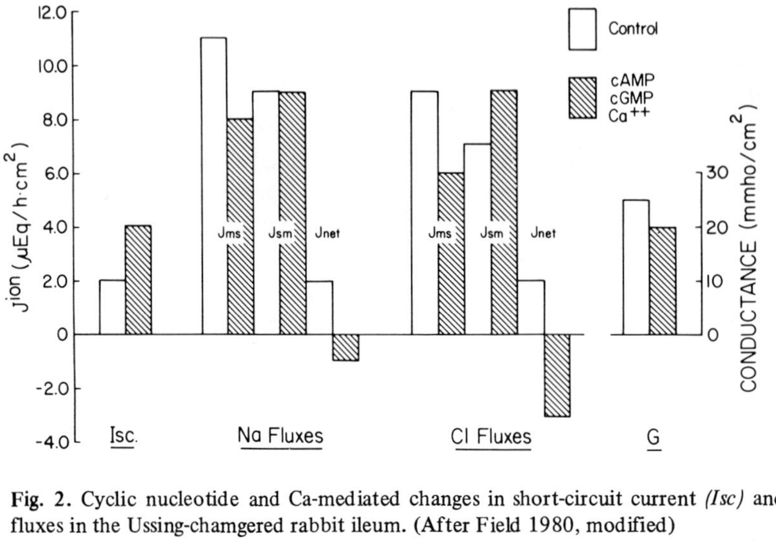

Fig. 2. Cyclic nucleotide and Ca-mediated changes in short-circuit current *(Isc)* and electrolyte fluxes in the Ussing-chamgered rabbit ileum. (After Field 1980, modified)

The evidence for Field's scheme is both direct and indirect. Indirect evidence includes the following observations: (1) Cyclic AMP inhibits NaCl absorption, but does not cause secretion in rabbit gallbladder, a tissue containing only NaCl absorptive processes. Conversely, cyclic AMP causes Cl secretion but no decrease in J_{ms}^{Na} and J_{ms}^{Cl} in Ussing chamber studies of descending rabbit colon, a tissue with Cl secretory but no NaCl absorptive mechanisms (Frizzell et al. 1979). (2) Brief exposure of rat small intestine to cholera toxin increases villous cell cyclic AMP content, but not crypt cell content, and only inhibits absorption. Prolonged exposure allows stimulation of crypt cell adenylate cyclase as well and produces actual secretion (de Jonge 1975). (3) Selective osmotic damage to villous cells, but not crypt cells, does not inhibit cholera toxin-induced secretion (Roggin et al. 1972). (4) Loop diuretics, such as ethacrynic acid and furosemide, thought to inhibit coupled NaCl transport across the basolateral cell membrane of Cl secreting cells, also inhbit cyclic nucleotide-stimulated secretion (Al-Awqati et al. 1974, Naftalin and Simmons 1979).

Direct evidence for the antiabsorptive effect of cyclic nucleotides comes from measured influxes of Na and Cl across the apical cell membrane as reported by Nellans et al. (1973) and Frizzell et al. (1973). As shown in Fig. 3, a component of Na influx across the brush border is dependent on the presence of Cl in the mucosal solution. Conversely, a similar obligatory need for Na in the mucosal solution is observed for part of the cell influx by Cl. Furthermore, theophylline inhibits the influx of both Na and Cl by the same amount as removal of the appropriate counter ion from the mucosal solution. Partial evidence for active Cl secretion is found in the demonstration that the intracellular content of Cl in various epithelia is above electrochemical potential (Frizzell and Duffey 1980).

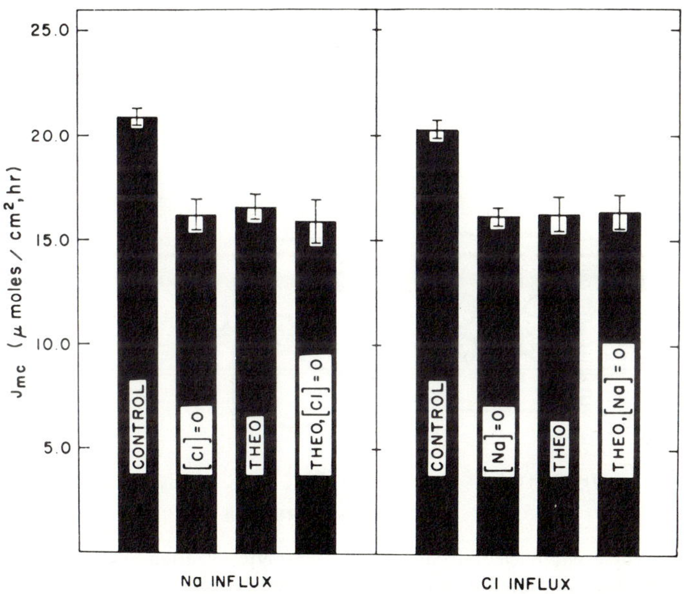

Fig. 3. Effects of removal of Cl or Na (mucosal solution) or addition of 10 mM theophylline (serosal solution) on Na and Cl influx across the apical cell membrane of rabbit ileum. (Nellans et al. 1973. Reproduced by permission of the Americal Physiological Society)

However, other explanations are also possible for the decrease in mucosal to serosal ion fluxes after stimulated secretion. For instance, in rabbit ileum, stimulated secretion is accompanied by a decrease in electrical conductance (Powell et al. 1974, Powell 1974). This change in conductance (G) may be due to cyclic AMP or Ca effects on the paracellular shunt path (see the review by Powell 1981) or to collapse of the intercellular space (Holman et al. 1979). It is possible that the decrease in passive fluxes of Na and Cl through the shunt path accounts for the observed reduction in mucosal to serosal fluxes. Additionally, it is likely that rabbit ileal mucosa is not uniformly short-circuited in conventional Ussing chambers (Tai and Tai 1981), raising the question whether some of the unidirectional flux changes might be experimental artifact.

Naftalin's Model

The most compelling alternative hypothesis comes from Naftalin and colleagues (Naftalin and Simmons 1979, Holman and Naftalin 1979, Holman et al. 1979). These investigators propose (Fig. 4) that secretion takes place throughout the ileal epithelium, villous as well as crypt cells, and that it occurs via cyclic nucleotide or calcium-induced regurgitation of Na and Cl from the intercellular space back across the cation-permeable tight junction. This recycling of Na and Cl would account for the observed decrease in J_{ms}^{Na} and J_{ms}^{Cl}. As in Field's hypothesis, Naftalin proposes a change in apical membrane Cl permeability after stimulated secretion, but his proposal does not necessitate an antiabsorptive effect in the villous cell.

The major proof for Naftalin's hypothesis comes from studies with triaminopyrimidine, a polyvalent, organic cation that is thought to block the junctional conductance of Na by titrating the negative charges that confer cation selectivity to many epithelial tight junctions (Moreno 1975). When this compound is used in conjunction with theophylline, it blocks the cyclic nucleotide-induced decrease in J_{ms}^{Na} and J_{ms}^{Cl},

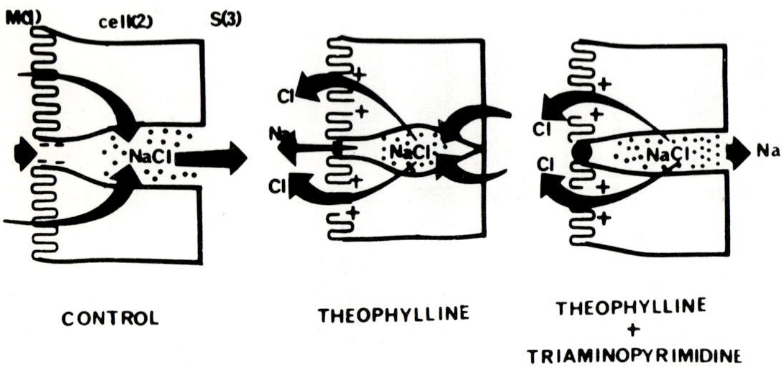

Fig. 4. Proposed scheme for cyclic nucleotide-stimulated electrolyte secretion in rabbit ileum. An increase in brush border Cl permeability throughout the mucosa leads to recycling of Na (and perhaps some Cl) from the lateral intercellular space back across the tight junction. Triaminopyrimidine blocks the tight junctional movement of Na, but has no effect on Cl secretion. (Naftalin and Simmons 1979. Reproduced by permission of The Physiological Society)

partially preventing the reduced Na absorption but not altering net Cl secretion at all (Naftalin and Simmons 1979). Thus this theory ascribes all of the changes in Na flux and the decreased m to s Cl fluxes to the properties of the tight junction-intercellular space complex. The idea that these cation-selected pathways might allow Na to recycle back to the mucosa solution is not unique to this hypothesis; a similar process has been advanced to explain the serosa-negative PD across rabbit gallbladder epithelium (Machen and Diamond 1969) and flounder small intestine (Field et al. 1978) and also the fact that the rate of active Cl absorption exceeds the rate of active Na absorption in the short-circuited flounder intestine (Field et al. 1978).

Use of Vesicles

Membrane vesicles are potential tools for testing these two hypotheses of Na and Cl transport in rabbit ileum. Even though the influx techniques of Nellans et al. (1973) are believed to differentiate between the movement of Na and Cl from mucosal solution into the cell as opposed to movement across the tight junction into the intercellular space, it is possible that their demonstration of coupled NaCl influx included movement into both compartments. Therefore confirmation of NaCl influx into villous cell brush border membrane vesicles would be important positive evidence for the proposed antiabsorptive mechanism. Similarly, demonstration of coupled NaCl influx across basolateral cell membranes of crypt cells would be useful evidence in favor of Field's proposed scheme of active chloride secretion. Furthermore, demonstrated inhibition of the brush border process by either cyclic nucleotide or calcium would represent compelling evidence in favor of Field's hypothesis. Conversely, the inability to demonstrate either the presence of coupled NaCl transport mechanisms or inhibition of these processes would not constitute compelling proof for Naftalin's hypothesis, but certainly would add to the attractiveness of that proposal. Since techniques to purify intestinal brush border membranes are more advanced than those for basolateral membranes, our laboratory set out first to study coupled NaCl transport processes in brush border vesicles.

NaCl Transport in Brush Border Vesicles

The first step in these studies was to decide on a method for purification of vesicles. While calcium-precipitated techniques have been favored by other investigators because of their simplicity, obviously such methods might interfere with subsequent studies of Ca inhibition of any putative coupled NaCl transport processes. Therefore, we used a modification (Fan et al. 1983) of a sucrose density gradient technique that we applied to epithelial cells harvested following their removal from the villus by vibration in a zero Ca-EGTA solution (Im et al. 1980). The brush border vesicles thus obtained from the villous cells were enriched 20-fold, as ascertained by comparison of maltase specific activity in the brush border fraction to that of the original cellular homogenate, and yet were not enriched at all in basolateral cell membranes as assayed by measurements of Na-K ATPase.

Using these purified villous cell brush border membrane vesicles, we determined the effect of various anions on Na uptake and, conversely, the effect of various cations on Cl uptake. Because it seemed likely that coupled NaCl influx processes might be but one of several entry mechanisms for either Na or Cl, and since the estimated K_m of the coupled process was quite low (11 mM) (Nellans et al. 1973), we studied uptake from solutions low in Na concentration (2 mM) and as low in Cl concentration (20 mM) as was possible considering the specific activities of ^{22}Na and ^{36}Cl that were commercially available. Furthermore, these uptake studies were performed in the absence of anion or cation gradients across the vesicle and in vesicles short-circuited with potassium and valinomycin, so that observed differences in uptake could not be ascribed to the passive permeability characteristic of the membrane.

Figure 5 demonstrates Na uptake in the presence of various anions. The rank order of anion-dependent Na uptake was Cl $\cong$ SCN $>$ NO_3 $>$ gluconate. Carrier mediation of Cl-stimulated uptake was suggested by the demonstration of competition for ^{22}Na uptake with increasing concentrations of unlabeled Na in the presence of Cl, but not when gluconate was the counter ion. By subtracting Na uptake in the presence of gluconate from that in the presence of Cl, a K_m of 4.5 mM was calculated for Cl-dependent Na uptake. Similar studies of ^{36}Cl uptake (Fig. 6) showed a rank order of cation dependence of Na $\cong$ Li $>$ K $>$ choline. Because ^{36}Cl specific activities precluded experiments with Cl concentration below 20 mM, kinetic studies of Na-dependent Cl transport were not done.

The rates of uptake of Na and Cl were not altered by potassium gradients across the vesicle in the presence of valinomycin, which is a technique that allows the experimenter to alter the transmembrane potential difference (PD). The lack of an effect of potential difference indicated that the NaCl coupled uptake was an electrically neutral process. However, the nature of the coupling of Na to Cl is still uncertain.

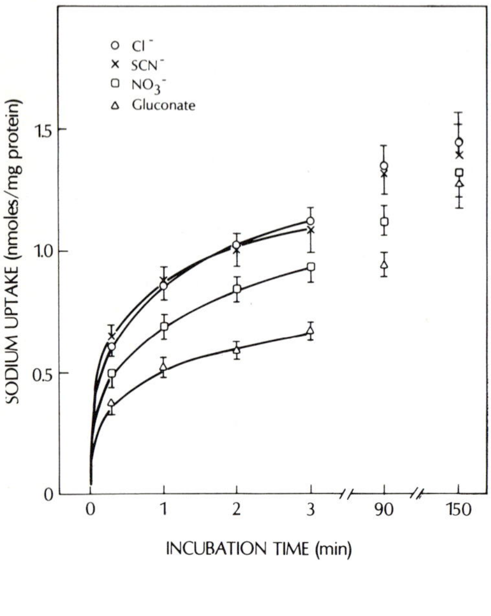

Fig. 5. Na uptake into rabbit ileal villous brush border membrane vesicles in the presence of various anions. The vesicles were preloaded with 100 mM mannitol, 100 mM K anion, valinomycin (30 μg mg^{-1} protein) and 20 mM Tris-Hepes (pH 7.4). The transport solution contained 100 mM mannitol, 2 mM ^{22}Na anion, 98 mM K anion, and 20 mM Tris-Hepes. Na uptake in the presence of Cl was significantly different ($p < 0.05$) from that with NO_3 or gluconate at all time periods. (Fan et al. 1983. Reproduced by permission of the American Physiological Society)

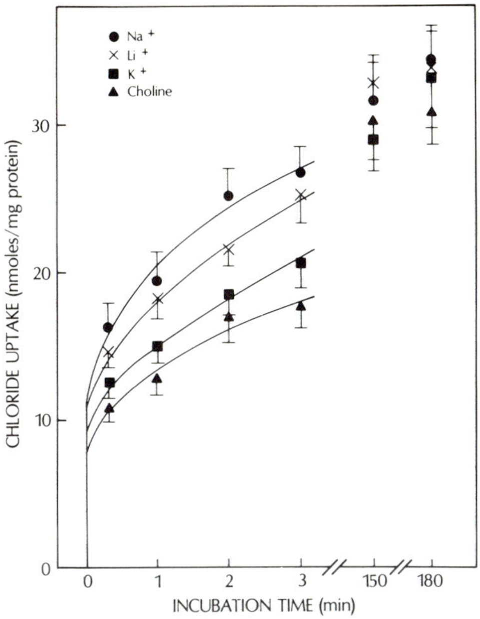

Fig. 6. Cl uptake by rabbit villous brush border membrane vesicles in the presence of various cations. The vesicles contained 100 mM mannitol, 20 mM Na, K, Li, or choline chloride, 80 mM K gluconate plus valinomycin, and 20 mM Tris-Hepes (ph 7.4). The transport medium had the same composition except ^{36}Cl salts of Na, K, Li, and choline were used. Cl uptake in the presence of Na was significantly different ($p < 0.05$) from that with K or choline. (Fan et al. 1983. Reproduced by permission of the American Physiological Society)

Apparent electrically neutral NaCl transport can come about by a process that is directly coupled (NaCl symport), or one that is indirectly coupled (Na:H antiport linked to simultaneous Cl:OH [or HCO_3] antiport) (Turnberg et al. 1970, Liedtke and Hopfer 1982a,b). In our vesicles, we could demonstrate Na:H exchange and Cl:OH (or Cl:HCO_3) exchange by studying ^{22}Na and ^{36}Cl uptake across vesicles with imposed H^+ gradients. Inhibitor studies did not settle the issue either. We found that furosemide, which is supposed to inhibit NaCl symport, would inhibit Na uptake in the presence of Cl but not in the presence of gluconate, and Cl uptake in the presence of Na but not in the presence of choline. However, we also could demonstrate that harmaline, an inhibitor of Na transport processes, and SITS and DIDS, inhibitors of anion exchange, would do so as well. Therefore, these studies would lead one to reach the same conclusions as Liedtke and Hopfer (1982a,b) that Na:H-Cl:HCO_3 exchange rather than NaCl cotransport is the mechanism of Na coupling to Cl across rabbit ileal intestinal brush borders. However, it is also possible that all three mechanisms reside in the intestinal cell membrane, as has been postulated by Spring and Ericson (1982) for *Necturus* gallbladder. It should be noted that similar processes have been proposed for the renal tubule, and the evidence for and against the various hypotheses has been discussed in detail by Warnock and Eveloff (1982).

Ca Inhibition of Coupled NaCl Uptake

Having demonstrated a "coupled" NaCl influx mechanism across brush border membrane vesicles from rabbit ileal villous cells, we next investigated the possible mechanisms of inhibition of this process. There remains a question whether cyclic nucleotides alter intestinal electrolyte transport directly or whether they do so through another intermediate messenger such as calcium. Thus cyclic nucleotides may release Ca from intercellular reservoirs rather than acting directly on cellular membrane transport processes for Na and Cl transport. Therefore, we decided to study the effect of calcium, since it was possible that this ion was involved in the final common pathway for nucleotide-induced secretion and certainly was involved in secretion induced by cholinergic agonist, serotonin, and by calcium ionophores which cause calcium gating across cell membranes.

In initial studies (Fan and Powell 1982), we compared the effect of Ca on Na uptake in the presence of Cl with that in the presence of gluconate. We also studied Cl uptake in the presence of Na and choline. Using Ca alone, we found that: (1) A concentration greater than 10^{-4} M inhibited Na and Cl uptake by 25%–30% when either Cl or Na were the counter ions respectively, but only by 10% when gluconate or choline were the counter ions present in the transport solution. (2) The inhibition was observed only when Ca was present inside the vesicle, that is after the vesicles had been preincutabed with calcium for 4 h at 4°C. If the vesicles were isolated with EGTA inside and then Ca was added to the outside of the vesicle, no inhibition was observed. (3) La, but not Mg or Ba, mimicked the affect of Ca on Na and Cl uptake. (4) Calcium at 1 mM also inhibited ^{3}H-mannitol uptake by 10% as did trifluoperazine (TFP) at concentrations greater than 10^{-5} M. (5) At concentrations of 10^{-6} M, neither Ca nor TFP had any effect on Na or Cl uptake.

These studies indicated that Ca inside the vesicle had an effect on general permeability, decreasing both ion and mannitol uptake by approximately 10%. High concentration of TFP, a known "membrane-stabilizing" drug, had a similar effect. However, Ca inside the vesicle had a greater effect on Na and Cl uptake when Cl and Na were the counter ions, and this property was shared by La, a cation known to mimic Ca effects on the membranes. The effects of Ca and La on Na and Cl uptake suggested an inhibitory effect on a coupled NaCl process. Of interest is that these results are almost opposite to those reported by Schulz and Heil (1979) in pancreatic cell membrane vesicles, where Ca outside the vesicle decreased Na permeability and Ca inside the vesicle increased Na permeability.

Since calmodulin appears to play a role in calcium-mediated intestinal secretion (Ilundain and Naftalin 1979, Naftalin 1981, Smith and Field 1980), we sought evidence for this in our vesicles. Calmodulin isolated from rat testes had no effect alone on electrolyte uptake by vesicles. Calcium at concentration of 10^{-5} M also had no effect by itself. However, as shown in Fig. 7, when calmodulin was present in concentrations of 33 μg ml^{-1} ($\sim$ 2 μM) together with 10^{-5} M Ca, the combination inhibited Na uptake by 20% when Cl was the counter ion. Ca plus calmodulin had no effect when gluconate was the counter ion. A similar Ca-calmodulin inhibition of Cl uptake could be demonstrated in the presence of Na but not choline. Furthermore, Ca-calmodulin inhibition of NaCl uptake was abolished by 50 μM TFP, a known inhibitor of the calmodulin-Ca complex (Fig. 8). Dose response studies showed a K_i of 0.2 μM calmodulin (when studied with 1 μM Ca) and a K_i of 0.5 μM Ca at 2 μM calmodulin.

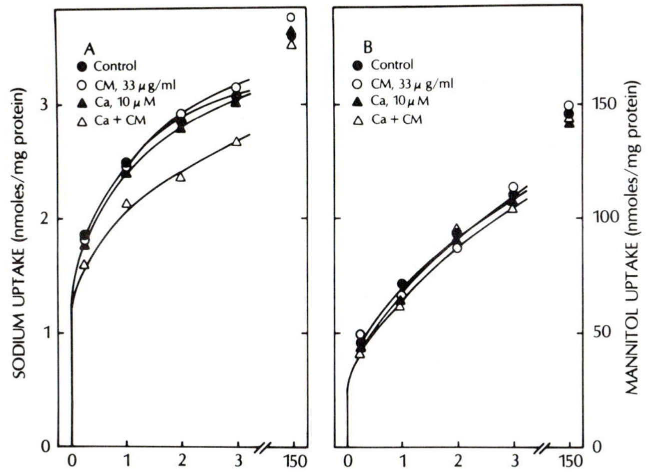

Fig. 7. A Inhibition of Na uptake by Ca-calmodulin (Ca 10 μM-calmodulin 33 μg ml^{-1} or ~ 2 μM). **B** Ca-calmodulin had no effect on mannitol uptake. The vesicles (**A**) were preloaded by incubating for 4 h at 4°C in a solution containing 100 mM mannitol, 2 mM NaCl, 98 mM KCl, 2 mM $MgSO_4$, 10 μM $CaCl_2$, and 20 mM Tris-Hepes (pH 7.4). The transport solution had the same composition except ^{22}NaCl replaced NaCl. In **B**, (^{3}H)-mannitol was present in the outside solution instead of ^{22}Na

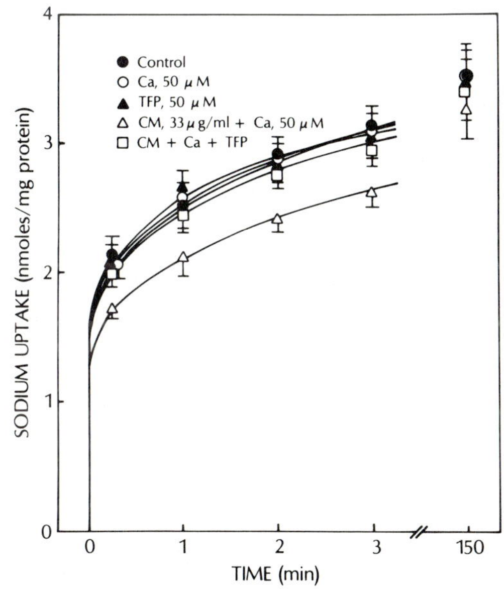

Fig. 8. Inhibition of Na uptake by Ca-calmodulin (50 μM Ca – ~ 2 μM calmodulin) and reversal of the inhibition by trifluoparazine (TFP 50 μM). Loading and transport solutions were the same as in Fig. 7A

Summary and Conclusions

These studies of brush border membrane vesicles indicate that a Ca-calmodulin complex mediates inhibition of coupled NaCl transport in apical cell membranes of rabbit villous enterocytes. As such, these experiments constitute evidence in favor of the Field-Frizzell-Schultz hypothesis of small intestinal secretion. A calmodulin-mediated inhibition of Na and Cl influx must be a significant part of the decrease in the mucosal to serosal fluxes in Na and Cl with calcium-stimulated intestinal secretion. Since calcium is possibly an intermediate for cyclic nucleotide-mediated secretion as well, these results may apply to secretion stimulation by cyclic AMP and cyclic GMP as well. This is not to say that recycling of Na across tight junctions, as proposed by Naftalin, does not occur. The evidence in favor of such recycling in gallbladder and flounder intestine make it likely that this occurs to a certain extent in rabbit intestine as well. Finally, the changes in shunt path conductance and subsequent decrease in passive Na and Cl fluxes across the paracellular pathway probably also contribute to the decrease in J_{ms}^{Na} and J_{ms}^{Cl}. The exact proportion that each of these processes play in the overall secretory process remains to be determined.

Perhaps more importantly, these studies indicate that Ca-calmodulin complexes play an important role in regulating intestinal electrolyte transport. Calcium-calmodulin is important in the metabolism of cyclic nucleotides which could then act through cyclic AMP-mediated protein kinases (Kuo 1980, Shlatz et al. 1978, 1979). Alternatively, Ca-calmodulin complexes might stimulate calcium-dependent protein kinases directly (Kuo et al. 1980, Kennedy and Greengard 1980, Taylor et al. 1981). A third hypothesis, recently proposed, is that all stimuli might work through synthesis of endogenous prostaglandins from membrane-bound arachidonic acid (Marshall et al. 1980, 1981). An understanding of the specific mechanisms whereby calcium-calmodulin complexes modulate intestinal secretion might delineate the intracellular control of intestinal electrolyte transport, but also might further understanding of stimulus-secretion coupling in all mammalian cells. Since intestinal secretion plays an important role in diarrhea diseases in animals and man, such information could have practical implications as well.

References

Al-Awqati Q, Field M, Greenough III WB (1974) Reversal of cyclic AMP-mediated intestinal secretion by ethacrynic acid. J Clin Invest 53:687–692

Binder HJ (1979) Mechanisms of intestinal secretion. Alan R Liss, New York

Bolton JE, Field M (1977) Ca ionophore-stimulated ion secretion in rabbit ileal mucosa: relation to actions of cyclic 3′,5′-AMP and carbamylcholine. J Membr Biol 35:159–173

Fan C-C, Powell DW (1982) Ca inhibition of NaCl uptake in rabbit ileal brush border membrane vesicles. Fed Proc 41:1265 (Abstr)

Fan C-C, Faust RG, Powell DW (1983) Coupled Na-Cl transport by rabbit ileal brush border membrane vesicles. Am J Physiol (in press)

Field M (1980) Regulation of small intestinal ion transport by cyclic nucleotides and calcium. In: Field M, Fordtran JS, Schultz SG (eds) Secretory diarrhea. Am Physiol Soc, Bethesda, pp 21–30

Field M, Karnaky KJ Jr, Smith PH, Bolton JE, Kinter WB (1978) Ion transport across the isolated intestinal mucosa of the winter flounder, *Pseudopleuronectes americanus*. I. Functional and structural properties of cellular and paracellular pathways for Na and Cl. J Membr Biol 41: 265–293

Frizzell RA, Duffey ME (1980) Chloride activities in epithelia. Fed Proc 39:2860–2864

Frizzell RA, Nellans HN, Rose RC, Markscheid-Kaspi L, Schultz SG (1973) Intracellular Cl concentrations and influxes across the brush border of rabbit ileum. Am J Physiol 224:328–337

Frizzell RA, Field M, Schultz SG (1979) Sodium-coupled chloride transport by epithelial tissues. Am J Physiol 236:F1–F8

Guandalini S, Migliavacca M, Campora E de, Rubino A (1982a) Cyclic guanosine monophosphate effects on nutrient and electrolyte transport in rabbit ileum. Gastroenterology 83:15–21

Guandalini S, Rao MC, Smith PL, Field M (1982b) cGMP modulation of ileal ion transport: in vitro effects of *Escherichia coli* heat-stable enterotoxin. Am J Physiol 243:G36–G41

Holman GD, Naftalin RJ (1979) Fluid movements across rabbit ileum coupled to passive paracellular ion movements. J Physiol 290:351–366

Holman GD, Naftalin RJ, Simmons NL, Walker M (1979) Electrophysiological and electron-microscopical correlations with fluid and electrolyte secretion in rabbit ileum. J Physiol 290: 367–386

Ilundain A, Naftalin RJ (1979) Role of Ca^{2+}-dependent regulator protein in intestinal secretion. Nature 279:446–448

Im WB, Misch DW, Powell DW, Faust RG (1980) Phenolphthalein- and harmaline-induced disturbances in the transport functions of isolated brush border and basolateral membrane vesicles from rat jejunum and kidney cortex. Biochem Pharmacol 29:2307–2317

Jonge HR de (1975) The response of small intestinal villous and crypt epithelium to choleratoxin in rat and guinea pig. Evidence against a specific role of the crypt cells in choleragen-induced secretion. Biochim Biophys Acta 381:128–143

Jonge JR de (1976) Cyclic nucleotide-dependent phosphorylation of intestinal epithelium proteins. Nature 256:590–593

Kennedy MB, Greengard P (1980) Two calcium/calmodulin-dependent protein kinases, which are highly concentrated in brain, phosphorylate protein I at distinct sites. Proc Natl Acad Sci USA 78:1293–1297

Kuo JF (1980) Cyclic nucleotide-dependent protein kinases: An overview. Prog Pharmacol 4:21–30

Kuo JF, Andersson RGG, Bradley CW, Mackerlova L, Salomonsson I, Brackett NL, Katoh N, Shoji M, Wrenn RW (1980) Calcium-dependent protein kinase: widespread occurrence in various tissues and phyla of the animal kingdom and comparison of effects of phospholipid, calmodulin, and trifluoperazine. Proc Natl Acad Sci USA 77:7039–7043

Liedtke CM, Hopfer U (1982a) Mechanism of Cl^- translocation across small intestinal brush-border membrane. I. Absence of Na^+-Cl^- cotransport. Am J Physiol 242:G263–G271

Liedtke CM, Hopfer U (1982b) Mechanism of Cl^- translocation across small intestinal brush-border membrane. II. Demonstration of Cl^--OH^- exchange and Cl^- conductance. Am J Physiol 242:G272–G286

Machen TE, Diamond JM (1969) An estimate of the salt concentration in the lateral intercellular spaces of rabbit gallbladder during maximal fluid transport. J Membr Biol 1:194–213

Marshall PJ, Dixon JF, Hokin LE (1980) Evidence for a role in stimulus-secretion coupling of prostaglandins derived from release of arachidonoyl residues as a result of phosphatidylinositol breakdown. Proc Natl Acad Sci USA 77:3292–3296

Marshall PJ, Boatman DE, Hokin LE (1981) Direct demonstration of the formation of prostaglandin E_2 due to phosphatidylinositol breakdown associated with stimulation of enzyme secretion in the pancreas. J Biol Chem 256:844–847

Moreno JH (1975) Blockage of gallbladder tight-junction cation selective channels by 2,4,6-triaminopyrimidine (TAP). J Gen Physiol 66:97–115

Naftalin RJ (1981) The role of intracellular calcium in the induction of intestinal secretion. In: Read NW (ed) Diarrhoea: new insights. Janssen Pharmaceutic, London, pp 63–71

Naftalin RJ, Simmons NL (1979) The effects of theophylline and choleragen on sodium and chloride ion movements within isolated rabbit ileum. J Physiol 290:331–350

Nellans HN, Frizzell RA, Schultz SG (1973) Coupled sodium-chloride influx across the brush border of rabbit ileum. Am J Physiol 225:467–475

Powell DW (1974) Intestinal conductance and permselectivity changes with theophylline and choleragen. Am J Physiol 227:1436–1443

Powell DW (1981) Barrier function of epithelia. Am J Physiol 241:G275–G288

Powell DW, Farris RK, Carbonetto ST (1974) Theophylline, cyclic AMP, choleragen, and electrolyte transport by rabbit ileum. Am J Physiol 227:1428–1435

Roggin GM, Banwell JG, Yardley JH, Hendrix TR (1972) Unimpaired response of rabbit jejunum to cholera toxin after selective damage to villus epithelium. Gastroenterology 63:981–989

Schulz I, Heil K (1979) Ca^{2+} control of electrolyte permeability in plasma membrane vesicles from cat pancreas. J Membr Biol 46:41–70

Shlatz LJ, Kimberg DV, Cattieu KA (1978) Cyclic nucleotide-dependent phosphorylation of rat intestinal microvillus and basal-lateral membrane proteins by an endogenous protein kinase. Gastroenterology 75:838–846

Shlatz LJ, Kimberg DV, Cattieu KA (1979) Phosphorylation of specific rat intestinal microvillus and basal-lateral membrane proteins by cyclic nucleotides. Gastroenterology 76:293–299

Smith PL, Field M (1980) In vitro antisecretory effects of trifluoperazine and other neuroleptics in rabbit and human small intestine. Gastroenterology 78:1545–1553

Spring KR, Ericson A-C (1982) Epithelial cell volume modulation and regulation. J Membr Biol 69:167–176

Tai Y-H, Tai C-Y (1981) The conventional short-circuiting technique under short-circuits most epithelia. J Membr Biol 59:173–177

Taylor L, Guerina VJ, Donowitz M, Cohen M, Sharp GWG (1981) Calcium and calmodulin-dependent protein phosphorylation in rabbit ileum. FEBS Lett 131:322–324

Turnberg LA, Bieberdorf FA, Morawski SG, Fordtran JS (1970) Interrelationships of chloride, bicarbonate, sodium, and hydrogen transport in the human ileum. J Clin Invest 49:557–567

Warnock DG, Eveloff J (1982) NaCl entry mechanisms in the luminal membrane of the renal tubule. Am J Physiol 242:F561–F574

Role of Calcium and Cyclic Nucleotides in the Regulation of Intestinal Ion Transport

M.C. RAO and M. FIELD [1]

Introduction

Studies in mammalian intestine reveal that a variety of humoral, microbial and pharmacological agents known to alter ion transport (either in the absorptive or secretory direction) appear to do so by altering the intracellular concentrations of calcium, cyclic AMP, or cyclic GMP (Field 1981). An increase in the intracellular concentration of any one of these mediators results in net secretion, whereas a decrease results in net absorption. Mammalian ileum can both absorb and secrete water and electrolytes. Net secretion can result from inhibition of an active absorptive process and/or the stimulation of an active secretory process. In contrast, the intestine of the marine teleost, *Pseudopleuronectes americanus* (winter flounder), only absorbs. This intestine is also devoid of crypts, which are believed to be the site of fluid secretion in mammalian intestine.

The effects of a number of modulators of secretion have been examined in our laboratory in both rabbit ileum and winter flounder intestine. These studies will be reviewed here with specific reference to the roles of the above three intracellular mediators in regulating ion transport.

Table 1 summarizes a list of humoral and microbial agents known to cause secretion in mammalian small intestine. They have been classified on the basis of their known or putative intracellular mediators and are listed with respect to the side of the epithelium on which they are maximally effective. In preliminary studies, the secretagogue action of a few of these agents (as measured by a change in transepithelial potential difference) has been tested by us (unpublished observations) and by Donowitz et al. (1981) in the flounder intestine (Table 1). Surprisingly, cholera toxin, *Escherichia coli* heat-stable enterotoxin and carbamylcholine, all potent secretagogues in mammalian small intestine, had no effect on flounder intestine. However, it remains to be established whether this tissue contains the appropriate cell membrane receptors for these agents.

1 The University of Chicago, Departments of Medicine and of Pharmacological and Physiological Sciences, 950 East 59th street, Chicago, IL 60637 and The Mount Desert Island Biological Laboratory, Salsbury Cove, ME 04672, USA

Intestinal Transport
(ed. by M. Gilles-Baillien and R. Gilles)

Table 1. Stimulus-secretion coupling in small intestine

Intracellular mediator	Extracellular stimulus	
	Luminal	Contraluminal
cAMP	Bacterial enterotoxins: cholera [a], heat-labile *E. coli*	Vasoactive intestinal peptide [b]
	Dihydroxy bile acids	Prostaglandins [b]
	Hydroxy fatty acids	Bradykinin
cGMP	Bacterial enterotoxins: heat-stable *E. coli* [a]	??
Ca^{2+}	? Detergents: bile salts, fatty acids	Acetylcholine [a]
		Serotonin [b]
		Substance P [b]
		Neurotensin [b]
Unknown	Other bacterial enterotoxins	Calcitonin
		Bombesin [b]
		Vasopressin

[a] Inactive in flounder
[b] Active in flounder

Role of Intracellular Mediators in Ion Transport in Mammalian Small Intestine

A current hypothesis for the regulation of mammalian intestinal ion transport (with specific reference to rabbit ileum) is shown in Fig. 1. Under normal circumstances, rabbit ileal epithelium absorbs Na and Cl and secretes HCO_3 and exhibits a serosa-positive transepithelial potential difference (PD). Secretory stimuli generally increase

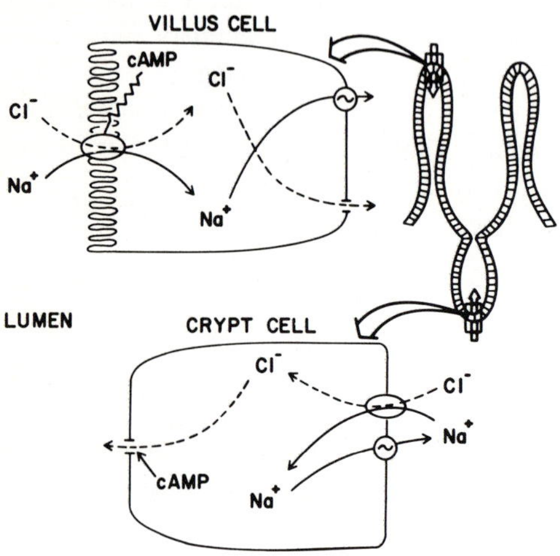

Fig. 1. Postulated model of cAMP action on ion transport in mammalian intestinal villus and crypt cells. There are two secretagogue-sensitive components which are spatially separated: an absorptive, electrically silent process localized to the villus cells and an electrogenic, secretory process localized to the crypt cells

this PD, inhibit Na absorption and reverse net Cl transport from absorption to secretion. In vivo, at least, they also stimulate HCO_3 secretion. There appear to be two secretagogue-sensitive components which are spatially separated in mammalian intestine: an absorptive process localized to the mature villus cells (which have a highly differentiated apical brush border membrane) and a secretory process localized to the crypt cells.

In the villus cell, in addition to providing the driving force for transport of sugars, amino acids and other solutes, the Na gradient (maintained by the Na/K ATPase located on the contraluminal membrane) is also responsible for uphill Cl entry across the brush border membrane. This NaCl cotransport appears to have a 1:1 stoichiometry and, like other NaCl cotransport systems, is inhibitiable by loop diuretics such as furosemide (Guandalini et al. 1982a). In both rabbit ileum and gall bladder this process is also inhibited by cAMP. The low intracellular Na concentration prevents the backflux of Cl through the cotransport system and Cl accumulates in the cell above electrochemical equilibrium. Most of the Cl appears to diffuse out across the contraluminal border. The nature of this serosal Cl permeability is unclear. This absorption of NaCl is electrically silent in the ileum as it is also in the gall bladder (Frizell et al. 1975). Thus an effect of cAMP on this process alone would not explain the associated increase in PD and short-circuit current (I_{sc}) (Field 1971).

In the crypt cell, the mechanism for Cl secretion resembles those present in a number of other secretory epithelia such as trachea (Shorofsky et al. 1982), cornea (Klyce and Wong 1977) and dogfish rectal gland (Silva et al. 1978). Cl enters these cells across the contraluminal membrane; this entry is coupled to that of Na and, therefore, Cl accumulates intracellularly above electrochemical equilibrium. Under nonsecreting conditions, most of the entering Cl simply recycles across the contraluminal membrane. The precise mechanisms by which Cl enters and exits across the basolateral membrane are poorly understood. In the presence of secretory stimuli, such as cAMP, there is a rapid increase in the conductive Cl permeability of the apical membrane, resulting in net Cl secretion. Most of the accompanying Na movement is passive, and probably via the paracellular pathway. Since secretion of Cl is electrogenic, any change in Cl flux across the apical membrane would change the transepithelial PD and therefore also change the electrical driving force for passive Na secretion.

Since Ca and cGMP also stimulate secretion, it is germane to compare their effects on each of the postulated secretagogue-sensitive components with those of cAMP. The effects of the three intracellular mediators on ion transport have been examined in vitro with respect to the following three parameters: (1) Effects on transepithelial potential difference (PD); this, however, does not account for changes in electroneutral processes. (2) Effects on active ion transport as measured by determining the transepithelial unidirectional and net fluxes of individual ions (Na and Cl) in the absence of a chemical or electrical gradient (i.e., under short-circuit conditions). (3) Effects on initial rate of uptake or influx of Cl from the luminal medium into the epithelium as a measurement of the NaCl coupled cotransport process. In our studies, the Ca ionophore A23187, *E. coli* heat-stable enterotoxin (ST) and theophylline have been used as Ca, cGMP and cAMP agonists respectively. (Although theophylline is a nonspecific phosphodiesterase inhibitor, the elevation in cGMP caused by this xanthine derivative [0.27 pmol mg^{-1} protein vs. 0.08 pmol mg^{-1} protein in control (Field et al. 1978a) is by itself insufficient to cause a change in transport (Guandalini et al. 1982a)].

Effect on PD and I_{sc}

Exposure of the serosal surface of stripped ileal preparations to cAMP, cGMP or Ca ionophore results in a rapid increase in PD which reaches a peak value within 5 min and then gradually (20–45 min) declines to and stablizes at about 50% of this peak value (Bolton and Field 1977, Field 1971). It must be noted that on exposure to cAMP and cGMP agonists such as cholera toxin and ST the peak PD values are sustained for prolonged periods of time, presumably due to a continued stimulation of intracellular cyclic nucleotide production (Field 1981). Neither cAMP nor cGMP is dependent on exogenous Ca for its effect and conversely A23187 does not alter intracellular concentrations of cAMP (Bolton and Field 1977) or cGMP (unpublished observations). Of the three intracellular mediators, cAMP appears to produce the largest changes in PD and I_{sc}. As shown in Fig. 2, although half-maximal effects were obtained at 10^{-5} M of each nucleotide, even at maximal doses, cGMP is only 70% as effective as cAMP. The tissue exposed to a maximal dose of cGMP remains sensitive to secretagogue action since subsequent addition of a maximal dose of cAMP results in a further increment in I_{sc}. However, it is important to note that the effects of the two nucleotides are non-additive, the increment caused by maximal doses of cAMP and cGMP together not being greater than that elicited by cAMP alone. Similarly although theophylline causes a greater stimulation of I_{sc} than does Ca ionophore (Bolton and Field 1977) their effects are nonadditive. These results suggest that the three mediators are acting on a common transport system.

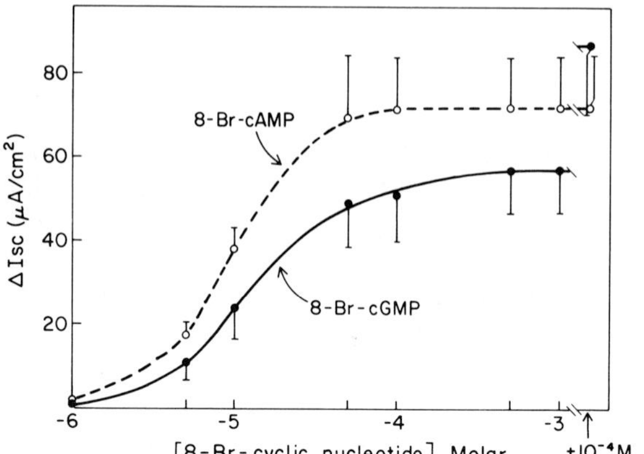

Fig. 2. Increments in short-circuit current (I_{sc}) produced by different concentrations of 8-bromo derivatives of cGMP and cAMP. Progressive doses of each agent were added to paired tissues from same animals (n = 4). *Brackets* represent 1 SE. After addition of 1.0 mM of each nucleotide, 0.1 μmol ml^{-1} of the other nucleotide was added. Although final mean increment in I_{sc} was greater for tissues exposed initially to 8-Br-cGMP and then to 8-BR-cAMP, this difference was not statistically significant. Nucleotides were added to serosal side only. (Guandalini et al. 1982a, reproduced by permission of the Americal Physiological Society)

Effects on Transepithelial Na and Cl Fluxes

All three mediators alter fluxes of Na as well as Cl under short-circuit conditions in the secretory direction. As shown in Table 2, theophylline actually causes net Na secretion by decreasing J^{Na}_{ms}. It causes a larger net Cl secretion by decreasing J^{Cl}_{ms} and increasing J^{Cl}_{sm}. Ca ionophore and *E. coli* ST cause similar changes in unidirectional Na and Cl fluxes (Table 2). In the case of ST these changes are associated with a decrease in Na absorption but not in net Na secretion (Table 2c). Experimental variability could account for this difference since, in the ST experiments, basline J^{Na}_{net} was higher than in the studies with theophylline and Ca ionophore. Recently Guandalini et al. (1982b) have reported that 8-Br-cGMP can cause net Na secretion.

Table 2. Effects of agonists of Ca, cAMP and cGMP on transepithelial ion fluxes in rabbit ileum

Condition	J^{Na}_{ms}	J^{Na}_{sm}	J^{Na}_{net}	J^{Cl}_{ms}	J^{Cl}_{sm}	J^{Cl}_{net}	I_{sc}	J^R	G_T
Control [a]	11.1	10.8	0.3	8.8	8.0	0.8	1.8	2.2	24
Theophylline	8.8 [c]	10.8	– 2.0 [c]	6.6 [c]	11.4 [c]	– 4.8 [c]	4.8 [c]	1.8	19 [c]
Control [a]	10.7	10.6	0.2	10.9	10.6	0.3	1.9	2.1	23
A23187	9.2 [c]	11.2	– 1.9 [c]	9.8	11.8	– 2.0 [c]	2.7 [c]	2.7	20 [c]
Control [b]	12.5	8.7	3.8	7.6	4.9	2.7	1.6	0.6	26
ST	10.4 [c]	10.6	– 0.1 [c]	5.1 [c]	6.5 [c]	– 1.5 [c]	2.9 [c]	1.6	25

Ion fluxes and I_{sc} are in μEq/h cm^{-2} and conductance (G_T) in ms cm^{-2}. J^R is the residual ion flux (I_{sc}-J^{Na}_{net} + J^{Cl}_{net}). Theophylline (5 μmol ml^{-1}) and A23187 (0.5 μg ml^{-1}) were added to the serosal bathing solution and ST (15 mouse units ml^{-1}) to the mucosal bathing solution. (I_{sc} = short circuit current)

[a] After Bolton and Field (1977), modified
[b] After Guandalini et al. (1982a), modified
[c] Different from control at $p < 0.05$

Theophylline and cAMP have been shown not to alter HCO_3 secretion in vitro both by direct measurements (Dietz and Field 1973, Sheerin and Field 1975) and when residual flux, J^R (I_{SC} – J^{Na}_{net} + J^{Cl}_{net}), is used as an index of HCO_3 secretion (Table 2). Ca ionophore and ST also do not appear to alter J^R (Table 2). Although Ca, cAMP or cGMP (or their agonists) do not alter HCO_3 transport in vitro the possibility that they do so in vivo cannot be excluded since cholera toxin has been shown to stimulate HCO_3 secretion in vivo (Leitch and Burrows 1968).

Of the three mediators, Ca and cAMP, but not cGMP, affected tissue conductance (G_T). A23187 caused a temporary decrease in G_T (120 min) which later reversed (Bolton and Field 1977). Cyclic AMP caused a more persistent decrease in G_T, an effect which in studies on *Necturus* gall bladder is accompanied by a reorientation of intramembranous junctional fibrils and has been correlated with decreases in the permeability of the paracellular pathway (Duffey et al. 1981).

As also suggested by the lack of additivity in PD change, all three mediators appear to act on the same transport processes since combined addition of Ca ionophore and theophylline or ST and theophylline does not result in ion flux changes

Table 3. Effects of combined additions of Ca or cGMP agonists with cAMP agonist on ion transport in rabbit ileum

Condition	J^{Na}_{ms}	J^{Na}_{sm}	J^{Na}_{net}	J^{Cl}_{ms}	J^{Cl}_{sm}	J^{Cl}_{net}	I_{sc}	J^{R}	G_T
A23187 (0.5 μg ml⁻¹)-treated [a]									
Baseline	8.6	11.5	− 2.9	7.1	11.1	− 4.0	3.7	2.6	22
+ Theophylline	9.2	11.1	− 2.0	7.2	12.6	− 5.4	5.3 [c]	1.9	21
ST (15 m.u. ml⁻¹)-treated [b]									
Baseline	Not determined			5.2	4.8	0.4	2.3	–	21
+ Theophylline	Not determined			4.0 [c]	6.6 [c]	− 2.5 [c]	3.5 [c]	–	21

Units are as expressed in Table 2. Theophylline (5 μmol ml⁻¹) was added to A23187 or ST-treated tissues.

[a] After Bolton and Field (1977), modified

[b] After Field et al. (1978a), modified

[c] Different from baseline at $p < 0.05$

greater than that seen with theophylline alone (i.e., no additive response) (Table 3). Theophylline, however, is a more effective secretagogue than ST or cGMP since, in ST-treated tissues, theophylline is capable of causing a further stimulation of electrogenic Cl secretion. The results with Ca ionophore are more equivocal: although the effects of theophylline in the presence of ionophore are not statistically different from those with ionophore alone, over a large range of experiments theophylline has a greater effect on mean J^{Cl}_{net} than does the Ca ionophore.

Effects on Initial Rate of Cl Influx

In contast to their differing capacities for stimulating electrogenic anion secretion, Ca, cAMP and cGMP appear to be equally effective in inhibiting Na coupled Cl influx across the brush border. When added on the luminal side, the loop diuretic furosemide inhibits Na-dependent Cl influx without stimulating electrogenic acion secretion. This is evident from the absence of an associated increase in I_{sc} (Table 4A). As also shown in Table 4A, the cGMP agonist, ST, inhibits the same fraction of Cl transport as furosemide (the larger mean change produced when the two agents were added together is not statistically significant) and causes a modest increase in I_{sc}. Theophylline and cAMP are known to maximally inhibit Cl influx, and, as shown in Tables 4B and 4C, these effects are equal to and not additive to the changes noted with ST or Ca ionophore. However, theophylline does cause a larger increase in I_{sc} than ST.

Table 4. Effects of Ca, cAMP and cGMP agonists on Cl influx in rabbit ileum

Condition	J^{Cl}_{me} Cl influx	ΔI_{sc}
A. Comparison of furosemide and ST		
Control	12.4	0.1
Furosemide	9.2 [a]	0.3
ST	9.3 [a]	1.3 [a]
Furosemide + ST	8.1 [a]	1.2 [a]
B. Comparison of ST and theophylline		
Control	11.8	0.2
ST	9.8 [a]	0.8 [a]
Theophylline	9.8 [a]	2.5 [a]
ST + Theophylline	9.6 [a]	2.1 [a]
C. Comparison of theophylline and A23187		
Control	4.9	1.8
Theophylline	4.1 [a]	3.3 [a]
A23187	4.3 [a]	3.1 [a]
Theophylline + A23187	3.7 [a]	3.9 [a]

Influx and I_{sc} are in μEq/h cm^{-2}. Experiments summarized in Panels A and B were conducted in 1/2 Cl Ringers (65 mM) and those in Panel C in 1/4 Cl Ringers (30 mM). The Cl in both cases was replaced with equimolar amounts of mannitol and SO_4^{2-}. The difference in baseline influx values in C (vs. A and B) is in part due to this use of different Cl concentrations. The lower concentrations of Cl were employed to reduce the contribution of the diffusional component (Smith et al. 1981). Values represent means from paired experiments. Panels A and B after Guandalini et al. (1982a), modified.

[a] Different from control at $p < 0.05$

Role of Intracellular Mediators in Ion Transport in Flounder Intestine

The intestinal epithelium of the winter flounder is less heterogeneous than the mammalian ileum: there are no morphologically distinct crypt regions and there are no differences at the ultrastructural level between the cells at the apex and base of the infoldings (Field et al. 1978b). A striking feature of this epithelium is that the absorptive cells are highly elongated (60 μ long × 3.5 μ wide at the apical end) with an apical brush border and a basolateral membrane with many infoldings. Adjoining cells are connected by junctional complexes at the apical end and the large intracellular spaces are bridged by intermittent desmosomes along the lateral surfaces. This unusual morphology may contribute to the unique transport properties of this epithelium: under open-circuit conditions, in the absence of a chemical gradient, the stripped flounder epithelium generates a serosa-negative PD and, under short-circuit conditions, although the epithelium absorbs almost 3 times as much Cl as Na, active Cl absorption ceases in the absence of Na, and occurs by a coupled cotransport driven by the Na gradient. Like rabbit ileum this cotransport process is inhibitable by loop diuretics such as furosemide (Frizzell et al. 1979b). Figure 3 illustrates a model proposed to explain these and some additional features of the transport properties of

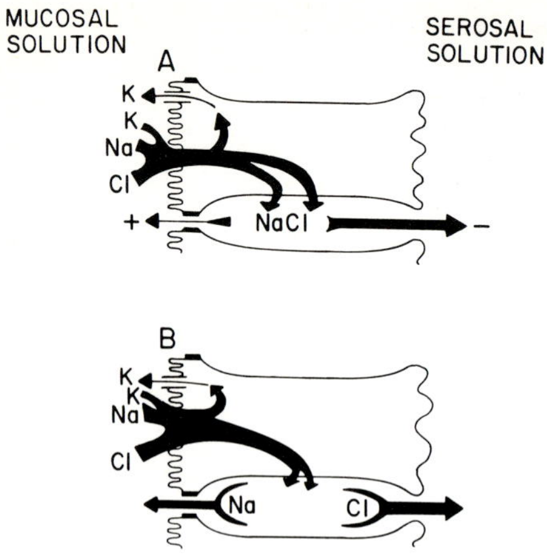

Fig. 3A, B. A postulated model of ion transport in flounder intestine. **A** Model to account for serosa-negative PD across the flounder under open-circuit conditions (modified from the model proposed for the rabbit gall bladder by Machen and Diamond, 1969): The serosa-negative PD is generated by a salt diffusion potential across the tight junction and is maintained by the cation selectivity of this junction. **B** Model to account for ion transport changes under short-circuit conditions: A threefold greater rate of Cl absorption than Na is explained by the presence of a highly cation-selective intercellular junctional complex that allows most of the transported Na to recycle to the luminal solution while preventing a similar movement of Cl

the flounder epithelium. The threefold difference in Na and Cl absorption has been explained on the basis of permselective and resistive properties of the paracellular shunt: a highly cation-selective junctional complex which is more permeant to Na than is the remainder of the lateral intercellular space. This results, under short-circuit conditions (Fig. 3B), in the recycling of most of the Na that is pumped out of the cell back to the luminal solution; in contrast, the relative Cl impermeability of the junctions causes virtually all of the Cl to enter the serosal solution. Under open-circuit conditions, the transepithelial serosa-negative PD is generated by a salt diffuison PD across the tight junction (since active transport is believed to increase salt concentrations in the lateral spaces) and is maintained by the cation selectivity of this junction. Since Cl mobility in free solution is greater than that of Na, a Cl diffusion potential along the lateral space may also make a small contribution to the transepithelial PD (Fig. 3A) (Field 1978, Frizzell et al. 1979a). Recent evidence (Musch et al. 1982a) shows that in addition to absorbing Na and Cl this tissue secretes K under short-circuit conditions. It has also been demonstrated that the brush border cotransport system is dependent on K in addition to Na and Cl and is in reality a Na/K/Cl cotransport process. A luminal, Ba-sensitive K conductive pathway is also present. Under short-circuit conditions (Fig. 3B), the K entering via the cotransport mechanism is recycled to the luminal solution via this K conductance pathway.

Since flounder intestine is an absorbing epithelium and possesses a Na/K/Cl cotransport mechanism, it provides an interesting model system to study regulation of absorptive processes. As mentioned earlier, calcium, cAMP and cGMP or their agonists have been tested for their ability to alter flounder intestinal ion transport (Field et al. 1980, Donowitz et al. 1981, Rao et al. 1982).

Effects on Transepithelial PD

All three intracellular mediators or their agonists cause a reduction in the absolute value of the transepithelial PD (Table 5). This change in PD develops more slowly

Table 5. Effects of Ca, cAMP and cGMP agonists on short-circuit current (I_{sc}) in flounder intestine

Condition	ΔI_{sc} μamp cm^{-2}	G_T mS cm^{-2}
Control	–	24
8-Br-cGMP (0.2 mM) [a]	84 [d]	26
8-Br-cAMP (0.2 mM) [a]	43 [d]	33 [d]
Serotonin (0.1 mM) [b]	37.0 [d]	22.1
Furosemide (1 mM) [c]	102.0 [d]	25
8-Br-cGMP (0.2 m) + 8-Br-cAMP (0.2 mM) [a]	90 [d]	33 [a]

Values represent peak I_{sc} changes. All agents with the exception of furosemide were added to the serosal bathing solution.
[a] After Rao et al. (1982), modified
[b] After Donowitz et al. (1981), modified.
[c] After Frizzell et al. (1979b), modified
[d] Different from control at $p < 0.05$

(over 10–15 min) than in mammalian intestine. In no case has the addition of a secretagogue resulted in the generation of serosa-positive PD, suggesting inhibition of absorption but no stimulation of secretion. As in mammalian ileum, the combined effects on PD of serotonin and cAMP (data not shown) and those of cAMP and cGMP are nonadditive in the flounder intestine, suggesting that the three mediators are acting on the same transport mechanism. In contrast to its effects on rabbit ileum, in the flounder cAMP caused a sustained increase in G_T, whereas, like in the mammalian ileum, Ca ionophore caused a transient decrease in G_T and cGMP had no effect. The effects of cAMP on G_T persist even in the presence of ouabain which reduces J^{Cl}_{net} and J^{Na}_{net} to zero. In contrast to rabbit ileum, in flounder intestine cGMP causes a larger reduction in transepithelial PD than does cAMP and can cause a further reduction in PD in tissues pretreated with maximal doses of cAMP; cGMP however does not reverse cAMP-induced changes in G_T. The loop diuretics furosemide and bumetanide also reduce transepithelial PD to near zero values without altering G_T.

Effects on Transepithelial Cl and K Fluxes

As suggested by their effects on PD, all 3 mediators inhibit Cl absorption, cGMP being the most effective (Table 6). In the presence of the calcium-dependent agonist, serotonin, there was a reduction in J^{Cl}_{ms} and no change in J^{Cl}_{sm} resulting in inhibition of Cl absorption. The effects of A23187 on the flounder are more ambigous and the ionophore may not be the optimal Ca agonist in this tissue. Only about 60% of the fish tested responded to the ionophore with a change in PD (Donowitz et al. 1981). Ca effects on K fluxes have not been examined.

Cyclic GMP and furosemide have similar effects on ion fluxes in the flounder: (1) both agents reduce J^{Cl}_{ms} resulting in an almost total inhibition of net Cl absorption.

Table 6. Effects of cAMP and cGMP on transepithelial ion fluxes in flounder intestine

Condition	J_{ms}^{Rb} [a]	J_{sm}^{Rb}	J_{net}^{Rb}	J_{ms}^{Cl}	J_{sm}^{Cl}	J_{net}^{Cl}
Control	0.3	1.0	– 0.7	7.6	3.2	4.4
cAMP (0.2 mM)	0.5 [b]	0.85	– 0.35 [b]	14.5 [b]	11.1 [b]	3.4
cGMP (0.2 mM)	0.56 [b]	0.79	– 0.23 [b]	3.0 [b]	2.2	0.8 [b]

Fluxes are in μEq/h cm^{-2} and represent average values. Cyclic nucleotides were added to the serosal bathing solution. After Rao et al. (1982), modified

[a] This tissue has been found not to distinguish between Rb and K

[b] Different from control at $p < 0.05$

(2) They both inhibit net K secretion by decreasing J_{sm}^{K} and causing a small increase in J_{ms}^{K}. (3) Their effects are not additive (Musch et al. 1982a, Rao et al. 1982).

In contrast to the Ca agonists and cGMP, cAMP produces its effects on net Cl transport by increasing both J_{ms}^{Cl} and J_{sm}^{Cl}. The effects of cAMP on J_{sm}^{Cl} persist even in the presence of ouabain or bumetanide when net Na and Cl absorption is abolished (Field et al. 1980, Rao et al. 1982). This suggests that cAMP is altering paracellular permeability to Cl. In recent studies, cAMP and not cGMP was found to inhibit the increase in PD (dilution potential) caused by imposing a NaCl concentration difference (in the absence of an osmotic gradient) across the tissue (Krasny and Frizzell 1982) providing further support for the notion that cAMP destroys the permselective nature of the paracellular pathway. It is clear that this cAMP-mediated increase in ionic permeability is restricted to anions since cAMP effects on J_{net}^{Rb} are similar to those caused by cGMP and are not accompanied by significant increases in unidirectional Rb fluxes (Table 6). Furthermore, cAMP has little effect on the magnitude of unidirectional Na fluxes (Field et al. 1980).

Effects on Influx

Consistent with their effects on transepithelial fluxes, cAMP, cGMP and furosemide (Ca effects on influx have not been determined) inhibited the initial rate of uptake of Cl and Rb into the epithelium (Table 7) (Frizzell et al. 1979b, Rao et al. 1982, Musch et al. 1982b). The effects of cAMP and cGMP were not additive to those of furosemide and again cGMP was more effective in inhibiting influx than cAMP. The disparity between the effects of cAMP in inhibiting J_{me}^{Cl} (Table 7) while increasing J_{ms}^{Cl} (Table 6) merits some consideration. It should be noted that, in order to reduce the paracellular component of J_{me}^{Cl}, the influx experiments were carried out at 50% of the Cl concentration used for transepithelial flux measurements. Although this optimizes assay conditions for the cotransport process. it may obscure other effects of cAMP on Cl permeability. Furthermore, cAMP, although it decreases the furosemide-sensitive component of Cl influx, appears to increase the furosemide-independent component of J_{me}^{Cl}. These results are consistent with simultaneous effects of cAMP on anion permeability and cotransport.

Table 7. Effects of cAMP and cGMP on influx in flounder intestine

Condition	J_{me}^{Cl} % control	J_{me}^{Rb} % control
A. [a]		
Furosemide	54% [d]	20% [b,d]
cAMP + theophylline	84%	39% [d]
cAMP + furosemide + theophylline	67% [d]	23% [d]
B. [c]		
cGMP	34% [d]	23% [d]
Bumetanide	23% [d]	n.d.
cGMP + bumetanide	41% [d]	n.d.

8-Br-cyclic GMP (0.2 mM); dibutyryl cAMP (0.25 mM); theophylline (5 mM); furosemide (1.0 mM) in a and 0.4 mM in b); bumetanide (10 μM)

[a] After Frizzell et al. (1979b), modified
[b] After Musch et al. (1982b), modified
[c] After Rao et al. (1982), modified
[d] Different from control at $p < 0.05$

Conclusions

Calcium and the cyclic nucleotides are intracellular mediators for the action of a number of secretagogues in the intestine. In mammalian systems, these mediators both inhibit absorption and directly stimulate secretion. In teleost intestine, they inhibit absorption but do not stimulate secretion. In both systems, the effects of all three mediators on net fluxes are nonadditive, i.e., they all appear to act on common transport processes. However there are species differences in the potencies of the three mediators. In rabbit ileum Ca, cAMP and cGMP are euqally effective in inhibiting NaCl cotransport (presumably a villus cell function) but cAMP is more effective than cGMP in stimulating electrogenic anion secretion (presumably a crypt cell function). Possibly, the smaller effect of cGMP on anion secretion could be due to a decreasing gradient of cGMP-dependent protein kinase activity from villus to crypt regions that may parallel the previously reported gradient of guanylate cyclase activity in these cells (de Jonge 1975). Although the actions of cAMP and cGMP do not require extracellular Ca for their effects it remains to be established whether they act by mobilizing intracellular pools of calcium. Frizzell (1977) showed that cAMP can increase Ca^{45} efflux from colonic epithelial cells. A key intracellular vehicle for Ca action is the small molecular weight protein calmodulin, whose actions are inhibited by neuroleptic drugs like trifluoperazine and chlorpromazine. These drugs can inhibit A23187, 8-Br-cAMP and ST-induced secretion (Smith and Field 1980, Ilundain and Naftalin 1979). However the phenothiazines are known to have other membrane effects and it is as yet unproven that this inhibition of secretion is due to inhibition of calmodulin action. Similarly, although A23187 does not alter total intracellular cAMP and cGMP concentrations, recent studies in other tissues on intracellular compartmentalization of cyclic nucleotides indicate another level of complexity that needs to be considered (Hayes and Brunton 1982).

In contrast to mammalian intestine, cGMP is more effective than cAMP in inhibiting brush border cotransport in the teleost intestine. Cyclic AMP appears to have the additional effect of markedly reducing the permselectivity of this epithelium most likely by increasing Cl permeability of the intercellular junctions.

The molecular mechanisms underlying these actions of cyclic nucleotides and calcium remain to be elucidated but one possible route is the phosphorylation of specific intracellular proteins, a process which the three mediators are known to affect in other tissues. It is perhaps naive to assume that all three mediators will regulate phosphorylation of a common protein. It is more likely that each will cause phosphorylation of different modulator proteins, each of which regulates a common transport mechanism. Recent studies in our laboratory on in vitro phosphorylation fail to reveal a common substrate for all three mediators in either mammalian or flounder intestine.

Acknowledgements. We are grateful to Ms. Susan Chang for secretarial assistance. This work was supported by United States National Institutes of Health Grants (USPHS) AM-21345, AI-15904 and AM-29778.

References

Bolton J, Field M (1977) Ca-ionophore-stimulated ion secretion in rabbit ileal mucosa: relation to actions of cyclic AMP and carbamylcholine. J Membr Biol 35:159–174

Dietz J, Field M (1973) Ion transport in rabbit ileal mucosa. IV. Bicarbonate secretion. Am J Physiol 225:858–861

Donowitz M, Battisti L, Madara JL, Trier JS, Cusolito S, Carlson S, Field M (1981) Calcium and active transport in intestine of the winter flounder, *Pseudopleuronectes americanus.* Bull Mt Desert Isl Biol Lab 21:22–26

Duffey ME, Hainau B, Ho S, Bentzel CJ (1981) Regulation of epithelial tight junction permeability by cyclic AMP. Nature 294:451–453

Field M (1971) Ion transport in rabbit ileal mucosa. II. Effects of cyclic 3′,5′-AMP. Am J Physiol 221:992–997

Field M (1978) Some speculations on the coupling between sodium and chloride transport processes in mammalian and teleost intestine. In: Hoffman JF (ed) Membrane transport processes, vol I. Raven Press, New York, pp 277–292

Field M (1981) Secretion of electrolytes and water by mammalian small intestine. In: Johnson LR (ed) Physiology of the gastrointestinal tract. Raven Press, New York, pp 963–982

Field M, Graf LH, Laird WJ, Smith PL (1978a) Heat stable enterotoxin of *Escherichia coli:* in vitro effects of guanylate cyclase activity, cyclic GMP concentration and ion transport in small intestine. Proc Natl Acad Sci USA 75:2800–2804

Field M, Karnaky KJ, Smith PL, Bolton JE, Kinter WB (1978b) Ion transport across the isolated intestinal mucosa of the winter flounder, *Pseudopleuronectes americanus.* I. Functional and structural properties of cellular and paracellular pathways for Na and Cl. J Membr Biol 41: 265–293

Field M, Smith PL, Bolton JE (1980) Ion transport across the isolated intestinal mucosa of the winter flounder, *Pseudopleuronectus americanus.* II. Effects of cyclic AMP. J Membr Biol 55: 157–163

Frizzell RA (1977) Active chloride secretion by rabbit colon: calcium-dependent stimulation by ionophore A23187. J Membr Biol 35:175–187

Frizzell RA, Dugas MC, Schultz SG (1975) Sodium chloride transport by rabbit gall bladder. Direct evidence for a coupled NaCl influx process. J Gen Physiol 65:769–795

Frizzell RA, Field M, Schultz SG (1979a) Sodium-coupled chloride transport by epithelial tissues. Am J Physiol 236:F1–F8

Frizzell RA, Smith PL, Vosburgh E, Field M (1979b) Coupled sodium-chloride influx across brush border of flounder intestine. J Membr Biol 46:27–39

Guandalini S, Rao MC, Smith PL, Field M (1982a) cGMP modulation of ileal ion transport: in vitro effects of *Escherichia coli* heat-stable enterotoxin. Am J Physiol 243:G36–G41

Guandalini S, Migliavacca M, Campora E de, Rubino A (1982b) Cyclic guanosine monophosphate effects on nutrient and electrolyte transport in rabbit ileum. Gastroenterology 83:15–21

Hayes JS, Brunton LL (1982) Functional compartments in cyclic nucleotide action. J Cyclic Nucl Res 8:1–16

Ilundain A, Naftalin RJ (1979) Role of Ca^{2+}-dependent regulator protein in intestinal secretion. Nature 279:446–448

Jonge HR de (1975) The localization of guanylate cyclase in rat small intestinal epithelium. FEBS Lett 53:237–242

Klyce SD, Wong RKS (1977) Site and mode of adrenaline action on chloride transport across the rabbit corneal epithelium. J Physiol (Lond) 266:777–799

Krasny EJ Jr, Frizzell RA (1982) Regulation of paracellular permselectivity in flounder intestine. Bull Mt Desert Isl Biol Lab 22: (in press)

Leitch GJ, Burrows W (1968) Experimental cholera in the rabbit ligated intestine: ion and water accumulation in the duodenum, ileum and colon. J Infect Dis 118:349–359

Machan TE, Diamond JE (1969) An estimate of the salt concentration in the lateral intercellular spaces of rabbit gall bladder during maximal fluid transport. J Membr Biol 1:194–213

Musch MW, Orellana SA, Kimberg LS, Field M, Halm DR, Krasny EJ, Frizzell RA (1982a) Na/K/Cl cotransport in the intestine of a marine teleost. Nature 300:351–353

Musch MW, Frizzell RA, Field M (1982b) Kinetics of Na, K, Cl cotransport in flounder intestine. Bull Mt Desert Isl Biol Lab 22: (in press)

Rao MC, Nash NT, Field M (1982) Cyclic GMP: the authentic inhibitor of Na, K, Cl cotransport in flounder intestine. Bull Mt Desert Isl Biol Lab 22: (in press)

Sheerin HE, Field M (1975) Ileal HCO_3 secretion: relationship to Na and Cl transport and effect of theophylline. Am J Physiol 228:1065–1074

Shorofsky S, Field M, Fozzard H (1982) Electrophysiology of Cl secretion in canine trachea. J Membr Biol (in press)

Silva P, Stoff JS, Field M, Fine L, Forrest JN, Epstein FH (1978) Mechanism of active chloride secretion by shark rectal gland: role of Na-K-ATPase in electrogenic chloride secretion. Am J Physiol 232:F298–F306

Smith PL, Field M (1980) In vitro antisecretory effects of trifluoperazine and other neuroleptics in rabbit and human small intestine. Gastroenterology 78:1545–1553

Smith PL, Blumberg JB, Stoff JS, Field M (1981) Antisecretory effects of indomethacin on rabbit ileal mucosa in vitro. Gastroenterology 80:356–365

Neuro Hormonal Control of Intestinal Transport

L.A. TURNBERG[1]

Introduction

An extremely complex picture is emerging of factors which influence intestinal transport and it is likely that final control is dependent on the complicated interplay of a variety of agents which influence the transporting epithelium. While a clear division between hormonal and neurological chemical messengers was originally discerned it is now clear that the division between these is becoming somewhat indistinct and artificial. The classical hormones, liberated from glands at a site removed from the point of action, such as thyroid, adrenal cortex and gonad are fairly clear-cut. However, the gut endocrine system made up of isolated cells in the gut mucosa specialised to produce one or more peptides not only subserves a hormonal role, by secreting their peptides into the bloodstream, but they also effect the local epithelial and sub-epithelial cells without recourse to the circulation. Such may be the case, for example, for gastrin, CCK, substance P and enteroglucagon. There has been an explosion of developments in the recognition of a variety of neurotransmitters other than the classical cholinergic and adrenergic agonists. Thus vasoactive intestinal peptide (VIP), substance P, serotonin, enkephalins and bombesin have been found in the enteric nervous system. Some nerves apparently contain more than one type of peptide. It is clear that some of these peptides are not restricted to nerves but some may also be found in paracrine cells. In addition, some of the neurotransmitters may escape into the bloodstream and subserve an endocrine role.

A further level of complexity also exists. It is now difficult to interpret the effect of a given agent on intestinal transport in isolation from other co-existing stimulating or inhibiting messengers. For example, a dose response curve to one hormone may give quite misleading information since it may bear little relationship to the concentration of that hormone which is effective in the presence of other potentiating agents. Furthermore, studies in which exogenous hormones are administered to produce plasma concentrations similar to those found under physiological circumstances, as for example after a meal, may also not be relevant since plasma concentrations may simply reflect the overflow of much higher local concentrations secreted adjacent to the mucosa. Thus attempts to mimic plasma concentrations by intravenous injection may not be relevant.

1 Department of Medicine, Hope Hospital, University of Manchester School of Medicine, Eccles Old Road, Salford M6 8HD, Great Britain

Intestinal Transport
(ed. by M. Gilles-Baillien and R. Gilles)

Table 1. Chemical messengers which have been shown to be capable of influencing intestinal transport experimentally

	Absorption	Secretion
Hormones	Cortisol Aldosterone Angiotensin (low dose) Prolactin	Gastrin Secretin CCK Enteroglucagon G.I.P.
Neuro-endocrine (including paracrine)	α_2 adrenergic β adrenergic Enkephalins Somatostatin	Acetyl choline V.I.P. Substance P Bombesin Neurotensin Gastrin/CCK tetrapeptide Serotonin ATP
Other local substances		Histamine Bradykinin Prostaglandins

A confused and confusing picture is emerging as an embarassingly large number of chemical messengers are discovered each of which may appear to have effects on intestinal transport under experimental conditions. Table 1 illustrates the range of chemical messengers which have been shown to have an effect on transport under some experimental conditions. It seems somewhat unlikely that all of these play a physiological role and in this chapter I wish to dissect out those which appear to have some potential for a physiological role and then attempt to piece the different elements into an overall picture, incomplete though this may be at present.

Hormones

The glucocorticoids have been shown to enhance sodium absorption in both the small and large intestine in a number of different experimental animals (Sellin and Field 1981, Binder 1978, Tai et al. 1981). Most of these observations have been made with high concentrations of corticosteroids. Nevertheless, there is some evidence that basal cortisol output from normal adrenal glands may influence transport, since in one series of studies, aminoglutethamide, which in that study inhibited endogenous cortisol production without affecting aldosterone output, inhibited sodium chloride absorption and induced secretion in rabbit ileum (Sellin and Field 1981). There is some dissociation between enhancement of activity of the basolateral enzyme sodium potassium ATPase and the enhancement of sodium absorption suggesting that enhanced absorption occurs independently of activation of sodium potassium ATPase. One important observation was the delay of at least three hours following administration of corticosteroids for an effect on transport to be observed suggesting that synthesis of specific proteins involved in transport is necessary.

The salt-retaining steroid, aldosterone, is also capable of enhancing sodium absorption although its effect appears to be limited to the colon, only trivial or no effects being observed in the small bowel (Frizzell and Schultz 1978, Dolman and Edmonds 1975). Since spironolactone inhibits sodium absorption it is likely that endogenous aldosterone output does exert some effect on colonic transport. Aldosterone has a more rapid action than cortisol although presumably it does require the mediation of increased protein production as in other transporting epithelia.

A physiological role for cortisol and/or aldosterone is not proved but it seems highly likely that they influence the diurnal variation in salt and water absorption which may be prominant in some species such as the rabbit colon. The rabbit produces soft edible faeces during the night and hard inedible faeces during the day. These hormones may also be important in retaining salt in certain birds (Thomas and Skadhauge 1979) since in some of these urine finds its way into the colon and control of sodium reabsorption will clearly be more important here than it is in mammalian species. Even in mammals, however, it may have some importance in conditions of salt excess or depletion. In normal man, salt and water output in the faeces is very small indeed, but in diarrhoeal disease there is some evidence that at least in children a secondary hyperaldosteronism modulates sodium excretion and encourages some salt retention (Rubens and Lambert 1972).

Angiotensin, in low dose, enhances small and large bowel sodium absorption (Poat et al. 1976, Levens et al. 1977) and prolactin when given over prolonged periods also enhances absorption (Mainoya et al. 1974, Mainoya 1975). However, a physiological role for either of these is uncertain although it is conceivable that like adrenal steroids they exert a relatively long-term, modest overall control of salt and water absorption.

Calcitonin in high concentrations has been shown to stimulate intestinal secretion in vivo in man (Kenney Gray et al. 1976), but a physiological role for this hormone seems extremely unlikely.

High doses of thyroxin have been shown to stimulate intestinal secretion in some experimental circumstances. Again a physiological role for this hormone in control of transport is unlikely, but it is conceivable that excess thyroid production in Grave's disease may provoke diarrhoea by this mechanism.

Of the gastrointestinal hormones, gastrin, secretin, CCK, enteroglucagon and G.I.P. have each been shown to be capable of inducing intestinal secretion when given in vivo in large doses in a variety of different mammalian species including man (Hicks and Turnberg 1974, Modligliani et al. 1976, Pansu et al. 1980, Matuchansky et al. 1972, Modligliani et al. 1971). However, at concentrations which are similar to those occurring under physiological circumstances they appear to be ineffective. In addition, where these have been examined in vitro on isolated intestinal mucosa they appear to have little or nor effect on transport even in high concentrations. Such is the case, for example, with glucagon (Isaacs and Turnberg 1977). However, we should not entirely discard the possibility that these hormones may have a physiological role for two reasons. Firstly, Professor Bernier and co-workers in Paris (Poitras et al. 1980) examined the effect of a combination of a number of these hormones (gastrin, secretin, G.I.P. and glucagon) given intravenously in doses just sufficient to raise plasma concentrations to those seen after a meal. This combination induced secretion in the jejunum in a group of normal human volunteers. Individually, none of the hormones were capable of exerting any effect at this concentration. They suggested that it is the combination of hormones which is important and presumably each of these has an additive or potentiating effect on the others.

Secondly, the concentration of these hormones appearing in plasma may be much smaller than the local concentration obtainable adjacent to the epithelium. A false idea may be gained from plasma concentrations. However, the observation that those hormones which have been tested in vitro do not apparently influence transport directly suggests that if this combination of hormones does have a physiological effect then it is likely to be mediated by some indirect mechanism, such as an effect on intestinal motility or blood flow.

Thus, there is modest evidence that the gastrointestinal hormones in combination may have a role in limiting absorption of salt and water after a meal, perhaps in order to maintain luminal fluidity while digestion continues.

Neuro-Endocrine Control

Acetyl choline and other cholinergic agonists stimulate intestinal secretion in both small and large bowel mucosa (Isaacs et al. 1976, Morris and Turnberg 1980). Since this effect is blocked by atropine it suggests that it is mediated by muscarinic receptors. Recently specific cholinergic binding receptors have been demonstrated on intestinal epithelial cells (Rimele et al. 1981, Issacs et al. 1982) and certainly there are a number of cholinesterase staining neurones adjacent to the intestinal epithelium particularly around the crypts (Isaacs et al. 1976). All these pieces of evidence support the possibility that the cholinergic innervation of the intestinal epithelium has a role in the control of intestinal transport. Experiments carried out over a century ago, in which the different branches of the autonomic innervation of the intestine were severed also supported the possibilities that cholinergic secretion was a physiological process (see review by Florey et al. 1941).

There is strong evidence too for an adrenergic control mechanism for intestinal transport. Alpha 2 adrenergic stimulation enhances absorption of sodium chloride in vitro (Chang et al. 1982) and furthermore selective binding receptors for alpha 2 adrenergic agonists have been demonstrated on isolated epithelial cells from rabbit ileum (Chang et al. 1982). The mechanism of action for adrenergic enhancement of absorption is not entirely clear but probably does not involve inhibition of adenylate cyclase or of cyclic AMP production. A role for calcium in adrenergic mechanisms is as yet unclear.

A large number of other putative neurotransmitters have been demonstrated in nerve and paracrine cells. Of these, VIP appears to be predominantly localised to nerve (Schultzberg et al. 1980). In the gut, it is clear that VIPergic neurones are distributed predominantly just beneath the mucosa in both large and small bowel. Since VIP is an extremely potent intestinal secretagogue and since specific binding receptors for VIP have been demonstrated on intestinal epithelial cells it seems likely that VIP does have a role in the control of intestinal transport (Davis et al. 1981,

Amiranoff et al. 1980). However, the way in which VIP is released under physiological circumstances remains to be elucidated. Lundgren and his co-workers in Sweden have produced indirect evidence to suggest that VIP may be involved in the secretory diarrhoea associated with cholera toxin. He found that VIP was liberated into the venous drainage of loops of intestine exposed to cholera toxin (Cassuto et al. 1982). Further support in favour of involvement of a neurological reflex arc is evidenced by inhibition of cholera toxin induced secretion by neuroblockade with tetrodotoxin and with ganglion blockers such as hexamethonium and also by local anaesthetics (Cassuto et al. 1981, Cassuto 1981). These intriguing observations require further work.

Substance P and bombesin both apparently induce secretion in isolated rabbit ileal mucosa although their mechanism of action is uncertain (Kachur et al. 1982a,b). They do apparently involve separate receptors. On the other hand, neurotensin which is also capable of provoking secretion probably mediates its effect through some other intermediary (Kachur et al. 1982a). Its action is blocked by tetrodotoxin and by prior desensitisation to substance P, suggesting that neurotensin may exert its actions through substance P. The possibility that any or all of these putative neurotransmitters have an important role in controlling transport remains an open question.

Of interest has been the recent demonstration by several groups of workers that the enteric nervous system contains endogenous opiates, met and leu enkephalins (Smith et al. 1976). It is now clear that not only can enkephalins influence motor activity in the gut, they are also capable of enhancing absorption in a dose related manner (McKay et al. 1981, Dobbins et al. 1980, Kachur et al. 1980). Competitive inhibition of the effects of opiates by the specific antagonist naloxone suggests that they exert their effects through specific receptor interactions (McKay et al. 1981). Enkephalins apparently are more potent than morphine in enhancing chloride absorption, supporting the suggestion that the type of receptor involved is a delta receptor (McKay et al. 1981, Kachur et al. 1980). The electrical effects of opiates are inhibited by tetrodotoxin indicating that they probably mediate their action indirectly through some neurological intermediary (Turnberg et al. 1982). The co-localisation of endogenous opiates with other neurotransmitters such as acetyl choline suggests the possibility that enkephalins are neuro-modulators rather than direct transmitters. The inhibition by calcium of opiate induced effects suggests that there may be some calcium/opiate antagonism (McKay et al. 1981). In animals made tolerant to morphine by repeated injections acute withdrawal provokes a variety of symptoms indlucing diarrhoea. This has recently been shown to be associated with intestinal secretion which supports the idea that endogenous opiates may have a physiological role (unpublished observations). However, much remains to be learnt about the place of opiates in the overall picture.

Somatostatin is found in a particular type of paracrine cell which has a diminutive process protruding from one aspect which apparently directs the liberation of somatostatin to a local group of one or more cells. This directed liberation of somatostatin may have a particular importance in the stomach where it has been shown to be important in regulating release of gastrin. Its role in the intestine is unclear but it has been shown to be capable of enhancing absorption and inhibiting secretion by a number of different secretagogues (Dharmsathaphorn et al. 1980, Guandalini et al. 1980). It is tempting to suggest that it exerts a rapid very local control over the effects of other secretagogues, modulating their effects.

Serotonin is a potent intestinal secretagogue which acts through liberation of calcium (Donowitz et al. 1980). It is found both in enteric neurones and enterochromaphin cells in the mucosa (Gershon 1981), but its physiological role if any is quite uncertain.

Other Compounds

Histamine is found in high concentrations in a number of subepithelial cells which are not mast cells (Lorenz et al. 1973). This pool of histamine turns over rapidly rather than being stored in granules. Histamine through a H1 receptor mechanism is capable of provoking intestinal secretion by a mechanism not involving cyclic nucleotide activation (Linaker et al. 1981). It is conceivable that histamine is liberated when the intestine is damaged or diseased and secretion may be provoked under these conditions by a histamine-mediated mechanism. The same may be true for prostaglandins which are produced rapidly by subepithelial tissues. A number of these (E and F groups) provoke intestinal secretion and they have been invoked as mediators of secretion provoked by a variety of diseases and laxatives for example (Beubler and Juan 1978, Rask-Madsen and Bukhave 1981).

Recently, bradykinin has also been shown to be capable of provoking intestinal secretion (Crocker and Willavoys 1975).

We thus have at least three substances which are almost certainly liberated in inflammatory reactions in the intestine and which are capable of provoking intestinal secretion and presumably diarrhoea under these conditions.

Summary

Of the large number of compounds which have been shown to be capable of influencing intestinal transport, corticosteroids and salt retaining steroids probably exert a prolonged overall control on salt and water absorption and are likely to be important in situations where there are diurnal variations in colonic absorption or in circumstances where salt depletion or salt overload are likely. It is conceivable but by no means proved that the gastrintestinal hormones, circulating in the plasma after a meal, may together limit the rate of absorption of salt and water from the gut. A more rapid short-term regulation of salt and water absorption in localised areas of the bowel is likely to be exerted by the enteric nervous system and paracrine secretions. Of these, reasonable evidence favours cholinergic and alpha 2 adrenergic control of secretion and absorption respectively. In addition, there is strong circumstantial evidence to favour a role for VIP in intestinal secretion, and this may also be relevant to secretion seen in certain diseases. A role for the other neuro-transmitters including enkephalins, neurotensin, substance P and bombesin remains uncertain. Finally, under pathological conditions where there is damage to or disease of the intestine the liberation of prostaglandin, histamine and bradykinin may provoke intestinal secretion and hence diarrhoea.

It is clear that much further work is required before these speculations can be confirmed or refuted and undoubtedly during the next few years the picture will become more confused as new peptides are discovered and their effects on the gut determined.

References

Amiranoff B, Laburthe M, Rosselin G (1980) Characterization of specific binding sites for vasoactive intestinal peptide in rat intestinal epithelial cell membranes. Biochim Biophys Acta 627:215–224

Beubler E, Juan H (1978) PGE-mediated laxative effect of diphenolic laxatives. Naunyn-Schmiedeberg's Arch Pharmacol 305:241–246

Binder HJ (1978) Effect of dexamethasone on electrolyte transport in the large intestine of the rat. Gastroenterology 75:212–217

Cassuto J (1981) Nervous mechanisms in cholera secretion. An experimental study in cats and rats. Ph D thesis, University of Göteborg

Cassuto J, Jodal M, Tuttle R, Lundgren O (1981) On the role of intramural nerves in the pathogenesis of cholera toxin-induced intestinal secretion. Scand J Gastroenterol 16:377–284

Cassuto J, Fahrenkrug J, Jodal M, Tuttle R, Lundgren O (1982) The release of vasoactive intestinal polypeptide from the cat small intestine exposed to cholera toxin. Gut (in press)

Chang EB, Field M, Miller RJ (1982) α_2 adrenergic receptor regulation of ion transport in rabbit ileum. Am J Physiol G237–G242

Chang EB, Field M, Miller RJ (1983) Enterocyte α_2 adrenergic receptors: Yohimbine and P-aminoclonidine binding relative to ion transport (in press)

Crocker AD, Willavoys SP (1975) Effect of bradykinin on transepithelial transfer of sodium and water in vitro. J Physiol (Lond) 253:401–410

Davis GR, Santa CA, Morawski SG, Fordtran JS (1981) Effect of vasoactive intestinal polypeptide on active and passive transport in the human jejunum. J Clin Invest 67:1687–1694

Dharmsathaphorn K, Sherwin RS, Dobbins JW (1980) Somatostatin inhibits fluid secretion in the rat jejunum. Gastroenterology 78:1554–1558

Dobbins J, Racusen L, Binder HJ (1980) Effect of D, alanine methionine enkephalin amide on ion transport in rabbit ileum. J Clin Invest 66:19–28

Dolman D, Edmonds CJ (1975) The effect of aldosterone and the renin angiotensin system on sodium, potassium and chloride transport by proximal and distal rat colon in vivo. J Physiol (Lond) 250:597–611

Donowitz M, Asarkof N, Pike G (1980) Calcium dependence of serotonin-induced changes in rabbit ileal electrolyte transport. J Clin Invest 66:341–352

Florey HW, Wright RD, Jennings MA (1941) The secretions of the intestines. Physiol Rev 21: 36–69

Frizzell RA, Schultz SG (1978) Effect of aldosterone on ion transport by rabbit colon in vitro. J Membr Biol 39:1–26

Gershon MD (1981) The enteric nervous system: an apparatus for intrinsic control of gastrointestinal motility. Viewpoints Digest Dis 13:13–16

Guandalini S, Kachur JF, Smith PL, Miller RJ, Field M (1980) In vitro effects of somatostatin on ion transport in rabbit intestine. Am J Physiol 238:G67–G74

Hicks T, Turnberg LA (1974) Influence of glucagon on the human jejunum. Gastroenterology 67:1114–1118

Isaacs PET, Turnberg LA (1977) Failure of glucagon to influence ion transport across human jejunal and ileal mucosa in vitro. Gut 18:1059–1061

Isaacs PET, Corbett Ch, Riley AK, Hawker PC, Turnberg LA (1976) In vitro behaviour of human intestinal mucosa. The influence of acetylcholine on ion transport. J Clin Invest 58:539–542

Isaacs PET, Whitehead JS, Kim YS (1982) Muscarinic acetylcholine receptors of the small intestine and pancreas of the rat: distribution and the effect of vagotomy. Clin Sci 62:203–207

Kachur JF, Miller RJ, Field M (1980) Control of guinea pig intestinal electrolyte secretion by δ-opiate receptor. Proc Natl Acad Sci USA 77:2753–2756

Kachur JF, Miller RJ, Field M, Rivier J (1982a) Neurohormonal control of ileal electrolyte transport. I. Bombesin and related peptides. J Pharmacol Exp Ther 220:449–455

Kachur JF, Miller RJ, Field M, Rivier J (1982b) Neurohormonal control of ileal electrolyte transport. II. Neurotensin and substance P. J Pharmacol Exp Ther 220:456–463

Kenney Gray T, Brannan P, Juan D, Morawski SG, Fordtran JS (1976) Ion transport changes during calcitonin-induced intestinal secretion in man. Gastroenterology 71:392–398

Levens NR, Munday KA, Stewart CP (1977) The effect of noradrenaline and angiotensin upon intestinal fluid transport in vivo in the rat. J Physiol (Lond) 270:77–78

Linaker BD, McKay JS, Higgs NB, Turnberg LA (1981) Mechanisms of histamine stimulated secretion in rabbit ileal mucosa. Gut 22:964–970

Lorenz W, Matejka E, Schmal A (1973) A phylogenetic study on the occurrence and distribution of histamine in the gastrointestinal tract and other tissues of man and various animals. Comp Gen Pharmacol 4:229–250

Mainoya JR (1975) Analysis of the role of endogenous prolactin on fluid and sodium chloride absorption by the rat jejunum. J Endocrinol 67:343–349

Mainoya JR, Bern HA, Regan JW (1974) Influence of ovine prolactin on transport of fluid and sodium chloride by the mammalian intestine and gall bladder. J Endocrinol 63:311–317

Matuchansky C, Huet PM, Mary JY, Rambaud JC, Bernier JJ (1972) Effects of cholecystokinin and metoclopramide on jejunal movements of water and electrolytes and on transit time of luminal fluid in man. Eur J Clin Invest 2:169–175

McKay JS, Linaker BD, Turnberg LA (1981) Influence of opiates on ion transport across rabbit ileal mucosa. Gastroenterology 80:279–284

Modligliani R, Huet PM, Rambaud JC, Bernier JJ (1971) Effect of secretin upon movements of water and electrolytes across the small intestine in man. Rev Eur Etud Clin Biol 16:361–364

Modligliani R, Mary JY, Bernier JJ (1976) Effect of synthetic human gastrin I on movements of water, electrolytes, and glucose across the human small intestine. Gastroenterology 71:978–984

Morris AI, Turnberg LA (with the technical assistance of L Hall and K Pimblett) (1980) The influence of a parasympathetic agonist and antagonist on human intestinal transport in vivo. Gastroenterology 79:861–866

Pansu D, Bosshard A, Dechelette MA, Vagne M (1980) Effect of pentagastrin, secretin and cholecystokinin on intestinal water and sodium absorption in the rat. Digestion 20:201–206

Poat JA, Parsons BJ, Munday KA (1976) Effects of angiotensin on transporting epithelia. J Endocrinol 68:2–3

Poitras P, Modigliani R, Bernier JJ (1980) Effect of a combination of gastrin, secretin, cholecystokinin, glucagon, and gastric inhibitory polypeptide on jejunal absorption in man. Gut 21: 299–304

Rask-Madsen J, Bukhave K (1981) The role of prostaglandins in diarrhoea. Clin Res Rev 1 (Suppl 1): 33–48

Rimele TJ, O'Dorisio MS, Gaginella TS (1981) Evidence for muscarinic receptors on rat colonic epithelial cells: binding of [^{3}H]Quinuclidinyl benzilate. J Pharmacol Exp Ther 218:426–434

Rubens RD, Lambert HP (1972) The homeostatic function of the colon in acute gastroenteritis. Gut 13:915–919

Schultzberg M, Hokfelt T, Nilsson G, Terenius L, Rehfeld JF, Brown M, Elde R, Goldstein M, Said S (1980) Distribution of peptide- and catecholamine-containing neurons in the gastro-intestinal tract of rat and guinea-pig: immunohistochemical studies with antisera to substance P, vasoactive intestinal polypeptide, enkephalins, somatostatin, gastrin/cholecystokinin, neurotensin and dopamine β-hydroxylase. Neuroscience 5:689–744

Sellin JH, Field M (1981) Physiologic and pharmacologic effects of glucocorticoids on ion transport across rabbit ileal mucosa in vitro. J Clin Invest 67:770–778

Smith TW, Hughes J, Kosterlitz HW et al. (1976) Enkephalins, isolation, distribution and function. In: Kosterlitz HW (ed) Opiates and endogenous opiate peptides. Elsevier/North Holland, Amsterdam New York, pp 57–62

Tai Yean-Heng, Decker RA, Marnane WG, Charney AN, Donowitz M (1981) Effects of methylprednisolone on electrolyte transport by in vitro rat ileum. Am J Physiol 240:G365–G370

Thomas DH, Skadhauge E (1979) Chronic aldosterone therapy and the control of transepithelial transport of ions and water by the colon and coprodeum of the domestic fowl *(Gallus domesticus)* in vivo. J Endocrinol 83:239–250

Turnberg LA, McKay J, Higgs N (1982) The role of opiates in the control of small intestinal transport. In: Case RM, Garner A, Turnberg LA, Young JA (eds) Electrolyte and water transport across gastrointestinal epithelia. Raven Press, New York, pp 287–294

Hormone Regulation of Intestinal Calcium and Phosphate Transport: Effects of Vitamin D, Parathyroid Hormone (PTH) and Calcitonin (CT)

T. DRÜEKE and B. LACOUR[1]

Introduction

Calcium (Ca) and phosphate (P_i) movement across the intestinal wall involves an active or facilitated, saturable, as well as a passive, non-saturable transport component. The transport proceeds almost certainly via both the transcellular and the paracellular routes, the latter process occurring across the tight junction. The former transcellular transport is at least in part coupled to transcellular sodium (Na) flux, i.e., secondary active, and subject to physiological and nutritional regulation, whereas the latter is thought to be a linear function of the transepithelial electrochemical gradient for each respective solute (Bonner and Peterlik 1981).

In contrast to Na, whose intraepithelial distribution appears to be homogeneous, the intracellular distribution of Ca and P_i is quite inhomogeneous and consists of several distinct pools. In addition, P_i undergoes metabolic conversion during its passage through the cell. These facts increase the difficulties in the analysis and the understanding of the mechanisms. Several experimental models have been used in order to elucidate the chain of events that are involved in the transepithelial transport of these ions, including in vivo perfusion of the intestine as well as in vitro techniques such as everted gut sacs, Ussing chambers, islolated enterocytes and enterocyte membrane vesicles.

Schematically, the entry step of Ca into the intestinal mucosa at the luminal side seems to involve a saturable, facilitated Na-dependent transport component as well as a non-saturable, diffusional component, at least in the duodenum (Bronner et al. 1981). The exit step at the basolateral membrane against a steep electrochemical potential seems to involve an ATP-dependent Ca extrusion process (Hildmann et al. 1982, Nellans and Popovitch 1981). In addition, a Ca/Na exchange mechanism could operate separately and contribute to Ca efflux across the basolateral membrane (Hildmann et al. 1982). The calcium movement in the opposite direction, i.e., directed from the serosal side to the cytosol, occurs by passive diffusion along a steep electrochemical gradient. Schematically, the luminal influx of Pi into the intestinal mucosa is a carrier-mediated uptake system whereas the P_i efflux in the opposite direction is passive (Peterlik and Wasserman 1977, 1978). The bulk of the transepithelial transport has been ascribed to a Na^+-dependent, secondary active transcellular pathway which is under the control of vitamin D (Murer and Hildmann 1981, Peterlik and

1 I.N.S.E.R.M. U 90, Hôpital Necker, Paris, France

Intestinal Transport
(ed. by M. Gilles-Baillien and R. Gilles)

Wasserman 1978, Peterlik et al. 1981). The Na-P_i cotransport mechanism across the brush border membrane seems to operate independently of Ca (Berner et al. 1976). No evidence for a coupled transport of Ca and P_i has been obtained. The nature of the exit step at the basolateral plasma membrane is not known (Murer and Hildmann 1981). Probably, P_i leave the cytosol down an electrochemical gradient (Peterlik and Wasserman 1978). A paracellular, low permeability route of diffusional P_i movement could also exist (Peterlik et al. 1981).

The role of specific enzymes involved in Ca and P_i transport such as Ca-ATPase for Ca (Bronner et al. 1981, Ghijsen and Van Os 1979, Murer and Hildmann 1981) and alkaline phosphatase for P_i (Birge and Avioli 1981) seems probable for the former but remains controversial for the latter (Bikle et al. 1978, Matsumoto et al. 1980). Moreover, specific transport and buffering proteins are certainly involved such as CaBP (Wasserman and Taylor 1968), calmodulin (Nellans and Popovitch 1981), IMCaL (Schachter and Kowarski 1982) and others (Bronner et al. 1981).

It must be noted that one of the important steps in Ca and P_i transport which has not yet been actively investigated is the stage of transcellular transport and the apparent maintenance of steep intracellular ion gradients. A subtle regulation of transported Ca and possibly also P_i between the free, ionised state and a bound form (binding proteins, sequestering organelles) must take place but its precise functioning is largely unknown.

Endocrine Regulation of Intestinal Transport

Vitamin D

1,25 diOH vitamin D_3 (1,25 diOH vit D), the most active metabolite of vit. D_3, stimulates intestinal Ca and P_i transport, and during vit D deficiency the intestinal absorption of these ions is impaired. Most experimental evidence favors the view that 1,25 diOH vit D enhances the active on carrier-mediated component of transepithelial Ca transport (Bronner et al. 1982, De Luca et al. 1982, Norman et al. 1982, Rasmussen et al. 1982). The precise site of the hormone action remains, however, unknown. In vitro studies using everted gut sacs (Lee et al. 1981) or enterocyte brush border membrane vesicles (BBMV) (Rasmussen et al. 1979) point to a stimulatory effect on the saturable, Na-dependent Ca entry at the luminal side of the cell. Generally, no such stimulation by 1,25 diOH vit D is believed to occur for the active exit step of Ca at the site of the basolateral membrane. However, a recent report favors the view that the hormone could also be involved in the regulation of ATP-dependent Ca transport at this side of Ca extrusion (Van Os and Ghijsen 1982). The hormone also stimulates the enterocyte uptake of P_i. The maximal velocity (V_{max}) of P_i entry into the epithelium across the luminal membrane is increased as demonstrated in in vitro experiments using everted gut or BBMV (Matsumoto et al. 1980, Peterlik and Wasserman 1978). In contrast to the increase in V_{max}, no change in carrier affinity for P_i (expressed in terms of K_m) could be found. The extrusion of P_i from the cell across its basolateral membrane seems not to be under the control of vit D (Peterlik and Wasserman 1978).

Recently numerous studies have been directed towards a better understanding of the cellular components of Ca and P_i transport in molecular terms. The classical and universally accepted mode of action of 1,25 diOH vit D involves cytosolic receptors, nuclear gene activation, and induction of specific carrier and binding proteins (De Luca et al. 1982). However, in several experimental conditions, no direct relationship could be established between de novo synthesis of Ca binding proteins and the change in Ca transport induced by 1,25 diOH vit D. Several arguments favor the view that genome activation is not the exclusive mechanism of the hormonal action on the intestinal cell (Norman et al. 1982, Rasmussen et al. 1982). These include a stimulation within 30 min of intestinal cAMP (Corradino 1973), alkaline phosphatase (Bachelet et al. 1979, Birge and Avioli 1981), and Ca uptake into Golgi vesicles (MacLaughlin et al. 1980), a rapid uptake of P_i within 1 h (Birge and Miller 1977), a change in fatty acid composition of phosphatidylcholine in BBMV (Max et al. 1978, Matsumoto et al. 1981), and an increase in luminal Ca entry preceding in time the increase in CaBP (Spencer et al. 1976) and occurring in spite of an efficacious blockade of RNA and protein synthesis by inhibitors such as actinomycin D and cycloheximide. Moreover, pretreatment with cycloheximide caused a greater than 95% inhibition of amino acid incorporation into BBMV proteins in 1,25 diOH vit D treated chicks as compared to the stimulation observed in the presence of the hormone alone. However, 1,25 diOH vit D conserved its stimulatory effect on Ca transport even in the presence of cycloheximide (Rasmussen et al. 1982).

For these many reasons, the group of H. Rasmussen and coworkers (Rasmussen et al. 1982) has proposed the hypothesis of a liponomic regulation of Ca transport. This hypothesis postulates a change in the lipid strucutre of the luminal membrane caused by 1,25 diOH vit D. The most striking positive arguments favoring this hypothesis are the following: (1) The polyene antibiotic filipin induced an increase of transcellular Ca transport in the intestine of vit D-deficient chicks, but not of vit D-repleted chicks (Adams et al. 1970, Rasmussen et al. 1979), indicating that filipin was not acting as a Ca ionophore but was modifying the properties of the Ca transport system. (2) The methyl ester of trans-vaccenic acid (MTVA) which decreases membrane fluidity inhibited the Ca effect of 1,25 diOH vit D at the level of intestinal BBMV, whereas the methyl ester of cis-vaccenic acid (MCVA) which increases membrane fluidity was also capable per se of increasing transmembranous Ca transport (Fontaine et al. 1981). It is therefore possible that the change in fluidity of the lipid domains of the membrane could be associated with a shift from a low to a high turnover state of the Ca transport system. (3) 1,25 diOH vit D stimulated Na-dependent P_i transport in enterocyte BBMV, even in the absence of a stimulation of alkaline phosphatase and of new protein synthesis (Matsumoto et al. 1980). (4) 1,25 diOH vit D treatment led to a change in BBMV phospholipid structure in that it increased the relative content of phosphatidylcholine in the membrane and it enhanced the relative content of polyunsaturated fatty acids in the phosphatidylcholine fraction (Matsumoto et al. 1980). The effect of the hormone on membrane lipids preceded in time its effects on Ca transport.

In personal studies (Karsenty et al. 1982) we recently demonstrated that the in vitro addition of 1,25 diOH vit D to the incubation medium led to a significant increase of P_i uptake velocity by isolated enterocytes of normal rats (Fig. 1). The effect was

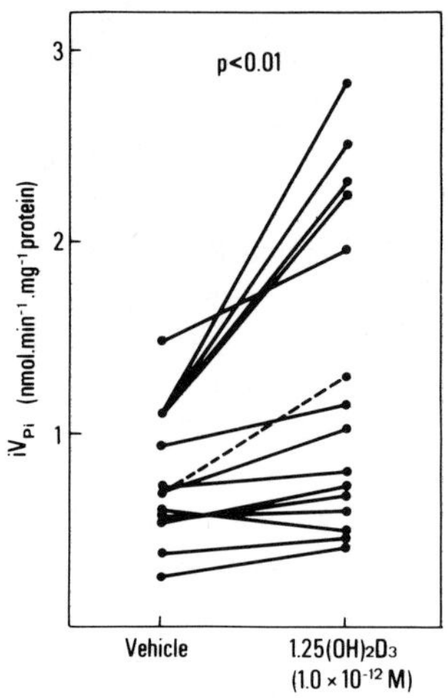

Fig. 1. P_i influx initial velocity (iVp_i) in isolated rat enterocytes, after in vitro exposition of the epithelial cells to either 1,25 diOH vit D (1 pM) or its ethanol vehicle during 20 min. Incubation medium P_i concentration is 3.0 mM, number of experiments = 14. The difference between mean ± SEM iVp_i of hormone and vehicle treated cells *(dashed line,* 1.30 ± 0.16 vs. 0.79 ± 0.09) was significant ($p < 0.01$). Student's paired t test

demonstrated as early as 20 min after cell exposure to the hormone and was inhibitable by prior treatment of the enterocytes with MTVA. These results also suggest a possible direct membrane action of 1,25 diOH vit D in the absence of nuclear gene activation, probably by modifying membrane fluidity.

In summary, vitamin D, via its active metabolite, 1,25 diOH vit D, controls intestinal Ca and P_i transport by the classical, steroidlike mode of genome activation but probably also by a direct, liponomic regulation of the enterocyte fluidity and permeability.

Parathyroid Hormone (PTH)

Earlier work has suggested that PTH could exert a direct action on mucosal cells of the small intestine (Birge et al. 1974). However, according to current views, PTH plays only an indirect role, whereas the influence of 1,25 diOH vit D on intestinal Ca and P_i absorption is primary. In fact, PTH stimulates the activity of 1α-hydroxylase, the renal enzyme effecting the final step in the generation of 1,25 diOH vit D. During recent years it has become apparent that PTH exerts direct effects on numerous organs and cell systems in addition to its well known action at the renal tubular and skeletal level (Massry and Goldstein 1977, Slatopolski et al. 1980). Several recent studies, including personal work, also point to a direct action of PTH on the intestinal mucosa. The proof of a direct effect requires the use of appropriate in vitro models in which the hormone is allowed to act directly at the cellular level in the absence of a previous stimulation of 1,25 diOH vit D synthesis and action.

Nemere and Szego (1981a,b) exposed isolated epithelial cells from rat intestine to the action of PTH in vitro. They were able to demonstrate a transient enhancement of enterocyte Ca uptake by PTH within 20 min of exposure. The effect of PTH and 1,25 diOH vit D on Ca uptake appeared to be additive. The authors speculated that the increased amounts of Ca found in the cells after PTH treatment might be bound by the plasma membrane, taken up by endocytotic vesicles that sequester also the hormonal ligand and function as a signal in the exocytosis of lysosomal enzymes. This hypothesis is based on their finding that PTH in concentrations as low as 10^{-16} M increased lysosomal enzyme release by isolated enterocytes. The increased liberation by PTH of enzymes such as hydrolase, cathepsin B, acid phosphatase and β-N-acetyl-D-glucosaminidase was even greater in enterocytes from parathyroidectomized rats and could be inhibited by indomethacin. Interestingly, 1,25 diOH vit D treatment of epithelial intestinal cells resulted in a similar response of lysosomal enzyme release. In contrast to their stimulatory effect of Na uptake, the enhancing effect of PTH and 1,25 diOH vit D on liberation of lysosomal enzymes was, however, not additive.

The results from several experimental studies done in our laboratory also pointed to an acute, possibly direct action of PTH on Na, Ca, and water transport by the jejunal mucosa of rats (Chanard et al. 1976, Drüeke et al. 1978, Drüeke et al. 1980). The stimulation of endogenous PTH secretion (Chanard et al. 1976, Drüeke et al. 1978) and the intravenous perfusion of PTH (Drüeke et al. 1980, Lacour et al. 1980) led to a decrease of net Na, Ca, and water absorption within 30 min as assessed by the technique of isolated intestinal loops perfused in vivo. Such effects were observed in the absence of increased generation of cAMP or cGMP by the intestinal mucosa (Lacour et al. 1980, 1981). In these in vivo experiments, however, no definite proof for a direct PTH effect could be provided since even in the acute situation the hormone could have exerted systemic or local effects resulting in the stimulation of other mediators.

Therefore, we decided to investigate a possible direct effect of PTH using the in vitro model of isolated intestinal cells. We found that in the presence of bovine PTH (1.2 I.U. ml^{-1}) in the incubation medium, the Na efflux rate constant (K_{Na}) of enterocytes was significantly reduced when compared to control experiments (Lacour et al. 1981). Figure 2 shows a typical experiment of enterocyte K_{Na} in the presence and the absence of PTH, respectively. The direct effect of PTH appeared to be inhibitory for the ouabain-sensitive, but not the ouabain-resistant Na pump (Table 1). The inhibitory effect of the hormone on K_{Na} could not be elicited in the absence of Ca in the incubation medium. Therefore, we speculated that PTH inhibition of entero-

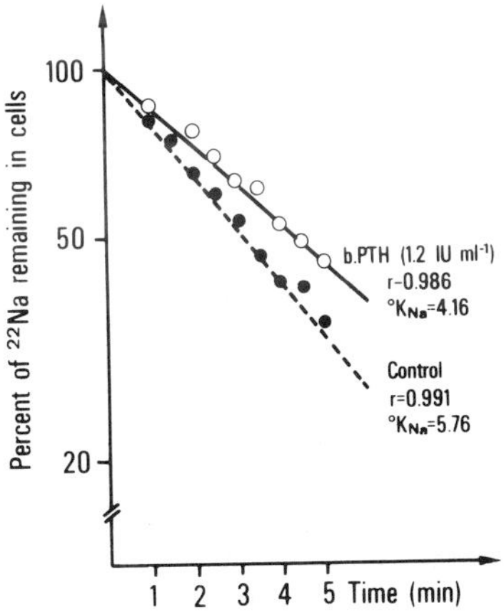

Fig. 2. A typical experiment of Na efflux rate constant (K_{Na}) in isolated rat enterocytes in the presence or absence of bovine parathyroid hormone (b.PTH) in the incubation medium

Table 1. Effect of PTH on total, ouabain-resistant, and oubain-sensitive efflux rate constant (K_{Na})

Hormone	Experimental conditions	Experimental values (h^{-1})		P	% Inhibition
		Hormone	Vehicle		
Bovine PTH (1.2 I.U. ml^{-1})	Total K_{Na}	3.01 ± 0.19 (n = 16)	4.06 ± 0.37 (n = 16)	< 0.02	26
	Ouabain-resistant K_{Na}	2.56 ± 0.27 (n = 7)	2.37 ± 0.21 (n = 7)	n.s.	–
	Ouabain-sensitive K_{Na}	0.45	1.69	–	73

n.s. = not significant

cyte K_{Na} could involve an increase in cytosolic Ca activity and thereby lead to a depression of Na-K-ATPase, similarly to the hormone action in other tissues (Proverbio and Del Castillo 1981). Similarly to 1,25 diOH vit D, the action of PTH on the brush border membrane of the enterocyte could involve a change of membrane fluidity via alterations of phospholipid structure such as phosphoinositides as has been described for kidney tubular epithelium (Bidot-Lopez et al. 1981).

Thus, Ca could be the second messenger of PTH in the intestinal epithelium. Further studies are clearly needed to sustain this hypothesis.

Calcitonin (CT)

It is well established that CT affects intestinal electrolyte and fluid transport. The watery diarrhea which has been reported to accompany approximately 30% of patients with medullary carcinoma of the thyroid (Hill et al. 1973, Steinfeld et al. 1973) has been attributed to the intestinal action of CT even though it could also be due to other hormones secreted by the tumor. Intravenously administered CT has been documented to induce a decrease in net absorption and/or an increase in net secretion of Na, Cl, HCO_3, Ca, P_i, and water in the jejunum and ileum of experimental animals (Drüeke et al. 1980, Kisloff and Moore 1977, Tanzer and Navia 1973) and man (Gray et al. 1973, 1976, Juan et al. 1976). However, these in vivo studies using CT doses higher than physiological levels did not prove a direct action of CT on the intestinal mucosa. CT could act only indirectly, and its effects could be mediated by the release of other hormones such as VIP (Gray et al. 1976) or 5-hydroxytryptamine (Nakhla and Latif 1978). The results of these studies are somewhat contradictory. Walling et al. (1977) reported a direct action of CT on rat ileum in vitro in that it induced a decrease in Na absorption and an increase in Cl secretion. However, extremely high, pharmacological doses of CT had to be used to obtain this effect. Gray et al. (1975) in another in vitro study using rabbit ileum failed to confirm a direct action of CT on water and electrolyte transport. They incriminated possible species differences in response to a foreign peptide hormone. In a personal study using isolated rat enterocytes, we were unable to observe a CT effect on Na uptake or K_{Na} (Lacour et al. 1981).

As to a possible direct effect of CT on intestinal Ca handling, there are also conflicting data from in vitro studies in the literature. Olson et al. (1972) reported a decrease in Ca absorption with low dose CT using isolated vascularly perfused rat small intestines. However, large doeses of CT led to an increase in Ca absorption. These effects could only obtained in vitamin D repleted, but not in vitamin D depleted rats. Walling et al. (1977) using stripped rat ileum in vitro demonstrated that high CT concentrations in the incubation medium led to an increase in bidirectional Ca fluxes but to no change in net Ca transport. In several in vitro studies a possible effect of CT on intestinal adenylate cyclase has been investigated (Kimberg et al. 1972, Lacour et al. 1981, Walling et al. 1977). However, these studies failed to demonstrate such an effect.

Thus, CT infusion into animals and man leads to numerous changes of intestinal electrolyte and water transport including Ca and P_i. Whether these effects are due to a direct action on the intestinal mucosa or only to indirect effects has still to be elucidated.

References

Adams TH, Wang RG, Norman AW (1970) Studies on the mechanism of calciferol. J Biol Chem 245:4432–4442

Bachelet M, Ulmann A, Lacour B (1979) Early stimulation of alkaline phosphatase activity in response to 1,25-dihydroxycholecalciferol. Biochem Biophys Res Commun 89:694–700

Berner W, Kinne R, Murer H (1976) Phosphate transport into brush border membrane vesicles isolated from rat small intestine. Biochem J 160:467–474

Bidot-Lopez P, Farese RV, Sabir MA (1981) Parathyroid hormone and adenosine-3′,5′-monophosphate acutely increase phospholipids of the phosphatidate-polyphosphoinositide pathway in rabbit kidney cortex tubule in vitro by a cycloheximide-sensitive process. Endocrinology 108:2078–2081

Bikle DD, Zolock DT, Morissey RL, Herman RH (1978) Independence of 1,25 dihydroxyvitamin D_3-mediated calcium transport from de novo RNA and protein synthesis. J Biol Chem 253:484–488

Birge SJ, Avioli RC (1981) Intestinal phosphate transport and alkaline phosphatase activity in the chick. Am J Physiol 240:E384–E390

Birge SJ, Miller R (1977) The role of Pi in the action of vitamin D on the intestine. J Clin Invest 60:980–988

Birge SJ, Switzer SA, Leonard DR (1974) Influence of sodium and parathyroid hormone on calcium release from intestinal mucosal cells. J Clin Invest 54:702–709

Bronner F, Peterlik M (1981) Calcium and phosphate transport across biomembranes. Academic Press, London New York

Bronner F, Pansu D, Buckley M, Singh RP, Lipton JH, Miller A III (1981) Intestinal calcium absorption. In: Bronner F, Peterlik M (eds) Calcium and phosphate transport across biomembranes. Academic Press, London New York, pp 135–142

Bronner F, Lipton J, Pansu D, Buckley M, Singh R, Miller A III (1982) Molecular and transport effects of 1,25-dihydroxyvitamin D_3 in rat duodenum. Fed Proc 41:61–65

Chanard J, Drüeke T, Pujade-Lauraine E, Lacour B, Funck-Brentano JL (1976) Effects of saline loading on jejunal absorption of calcium, sodium, and water and on parathyroid hormone secretion in the rat. Pfluegers Arch 367:169–175

Corradino RA (1973) Embryonic chick intestine in organ culture. A unique system for the study of the intestinal calcium absorptive mechanism. J Cell Biol 58:64–78

Drüeke T, Chanard J, Lacour B, Pujade-Lauraine E, Funck-Brentano JL (1978) Effects of hyperoncotic albumin and parathyroid hormone infusion on jejunal electrolyte and water absorption in the rat. Pfluegers Arch 373:249–257

Drüeke T, Lacour B, Chanard J (1980) Acute inhibitory effects of parathyroid hormone and calcitonin on jejunal electrolyte and water transport in normal and thyroparathyroidectomized rats. Min Electrol Metab 4:68–80

Fontaine O, Matsumoto T, Goodman DBP, Rasmussen H (1981) Liponomic control of Ca^{2+} transport: relationship to mechanism of action of 1,25-dihydroxyvitamin D_3. Proc Natl Acad Sci USA 78:1751–1754

Ghijsen WEJM, Os CH Van (1979) Ca^{2+} stimulated ATPase in brush border and basolateral membranes of rat duodenum with high affinity sites for Ca^{2+} ions. Nature 279:802–803

Gray TK, Bieberdorf FA, Fordtran JS (1973) Thyrocalcitonin and the jejunal absorption of calcium, water and electrolytes in normal subjects. J Clin Invest 52:3084–3088

Gray TK, Juan D, Powell DW (1975) Salmon calcitonin and water and electrolyte transport in rabbit ileum. Proc Soc Exp Biol Med 150:151–154

Gray TK, Brannan P, Juan D, Morawski SG, Fordtran JS (1976) Ion transport changes during calcitonin-induced intestinal secretion in man. Gastroenterology 71:392–398

Hildmann B, Schmidt A, Murer H (1982) Ca^{++}-transport across basolateral plasma membranes from rat small intestinal epithelial cells. J Membr Biol 65:55–62

Hill CS, Ibanez ML, Samaan NA, Ahearn MJ, Clark RL (1973) Medullary (solid) carcinoma of the thyroid gland: an analysis of the M.D. Anderson Hospital experience with patients with tumor, its special features, and its histogenesis. Medicine 52:141–171

Juan D, Liptak P, Gray TK (1976) Absorption of inorganic phosphate in the human jejunum and its inhibition by salmon calcitonin. J Clin Endocrinol Metab 43:517–522

Karsenty G, Ulmann A, Lacour B, Pierandreï E, Drüeke T (1982) Early stimulation by 1,25 $(OH)_2$ D_3 of $^{33}P_i$ uptake by isolated enterocytes. In: Norman AW, Schaefer K, Herrath DV, Grigoleit HG (eds) Vitamin D, chemical, biochemical and clinical endocrinology of calcium metabolism. De Gruyter, Berlin New York, pp 345–347

Kimberg DV, Field M, Johnson J, Handerson A, Gershon E (1972) Stimulation of intestinal mucosal adenyl cyclase by cholera enterotoxin and prostaglandins. J Clin Invest 50:1218–1230

Kisloff B, Moore EW (1977) Effects of intravenous calcitonin on water, electrolyte, and calcium movement across in vitro rabbit jejunum and ileum. Gastroenterology 72:462–468

Lacour B, Drüeke T, Chanard J (1980) Effect of parathyroid hormone on jejunal electrolyte and water transport and cyclic nucleotide formation in the rat. Horm Metab Res 12:624–628

Lacour B, Drücke T, Pirandrei E, Nabarra B, Funck-Brentano JL (1981) Rat enterocyte sodium transport in vitro: action of parathyroid hormone and calcitonin. Biochim Biophys Acta 648: 151–161

Lee DBN, Walling MM, Levine BS, Gafter U, Silis V, Hodsman A, Coburn JW (1981) Intestinal and metabolic effect of 1,25-dihydroxyvitamin D_3 in normal adult rat. Am J Physiol 240: G90–G96

Luca HF De, Franceschi RT, Halloran BP, Massaro ER (1982) Molecular events involved in 1,25-dihydroxyvitamin D_3 stimulation of intestinal calcium transport. Fed Proc 41:66–71

MacLaughlin JA, Weiser MM, Freedman RA (1980) Biphasic recovery of vitamin D-dependent Ca^{2+}-uptake by rat intestinal Golgi membranes. Gastroenterology 78:325–332

Massry SG, Goldstein DA (1977) The search for uremic toxin(s) "X". "X" = PTH. Clin Nephrol 11:181–189

Matsumoto T, Fontaine O, Rasmussen H (1980) Effect of 1,25-dihydroxyvitamin D_3 on phosphate uptake into chick intestinal brush border membrane vesicles. Biochim Biophys Acta 599:13–23

Matsumoto T, Fontaine O, Rasmussen H (1981) Effect of 1,25-dihydroxyvitamin D_3 on phospholipid metabolism in chick duodenal mucosal cells. J Biol Chem 256:3354–3360

Max E, Goodman D, Rasmussen H (1978) Purification and characterization of chick intestine brush border membrane. Biochim Biophys Acta 511:224–239

Murer H, Hildmann B (1981) Transcellular transport of calcium and inorganic phosphate in the small intestinal epithelium. Am J Physiol 240:G409–G416

Nakhla AM, Latif A (1978) A possible role for 5-hydroxytryptamine as a mediator for calcitonin actions on the gastrointestinal tract and pancreas in rats. Biochem J 176:339–342

Nellans HN, Popovitch JR (1981) Calmodulin-sensitive, ATP-dependent calcium transport by plasma membrane of small intestinal epithelium. In: Bronner F, Peterlik M (eds) Calcium and phosphate transport across biomembranes. Academic Press, London New York, pp 163–166

Nemere I, Szego CM (1981a) Early actions of parathyroid hormone and 1,25-dihydroxycholecalciferol on isolated epithelial cells from rat intestine: I. Limited lysosomal enzyme release and calcium uptake. Endocrinology 108:1450–1462

Nemere I, Szego CM (1981b) Early actions of parathyroid hormone and 1,25-dihydroxycholecalciferol on isolated epithelial cells from rat intestine: II. Analyses of additivity, contribution of calcium, and modulatory influence of indomethacin. Endocrinology 109:2180–2187

Norman AW, Putkey JA, Nemere I (1982) Intestinal calcium transport: pleiotropic effects mediated by vitamin D. Fed Proc 41:78–83

Olson BE, Luca H De, Potts JT Jr (1972) Calcitonin inhibition of vitamin D-induced intestinal calcium absorption. Endocrinology 90:151–157

Os, CH Van, Ghijsen WEJM (1982) Calcium transport mechanisms in rat duodenal basolateral plasma membranes: effects of 1,25 $(OH)_2D_3$. In: Norman AW, Schaefer K, Herrath DV, Grigoleit HG (eds) Vitamin D, chemical, biochemical and clinical endocrinology of calcium metabolism. De Gruyter, Berlin New York, pp 295–297

Peterlik M, Wasserman RH (1977) Effect of vitamin D_3 and 1,25-dihydroxyvitamin D_3 on intestinal transport of phosphate. Adv Exp Med Biol 81:323–332

Peterlik M, Wasserman RH (1978) Effect of vitamin D on transepithelial phosphate transport in chick intestine. Am J Physiol 234:E379–E388

Peterlik M, Fuchs R, Sing Cross H (1981) Phosphate transport in the intestine: cellular pathways and hormonal regulation. In: Bronner F, Peterlik M (eds) Calcium and phosphate transport across biomembranes. Academic Press, London New York, pp 173–179

Proverbio F, Castillo JR Del (1981) Na^+-stimulated ATPase activities in kidney basal-lateral plasma membranes. Biochim Biophys Acta 646:99–108

Rasmussen H, Fontaine O, Max E, Goodman DBP (1979) The effect of 1α-hydroxyvitamin D_3-administration on calcium transport in chick intestine brush border membrane vesicles. J Biol Chem 254:2993–2999

Rasmussen H, Matsumoto T, Fontaine O, Goodman DBP (1982) Role of changes in membrane lipid structure in the action of 1,25-dihydroxyvitamin D_3. Fed Proc 41:72–77

Schachter D, Kowarski S (1982) Isolation of the protein IMCal, a vitamin D-dependent membrane component of the intestinal transport mechanism for calcium. Fed Proc 41:84–87

Slatopolski E, Martin K, Hruska K (1980) Parathyroid hormone metabolism and its potential as a uremic toxin. Am J Physiol 239:F1–F12

Spencer R, Charman M, Wilson P, Lawson E (1976) Vitamin D-stimulated intestinal calcium absorption may not involve calcium-binding protein directly. Nature 263:161–163

Steinfeld CM, Moertel CG, Woolner LB (1973) Diarrhea and medullary carcinoma of the thyroid. Cancer 31:1237–1239

Tanzer FS, Navia JM (1973) Calcitonin inhibition of intestinal phosphate absorption. Nature 242:221–222

Walling MW, Brasitus TA, Kimberg DV (1977) Effects of calcitonin and substance P on the transport of Ca, Na, and Cl across rat ileum in vitro. Gastroenterology 73:89–94

Wasserman RH, Taylor AM (1968) Vitamin D-dependent calcium-binding protein. J Biol Chem 243:3987–3993

Part 4
Comparative Aspects of Intestinal Transport

Comparative Aspects of Amino Acid Transport in Guinea Pig, Rabbit and Rat Small Intestine

B.G. MUNCK [1]

Introduction

This chapter will deal mainly with the transport of cationic amino acids and imino acids, including non-α-monoaminocarboxylic acids across the brush-border membrane of the guinea pig, rabbit and rat small intestine. It will be based on data from ongoing research in my laboratory. The aim of this research is to resolve a number of the differences between current views on intestinal transport of amino acids. These differences exist mainly regarding the number of transport mechanisms for cationic and neutral amino acids both within and between species (Munck and Schultz 1969a,b, Preston et al. 1974, Munck and Rasmussen 1975, 1979, Paterson et al. 1979, 1980, 1981, Sepulveda and Smith 1978, Robinson and van Melle 1982), on the question of the stimulating effects of a number of neutral amino acids on the transport of cationic amino acids (Robinson and Felber 1964, Munck 1965, 1966b, 1980b, Munck and Schultz 1969b, Robinson and Alvarado 1977), and on the question of the transport of imino acids and non-α-neutral amino acids (Hagihira et al. 1961, Munck 1966a, Peterson et al. 1970, Sepulveda and Smith 1978, Stevens et al. 1982).

Symbols, Abbreviations, Methods

The unidirectional fluxes involved in transepithelial fluxes are indicated by a J where the symbols m, c, and s for the mucosal, cellular, and serosal side compartments are arranged as subscripts with the direction of the flux indicated by the order of the symbols, as in J_{mc} for influx across the brush-border membrane; the generally accepted abbreviations for sugars and amino acids are used as indices to show which substance is under consideration, as in J_{mc}^{Lys} for lysine influx across the brush-border membrane. The concentration of a substance is symbolized by its abbreviation in rectangular parentheses as in $[Lys]_m$ for the concentration of lysine in the fluid bathing the mucosal surface of the epithelium. A compartment's electrical potential is symbolized by ψ with the appropriate subscript. The electrical potential difference across the brush-border membrane: $PD_a = \psi_c - \psi_m$.

1 Institute of Medical Physiology A, The Panum Institute Blegdamsvej 3C, DK-2200 Cøpenhagen N, Denmark

Intestinal Transport
(ed. by M. Gilles-Baillien and R. Gilles)

ABA, β-ABA, and GABA symbolize respectively α-, β-, and γ-amino-n-butyric acid, MeAIB, Meala and Megly are the N-methyl derivatives of amino-isobutyric acid, alanine, and glycine (Megly = sarcosine).

For measurements of J_{mc} the technique of Schultz et al. (1967) has been applied as previously described using incubation periods of 0.5 min. The steady state transmural fluxes, J_{ms} and J_{sm} have been measured under short circuit conditions, using the Ussing technique (Ussing and Zerahn 1951).

It is assumed that transport across the brush border membrane is the sum of one or two Michaelis-Menten processes and free diffusion:

$$J^A_{mc} = MM_1 \, (+ MM_2) + P[A]_m, \text{ where} \tag{1}$$

$$MM = \frac{J_{max} \, [A]_m}{K_t + [A]_m + \frac{K_t \, [I]}{K_i}}$$

P = passive permeability of A in μmol cm^{-2} (serosal area) · h^{-1} · $[A]_m^{-1}$

J_{mc} = μmol cm^{-2} (serosal area) h^{-1}

K_t and K_i are in mM.

Consequently the estimates of transport kinetics presented in Eq. (1)–(4) and (6)–(8) were obtained by non-linear least-squares fitting of the model to the experimentally determined relationships between J^A_{mc} and $[A]_m$. The errors of these estimates are S.D., and they are evaluated by the chi-square test.

The K_i values of tables are, as described in the tables, calculated from ratios between inhibited and uninhibited fluxes assuming that these are described by Eq. (1). Errors on fluxes and K_i are S.E. For these data P values below 0.05 by Student's t-test are taken as evidence of statistical significance.

PD_a, Unstirred Layers, Diffusion

The characterization of the several carriers of amino acids in the intestinal brush border membrane rests on interpretations of studies of inhibitory interactions, fully or partially competitive, including studies of selfinhibition. The manifestations of these interactions are influenced by the thickness of unstirred layers, by the PD_a and its possible partial depolarization by the amino acids, and by the contribution to the J_{mc} by free diffusion. These aspects of the transport across the brush-border membrane must be considered before the transport of individual amino acids can be evaluated.

Unstirred Layers

The influence of the thickness of unstirred layers on the apparent kinetics of intestinal transport has been treated in theoretical papers (Winne 1973, Thomson and Dietschy 1977). In experiments where the rate of stirring has been graduated by means of magnetic stirring, a dramatic difference between stirred and unstirred conditions has also been demonstrated, leading to severe criticism of studies in which magnetic stirring has not been used (Thomson and Dietschy 1980). However, the influx technique of Schultz et al. (1967) was designed in a way which assured an unstirred layer of moderate thickness (Preston et al. 1974), and in the version used in the author's laboratory its thickness has been further reduced by placing septa between the wells of the apparatus, allowing a much higher rate of oxygen flow into the wells, and by removing surplus water from the mucosal surface by touching it with soft paper before the test solution is injected. To allow a comparison with the quantitative studies (Thomson and Dietschy 1980) on which the criticism of most previous studies was based, the kinetics of glucose transport in the rabbit small intestine was reexamined. J_{mc}^{Glu} was measured in paired experiments at 7 different concentrations between 0.5 and 100 mM D-glucose. The results are well described as

$$J_{mc}^{Glu} = \frac{4.8 \pm 0.2\,[Glu]_m}{(1.9 \pm 0.1) + [Glu]_m} + (0.033 \pm 0.003)\,[Glu]_m \tag{2}$$

By the chi-square test the fit between this equation and the experimental data is characterized by a P-value of 0.9.

With a dry weight (d.w.) of 6.79 ± 0.20 mg cm^{-2} = 14.7 cm^2/100 mg d.w., 4.8 μmol cm^{-2} h^{-1} corresponds to 1.18 μmol/100 mg d.w. min^{-1}, which is 5 times the value of 0.23 μmol/100 mg d.w. · min, observed by Thomson and Dietschy (1980). In spite of this much higher rate of transport, the Schultz technique provides an estimate of K_t identical to that observed by these authors at their maximal rate of stirring. This confirms the conclusion of Preston et al. (1974) that with the influx technique described by Schultz et al. (1967) only moderate unstirred layers exist.

Partial Depolarization of PD_a

The coupled transport of Na^+ and sugars or amino acids across the intestinal brush border membrane is rheogenic (Gunter-Smith et al. 1982, White and Armstrong 1971, Okada et al. 1977, Rose and Schultz 1971). Sugars and amino acids as well as two amino acids restricted to use one carrier each will therefore induce some Na^+-dependent mutual inhibition (Murer et al. 1975, Munck 1980a). The effect on PD_a and consequently the degree of mutual inhibition will increase with increasing rate of the rheogenic influx and decrease with increasing conductance of the luminal membrane (Schultz 1980). Therefore a certain rate of influx through one rheogenic, Na^+-nonelectrolyte pathway will have a smaller effect, if another such pathway has already been activated. In addition any effect on PD_a will be attenuated if the basolateral Na^+-K^+-pump is activated and/or the passive permeabilities of the limiting cell membrane are increased (Gunter-Smith et al. 1982, Schultz 1977, 1981). The earlier failure to demonstrate mutual inhibition between sugars and amino acids in rabbit (Chez et al. 1966) and rat (Munck 1972) and the only 10%–20% inhibition seen in the guinea pig (Robinson and Alvarado 1977), Munck 1980a) is therefore consistent with the relative small effects of sugars and amino acids on PD_a.

As a prelude to the studies on guinea pig, rabbit, and rat intestine, the magnitude of the PD_a-related fraction of the inhibitory effect of lysine on the transport of leucine (guinea pig, rat) or alanine (rabbit) was evaluated through the effect of lysine on the influx of D-galactose as measured at 1 mM D-glalactose in the presence of 10 mM leucine (guinea pig and rat) or 90 mM alanine (rabbit). In the guinea pig J_{mc}^{Gal} was reduced by 13% from 0.20 ± 0.13 (8) to 0.17 ± 0.14 (8) μmol cm^{-2} h^{-1}; in the rabbit by 13% from 0.33 ± 0.028 (16) to 0.29 ± 0.020 (16) μmol cm^{-2} h^{-1}, and in the rat by 15% from 0.14 ± 0.026 (8) to o.12 ± 0.019 (8) μmol cm^{-2} h^{-1}. For all three species the inhibition of J_{mc}^{Gal} by 200 mM lysine was statistically insignificant. The present studies have all been performed in the presence of 5 mM glucose which represents 70% saturation of the sugar carrier [Eq. (2)], whereas 1 mM galactose represents 17% saturation (Goldner et al. 1977). Therefore, the observed lysine-inhibitions of J_{mc}^{Gal} overestimate the PD_a related fraction of the lysine inhibition of the transport of neutral amino acids.

Diffusion

Several procedures have been used to estimate the diffusional contribution to the intestinal transport of sugars and amino acids, such as the rate of transport of D-amino acid or L-sugars (Murer et al. 1975, Mircheff et al. 1980), or the rate of transport observed in the presence of high concentrations of competitive inhibitors (Lerner and Karcher 1978, Stevens et al. 1982). The advantage of these procedures is that unquestionably the probe has the right passive permeability. But they are misleading if D-amino acids and L-sugars are transported or if low affinity carriers are operating in parallel with the one being studied.

It has been shown that tetraethylammonium (TEA) and lysine have identical passive permeabilities in the rat small intestine (Munck and Rasmussen 1979). We have therefore measured J_{mc}^{TEA} also in the guinea pig and rabbit. In the guinea pig, rabbit, and rat J_{mc}^{TEA} was respectively 0.026 ± 0.003 (16), 0.015 ± 0.005 (16), and 0.016 ± 0.006 (8) μmol cm^{-2} h^{-1}. As a possible estimate of the passive permeability of the neutral amino acids J_{mc} of mannitol was measured at 1 mM mannitol. In the guinea pig, rabbit, and rat J_{mc}^{Man} was respectively 0.061 ± 0.005 (16), 0.034 ± 0.002 (13), and 0.021 ± 0.002 (16) μmol cm^{-2} h^{-1}. When in rabbit and guinea pig J_{mc}^{Gal} was measured at 1 mM galactose + 200 mM glucose, it was reduced to respectively 0.025 ± 0.003 (16) and 0.028 ± 0.006 (16) indicating that passive permeabilities of molecules of the size of sugars and amino acids are overestimated by J_{mc} of mannitol. Therefore, unless independent estimates of the amino acid passive permeabilities are available a value of 0.02 μmol cm^{-2} h^{-1} mMol^{-1} has been used.

Transport of Cationic Amino Acids

The first description (Hagihira et al. 1961) of a separate transport system for cationic amino acids contained some evidence of overlapping specificity with neutral amino acids. Subsequently, mutual inhibition between neutral and cationic amino acids was demonstrated in sacs (Munck 1966b, 1968, Reiser and Christiansen 1969) and rings (Robinson 1968) of everted rat intestine, and later in isolated rat enterocytes (Reiser and Christiansen 1971a,b, 1972, 1973). These studies also dealt in detail with the phenomenon of stimulation of tissue uptake (Robinson and Felber 1964) and transepithelial net transport of cationic amino acids by several neutral amino acids. However, for methodological reasons (too long periods of incubation, measurements of net transepithelial transport, or net tissue uptake, exposure of both luminal and basolateral membranes to the bathing solutions, and use of too narrow ranges of concentrations) is was impossible to interpret the results of these studies in terms of unidirectional fluxes across the two limiting cell membranes.

The first unequivocal evidence of mutual inhibition between neutral and cationic amino acids and sharing of one or more transport systems in the brush border membrane was reported for the rabbit small intestine by Munck and Schultz (1969a,b), who resolved the transport of both neutral and cationic amino acids into two saturable processes. In these studies it was established that lysine was transported by a high affinity-low capacity carrier which appeared sodium-independent and by a sodium-dependent low affinity-high capacity carrier; but it was not determined which of these was shared with neutral amino acids. Preston et al. (1974) observed a 30% inhibition of J_{mc}^{Met} by a series of cationic amino acids, but did not incorporate this observation in their evaluation of the characteristics of the transport of neutral amino acids across the rabbit small intestine. The study by Paterson et al. (1981) is the first follow-up on the results of Munck and Schultz (1969a,b). In this report it is shown that in the absence of sodium alanine is a competitive inhibitor of a fraction of J_{mc}^{Lys}, and lysine a competitive inhibitor of a fraction of J_{mc}^{Ala}. The data indicate that it is a high affinity-low capacity carrier for both lysine and alanine which is involved in these interactions. The situation at normal concentrations of sodium was not examined.

In the rabbit small intestine Sepulveda and Smith (1978) and Paterson et al. (1979, 1980) have also studied in details the transport of neutral amino acids. Following the terminology of Oxender and Christensen (1963) these authors propose that neutral amino acids cross the brush border membrane by a low affinity-high capacity carrier, which is totally sodium-independent and a high affinity-low capacity carrier, which is completely sodium-dependent. However, the results of Curran et al. (1967), Munck and Schultz (1969a,b), Paterson et al. (1981) and those of Christensen et al. (1969) suggest for the rabbit intestine an alternative model with three carriers. *Carrier 1* is the principal carrier of neutral amino acids. At 140 mM sodium it is a high affinity-high capacity carrier and at 0 mM sodium a low affinity-high capacity carrier of only neutral amino acids. *Carrier 2* is both at 140 and 0 mM sodium a high affinity-low capacity carrier of both neutral and cationic amino acids. *Carrier 3* is the low affinity-high capacity carrier of lysine, which at 140 mM sodium is a low affinity-high capacity carrier also of neutral amino acids, but at 0 mM sodium a low affinity-high capacity carrier of lysine, which is inaccessible to neutral amino acids.

In the rat lysine appears to be transported by only one carrier (Munck and Rasmussen 1975, 1979), which can be completely inhibited by neutral amino acids (Munck and Rasmussen 1975, Munck 1980b). It has not yet been determined whether this carrier also transports neutral amino acids. In the guinea pig these problems have not yet been studied.

The stimulating effect of neutral amino acids on the transepithelial transport of cationic amino acids (Munck 1965, 1966b) was confirmed for the rabbit and resolved into its effects on the two steps of transepithelial passage, the first across the brush border and the second across the basolateral membrane. It was found to consist of an inhibition of the former and a stimulation of the latter (Munck and Schultz 1969b). In addition preloading of the epithelium with neutral amino acids enhanced the influx of cationic amino acids (Munck and Schultz 1969b). These are also the effects of neutral amino acids on the transport of cationic amino acids in the rat (Munck 1980b). In contrast Robinson and Alvarado (1977) have observed that at low concentrations of cationic amino acids low concentrations of neutral amino acids stimulate the initial rate of uptake of the cationic amino acids, which is inhibited by higher concentrations of neutral amino acids. These differences have been ascribed to methodological weaknesses (Robinson and Alvarado 1977, Munck 1980b); and the more likely possibilities that either too narrow concentration ranges were used, or that real species differences were involved have not been seriously considered.

Kinetics of Influx of Lysine Across the Brush-Border Membrane

The Guinea Pig. For this species the kinetics of the transport of cationic amino acids have not previously been published. Therefore J_{mc}^{Lys} has been studied using unfasted, 300–400 g animals of either sex. Using the mid small intestine for paired measurements J_{mc}^{Lys} was measured at concentrations between 0.5 and 100 mM lysine in the presence of 5 mM glucose. The results were analyzed assuming that the J_{mc}^{TEA} of 0.026 ± 0.003 μmol cm^{-2} h^{-1} described the passive permeability to lysine. It was found that

$$J_{mc}^{Lys} = \frac{(0.20 \pm 0.04)\,[Lys]_m}{(0.31 \pm 0.16) + [Lys]_m} + \frac{(8.92 \pm 1.39)\,[Lys]_m}{(92 \pm 21) + [Lys]_m} + 0.026 \pm 0.003\,[Lys]_m \quad (3)$$

provided the best description of this transport function. By the chi-square test the fit between this equation and the experimental data is characterized by a P value of 0.8. Thus in the guinea pig as in the rabbit, lysine uses at least two transport systems. When compared with data for the rabbit Paterson et al. (1981, and below), the most striking difference is the very low capacity of the high affinity carrier.

The Rabbit. The data of Paterson et al. (1981) confirmed those of Munck and Schultz (1969a) with respect to the existence of two carriers, of which one is sodium-independent. These results also indicated that the previous estimates of the kinetics for lysine transport (Munck and Schultz 1969a) could be too inaccurate. In order to get a better estimate of the kinetics and a set of directly comparable data for the further study of the interaction between neutral and cationic amino acids, J_{mc}^{Lys} was measured at concentrations between 0.5 and 200 mM at 5 mM glucose, and both at 140 and 0 mM sodium. For this purpose the distal 20–30 cm small intestine from unfasted 2500–3000 g female albino rabbits were used. Assuming again that the diffusional contribution to the transport of lysine is well described by J_{mc}^{TEA}, this process was best described at 140 mM sodium as

$$J_{mc}^{Lys} = \frac{(2.33 \pm 0.34)\,[Lys]_m}{(0.56 \pm 0.20) + [Lys]_m} + \frac{(6.87 \pm 1.60)\,[Lys]_m}{(64.4 \pm 34.5) + [Lys]_m} + 0.015 \pm 0.005\,[Lys]_m \quad (4)$$

by the chi-square test P = 0.05, and at 0 mM sodium as

$$J_{mc}^{Lys} = \frac{(1.19 \pm 0.15)\,[Lys]_m}{(1.23 \pm 0.28)\,[Lys]_m} + \frac{(8.8 \pm 3.5)\,[Lys]_m}{(231 \pm 149) + [Lys]_m} + 0.015 \pm 0.005\,[Lys]_m \quad (5)$$

by the chi-square test P = 0.85.

In principle these results confirm those of Munck and Schultz (1969a) and Paterson et al. (1981). They do, however, demonstrate that although the high affinity-low capcity carrier is sodium-independent with respect to its K_t, its maximum rate of transport is reduced by 50% under sodium-free conditions.

The Rat. For the rat mid small intestine J_{mc}^{Lys} has previously been examined (Munck and Rasmussen 1975, 1979). Using concentrations between 0.1 and 60 mM lysine no evidence was procued of the involvement of more than one transport system in addition to diffusion, and the transport could be adequately described as

$$J_{mc}^{Lys} = \frac{2.25\,[Lys]_m}{3 + [Lys]_m} + 0.012\,[Lys]_m \quad (6)$$

Transport Mechanisms Common to Neutral and Cationic Amino Acids

The Guinea Pig. In the guinea pig cationic amino acids are transported by two carriers one with high and one with low affinity [Eq. (3)]. Consequently a neutral amino acid, leucine, was tested as inhibitor both at a low (1 mM lysine) and a high (20 mM lysine) concentration of a cationic amino acid. The effect of low concentrations of leucine on J_{mc}^{Lys} at 1 mM lysine was also examined under sodium-free conditions. Finally, lysine was tested as an inhibitor of J_{mc}^{Leu} at 10 mM leucine. As demonstrated by Fig. 1 the results of these experiments provided important information on the characteristics of the guinea pig intestine. (1) Figure 1A and B shows that leucine is only a partial inhibitor of J_{mc}^{Lys}, with the maximum degree of inhibition suggesting that it is the low affinity-high capacity lysine carrier, which is insensitive to neutral amino acids. (2) Confirming the observations of Robinson and Alvarado (1977) Fig. 1C demonstrates that at low concentrations leucine strongly stimulates J_{mc}^{Lys}. This figure also shows that the stimulating effect is sodium-dependent, and at 0 mM

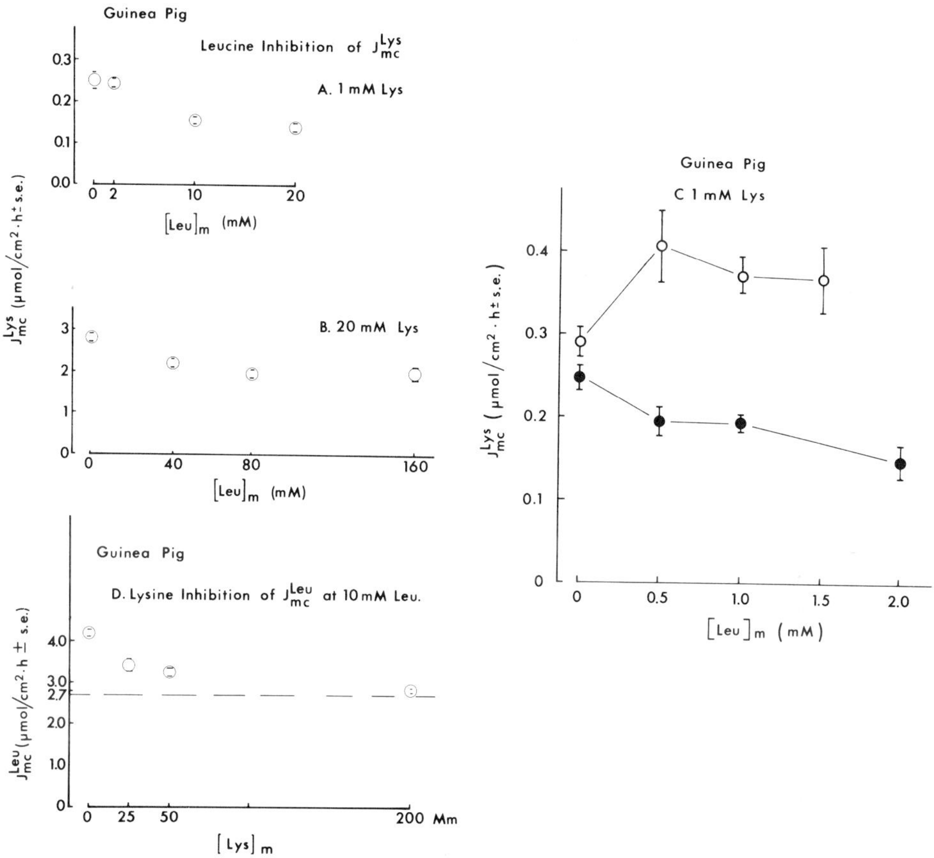

Fig. 1. A Leucine inhibition of lysine influx across guinea pig small intestine measured at 1 mM lysine in the presence of 5 mM D-glucose. B Leucine inhibition of lysine influx across guinea pig small intestine measured at 20 mM lysine in the presence of 5 mM D-glucose. C Na^+-dependence of cis-stimulation of lysine influx in guinea pig small intestine measured at 1 mM lysine in the presence of 5 mM D-glucose. 140 mM Na^+ (○); 0 mM Na^+ (•). 0 mM Na^+. D Lysine inhibition of leucine influx across guinea pig small intestine measured at 10 mM leucine in the presence of 5 mM D-glucose. *Dashed line* indicates the magnitude, 2.7 μmol cm^{-2} h^{-1} of the lysine resistant fraction of leucine influx obtained by substituting choline chloride for sodium choride and by repeated changing of the sodium-free preincubation fluid

sodium turned into inhibition. This latter type of interaction suffices to rule out that the partial inhibition illustrated by Fig. 1A and B could be caused by a leucine induced depolarization of PD_a. (3) By comparing the degree of lysine inhibition of J_{mc}^{Leu} (Fig. 1D) with the inhibitory effect of 200 mM lysine on J_{mc}^{Gal}, it is seen that leucine must be transported by at least one of the carriers of lysine, most likely that with a high affinity and a low capacity for lysine.

Rabbit. Also in the rabbit lysine is transported by a high affinity-low capacity and by a low affinity-high capacity carrier. This makes it necessary that neutral amino acids be used as inhibitors both at a relatively low and a relatively high concentration of a cationic amino acid. It will be clear that neutral amino acids also have high affinities for the high affinity lysine carrier and relatively low affinities for the low affinity lysine carrier. Therefore transport of neutral amino acids by the lysine carriers must also be examined at low and high concentrations. In order to allow direct comparison with the work of Sepulveda and Smith (1978), Paterson et al. (1979, 1980, 1981) and Munck and Schultz (1969b), data will be presented for the interaction of lysine with both leucine and alanine.

The interactions of neutral amino acids with the carriers of cationic amino acids are characterized by Figs. 2–7, of which Fig. 6 characterizes the lysine-insensitive transport of alanine. In Table 1 K_i values for alanine, leucine, methionine and lysine from the present study are compared with previously reported estimates. The present estimates are reached as described in the legends for the figures.

Table 1. Kinetics of mutual inhibition between lysine, leucine, and alanine (K_i in mM) for transport across the brush border membrane of rabbit small intestine. The K_i values for the mutual inhibition between neutral and cationic amino acids were calculated by applying Eq. (1) to the data described in Figs. 4–6, using the procedures described in the text and in the legends for these figures

Inhibitors	Substrates				
	1 mM lys	20 mM lys	1 mM leu	1 mM ala	1 mM ala (+ 40 mM lys)
Leu at 140 mM Na^+	1.1 ± 0.1 (6)	36 ± 6 (3) [a]			
Leu at 0 mM Na^+	1.2 ± 0.1 (3)				
Lys at 140 mM Na^+			0.9; 1.5	5 ± 3 (3)	
Lys at 0 mm Na^+			0.5; 1.9	0.9; 2.6	
Ala at 140 mM Na^+	7.7 ± 0.3 (3)	85 ± 6 (3)			12.6 ± 1.1 (3) [b]
Ala at 0 mM Na^+	8.9 ± 0.7 (3) [d]				76 ± 5 (3) [c]

[a] Preston et al. (1974): $K_t^{ala; 0\ Na^+}$: 59 ± 6 mM, $K_t^{leu; 0\ Na}$: 27 ± 6 mM
[b] Paterson et al. (1979): $K_t^{ala; 140\ Na^+}$: 4.6 mM
[c] Paterson et al. (1980): $K_t^{ala; 0\ Na^+}$: 77 ± 27 mM
[d] Paterson et al. (1980): $K_i^{ala; 0\ Na^+}$: 14.3 mM

From Fig. 2A and B it is evident that alalnine and leucine at 140 as at 0 mM sodium are fully competitive inhibitors of the high affinity-low capacity carrier of lysine. The data also demonstrate that for both neutral amino acids K_i is sodium-independent. Figure 3 shows that alanine fully inhibits the transport of lysine by the low affinity-high capacity carrier, but does so with a K_i more than 10 times the K_i for the high affinity carrier. Figure 4 describes the inhibitory effect of leucine on J_{mc}^{Lys} measured at 20 mM lysine. In addition to allowing the same conclusion as for alanine with respect to the high capacity lysine carrier the data of this figure show that at 0 mM sodium the low affinity-high capacity carrier of lysine becomes insen-

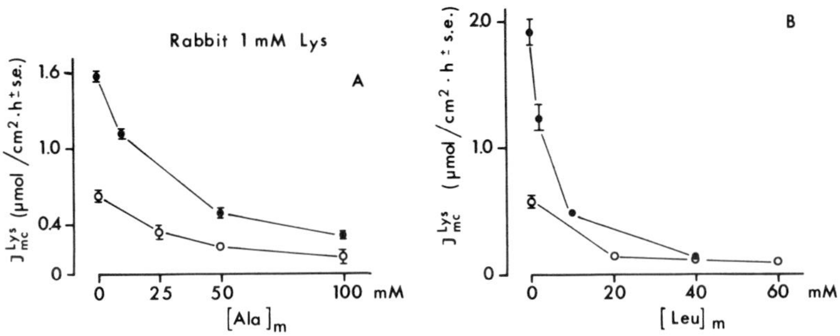

Fig. 2A, B. Influx of lysine across the brush border membrane of rabbit small intestine measured at 1 mM lysine + 5 mM D-glucose at 140 mM Na^+ (•); or at 0 mM Na^+ (○). **A** Inhibitory effect of alanine. Calculated from the mean values as described in Eq. (1) assuming a diffusional contribution of 0.015 µmol cm^{-2} h^{-1} and a K_t^{lys} of 0.5 mM the K_i^{ala} was found to be 7.7 ± 0.3 mM at 140 mM Na^+, and 8.9 ± 0.7 mM at 0 mM Na^+. **B** Inhibitory effect of leucine, calculated as for alanine K_i^{leu} = 1.1 ± 0.1 mM at 140 mM Na^+ and 1.2 ± 0.1 at 0 mM Na^+

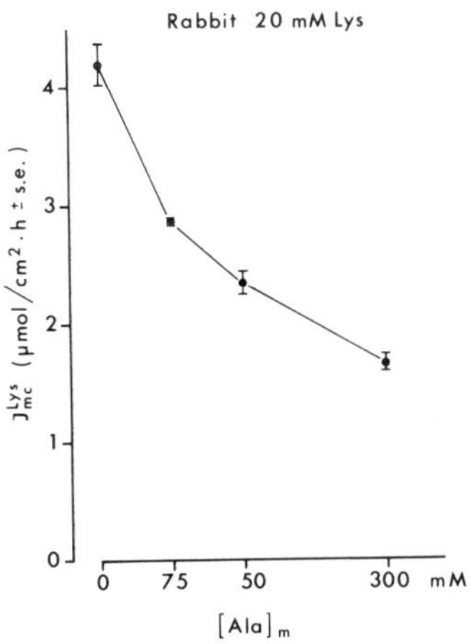

Fig. 3

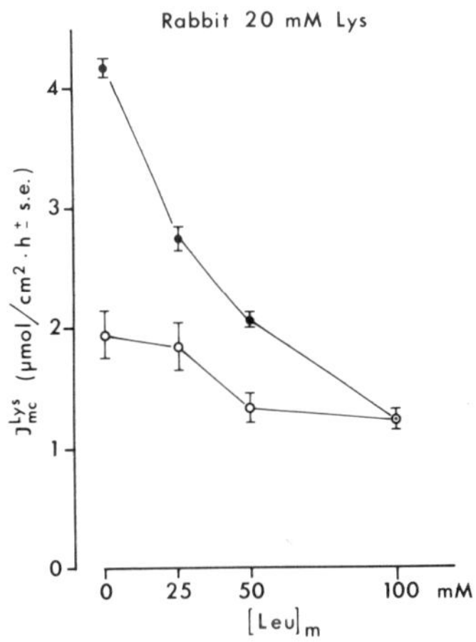

Fig. 4

Fig. 3. Alanine inhibition of lysine transport across the brush border membrane of the rabbit small intestine measured at 20 mM lysine + 5 mM D-glucose. Assuming a J_{max} of 2.25 µmol cm^{-2} h^{-1} at K_t^{lys} of 0.5 mM and a K_i^{ala} of 7.7 mM with respect to the high affinity lysine carrier the data of the figure were corrected for transport by this carrier; assuming then a K_t^{lys} of 70 mM and using the procedure of calculation described in Fig. 4, the K_i^{ala} for the low affinity carrier of lysine was 85.6 mM

Fig. 4. Leucine inhibition of lysine transport in rabbit small intestine measured at 20 mM lysine + 5 mM D-glucose at 140 mM Na^+ (•); or at 0 mM Na^+ (○). Making assumptions as described for Fig. 5 except that at 140 mM Na^+ the K_i^{leu} = 1.1 mM the K_i^{leu} for the low affinity carrier of lysine was 36 ± 6 mM

sitive to leucine. Identical results obtained using methionine as inhibitor indicate that in general, in the absence of sodium, neutral amino acids have access only to the high affinity-low capacity carrier of lysine. These results agree with those of Paterson et al. (1981) who observed that at 0 mM sodium alanine could only eliminate a fraction of J^{Lys}_{mc}.

The inhibitory effect of lysine on J^{Ala}_{mc} (1 mM) and J^{Leu}_{mc} (1 mM) is described by Fig. 5. Both at 140 and at 0 mM sodium lysine is a partial inhibitor of the transport of alanine and leucine with a sodium-independent K_i which is in the range of its K_t for the high affinity-low capacity carrier of lysine. From Fig. 2 it appears that alanine and leucine have sodium-independent affinities for the high affinity carrier of lysine. Here we see that in the absence of sodium the lysine sensitive fraction of J^{Ala}_{mc} and J^{Leu}_{mc} are both reduced by a factor of approximately 10. Thus, as for lysine, the maximum rates of transport of alanine and leucine by the high affinity carrier of lysine are sodium-dependent.

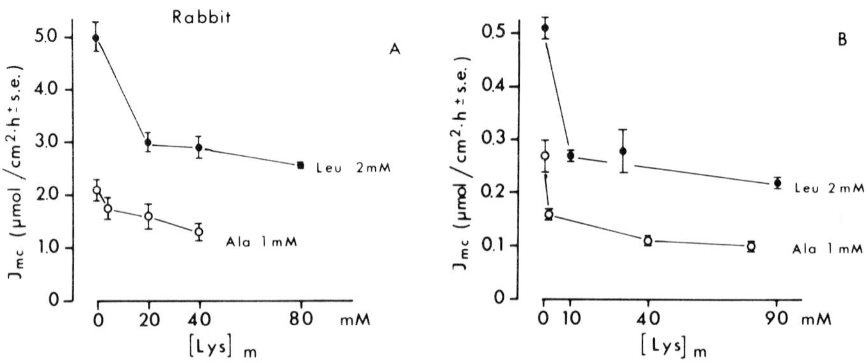

Fig. 5A, B. Lysine inhibition of influx of leucine (•) and alanine (○) across the rabbit brush-border membrane measured at 1 mM alanine or 2 mM leucine at 140 mM Na^+ (**A**) and at 0 mM Na^+ (**B**). K_i^{lys} was calculated as described for Fig. 4 assuming that $K_i = K_t$ for both alanine and leucine, that at 140 mM Na^+, the lysine resistant fractions of J^{ala}_{mc} and J^{leu}_{mc} are 1.3 and 2.7 μmol cm^{-2} h^{-1} respectively, and at 0 mM Na^+ respectively 0.1 and 0.2 μmol cm^{-2} h^{-1}. The results are included in Table 3

This discussion has shown beyond doubt that neutral amino acids are transported by the high affinity carrier of lysine, but only demonstrated that they are perfect competitive inhibitors of the low affinity lysine carrier. For intestinal transport of amino acids, a dissociation between fully competitive interaction with a carrier and transport by that carrier has not yet been demonstrated with certainty. Therefore, the most likely conclusion is that at 140 mM sodium neutral amino acids are also transported by the low affinity carrier of lysine. Sepulveda and Smith (1978, 1979) and Paterson et al. (1979, 1980) observed that at 140 mM sodium neutral amino acids are transported by a low affinity carrier in addition to the classical high affinity carrier of neutral amino acids. Figure 6 demonstrates that in the presence of a high concentration of lysine and at 140 mM sodium, alanine is transported by a single

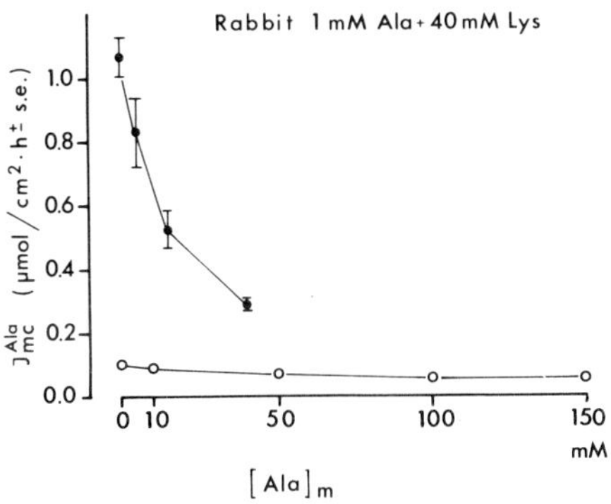

Fig. 6. Alanine inhibition of influx of alanine across the rabbit brush border membrane measured at 1 mM alanine + 5 mM D-glucose and 40 mM lysine, at 140 mM Na^+ (•) and at 0 mM Na^+ (○). Calculated as described in Fig. 4. $K_i^{ala} = K_t^{ala}$ = 12.6 ± 1.1 mM and 76 ± 5 mM at 140 mM Na^+ and 0 mM Na^+ respectively

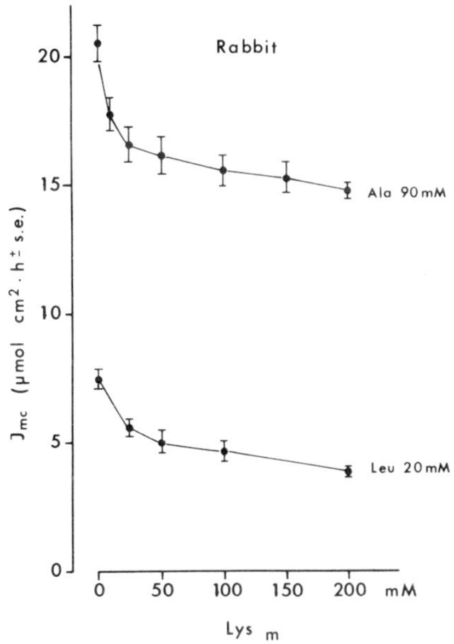

Fig. 7. Lysine inhibition of alanine and leucine influx across the rabbit brush-border membrane measured at 90 mM alanine or 20 mM leucine in the presence of 5 mM D-glucose

high affinity carrier, and at 0 mM sodium by a single low affinity carrier. Together these observations strongly support that neutral amino acids use the low affinity carrier of lysine as a low affinity carrier when sodium is present.

Using a concentration at which the low affinity carrier of lysine should significantly contribute to their transport, the inhibitory effect of lysine on the transport of alanine (90 mM), leucine (20 mM), and metionine (20 mM), (not shown) was examined. As shown for alanine and leucine (Fig. 7) 200 mM lysine reduced J_{mc} of all three amino acids by more than corresponding to the maximum rate of the high affinity-low capacity lysine carrier, the data of both Figs. 5 and 7 can be accounted for, if the transport of alanine and leucine is governed by the affinities stated in Table 3, and if alanine has a J_{max} of 4 μmol cm^{-2} h^{-1} for both lysine carriers and leucine a J_{max} of 2.7 and 4 μmol cm^{-2} h^{-1} on respectively the high and the low affinity lysine carriers. I believe that together with the data of Sepulveda and Smith (1978, 1979) and Paterson et al. (1979, 1980, 1981) the data presented here warrant the conclusion that at 140 mM sodium neutral amino acids are transported by both the carriers of lysine, and that the low affinity transport of neutral amino acids are effected by different carriers at 140 and at 0 mM sodium.

Rat. The effect of neutral amino acids on influx of a cationic amino acid across the brush border membrane of the rat small intestine has been previously described in some detail. In all cases the data were consistent with competition for a single carrier of lysine (Munck 1980b).

As discussed above no conclusive evidence has been published in favor of transport of neutral amino acids by the rat lysine carrier. The considerable suggestive evidence (Munck 1968, Reiser and Christiansen 1971a, Robinson 1968) has, however, now been confirmed by measurements of the inhibitory effect of lysine on J^{Leu}_{mc} at 10 mM leucine. It is seen (Fig. 8) that 200 mM lysine eliminates almost one third of J^{Leu}_{mc}. In view of the statistically insignificant 15% inhibition which can be attributed to an effect of PD_a on J^{Leu}_{mc}, Fig. 8 unequivocally demonstrates that leucine in the rat is transported both by the carrier of neutral amino acids and by the lysine carrier. The data also confirm the previous analysis (Munck and Rasmussen 1975) which ascribed one third of the total transport of leucine to the carrier of lysine.

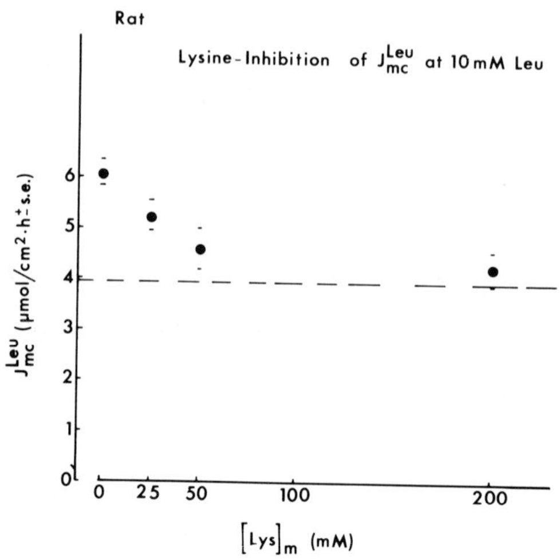

Fig. 8. Lysine inhibition of influx of leucine across the brush-border membrane of the rat small intestine. The experiments were carried out at 10 mM leucine in the presence of 5 mM D-glucose. *Dashed line* indicates the magnitude of the lysine-resistant fraction of J^{leu}_{mc}

Stimulating Effects of Neutral Amino Acids on the Transport of Cationic Amino Acids

In rat and rabbit small intestine neutral amino acids inhibit J^{Lys}_{mc}, enhance the steady state parameter J^{Lys}_{ms} without affecting J^{Lys}_{sm}, and reduce the steady-state epithelial uptake of lysine (Munck and Schultz 1969b, Munck 1980b). In addition intracellularly accumulated neutral amino acids stimulate J^{Lys}_{mc}. These observations have been summarized in a model of a common carrier of neutral and cationic amino acids in the brush border membrane, a possibly allosteric stimulation by neutral amino acids of efflux of cationic amino acids across the basolateral membrane, and, under the condition of normal intracellular sodium, an acceleration of J_{mc}^{Lys} caused by a hyperpolarization of the brush border membrane by coupled efflux of sodium and neutral amino acids across this membrane (Munck 1981).

With rings of everted guinea pig small intestine strong evidence was found (Robinson and Alvarado 1977) for a stimulating effect on J_{mc}^{Lys} of neutral amino acids at the outside of the brush border membrane. This observation was confirmed by the data of Fig. 3C. In reviewing previously published data (Munck and Schultz 1969b, Munck 1980b), it became clear that because of the choice of concentrations used for the studies on rabbit and rat, it could not be ruled out with certainty that at low concentrations neutral amino acids would stimulate J_{mc}^{Lys}. In the rat the K_i of leucine against 1 mM lysine appeared lower at 2 than at 10 and 20 mM (Munck and Rasmussen 1975), indicating that leucine might be the best candidate. For the rabbit unpublished results have similarly pointed to methionine. Consequently J_{mc}^{Lys} was measured at 1 mM lysine with 0, 0.5, 1.0, and 1.5 mM leucine in the rat experiments or methionine in the rabbit experiments. In the guinea pig J_{mc}^{Lys} was again measured at 1 mM lysine in the presence of 0, 05., 1.0 or 1.5 mM leucine but reducing the incubation period from 0.5 to 0.25 min. In both rabbit and rat all three concentrations of the neutral amino acids inhibited J_{mc}^{Lys}, demonstrating that in these two species neutral amino acids act only as cis-inhibitors of J_{mc}^{Lys}. For the guinea pig the information of Fig. 1C was confirmed establishing beyond doubt that in this species neutral amino acids can act as cis-stimulators of J_{mc}^{Lys}.

Transport of Imino Acids and of Non-α Neutral Amino Acids

Transport of imino acids by a separate intestinal transport system was first described for the hamster small intestine, where it appeared to transport N-mono-, di-, and trimethyl-glycine, proline, HO-proline (Hagihira et al. 1962), azetidine-2-carboxylic acid, and piperidine-2-carboxylic acid (Spencer and Brody 1964). β-alanine appeared not to be transported by the hamster imino acid carrier (Hagihira et al. 1962).

Most extensively the transport of imino acids and non-α neutral amino acids has been studied using various preparations of the rat small intestine (Daniels et al. 1969a,b, Munck 1966a, 1977, 1981, Newey and Smyth 1964). It is the results from this species (Daniels et al. 1969a, Munck 1977, 1981) which with respect to transport led to the grouping of the non-α neutral amino acids with the imino acids (Munck 1981).

The Guinea Pig. The transport of imino acids and non-α neutral amino acids by the guinea pig has not previously been studied.

An overview of the specificity of a transport system for MeAIB can be gained from Table 2. This table contains data for the inhibition of J_{mc}^{MeAIB} and $J_{mc}^{\beta\text{-ala}}$ measured using 1 mM of these two amino acids and 40 mM of the inhibitors.

In Table 2, from rows 1 and 2, it is seen that neither imino acids nor β-alanine itself inhibit $J_{mc}^{\beta\text{-ala}}$. This strongly suggests absence of carrier mediated transport of

Table 2. Characteristics of imino acid transport in the guinea pig small intestine. Influx of MeAIB or β-alanine was measured at 1 mM of these amino acids in the presence of 5 mM D-glucose, with or without 40 mM of the inhibitors. J_{mc} is expressed in μmol cm^{-2} h^{-1} ± S.E. with the number of observations in parentheses. Within each row the data are the results of paired experiments

Substrates	Inhibitors				
	None	MeAIB	Me-DL-ala	β-ala	
1 β-alanine (1 mM)	0.04 ± 0.01 (4)	0.06 ± 0.01 (4)	0.05 ± 0.01 (4)		
2 β-alanine (1 mM)	0.07 ± 0.01 (8)			0.07 ± 0.01 (8)	
	None	MeAIB	Me-DL-ala	Me-D-ala	
3 MeAIB (1 mm)	0.41 ± 0.02 (7)	0.06 ± 0.01 (7)	0.11 ± 0.02 (7)	0.16 ± 0.03 (7)	
	None	ABA	β-ABA	GABA	
4 MeAIB (1 mM)	0.50 ± 0.02 (4)	0.30 ± 0.02 (4)	0.42 ± 0.01 (4)	0.51 ± 0.03 (4)	
	None	L-pro	D-pro	HO-L-pro	HO-D-pro
5 MeAIB (1 mM)	0.52 ± 0.04 (6)	0.06 ± 0.01 (5)	0.15 ± 0.03 (5)	0.06 ± 0.01 (5)	0.32 ± 0.04 (5)
	None	Piperidine-2-carboxylic acid	Piperidine-3-carbocylic acid	Piperidine-4-carboxylic acid	
6 MeAIB (1 mM)	0.46 ± 0.03 (4)	0.05 ± 0.02 (4)	0.30 ± 0.02 (4)	0.38 ± 0.02 (4)	
	None	β-ala	L-leu	L-lys	L-ala [a]
7 MeAIB (1 mM)	0.59 ± 0.03 (8)	0.56 ± 0.01 (8)	0.33 ± 0.02 (8)	0.59 ± 0.06 (8)	0.39 ± 0.02 (4)

[a] Results from another series of experiments normalized to the control value of 0.59 μmol cm^{-2} h^{-1}

β-alanine. The data of row 1 demonstrate that β-ala does not significantly inhibit J_{mc}^{MeAIB}. In row 3 Me-DL-ala and Me-D-ala are shown to exert inhibitory effects similar to their effects on $J^{\beta\text{-ala}}$ in the rat. From row 4 it can be seen that ABA and β-ABA but not GABA inhibit J_{mc}^{MeAIB}. The effect of ABA is comparable to the effect seen in the rat; but the effects of β-ABA and GABA are smaller. The data of row 5 show that in the guinea pig L-proline and HO-L-proline are as effective inhibitors of J_{mc}^{MeAIB} as they are of $J_{mc}^{\beta\text{-ala}}$ in the rat; the data on the effects of D-proline and HO-D-proline show more clearly than the data of row 3 that in the guinea pig the carrier of imino acids is stereospecific. From row 6 it is seen that the piperidine carboxylic acids all inhibit J_{mc}^{MeAIB}. The inhibitory effect declines as the carboxylic acid group is moved from α- through β- to γ-position relative to the imino group. In the rat (Table 5) the inhibitory effect against β-alanine increased with the same configurational change. The data of row 7 show that leucine inhibits J_{mc}^{MeAIB}, but that lysine does not. In the rat neutral amino acids with side chains longer than that of 2-amino-n-butyric acid did not inhibit $J_{mc}^{\beta\text{-ala}}$ (Munck 1981).

The transport of MeAIB was studied by paired measurements of J_{mc}^{MeAIB} at 7 concentrations between 0.5 and 60 mM MeAIB. The results of these experiments are well described as

$$J_{mc}^{MeAIB} = \frac{(1.14 \pm 0.10)\,[\mathrm{MeAIB}]_m}{(1.17 \pm 0.23) + [\mathrm{MeAIB}]_m} + (0.046 \pm 0.004)\,[\mathrm{MeAIB}]_m \qquad (7)$$

Evaluated by the chi-square test the fit of the experimental results to Eq. (7) is characterized by a P-value of 0.92. The inhibitory effect of leucine on J_{mc}^{MeAIB} was further examined by paired measurements at 1 mM MeAIB with 0, 10, 20, 40, and 80 mM leucine. Indicating regular competitive inhibition the K_i for leucine was independent of the concentration of leucine; and using the data of Eq. (7) for its estimate the K_i was 21 ± 1 mM (n = 4). These data suffice to distinguish the carrier of MeAIB from that of lysine. The K_i = 21 mM also differs by an order of magnitude from the estimate of 1.70 mM for the K_t of leucine for the carrier of neutral amino acids (Robinson and Van Melle 1982), indicating that these are different transport mechanisms.

It is clear now that the guinea pig small intestine is equipped with a separate carrier of imino acids which differs from that of the rat in having a higher degree of stereo-specificity, in preferring the imino group in an α-position, in being inhibitable by long chain neutral amino acids, and in not transporting β-alanine.

Rabbit. In a study of amino acid transport in the rabbit Peterson et al. (1970) concluded that a small contribution to the transport of proline might come from an equivalent of the rat's imino acid carrier, which, however, was not used by glycine. In contrast Sepulveda and Smith (1978) reported that both proline and MeAIB significantly inhibited the transport of glycine. In a study of the transport of neutral amino acids by microvesicles of rabbit intestinal brush border membranes Stevens et al. (1982) observed complete mutual inhibition between proline and MeAIB, complete inhibition of the transport of proline and MeAIB by phenylalanine but no inhibition of the transport of phenylalanine by MeAIB and only 33% inhibition by proline. The transport of proline and MeAIB appeared to be completely sodium-dependent, but both were weak inhibitors of the sodium-independent transport of glycine and phenylalanine. MeAIB appeared to stimulate the sodium-dependent transport of glycine and the sodium-independent transport of alanine. β-alanine was not subject to mediated transport but did weakly inhibit sodium-dependent transport of alanine, glycine, and phenylalanine without having any effect on the transport of MeAIB. Lysine did not inhibit the transport of any of these amino acids (alanine, glycine, phenylalanine, proline, MeAIB). In the case of lysine these results differ from those discussed in a previous section. As it will be described now, microvesicles appear also to have different characteristics with respect to the transport of β-alanine and MeAIB than the intact epithelium as determined by measurements of influx across the brush border membrane.

Some aspects of the transport of β-alanine are illustrated by the data of Table 3A. In this table the inhibitory action of 40 mM of an amino acid on the transport of β-alanine at 1 mM is shown. Clearly β-alanine is a strong inhibitor of its own transport. This indicates that $J_{mc}^{\beta\text{-ala}}$ is accomplished by a saturable process. $J_{mc}^{\beta\text{-ala}}$ is also inhibited by MeAIB, Me-DL-alanine and MeGly. Strikingly, alanine, leucine and lysine reduce $J_{mc}^{\beta\text{-ala}}$ to the level of free diffusion. In addition the data of Table 3A show for both the amino-butyric acids and the piperidine-carboxylic acids that the degree of inhibitory effect decreases as the amino or imino group moves from α- through β- to γ-position.

Table 3. Inhibition of influx (μmol cm^{-2} h^{-1} ± S.E.) of β-alanine (Part A) and MeAIB (Part B) across the brush border membrane of rabbit small intestine. The data of each row of the table represent results of paired measurements of influx across the brush-border membrane. Fluxes were measured at 1 mM β-alanine or at 1 mM MeAIB without or with 40 mM of the inhibitors. The numbers in parentheses are numbers of observations. Ala* and Ala** are from different series of experiments, the values are normlaized to the control values with which they are listed

Part	Row					
		None	β-ala	Leu	Ala*	
	1.	0.44 ± 0.01 (6)	0.05 ± 0.01 (5)	0.03 ± 0.01 (5)	0.02 ± 0.02 (3)	
		None	MeGly	Me-DL-ala	Lys	
	2.	0.26 ± 0.04 (4)	0.12 ± 0.01 (4)	0.08 ± 0.01 (4)	0.04 ± 0.01 (4)	
		None	MeAIB			
A	3.	0.23 ± 0.04 (4)	0.10 ± 0.01 (5)			
		None	α-ABA	β-ABA	γ-ABA	
	4.	0.31 ± 0.08 (4)	0.02 ± 0.01 (4)	0.06 ± 0.01 (4)	0.10 ± 0.02 (4)	
		None	Piperidine-2-carboxylic acid	Piperidine-3-carboxylic acid	Piperidine-4-carboxylic acid	
	5.	0.23 ± 0.04 (4)	0.03 ± 0.02 (4)	0.11 ± 0.02	0.13 ± 0.02	
		None	MeAIB	β-ala	Lys	
	6.	1.16 ± 0.01 (4)	0.14 ± 0.01 (4)	0.90 ± 0.10 (4)	1.14 ± 0.10 (4)	
B		None	Me-DL-ala	Leu	Ala**	
	7.	0.85 ± 0.08 (5)	0.13 ± 0.01 (6)	0.38 ± 0.01 (5)	0.70 ± 0.03 (8)	
		None	L-pro	D-pro	HO-L-pro	HO-D-pro
	8.	1.10 ± 0.22 (4)	0.05 ± 0.02 (3)	0.24 ± 0.01 (3)	0.05 ± 0.01 (3)	0.50 ± 0.01 (3)

Similarly in Table 3B the transport of MeAIB is partly characterized. In agreement with the data of Stevens et al. (1982) MeAIB strongly inhibits its own transport. J_{mc}^{MeAIB} is similarly effectively inhibited by Me-DL-ala. It is moderately inhibited by β-alanine. This indicates that a fraction of J_{mc}^{MeAIB} may be transported by the carrier of β-alanine. The data of Table 3B on the inhibitory effects of proline and HO-proline demonstrate a considerable degree of stereospecificity. J_{mc}^{MeAIB} is moderately inhibited by alanine and leucine but not at all by lysine. The moderate effect of alanine and leucine suffices to distinguish the carrier of MeAIB from those described previously for neutral and cationic amino acids; the lack of inhibition by lysine clearly shows that only a very small fraction of J_{mc}^{MeAIB} can be by the carrier of β-alanine.

The transport of β-alanine was further characterized in studies of the dependence of $J_{mc}^{\beta\text{-ala}}$ on $[\beta\text{-ala}]_m$, and by measurements of its inhibition by lower concentrations of alanine, leucine and lysine.

The results of paired measurements of $J_{mc}^{\beta\text{-ala}}$ at 8 concentrations between 0.5 and 100 mM β-alanine were well described as

$$J_{mc}^{\beta\text{-ala}} = \frac{(0.79 \pm 0.09)\,[\beta\text{-ala}]_m}{(2.05 \pm 0.37) + [\beta\text{-ala}]_m} + (0.018 \pm 0.004)\,[\beta\text{-ala}]_m \qquad (8)$$

Evaluated by the chi-square test the fit between this equation and the observed flux is characterized by a P-value of 0.85. In Fig. 9 the inhibitory effect of alanine, leucine and lysine are examined at low inhibitor concentrations. Using the data of Eq. (8) the estimates of K_i of lysine are constant with increasing inhibitor concentration with a mean value of 0.38 ± 0.06 mM (4). For alanine and leucine the K_i is about 0.1 mM.

Clearly the rabbit small intestine possesses a carrier for β-alanine with a rather low transport capacity. This carrier can be distinguished from previously described carriers by the very high affinities for alanine, leucine and lysine. However, even if this

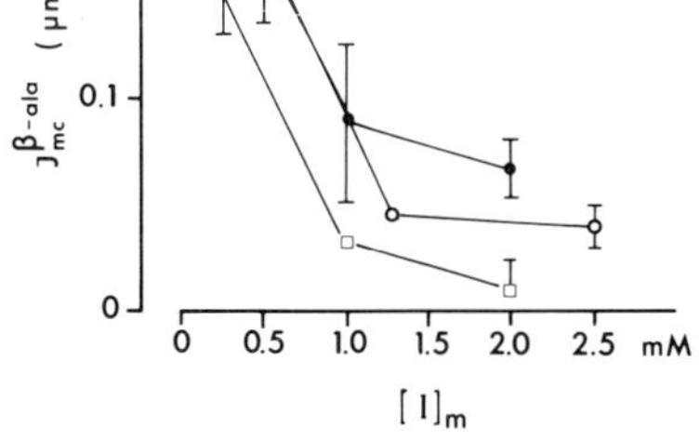

Fig. 9. Influx of β-alanine across the brush border membrane of the rabbit small intestine measured at 1 mM β-alanine + 5 mM D-glucose and inhibited by lysine (•), alanine (□), or leucine (○). Calculated as described for Fig. 4, using the data of Eq. (7), K_i^{lys} was found to be 0.38 ± 0.006 mM. For both alanine and leucine K_i is about 0.1 mM

transport mechanism clearly accepts β-alanine, the much higher affinity for alanine (Fig. 9) and the data on the effects of the amino-butyric acids and the piperidine-carbocylic acids amply demonstrate a preference for α-amino acids over non-α amino acids.

Lysine's K_i of 0.4 mM against $J_{mc}^{\beta\text{-ala}}$ raised the question whether the carrier of β-alanine could be identical with the high affinity carrier of lyseine [Eq. (4)]. This possiblity could be ruled out by the very low K_i values for alanine and leucine and by the inability of 40 mM β-alanine to inhibit J_{mc}^{Lys} as measured at 1 mM lysine. Nevertheless, when the concentration of lysine was reduced to 0.1 mM significant degrees of inhibition of J_{mc}^{Lys} were observed with 20 and 40 mM β-alanine. 40 mM β-alanine reduced J_{mc}^{Lys} from 0.40 ± 0.06 (8) to 0.26 ± 0.02 (8) μmol cm^{-2} h^{-1}. It must therefore be concluded that the high affinity transport of lysine described by Eq. (4) comprises at least two high-affinity mechanisms of lysine transport.

The transport of MeAIB was similarly measured at 8 concentrations between 0.5 and 100 mM MeAIB. The data on the concentration-dependence of J_{mc}^{MeAIB} can be adequately described as

$$J_{mc}^{MeAIB} = \frac{(4.73 \pm 0.22)\,[MeAIB]_m}{(3.6 \pm 0.2) + [MeAIB]_m} + 0.003 \pm 0.003)\,[MeAIB]_m \qquad (9)$$

with a P-value by the chi-square test of 0.95.

In paired experiments J_{mc}^{MeAIB} was measured at 1 mM MeAIB with 0, 50, 100, 200 and 300 mM alanine in one series and with 5, 25, 50, and 100 mM leucine in another series of experiments. It was found that alanine only partially inhibited J_{mc}^{MeAIB} to an extent which could be interpreted as a sum of the elimination of transport of MeAIB by the carrier of β-alanine and a reduction of J_{mc}^{MeAIB} on its own carrier by a partial depolarization of PD_a. At all concentrations of leucine its effect was described by the same K_i which had a mean value of 23 ± 1 mM (4). These data are supplemented by data from paired experiments on the inhibitor action of 40 mM MeAIB on the transport of alanine, leucine and lysine measured at 1 mM of these amino acids (Table 4). J_{mc}^{Lys} is not affected by MeAIB. J_{mc}^{Leu} is significantly inhibited by MeAIB indicating that leucine ias also transported by the carrier of MeAIB. The reduction of J_{mc}^{ala} is not statistically significant. These results confirm those of Stevens et al. (1982), with respect to the existence of a carrier of MeAIB and to the inhibitory action of long and short side-chain neutral amino acids. In addition it is demonstrated that the fully inhibitory neutral amino acid appears also to be transported by this carrier. The carrier of MeAIB can be distinguished from previously described carriers by the combination of a high K_i for leucine and lack of inhibitory effect of lysine.

Table 4. Effect of MeAIB on influx (μmol cm^{-2} h^{-1} S.E.) of alanine, leucine and lysine across the brush border membrane of rabbit small intestine

	Control	+ 40 mM MeAIB
Lys (1 mM)	1.32 ± 0.03 (4)	1.45 ± 0.11 (4)
Leu (1 mM)	3.90 ± 0.23 (8)	2.45 ± 0.16 (8)
Ala (1 mM)	3.02 ± 0.39 (4)	2.76 ± 0.27 (4)

Rat. For the rat Newey and Smyth (1964) demonstrated that in sacs of everted small intestine glycine, methionine, and proline shared a transport mechanism, while proline and glycine in addition were transported by another methionine insensitive system. It was also shown in everted sacs that glycine, proline and MeGly (sarcosine) shared a transport mechanism (Munck 1966a), that D- and L-enanthiomorphs of alanine acetidine-2-carboxylic acid, HO-proline, proline, and pipecolic acid were equally effective as inhibitors of the transport of sarcosine (Daniels et al. 1969a), and that β- and γ- but not α-aminobutyric acid, and α- and β-isobutyric acid equally inhibited this transport (Daniels et al. 1969b). These studies were all of 30–60 min net transport across the wall of everted sacs. It is, therefore, not possible unambiguously to interpret the results in terms of unidirectional transport across one or the other of the limiting cell membranes. Nevertheless these results from everted sac studies are in principle confirmed by the following data on influx across the brush-border membrane (Munck 1977, 1981, Munck and Rasmussen, unpublished data).

In Table 5 the specificity of the imino- non-α neutral amino acid carrier of the rat is characterized through the values of K_i as determined by the inhibitory effect on $J_{mc}^{\beta\text{-ala}}$ or J_{mc}^{Sar}, and in some cases on both; β-alanine and sarcosine were ascribed K_t values of respectively 15 and 11 mM. This procedure was based on the observation that both β-alanine and sarcosine were mutually fully competitive and each apparently transported by only one carrier. In addition proline has the same K_i against these two amino acids.

The data from Table 5 for glycine, L-alanine, L-α-aminobutyric acid, and L-n-valine show that for aliphatic neutral amino acids a side chain larger than the ethyl group is not tolerated by the rat imino acid carrier. This is in contrast with the data for the rabbit (Table 3) which accepts leucine, and according to Stevens et al. (1982) also phenylalanine, and to the data for the guinea pig (Table 3) where J_{mc}^{MeAIB} is competitively inhibited by leucine.

The question of stereospecificity is answered differently on the basis of data for alanine, Meala, and proline, which indicate a moderate stereospecificity than of data for HO-proline and serine, which indicate preference for the D-configuration. This difference may suggest that the disadvantage of a polar group on the side chain is greater than that of D-configuration. The stereospecificity seems less developed than in the guinea pig (Table 3).

For alanine, AIB, amino-n-butyric acid, piperidine carboxylic acid, and proline the affinity is higher with the amino group in β-position, and this advantage is even greater with the amino group in γ-position as in γ-amino-butyric acid and piperidine-4-carboxylic acid (Table 5). This is in contrast to the situation in the rabbit where alanine has a much higher affinity for the β-alanine carrier than β-alanine itself [Fig. 9, Eq. (8)]. It also differs from the guinea pig, where, in contrat to alanine, β-alanine has no effect on J_{mc}^{MeAIB}, and the order of the affinities for the piperidine carboxylic acids is reversed (Table 2).

Most of the data of Table 5 were collected before ^{14}C-labeled MeAIB became commercially available. However, the transport of MeAIB is partially characterized by the results of paired measurements of J_{mc}^{MeAIB} at concentrations between 0.94 and 300 mM. With no indication of involvement of more than one carrier, the data on the dependence of J_{mc}^{MeAIB} on [MeAIB] are well described as

Table 5. Characteristics of the imino acid carrier in rat small intestine. Affinities of amino acids for the imino acid carrier of the brush border membrane of the rat small intestine were evaluated through estimates of their K_i against sarcosine or β-alanine. The K_i values were calculated using data from paired experiments with 1 mM alanine or 1 mM sarcosine 40 mM inhibitor in the presence of 5 mM D-glucose. For the calculations Eq. (1) was used assuming a K_t of 11 mM and 15 mM for respectively sarcosine and β-alanine. Means ± S.E. of 4–5 observations are given (Munck 1977, 1981)

Effect of side chain length, steric configuration, and N-methylation of inhibitory efficiency

			N-CH_3-derivative	
	L	D	L	D
Glycine	37 ± 7		15 ± 2	
Alanine	24 ± 5	31 ± 2	7 ± 1	24 ± 3
α-ABA	93	155		
α-AIB acid	81 ± 14		17 ± 1	
N-valine	∞	∞	∞	
Proline	12 ± 1 [a]	14 ± 1	∞	
HO-proline	16 ± 2	12 ± 1		
Iso-nipecotic acid	22 ± 4		57 ± 9	
Taurine	36 ± 8		83 ± 14	
Serine	∞	58 ± 8	221 ± 38	
γ-ABA	17 ± 4		∞	

Effect of the position of the amino/imino group on inhibitory efficiency

	α	β	γ	δ	ε
Alanine	24 ± 5	14 ± 1			
AIB acid	81 ± 14	22 ± 3 [b]			
ABA	93	19 ± 3 [b]	17 ± 4		
N-valine	∞			∞	
N-leucine	∞			∞	∞
Proline	10 ± 1	9 ± 1			∞
Piperidine carboxylic acids	50 ± 12 [b]	33 ± 4 [b]	22 ± 4		

[a] Determined against β-ala; against sarcosine K_i was 11 ± 2 (4)
[b] The inhibitors were DL-amino acids

$$J_{mc}^{MeAIB} = \frac{(3.53 \pm 0.22)\,[MeAIB]_m}{(8.35 \pm 0.75) + [MeAIB]_m} + 0.061 \pm 0.002\,[MeAIB]_m \tag{10}$$

Evaluated by the chi-square test the fit between this equation and the experimental data is characterized by $P = 0.95$. Nevertheless, there are some indications that a second carrier may be involved. Thus, methionine at 13 times its K_t reduces J_{mc}^{MeAIB} by 22%. Also a K_i of 32 ± 4 mM for β-alanine against MeAIB is much higher than against the imino acids in general (Table 5); but if control and inhibited fluxes are both reduced by the methionine sensitive fraction of J_{mc}^{MeAIB} the K_i is reduced to its usual level (Table 5).

Acknowledgements. The author's research was supported by grants from the Novo Foundation, the P. Carl Petersen Foundation, the Nordic Insulin Foundation, and the Danish Medical Research Council.

References

Chez RA, Schultz SG, Curran PF (1966) Effect of sugars on transport of alanine in intestine. Science 153:1012–1013

Christensen HN, Handlogten ME, Thomas El (1969) Na-facilitated reactions of neutral amino acids with a cationic amino acid transport system. Proc Natl Acad Sci USA 63:948–955

Curran PF, Schultz SG, Chez RA, Fuisz RE (1967) Kinetic relations of the Na-amino acid interaction at the mucosal border of intestine. J Gen Physiol 50:1261–1286

Daniels VG, Dawson AG, Newey H, Smyth DH (1969a) Effect of carbon chain length and amino group position on neutral amino acid transport systems in rat small intestine. Biochim Biophys Acta 173:575–577

Daniels VG, Newey H, Smyth DH (1969b) Stereochemical specificity of neutral amino acid transfer systems in rat small intestine. Biochim Biophys Acta 183:637–639

Goldner AM, Sarracino S, Estep JA (1977) Coupling of sodium and substrate fluxes in the intestine: galactose transport in the jejunum. In: Kramer M, Lauterbach F (eds) Intestinal permeation. Excerpta Medica, Amsterdam Oxford, p 282

Gunter-Smith PJ, Grasset E, Schultz SG (1982) Sodium-coupled amino acid and sugar transport by *Necturus* small intestine: An equivalent electrical circuit analysis of a rheogenic cotransport system. J Membr Biol 66:25–39

Hagihira H, Lin ECC, Samiy AH, Wilson TH (1961) Active transport of lysine, ornithine, arginine and cystine by the intestine. Biochem Biophys Res Commun 4:478–481

Hagihira H, Wilson TH, Lin ECC (1962) Intestinal transport of certain N-substituted amino acids. Am J Physiol 203:637–640

Lerner J, Karcher CA (1978) Kinetic properties of imino acid transport system in the chicken intestine. Comp Biochem Physiol 60A:503–505

Mircheff AK, Os CH van, Wright EM (1980) Pathways for alanine transport in intestinal basal lateral membrane vesicles. J Membr Biol 52:83–92

Munck BG (1965) Amino acid transport by the small intestine of the rat. The effect of amino acid pre-loading on the transintestinal amino acid transport by the everted sac preparation. Biochim Biophys Acta 109:142–150

Munck BG (1966a) Amino acid transport by the small intestine of the rat. The existence and specificity of the transport mechanism of imino acid and its relation to the transport of glycine. Biochim Biophys Acta 120:97–103

Munck BG (1966b) Amino acid transport by the small intestine of the rat. On the counterflow phenomenon as a cause of the accelerating effect of leucine on the trans-intestinal transport of diamino acids. Biochim Biophys Acta 120:282–291

Munck BG (1968) Amino acid transport by the small intestine of the rat. Evidence against interactions between sugars and amino acids at the carrier level. Biochim Biophys Acta 156:192–194

Munck BG (1972) Effects of sugar and amino acid transport on transepithelial fluxes of sodium and chloride of short circuited rat jejunum. J Physiol (Lond) 223:699–717

Munck BG (1977) Intestinal transport of amino acids. In: Kramer M, Lauterbach F (eds) Intestinal permeation, vol IV. Excerpta Medica, Amsterdam, p 123

Munck BG (1980a) Transport of sugars and amino acids across guinea pig small intestine. Biochim Biophys Acta 597:411–417

Munck BG (1980b) Lysine transport across the small intestine. Stimulating and inhibitory effects of neutral amino acids. J Membr Biol 53:45–53

Munck BG (1981) Intestinal absorption of amino acids. In: Johnson LR (ed) Physiology of the gastrointestinal tract. Raven Press, New York, p 1097

Munck BG, Rasmussen SN (1975) Characteristics of rat jejunal transport of tryptophan. Biochim Biophys Acta 389:261–280

Munck BG, Rasmussen SN (1979) Lysine transport across rat jejunum: Distribution between the transcellular and the paracellular routes. J Physiol (Lond) 291:291–303

Munck BG, Schultz SG (1969a) Lysine transport across isolated rabbit ileum. J Gen Physiol 153: 157–182
Munck BG, Schultz SG (1969b) Interactions between leucine and lysine transport in rabbit ileum. Biochim Biophys Acta 183:182–193
Murer H, Sigrist-Nelson K, Hopfer U (1975) On the mechanism of sugar and amino acid interaction in intestinal transport. J Biol Chem 250:7392–7396
Newey H, Smyth DH (1964) The transfer system for neutral amino acids in the rat small intestine. J Physiol 170:328–343
Okada Y, Tsuchiya W, Irimajiri A, Inouye A (1977) Electrical properties and active solute transport in rat small intestine. I. Potential profile changes associated with sugar and amino acid transport. J Membr Biol 31:205–219
Oxender DL, Christensen HN (1963) Distinct mediating systems for the transport of neutral amino acids by the Ehrlich cell. J Biol Chem 238:3686–3699
Paterson JYF, Sepulveda FV, Smith MW (1979) Two carrier influx of neutral amino acids into rabbit ileal mucosa. J Physiol (Lond) 292:339–350
Paterson JYF, Sepulveda FV, Smith MW (1980) A sodium-independent low affinity transport system for neutral amino acids in rabbit ileal mucosa. J Physiol (Lond) 298:333–346
Paterson JYF, Sepulveda FV, Smith MW (1981) Distinguishing transport systems having overlapping specificities for neutral and basic amino acids in the rabbit ileum. J Physiol 319:345–354
Peterson SC, Goldner AM, Curran PF (1970) Glycine transport in rabbit ileum. Am J Physiol 219:1027–1032
Preston RL, Schaeffer J-F, Curran PF (1974) Structure-affinity relationships of substrates for the neutral amino acid transport system in rabbit ileum. J Gen Physiol 64:443–467
Reiser S, Christiansen PA (1969) A cross-inhibition of basic amino acid transport by neutral amino acids. Biochim Biophys Acta 183:611–624
Reiser S, Christiansen PA (1971a) The properties of the preferential uptake of L-leucine by isolated intestinal epithelial cells. Biochim Biophys Acta 225:123–139
Reiser S, Christiansen PA (1971b) Stimulation of basic amino acid uptake by certain neutral amino acids in isolated intestinal epithelial cells. Biochim Biophys Acta 241:102–113
Reiser S, Christiansen A (1972) A basis for the difference in the inhibition of the uptake of various neutral amino acids by lysine in intestinal epithelial cells. Biochim BiophysActa 266: 217–229
Reiser S, Christiansen PA (1973) Exchange transport and amino acid charge as the basis for Na^+-independent lyseine uptake by isolated intestinal epithelial cells. Biochim Biophys Acta 307: 223–233
Robinson JWL (1968) Interactions between neutral and dibasic amino acids for uptake by the rat intestine. Eur J Biochem 7:78–89
Robinson JWL, Alvarado F (1977) Comparative aspects of the interactions between sugar and amino acid transport systems. In: Kramer M, Lauterbach F (eds) Intestinal permeation,vol IV. Excerpta Medica, Amsterdam Oxford, p 145
Robinson JWL, Felber JP (1964) A survey of the effect of other amino-acids on the absorption of L-arginine and L-lysine by the rat intestine. Gastroenterology 101:330–338
Robinson JWL, Melle G van (1982) Single-site uptake of neutral amino acids into guinea-pig intestinal rings. J Physiol (Lond) 323:569–587
Rose RC, Schultz SG (1971) Studies on the electrical potential profile across rabbit ileum. Effects of sugars and amino acids on transmural and transmucosal electrical potential differences. J Gen Physiol 57:639–663
Schultz SG (1977) Sodium-coupled solute transport by small intestine: a status report. Am J Physiol 233:E249–E254
Schultz SG (1980) Basic principles of membrane transport. Cambridge Univ Press, New York
Schultz SG (1981) Homocellular regulatory mechanisms in sodium-transporting epithelia: avoidance of extinction by "flush-through". Am J Physiol 241:F579–F590
Schultz SG, Curran PF, Chez RA, Fuisz RE (1967) Alanine and sodium fluxes across mucosal border of rabbit ileum. J Gen Physiol 50:1241–1260

Sepulveda FV, Smith MW (1978) Discrimination between different entry mechanisms for neutral amino acids in rabbit ileal mucosa. J Physiol (Lond) 282:73–90

Sepulveda FV, Smith MW (1979) Different mechanisms for neutral amino acid uptake by new-born pig colon. J Physiol (Lond) 286:479–490

Spencer RP, Brody KR (1964) Intestinal transport of cyclic and noncyclic amino acids. Biochim Biophys Acta 88:400–406

Stevens BR, Ross HJ, Wright EM (1982) Multiple transport pathways for neutral amino acids in rabbit jejunal brush border vesicles. J Membr Biol 66:213–225

Thomson ABR, Dietschy JM (1977) Derivation of the equations that describe the effects of unstirred water layers on the kinetic parameters of active transport processes in the intestine. J Theoret Biol 64:277–294

Thomson ABR, Dietschy JM (1980) Experimental demonstration of the effect of the unstirred water layer on the kinetic constants of the membrane transport of D-glucose in rabbit ileum. J Membr Biol 54:221–229

Ussing HH, Zerahn K (1951) Active transport of sodium as the source of electric current in the short-circuited frog skin. Acta Physiol Scand 214:110–127

White JF, Armstrong WMcD (1971) Effect of transported solutes on membrane potentials in bullfrog small intestine. Am J Physiol 221:194–201

Winne D (1973) Unstirred layer, source of biased Michaelis constant in membrane transport. Biochim Biophys Acta 298:27–31

Temporal Adaptation and Hormonal Regulation of Sodium Transport in the Avian Intestine

E. SKADHAUGE[1]

Introduction

The lower intestine of birds (coprodeum and colon) is a common storage organ for faeces and urine. Resorption of electrolytes, some organic nutrients, and water takes place during the intestinal sojourn of the excreta. Urine is regurgitated into coprodeum and colon from the urodeum, and may even reach the caeca (Skadhauge 1981).

We have in a number of studies (summarized in Skadhauge 1982, Thomas 1982) in vivo and in vitro investigated the transport characteristics of coprodeum, colon, and caecum, largely in the domestic fowl but also in other species (Choshniak et al. 1977, Lyngdorf-Henriksen et al. 1978, Lind et al. 1980a, Skadhauge and Dawson 1980, Holtug and Skadhauge 1982, Rice and Skadhauge 1982a,b,c). Important changes in transport pattern were observed, both in coprodeum and colon, when birds were switched from a high-NaCl to a low-NaCl diet and vice versa.

This paper outlines first the transport characteristics after full adaptation to either a high-NaCl or a low-NaCl diet, second the temporal development of transport parameters and plasma hormone concentrations, and third the effects of hormone injections.

Transport Pattern on Constant Intake of a Low-NaCl or a High-NaCl Diet

As described by Thomas and Skadhauge (1982a), the low-NaCl diet, based largely on wheat and barley, contained around 5 mmol Na kg^{-1} whereas the high-NaCl diet contained around 176 mmol Na kg^{-1}. This corresponds to a daily intake of 0.2 mmol kg^{-1} body weight respectively. In other experiments the NaCl intake was further augmented by offering 0.5% NaCl as the only drinking fluid in addition to the high-NaCl diet (25 mmol Na kg^{-1} body weight day).

On the *low-NaCl diet* net sodium absorption was in the range of 10–20 μmol cm^{-2} h^{-1} both in coprodeum and colon as investigated in vitro (Choshniak et al. 1977, Lind et al. 1980a) and in vivo (Rice and Skadhauge 1982a,b). The in vitro measurements, carried out in the Ussing chamber, demonstrated that the net rate of sodium

1 Department of Veterinary Physiology and Biochemistry, The Royal Veterinary and Agricultural University, Bülowsveg 13, DK-1870 Copenhagen, Denmark

Intestinal Transport
(ed. by M. Gilles-Baillien and R. Gilles)

absorption was nearly equal to the short-circuit current (SCC); the flux ratio for chloride was unity, and there was a slight secretion of potassium. The transport of sodium was not stimulated by the presence of amino acids in the luminal bathing fluid but was fully inhibited by amiloride. In the Ussing chamber the electrical potential difference (PD) was recorded under open circuit and the epithelial resistance calculated. The colon had a lower resistance than coprodeum; the PD was around 20 mV on the low-NaCl diet whereas coprodeum reached 50–60 mV.

On the *high-NaCl diet* an entirely different pattern was observed. In the coprodeum the net rate of sodium transport was nearly suppressed due to a sharply reduced sodium influx across the luminal membrane (Bindslev 1979). In the colon, sodium transport remained high (around 8 μmol cm^{-2} h^{-1}), but only after stimulation with amino acids in the luminal solution. Leucine and lysine (4 mM) gave a maximal increase in the SCC (Lind et al. 1980b). The kinetics of sodium absorption changed also. Both in coprodeum and colon the K_m for sodium absorption was augmented considerably from low- to high-NaCl diet (Lyngdorf-Henriksen et al. 1978, Holtug and Skadhauge 1982). Binding studies were done with the amiloride derivative benzamil (Cuthbert et al. 1982). This compound showed high binding capacity to homogenized epithelial tissue from hen coprodeum both from birds on high- and low-NaCl diets, but higher binding from intact tissue from low-NaCl diet birds. This may be interpreted to indicate that sodium depletion augments the luminal permeability by insertion of a preformed cytoplasmatic component into the membrane.

It may thus be concluded that adaptation to the low-NaCl diet involves induction of amiloride-sensitive sodium channels in the mucosal membrane of the epithelial cells, and it is likely that these channels limit sodium transport (Bindslev 1979, Skadhauge 1980). In contrast high-NaCl adapted birds have substantial capacity in colon for the co-transport of sodium with hexoses and amino acids. This transport mechanism disappears and is replaced by the amiloride-sensitive mechanism for sodium transfer at adaptation to low-NaCl diet. The increase in colonic SCC shows that the sodium channels more than compensate for the loss of the system for sodium hexose/amino acid co-transport by adaptation to low-NaCl diets (Thomas and Skadhauge 1982b).

Recent studies in vivo (Rice and Skadhauge 1982a,b) have confirmed the modifications in transport with diet as observed in vitro, both in the coprodeum and colon, and have also demonstrated that although the final output of sodium is regulated by the coprodeum the main electrolyte absorption takes place in the colon. This applies both to the domestic fowl (see Skadhauge 1982) and to the domestic duck (Rice et al. 1981).

Several other transport functions have been investigated in coprodeum and colon. It should be noted that the Na/K-stimulated ATPase which is ouabain-inhibited did not change in the colon with diet, but was doubled in the coprodeum (Skadhauge 1980) from high- to low-NaCl adaptation. Staining of ultrathin sections, 1 μ, with toluidine blue showed the presence of so-called dark cells which increased from 2% of the total number of epithelial cells on high-NaCl diet, to 20% on the low-NaCl diet (Eldrup et al. 1979). These cells are believed to be mitochondrion-rich cells.

The plasma concentrations of aldosterone, corticosterone, arginine vasotocin (AVT) and prolactin were measured using a constant intake of the two diets. A significant difference was not observed for corticosterone, but aldosterone was significantly higher on the low-NaCl diet (Thomas et al. 1980). The plasma concentrations of AVT was tripled and prolactin doubled on the high-NaCl diet in association with an 8% increase in plasma osmolality (Skadhauge et al. 1982a, Rice, Árnason, Chadwick and Skadhauge, unpublished data). A higher plasma concentration of prolactin was previously reported after NaCl loading (Scanes et al. 1976). Plasma renin activity was measured in several batches of high-NaCl and low-NaCl adapted birds but, surprisingly, a significant difference was not observed (Nishimura and Skadhauge, unpublished data).

Temporal Development of Plasma Hormone Concentrations and Transport Parameters of the Gut Epithelia

In two recent studies (Thomas and Skadhauge 1982b, Skadhauge et al. 1983a,b) both the NaCl balance, the electrical parameters of the epithelia and the plasma hormone concentrations (aldosterone, corticosterone and prolactin) were measured after sudden change from the high- to the low-NaCl diet and after renewed "resalination". Initial experiments showed that the full turn-over from high- to low-NaCl pattern took approximately a week, whereas the switch-back to high-NaCl pattern was nearly complete within 24 h. Subsequent studies were therefore made at 1, 2, 4, and 8 days after change from high- to low-NaCl diet and 8, 16, 24 h after start of resalination. The cumulative balance of Na (and Cl) indicated a net gain or loss of about 12–13 mmol Na kg^{-1} body weight during NaCl depletion or repletion respectively, as observed by the lag of change in rate of excretion behind change in rate of intake. During NaCl depletion the time taken for the rate of sodium excretion to match the rate of intake was about 6 days with a half-time of 1–1.5 day. During NaCl repletion the rate of sodium excretion matched the rate of intake by 2–3 days of NaCl repletion with a half-time of 0.5–1 day.

The time course of the change in the electrical variables, SCC, PD, and resistance was followed over the intervals previously indicated. NaCl-depletion increased SCC and PD fully by 4 days in colon but took 8 days in coprodeum (half-time: 1.2 day and 4 days, respectively). The SCC in the colon that was sensitive to amino acids was abolished while the SCC sensitive to amiloride became established by 4 days (half-time: 1 day). Changes in the resistance were transient and relatively small in the colon but the resistance of the coprodeum decreased. Resalination reversed these changes very rapidly in both tissues, within 24–48 h (half-times: 10–20 h). The mean SCC's of the coprodeum and colon are shown in Fig. 1.

The plasma concentrations of sodium and chloride ions and the osmolality were reduced in birds which were adapted chronically to the low-NaCl compared to those on the high-NaCl diet. When birds were switched from high-NaCl to low-NaCl diet reductions in sodium, chloride and osmotic concentrations occurred quickly (within 1–2 days) to levels found in chronically maintained birds. After resalination the concentrations of sodium, chloride and the osmolality were augmented in 8 h, overshooting, and then subsequently declining to the levels found in birds kept chronically on the high-NaCl diet.

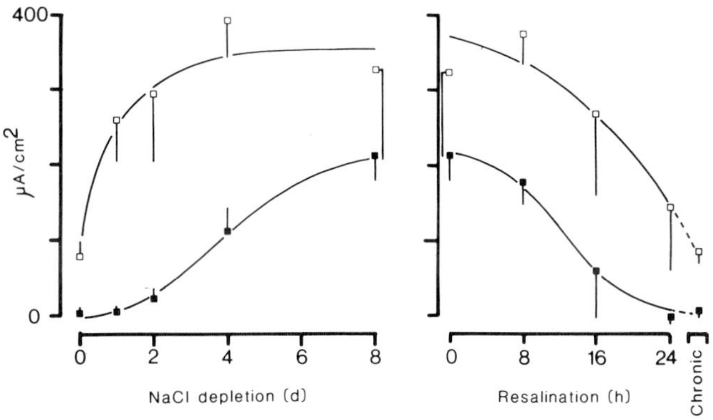

Fig. 1. Short-circuit current after sudden changes of NaCl intake. *Left* short-circuit current (SCC) of colon *(open squares)* and coprodeum *(closed squares)* 1, 2, 4, and 8 days after switch-over from high-NaCl to low-NaCl diet. *Right* SCC 8, 16, and 24 h after acute resalination. *0 (left)* and *chronic (right)* indicate values in the fully adapted high-NaCl state. Means + S.E. are reported. (After Thomas and Skadhauge 1982a, modified)

The levels of aldosterone in the plasma rose approximately hyperbolically during NaCl depletion taking 2 days for half the complete change. After resalination there was a much more rapid decrease in aldosterone concentrations which was complete in less than 8 h. Epithelial SCC's plotted against plasma aldosterone concentration during NaCl depletion indicate that the coprodeum is less sensitive than colon to this hormone (Thomas and Skadhauge 1982b).

Corticosterone concentration in plasma showed a peak about one day after the change in diets from high- to low-NaCl intake, but otherwise showed no change during NaCl-depletion. The plasma corticosterone concentrations 16–24 h after resalination were similar to those after 1 day of NaCl depletion.

Prolactin concentrations showed a steady decline during NaCl depletion, whereas resalination produced a large increase in prolactin concentrations to plateau levels at 8–24 h, well above the levels in birds chronically adapted to high-NaCl intake. The concentrations of these hormones are shown in Fig. 2. The plasma concentrations of AVT were halved 1 day after switching to the low-NaCl diet and reached the constant, low-NaCl level at the 6th day. Acute resalination tripled the concentration after 16 h, but fluctuating concentrations were observed the following days (Rice, Árnason and Skadhauge, unpublished experiments).

The brief rise in corticosterone concentrations following both NaCl depletion and resalination may probably be due to non-specific "stress".

It is concluded that the lower intestinal adaptation to dietary NaCl levels may be controlled by the change of NaCl balance via its effect on plasma aldosterone concentration, while the difference in adaptation of coprodeum and colon may depend on differential sensitivities to aldosterone. The inverse changes in plasma concentrations of aldosterone and prolactin (and AVT) during NaCl depletion/resalination suggest that prolactin and AVT may be cadidate antagonists to aldosterone actions during states of high-NaCl balance.

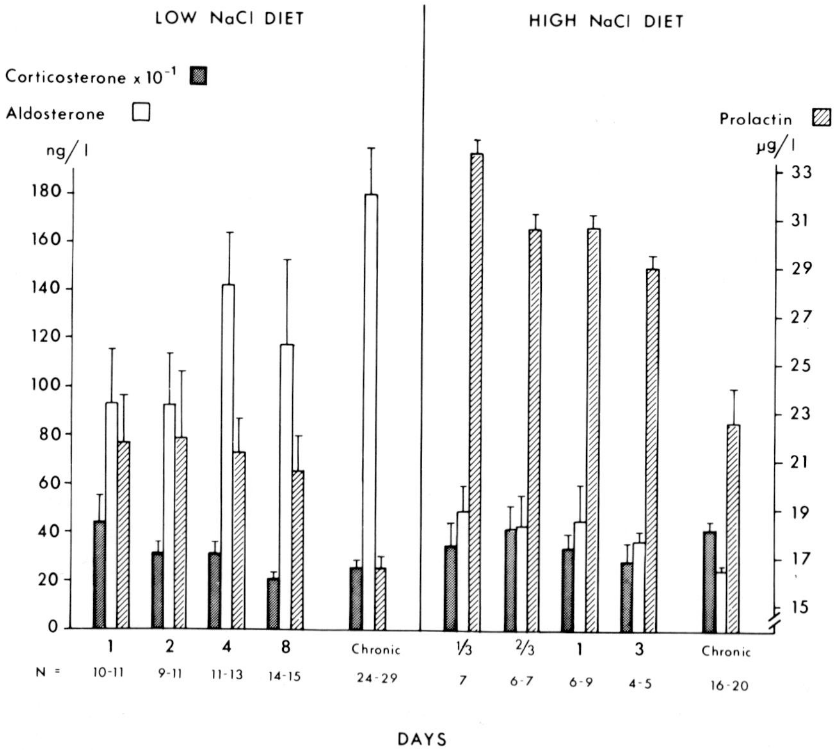

Fig. 2. Plasma hormone concentrations after sudden changes of NaCl intake. *Left* plasma concentrations of aldosterone, corticosterone and prolactin 1, 2, 4, and 8 days after switch-over from the high-NaCl to the low-NaCl diet. *Right* hormone concentrations 1/3 day (8 h), 2/3 day (16 h), 1, and 3 days after acute resalination. On both diets *Chronic* indicates the time-independent hormone concentrations measured after a minimum of 14 days' adaptation. Means + S.E. are reported. (Skadhauge et al. 1982a,b, including unpublished values. Hormone analyses by courtesy of Drs. M. Jallagéas and A. Chadwick)

Recent studies (Árnason and Skadhauge, unpublished data) of epithelial electrical parameters and plasma hormone concentrations during constant intake of NaCl in the range from low- to high-NaCl intake shed further light on the different speed of adaptation after NaCl depletion and repletion.

There was a well defined transition zone (turning point) from low-NaCl pattern to high-NaCl pattern. Both the SCC/PD, the amiloride inhibition and amino acid stimulation, and the plasma hormone concentrations (aldosterone and AVT/prolactin) changed over a range from 1–10 mmol Na kg^{-1} body weight · day. The switch-over started close to the intake of low-NaCl diet and was equal to the high-NaCl pattern at less than 50% of the high-NaCl intake (10 vs. 25 mmol Na kg^{-1} day). This means that practically complete NaCl depletion is necessary to induce the low-NaCl pattern, whereas a smaller amount of NaCl suffices to suppress aldosterone completely and bring about the high-NaCl adaptation.

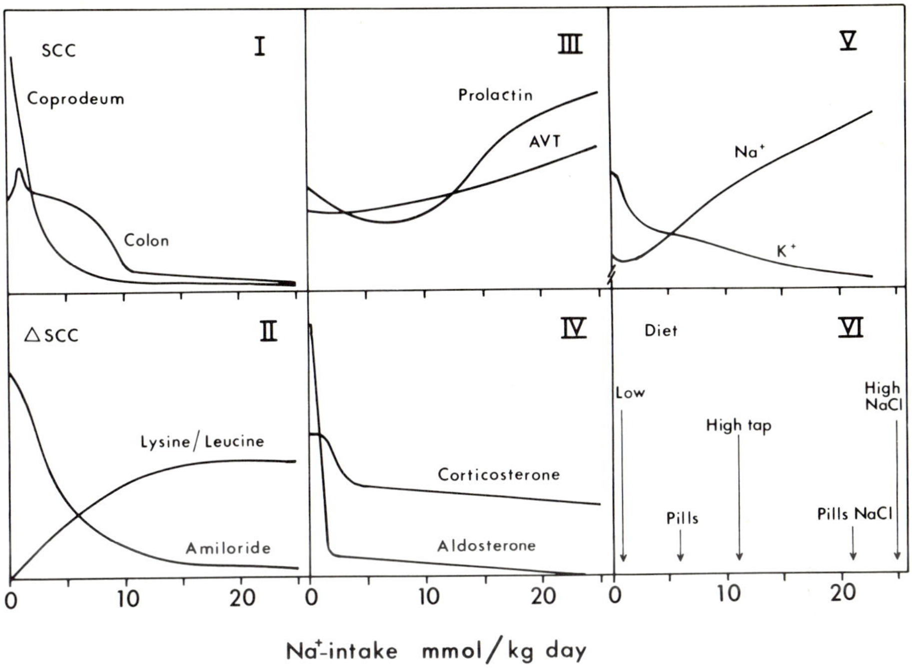

Fig. 3. A survey of short-circuit currents and plasma concentrations of constant intake of diets varying in NaCl-content from the low- to the high-NaCl diet. *I* short-circuit currents (SCC) of coprodeum and colon (without glucose on the mucosal side before addition of amino acids); *II* the changes in short-circuit currents after addition of first amino acids and later amiloride to colon; *III* plasma concentrations of prolactin and arginine vasotocin (AVT); *IV* plasma concentrations of aldosterone and corticosterone; *V* plasma concentrations of sodium and potassium; and *VI* the Na intakes resulting from the dietary regimes in use. It will appear that the change from high-NaCl to low-NaCl pattern occurs over a range of 1–10 mmol kg^{-1} day. The high-NaCl diet + tap water (around 11 mmol kg^{-1} day) induces the complete high-NaCl pattern. Arginine vasotocin and prolactin rise further on the high-NaCl diet + 0.5% NaCl drinking solution (25 mmol kg^{-1} day) parallel to the increase in plasma sodium concentration (and osmolarity). (S.S. Árnason and E. Skadhauge, unpublished data)

The measured variables as functions of the NaCl intake are shown in Fig. 3.

The relatively slow adaptation during NaCl-depletion in principle allows a morphological change such as the induction of a new cell population to be important for the switch-over of the sodium transport pattern, but the fast response during resalination seems to rule out such a mechanism. Further studies of the dark cells (M. Bundgaard and E. Skadhauge, unpublished data) suggest that there is a continuous range in staining behavior of the principal cells of the mucosal layer presumably depending on the mitochondrion content. This causes these cells to be grouped as dark cells or not rather haphazardly according to minor changes in staining. Although a qualitative assessment confirms that there is a larger number of dark cells in the colon and coprodeum of low-NaCl adapted birds we also observed a large difference from area to area. This patchiness unfortunately precludes a regular morphometric analysis of the relative dark-cell number.

The studies on temporal adaptation as well as the turning-point experiments confirm the observations of Thomas et al. (1980) that amiloride inhibition and amino acid stimulation gradually replace each other (Fig. 3 II). Although the difference in patterns of colonic and coprodeal adaptation to different levels of NaCl intake is largely due to the waxing and waning of the co-transport component in colon and to the more sensitive response of the sodium channel blocked by amiloride it appears that the sodium co-transport mechanism and the amiloride channel are not alternative modes of a common structure. If this was so strict, reciprocity between modes would be expected. There is clearly some departure from reciprocity and there is some hysteresis between adaptation to, and de-adaptation from, the low-NaCl diet (Thomas and Skadhauge 1982b). It would thus seem that as a function of presence or absence of aldosterone (and possibly other hormones) the amiloride-sensitive sodium channel, and the amino acid/hexose-linked sodium system in the luminal membrane are induced and inactivated at independent sites.

Effects of Hormone Injections

The main conclusion from the foregoing sections, that the adaptation of colonic and coprodeal epithelium to NaCl is largely induced by aldosterone is further supported by a number of hormone injection experiments. Work in progress suggests, however, a rather complicated function for aldosterone.

Early studies of the sodium and chloride transport in coprodeum and colon by in vivo perfusion showed pronounced effects of aldosterone both after acute injections (Thomas et al. 1979) and chronic (2 days) treatment (Thomas and Skadhauge 1979). The main result of the acute injections was an increase after 4 h of the rate of sodium and chloride absorption and secretion of potassium. The chronic injections also augmented the net rate of sodium absorption and typical saturation kinetics were observed as function of luminal sodium concentration. In later studies the electrical parameters of coprodeum and colon were measured in the Ussing chamber after previous aldosterone injections (Thomas et al. 1980). The hormone caused a conspicuous increase in coprodeal PD and SCC, and an increase of the amiloride inhibition and suppression of amino acid stimulation in the colon. It was obvious, however, both in vitro and in vivo, that the fully developed quantitative effects of a chronic low-NaCl diet were not obtained. Only 25%–50% of the effects were observed (see Thomas and Skadhauge 1982a). In addition, qualitative differences were noted (Thomas et al. 1979, Thomas and Skadhauge 1979). These observations led to attempts to find effects of hormones which might influence electrolyte transport of the intestine and were likely to be affected by the intake of NaCl. Both angiotensin (Hypertensin Ciba) and AVT were tested in vitro but no effects on SCC or PD were observed (Skadhauge and Thomas, unpublished data). Bovine prolactin was also without effect (Árnason and Skadhauge, unpublished data).

It might be noted in passing that vasoactive intestinal peptide (VIP), both chicken and porcine, induces chloride secretion and somatostatin induces a slight augmentation of the rate of sodium absorption (Andersen, Munck and Skadhauge, unpublished data).

A possible explanation for the limited effects of aldosterone may be found in the following observations. The injection studies noted above in which stimulation of SCC/PD or induction of amiloride inhibition partly failed were all carried out in birds chronically exposed to the high-NaCl diet (see right columns in Fig. 4). This pattern was, however, not observed if the aldosterone injections were given instead simultaneously with resalination and the Ussing-chamber experiments performed from tissues removed from birds slaughtered after 24 h (see left columns in Fig. 4). In these experiments the typical pattern of the low-NaCl diet birds remained: the SSC was over 300 μÅ cm^{-2}, the PD over 40 mV, and amiloride fully suppressed the SCC. The PD was even reversed as typically encountered in low-NaCl diet birds (Lyngdorf-Henriksen et al. 1978, Lind et al. 1980a). One interpretation of these findings is that the tissues seem to have a *different sensitivity* to aldosterone depending on the length of exposure to the diets. This tentative conclusion is supported by injections

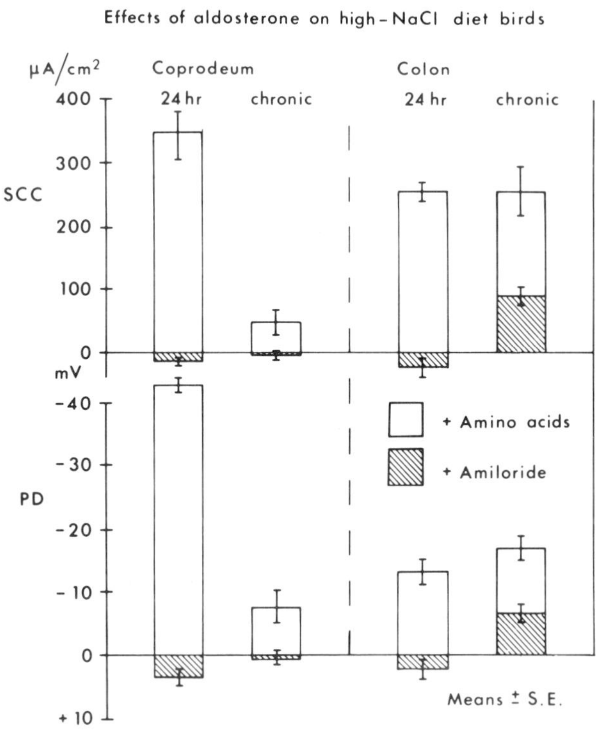

Fig. 4. Effects of aldosterone on electrical parameters of epithelia from birds on high-NaCl diet. Short-circuit current (SCC) and electrical potential difference (PD) of coprodeum *(left)* and colon *(right)* mounted in the Ussing chamber after injections of aldosterone (methods described in Thomas et al. 1980). The colon values are recorded after amino acid stimulation before and after amiloride inhibition, the coprodeum values before and after amiloride inhibition. *Left columns* of each half figure indicate, both for coprodeum and colon, aldosterone injections simultaneously with acute resalination (Árnason and Skadhauge, unpublished data); *right columns* values after aldosterone injections to birds chronically adapted to the high-NaCl diet (Thomas et al. 1980)

of spironolactone (Árnason and Skadhauge, unpublished data). This aldosterone antagonist was ineffective when given to birds fully adapted to the low-NaCl diet, but inhibited the effect of aldosterone when the hormone and spironolactone were injected simultaneously with acute resalination. It is suggested that aldosterone may act more as a permissive agent than as a direct quantitative mediator in a feed-back loop. It remains to be established to what extent aldosterone produces a "snowball effect" augmenting its own action on low-NaCl adapted epithelia, or other hormones are involved in regulating epithelial sensitivity to aldosterone. The large effect of aldosterone given simultaneously with the first day of resalination seems to rule out other variables such as extracellular fluid volume and electrolyte or protein concentrations of plasma as directly responsible for the adaptation. The increase in mitochondrion content might both be induced by aldosterone and responsible for the change in effect of the hormone.

Summary

The sodium transport of the lower intestine (coprodeum and colon) of birds varies with the NaCl intake. When birds are switched from a high-NaCl to a low-NaCl diet the adaptation takes nearly a full week, but the intestinal transport pattern is reversed only 24 h after resalination. The main explanation for this difference seems to be that a nearly complete NaCl depletion is necessary to switch the transport parameters into the low-NaCl mode, wheras a smaller amount of NaCl is sufficient to reverse the pattern. The speed of change after acute resalination seems to rule out a change of epithelial cell population as responsible for the augmented sodium transport.

The changes in intestinal sodium transport follow the plasma aldosterone concentrations closely with colon more sensitive than coprodeum. Aldosterone injections to birds chronically exposed to the high-NaCl diet simulate only partially (25%–50%) the effects of a low-NaCl diet, but are fully effective when given simultaneously with acute resalination. This finding may suggest that aldosterone acts more as a permissive agent than as a quantitative mediator in a feed-back loop.

Other studies have shown that the event directly responsible for augmented sodium transport is an increased permeability of the luminal membrane. Studies on the binding of amiloride derivatives suggest that the permeation effect is caused by insertion into the membrane of preformed cytoplasmatic material. The remaining high sensitivity to aldosterone after acute resalination may be due to increase of the mitochondrion content of the transporting cells.

Although the plasma concentrations of both AVT and prolactin are higher in birds on high-NaCl diet neither of these hormones independently affects electrolyte transport of the lower intestine. Attempts to demonstrate a role for the renin-angiotensin system have failed so far.

Acknowledgements. The collaroration of S.S. Árnason, G.E. Rice and D.H. Thomas is gratefully acknowledged. Major support came from the Danish Natural and Medial Science Research Councils, NOVO's Fond and NATO Research Grants No. 1795 and 11882. Aldosterone was a gift from Ciba-Geigy.

References

Bindslev N (1979) Sodium transport in the hen lower intestine. Induction of sodium sites in the brush border by a low sodium diet. J Physiol (Lond) 288:449–466

Choshniak I, Munck BG, Skadhauge E (1977) Sodium chloride transport across the chicken coprodeum. Basic characteristics and dependence on sodium chloride intake. J Physiol (Lond) 271:489–504

Cuthbert AW, Edwardson JM, Bindslev N, Skadhauge E (1982) Identification of potential components of the transport mechanism for Na^+ in the hen colon and coprodeum. Pfluegers Arch 392:347–351

Eldrup E, Møllgård K, Bindslev N (1979) Possible sodium channels in the luminal membrane of the hen lower intestine visualized by freeze fracture. INSERM 85:253–262

Holtug K, Skadhauge E (1982) NaCl transport across hen colon. Dependence on electrochemical driving force. Pfluegers Arch 394:222–225

Lind J, Munck BG, Olsen O, Skadhauge E (1980a) Effects of sugars, amino acids and inhibitors on electrolyte transport across hen colon at different sodium chloride intakes. J Physiol (Lond) 305:315–325

Lind J, Munck BG, Olsen O (1980b) Effects of dietary intake of sodium chloride on sugar and amino acid transport across isolated hen colon. J Physiol (Lond) 305:327–336

Lyngdorf-Henriksen P, Munck BG, Skadhauge E (1978) Sodium chloride transport across the lower intestine of the chicken. Dependence on sodium chloride concentration and effect of inhibitors. Pfluegers Arch 378:161–165

Rice GE, Skadhauge E (1982a) The in vivo dissociation of colonic and coprodeal transepithelial transport in NaCl-depleted domestic fowl. J Comp Physiol B 146:51–56

Rice GE, Skadhauge E (1982b) Colonic and coprodeal transepithelial transport parameters in NaCl-loaded domestic fowl. J Comp Physiol B 147:65–69

Rice GE, Skadhauge E (1982c) Caecal water and electrolyte absorption and the effects of acetate and glucose in the dehydrated NaCl-depleted domestic fowl. J Comp Physiol B 147: 61–64

Rice GE, Munck BG, Skadhauge E (1981) Water and electrolyte absorption in duck colon: Effects of high and low NaCl diets. Abstr 9th Int Symp Endocrinol, HongKong, p 152

Scanes CG, Chadwick A, Bolton NJ (1976) Radioimmunoassay of prolactin in the plasma of the domestic fowl. Gen Comp Endocrinol 30:12–20

Skadhauge E (1980) Intestinal osmoregulation. In: Epple A, Stetson MH (eds) Avian endocrinology. Academic Press, London New York, p 481

Skadhauge E (1981) Osmoregulation in birds. Springer, Berlin Heidelberg New York

Skadhauge E (1982) A quantitative survey of salt and water excretion. Comp Biochem Physiol A 74:481–483

Skadhauge E, Dawson TJ (1980) In vitro studies of sodium transport across the lower intestine of a desert parrot. Am J Physiol 239:R285–R290

Skadhauge E, Arnason SS, Rice GE (1983a) Hormonal regulation of excretory functions in birds. Proc 9th Int Symp Comp Endocrinol (in press)

Skadhauge E, Thomas DH, Chadwick A, Jallageas M (1983b) Time course of adaptation to low and high NaCl diets in the domestic fowl: Effects on electrolyte excretion and on plasma hormone levels (aldosterone, corticosterone and prolactin). Pfluegers Arch 396:301–307

Thomas DH (1982) Salt and water excretion by birds: the lower intestine as an integrator of renal and intestinal excretion. Comp Biochem Physiol A 71:527–535

Thomas DH, Skadhauge E (1979) Chronic aldosterone therapy and the control of transepithelial transport of ions and water by the colon and coprodeum of the domestic fowl *(Gallus domesticus)* in vivo. J Endocrinol 83:239–250

Thomas DH, Skadhauge E (1982a) Regulation of electrolyte transport in the lower intestine of birds. In: Case M, Garner A, Turnberg L (eds) Electrolyte and water transport across gastrointestinal epithelia. Raven Press, New York, pp 295–303

Thomas DH, Skadhauge E (1982b) Time course of adaptation to low and high NaCl diets in the domestic fowl: Effects on elctrical behaviour of isolated epithelia from the lower intestine. Pfluegers Arch 395:165–170

Thomas DH, Skadhauge E, Read MW (1979) Acute effects of aldosterone on water and electrolyte transport in vivo by the colon and coprodeum of the domestic fowl *(Gallus domesticus).* J Endrocinol 83:229–237

Thomas DH, Jallageas M, Munck BG, Skadhauge E (1980) Aldosterone effects on electrolyte transport of the lower intestine (coprodeum and colon) of the fowl *(Gallus domesticus)* in vitro. Gen Comp Endocrinol 40:44–51

Effect of Galactose on Intracellular Potential and Sodium Activity in Urodele Small Intestine. Evidence for Basolateral Electrogenic Transport

J.F. WHITE and M.A. IMON [1]

Introduction

It has been appreciated for many years that the presence of Na^+ in the intestinal lumen enhances and indeed is required for active sugar absorption by the small intestinal mucosa. Since the seminal observation (Riklis and Quastel 1958, Czaky and Thale 1960, Crane et al. 1961) was made there have been numerous studies aimed at characterizing the transport mechanism in the brush border. There is now wide acceptance for the view that Na^+ and sugar are bound and cotransported on the same apical membrane carrier (Fig. 1a). More recently the way in which sugars are translocated across the basolateral membrane have received attention and these studies, pointing to facilitated diffusion of sugar, are serving to round out our view of transcellular sugar transport and sugar absorption. It has also been known for a long time that actively transported sugars stimulate the absorption of Na^+ (Fig. 1b). For example actively transported sugars increase the transepithelial electrical potential difference (ψ_{ms}) which is measurable across the wall of the intestine (Barry et al. 1961, Baillien and Schoffeniels 1961, Clarkson et al. 1961, Schachter and Britten 1961). The short-circuit current (I_{sc}) is similarly stimulated as illustrated in Fig. 2 without any early changes in tissues resistance. Phloridzin, which blocks sugar absorption (Newey et al. 1959) blocks the stimulation of I_{sc}. The simplest explanation for sugar-stimulated Na^+ absorption is suggested by Fig. 1b, i.e., luminal Na^+-sugar cotransport promotes Na^+ entry and raises cytoplasmic Na^+ providing more cations for a basolateral

Fig. 1a, b. Models of intestinal sodium and sugar absorption

1 Department of Physiology, Emory University, Atlanta, GA 30322, USA

Intestinal Transport
(ed. by M. Gilles-Baillien and R. Gilles)

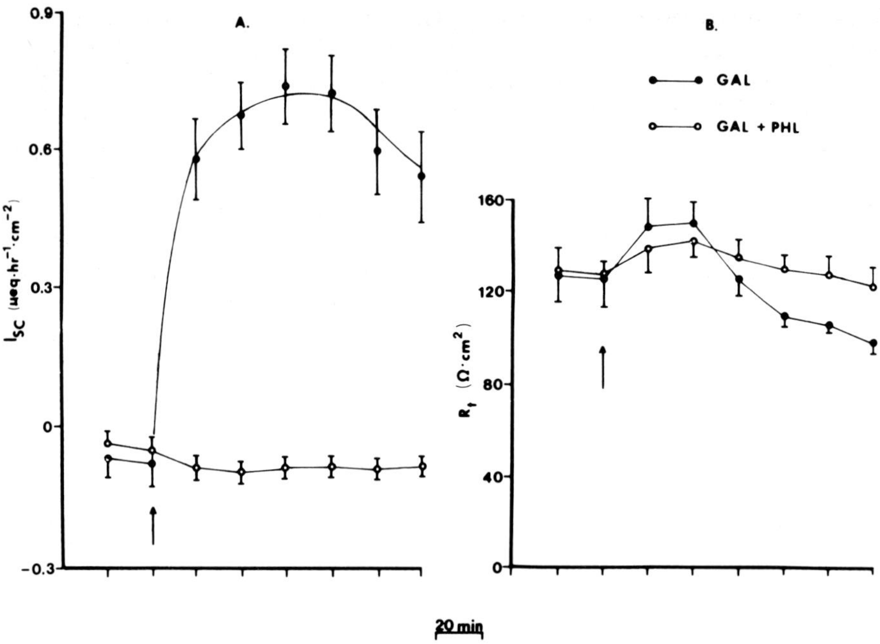

Fig. 2A, B. Time course of effect of 65 mM galactose (GAL) on short-circuit current (I_{SC}) and tissue resistance (R_t) compared in the presence and absence of 1 mM phlorizin (PHL). Additions at *arrow*

rheogenic Na^+ pump. Unlike the model for Na-dependent sugar transport (Fig. 1a) which has much experimental support, the explanation for sugar-stimulated Na^+ absorption, though elegant in its simplicity, does not have strong experimental support.

In this paper we will show evidence, derived from amphibian and mammalian small intestine, that the simplest prediction from Fig. 1b, namely that intracellular Na^+ concentration is increased by actively transported sugars, is contradicted by direct and indirect observations. Nevertheless Na^+ transport is enhanced. We will show evidence that the Na^+ pump is electrogenic and suggest possible models by which these observations can be reconciled.

Effect of Sugars on Intestinal Electrophysiology

Changes in the Luminal Membrane

If actively absorbed sugars can stimulate the transepithelial electrical potential consistent with greater basolateral (serosal) electrogenic Na^+ transport it would be expected that the serosal membrane potential (ψ_s) would be hyperpolarized. Although early microelectrode studies reported this finding (Barry and Eggenton 1972, Gilles-Baillien and Schoffeniels 1965, Wright 1966) it was subsequently reported for bullfrog intestine

(White and Armstrong 1971) and rabbit ileum (Rose and Schultz 1971) that both mucosal and serosal membranes of the absorptive cell are depolarized by actively transported sugars. The electrical response of a bullfrog intestinal cell to galactose is seen in Fig. 3a. There was a rapid and sustained depolarization of the luminal membrane concommitant with a smaller increase in the transepithelial potential (ψ_{ms}). The potential profile illustrated in Fig. 3b shows that the serosal membrane was also depolarized but by less. It was proposed that luminal membrane depolarization was due to increased permeability to Na^+ resulting from coupled Na^+-sugar cotransport (White and Armstrong 1971). (An additional factor which may promote luminal depolarization, a decrease in cell K^+, is discussed later in this chapter.)

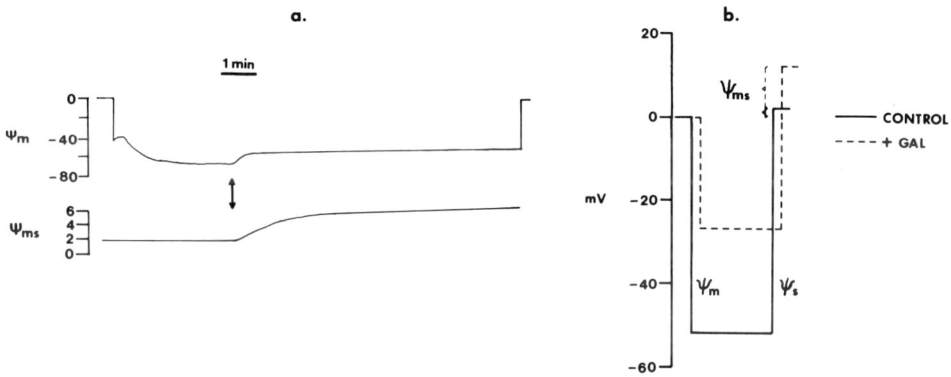

Fig. 3. a Time course of effect of galactose on mucosal membrane potential (ψ_m). Galactose-containing medium infused at *arrow*. (After White and Armstrong 1971, modified). b Potential profile of mucosal and serosal (ψ_s) membrane response to galactose

Changes in the Serosal Membrane

Rather than becoming hyperpolarized the serosal membrane was also depolarized by sugars as illustrated in Fig. 3b. Most likely ψ_s declines because of an IR drop across the serosal membrane, current being induced over low resistance paracellular (shunt) pathways. In other words, the mucosal and serosal membranes are electrically coupled. From similar observations Rose and Schultz (1971) proposed that serosal electrogenic Na^+ transport was stimulated in rabbit ileum, the effect being overshadowed by a decline in ψ_m due to the IR drop.

In subsequent years identical effects of sugar and amino acids have been described for several other intestinal preparations and in the renal tubule (e.g., Frömter 1982). Data from intestinal tissues is listed in Table 1. In most of these cases ψ_m declined to a new stable value upon exposure to actively transported solute. An exception, the response of *Amphiuma* intestine is described below.

Table 1. Effect of actively transported solutes on membrane potentials in intestinal tissue

Animal	Peak $\Delta\psi_m$ (mV)	Steady state $\Delta\psi_m$ (mV)	$\Delta\psi_{ms}$ (mV)	$\Delta\psi_s$ (mV)	Solute (mM)	
Amphiuma	19.4	13.9	3.3	10.6	Gal (65)	White and Imon, this chapter
Aplysia		6.1	1.1	5.0	Glu (50)	Gerencser and White (1980)
Bullfrog		25	10	15	Gal (65)	White and Armstrong (1971)
Necturus	50	12	3.8	8.4	Ala (10)	Gunter-Smith et al. (1982)
Rabbit ileum		10	3	7	Ala (20)	Rose and Schultz (1971)
Rat		7.7	1.3	6.4	Glu (5.5)	Dinno and Huang (1977)
Rat duodenum		3.5			Glu (20)	Okada et al. (1977)

Response of Amphiuma Intestine to Galactose – the Biphasic Response

Recent observations in the small intestine of *Amphiuma* (order: urodeles) have provided evidence that the Na^+-K^+ pump is electrogenic and is stimulated by actively transported sugar. Exposure of the mucosal surface of isolated stripped segments of *Amphiuma* intestine to galactose was followed by a reduction of ψ_m, measured with microelectrodes, which was not sustained. In Fig. 4 two recordings illustrating the effect are shown. When the medium containing galactose (20 mM) was infused ψ_m was depolarized over 2–3 min as reported for bullfrog and other intestinal preparations but then became partially repolarized over several minutes before stabilizing. The transmural potential began to climb only after ψ_m was reduced. After several minutes ψ_{ms} reached a new steady state and did not exhibit any change that could be associated with repolarization of ψ_m. As seen in Table 2 in 10 of 11 recordings from cells lining the villus glucose or galactose reduced ψ_m an average of 6.1 mV. In 7 of 10 recordings the mucosal membrane was subsequently repolarized an average of 3.1 mV.

In a limited number of measurements in intervillus cells (the epithelial cells lying between the villi) depolarization by actively transported sugar was also observed (Table 2). Furthermore repolarization was seen in 3 of 4 recordings.

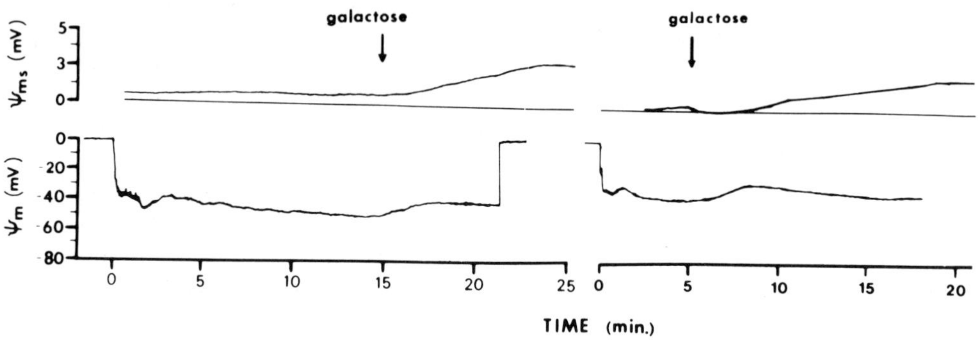

Fig. 4. Effect of galactose (20 mM) added at *arrow* on electrical potentials in *Amphiuma* intestine

Table 2. Effect of transported sugars on mucosal membrane potential (ψ_m) and transepithelial potential (ψ_{ms}) in villus and intervillus cells

Major anion	Sugar (mM)	ψ_m (mV)	max $\Delta\psi_m$ (mV)	$\Delta\psi_{ms}$ (mV)
Villus cells				
Cl	Glucose (10)	– 42	+ 2	+ 0.9
Cl	Glucose (10)	– 38	+ 3	+ 1.1
Cl	Galactose (10)	– 50	0	+ 1.5
SO_4^{2-}	Glucose (10)	– 52	+ 7	–
SO_4^{2-}	Glucose (10)	– 37	+ 10	+ 2.3
SO_4^{2-}	Glucose (10)	– 72	+ 4	+ 2.1
SO_4^{2-}	Galactose (20)	– 53	+ 6	+ 4.3
SO_4^{2-}	Galactose (20)	– 35	+ 11	+ 1.8
SO_4^{2-}	Galactose (20)	– 32	+ 8	+ 1.6
SO_4^{2-}	Galactose (20)	– 30	+ 9	+ 2.1
	Average:	– 43 ± 4	+ 6 ± 1	+ 2.1 ± 0.3
Intervillus cells				
SO_4^{2-}	Galactose (20)	– 29	+ 5	1.5
SO_4^{2-}	Galactose (20)	– 39	+ 11	2.2
SO_4^{2-}	Galactose (20)	– 41	+ 8	–
SO_4^{2-}	Galactose (20)	– 52	+ 5	–
	Average:	– 40 ± 5	+ 7 ± 1	1.8

One explanation for depolarization of the luminal membrane mentioned above is that the entry in cotransport with sugar serves to increase the passive permeability of the membrane for Na^+, i.e., that ψ_m approaches E_{Na}^m, the Nernst potential for Na^+ across the luminal membrane. Indeed in a limited number of measurements, described later in this chapter, the polarity of the luminal membrane was transiently reversed (e.g., Fig. 7). Another factor which may contribute to the reduction in ψ_m is the decrease in the intracellular concentration of K^+ which follows exposure to actively transported solutes. This effect, described later in this chapter, would lower E_K, the Nernst potential for K^+ at the luminal membrane and therefore reduce ψ_m as pointed out by White and Armstrong (1971) and Okada et al (1976). The influence of these factors on ψ_m would be determined by the relative permeability of the membrane for Na^+ and K^+. A definitive study of these factors has not been conducted to date.

Repolarization of the mucosal membrane following depolarization by actively absorbed solutes is an unusual response compared with that in other animals (see Table 1). A similar observation was made in *Necturus* intestine where it was shown that repolarization of ψ_m is blocked upon exposure to cyanide and iodoacetamide (Gunter-Smith et al. 1982). If the initial depolarization is due to enhanced Na^+ entry then it seems unlikely that the repolarization of the mucosal membrane is due to a subsequent reduction of mucosal Na^+ permeability because the stimulation of ψ_{ms} and I_{sc} are sustained. A more likely explanation of the repolarization is that hyperpolarization of the serosal membrane occurs because of greater electrogenic Na transport, this electrical effect being reflected in the mucosal membrane as a result of the electrical coupling that exists between the two membranes which are in series.

Evidence for Basolateral Rheogenic Na Transport

Some support for the notion that the serosal Na pump is electrogenic is illustrated in Fig. 5. Intracellular K^+ activity (a^i_K) was measured in cells of urodele jeunum following exposure of the serosal surface to ouabain (1 mM), the inhibitor of Na^+-K^+ ATPase. Each data point reflects the mean ± S.E.M. of at least 37 impalements in four tissues. After 1 h exposure to ouabain ψ_m had declined 11 mV while a^i_K had declined only slightly. Therefore this change could not be due to a change in the transmembrane K^+ diffusion potential. Only in subsequent time periods was the decline in ψ_m matched by a drop in a^i_K. Assuming that the change in ψ_m reflects via coupling a primary change in the serosal membrane potential, then the simplest explanation for this observation is that the electrogenic Na^+ pump contributes about 11 mV to the membrane potential. Possibly the Na^+ pump is rheogenic, e.g., exchanging 3 Na^+ for 2 K^+. Evidence for rheogenic Na^+ transport has been reported for rat duodenum (Okada et al. 1977), for choroid plexus (where ouabain rapidly depolarized the membrane by 10.2 mV, Zeuthen and Wright 1981) and for rat kidney tubule cells (which were depolarized 10.6 mV by oubain, Frömter and Gessner 1975). These results provide circumstantial evidence that the Na pump in the small intestine is electrogenic.

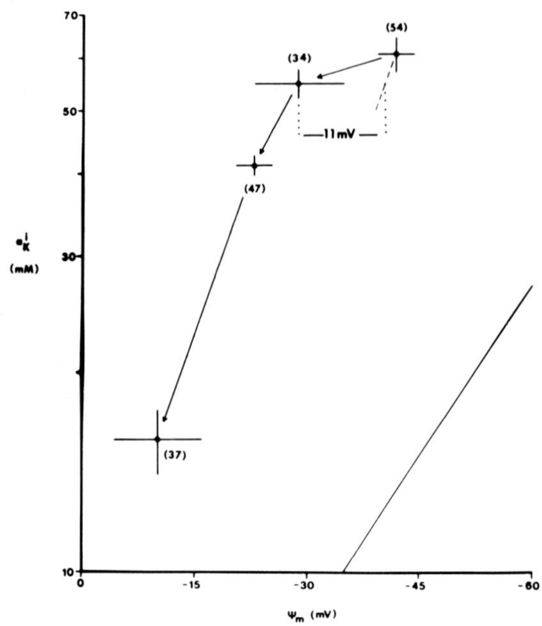

Fig. 5. Relationship between a^i_K and ψ_m in 4 tissues. The *arrows* indicate the time course of decline of a^i_K and ψ_m over 3 h after exposure to ouabain. *Solid line* relationship if K^+ was in electrochemical equilibrium. *Dashed line* illustrates the expected decline of ψ_m due simply to a loss of K^+ from the cell

The Effect of Sugars on Intracellular Na^+

Cytoplasmic Na^+ Activity in the Urodele

If galactose stimulates Na^+ transport by providing Na^+ in greater concentration for a serosal Na^+ pump we should be able to demonstrate an increase in cell Na^+ concentration (or activity) in the presence of galactose. This was examined using Na-sensitive microelectrodes in single and double-barreled configuration. Representative recordings are illustrated in Fig. 6. The potential of the Na^+-sensing microelectrode (ψ_{Na}) when corrected for ψ_m, was equivalent to the intracellular Na^+ activity (a_{Na}). In Table 3 the average value of a^i_{Na} in *Amphiuma* intestine bathed in Tris-buffered Cl^--free medium is compared with similar measurements in other epithelial preparations and is seen to be in the mid range. The Na^+-exchanger used here, as reported on by Steiner et al. (1979), is sensitive to calcium at concentrations found in the extracellular fluid. However, measurements in the cytoplasm, where a^i_{Ca} is low (e.g., O'Doherty and Stark 1981), should be free of this interference.

When the small intestine is exposed to galactose, intracellular Na^+ activity is not increased. In Fig. 7 a representative experiment is illustrated in which a medium containing galactose (65 mM) was infused while a^i_{Na} and ψ_m were measured in the same villus cell by using a double-barreled microelectrode. Galactose infusion produced a rapid depolarization of such a magnitude as to reverse the polarity of the mucosal membrane. Repolarization of the membrane followed within 25 s but in this cell the original polarity (cell interior negative) was not restored. Intracellular a_{Na} declined from 8.1 to 5.0 mM, all of the drop occurring during the period in which ψ_m was changing rapidly as indicated by the trace of ψ_{Na}-ψ_m. Since the response of the membrane potential-sensing barrel is essentially instantaneous while the response of

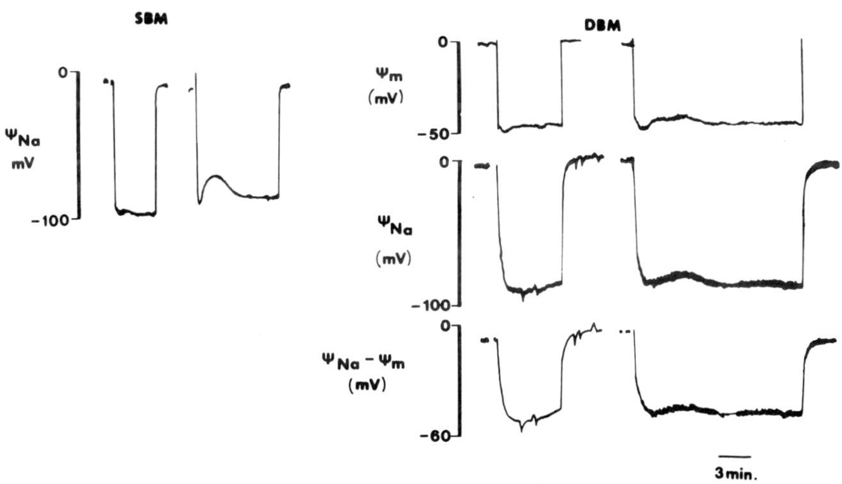

Fig. 6. Recordings obtained with Na^+-sensitive liquid ion-exchanger microelectrodes. *Left* two recordings of Na potential (ψ_{Na}) with single-barreled microelectrodes *(SBM)*; *right* with double-barreled microelectrodes *(DBM)*. The difference, ψ_{Na}-ψ_m is equivalent to cell Na activity

Table 3. Values of a^i_{Na} in cells from various epithelia

	a^i_{Na} (mM)	Reference
Necturus proximal tubule	7.4	Cemerikic and Giebisch (1980)
Necturus small intestine	9	O'Doherty et al. (1979)
Frog skin	14	Nagel et al. (1981)
Bullfrog proximal tubule	16.8	Fujimoto and Honda (1980)
Bullfrog small intestine	18	Armstrong et al. (1979)
Amphiuma small intestine	19	White and Imon (this chapter)
Bullfrog proximal tubule	20.7	Kotera et al. (1979)
Necturus gallbladder	22.2	Reuss and Weinman (1979)
Necturus proximal tubule	29.7	Kimura and Spring (1979)

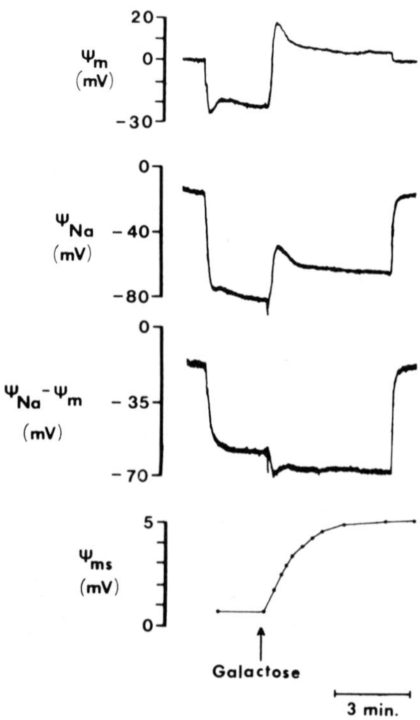

Fig. 7. Effect of galactose (65 mM) added at *arrow* on Na activity and electrical properties of intestinal cells

the liquid ion-exchanger barrel is much slower (about 5 s for 90% response) the initial change in ψ_{Na}-ψ_m is due largely to the change in ψ_m. In any case in the new steady state intracellular Na^+ activity was lower. In this way the cell Na^+ activity was successfully monitored in 9 cells from separate tissues during galactose perfusion. The averaged results are tabulated in Table 4. On average a^i_{Na} declined from 15.0 to 12.5 mM. This change was insignificant ($P < 0.05$). At the same time ψ_m was depolarized on average from -29.3 to -9.9 mV before repolarizing to -15.4 mV. Thus upon exposure to galactose the activity of Na^+ in the cytoplasm is not increased

Table 4. Effect of galactose on intracellular Na^+ activity (a^i_{Na}) and electrical parameters

	a^i_{Na} (mM)	ψ_m (mV)	ψ_{ms} (mV)	ψ_s (mV)
Control	15.0 ± 2.3	– 29.3 ± 3.0 (– 9.9 ± 5.0)	2.4 ± 1.5	31.7 ± 3.1
+ Galactose (65 mM)	12.5 ± 2.5	– 15.4 ± 4.0	5.7 ± 1.2	21.1 ± 4.1
Δ	– 2.5 ± 1.4	+ 13.9 ± 2.3	+ 3.3 ± 0.4	– 10.6 ± 2.1
P	N.S.	< 0.001	< 0.001	< 0.005

Galactose was added at 65 mM. n = 9 tissues. N.S. = not significant at $P < 0.05$. The value of ψ_m in parentheses is the average lowest value obtained after addition of sugar

but rather remains essentially unchanged despite the fact that galactose produced electrical changes suggestive of enhanced mucosal Na^+ entry and enhanced serosal Na^+ extrusion. For this reason some stimulus other than increased intracellular Na^+ concentration must be responsible for stimulating serosal Na^+ extrusion.

The results tabulated in Table 4 are in very good agreement with those reported by Lee and Armstrong (1972) for bullfrog small intestine, using pairs of single-barreled K^+ selective glass microelectrodes of differing selectivity for potassium. They found that a^i_{Na} declined from 14.4 mM to 12.7 mM after exposure to 3-O-methylglucose (26 mM), the change being statistically significant. In the bullfrog small intestine the change in ψ_m produced by sugar is monophasic (White and Armstrong 1971).

Effect of Sugar on Cell Sodium Concentration

The lack of effect of galactose on intracellular Na^+, as measured with ion-sensitive microelectrodes, also agrees with earlier photometric measurements of the sodium concentration of the cell (c^i_{Na}). These are listed in Table 5. It is seen that 3-O-methylglucose, a non-metabolized but actively absorbed sugar, which stimulates Na^+ absorption in the intestine, did not elevate the cytoplasmic Na^+ concentration in several in vitro preparations. A possible explanation for the observed deline in cell Na^+ is also seen in Table 5. After exposure to 3-O-methylglucose cell K^+ undergoes a significant decline. The validity of this observation is supported by the data of Lee and Armstrong (1972) who observed a 24% decline in intracellular K^+ activity after exposure to 3-O-methylglucose. The decline in cell K^+ activity is most likely due to cell swelling which is seen to occur from the data for cell water of Czaky and Esposito (1969) and Armstrong et al. (1970) listed in Table 5.

Entry of sugar, Na^+ and associated anion into the cytoplasm should promote an osmotic influx of water and may be the cause of the gain in cell water and the decline in cell K^+. For intracellular Na^+ the expected increase in concentration due to coupled entry with sugar may be offset or even negated by a concommitant entry of water. Okada et al. (1976) computed that a net gain of Na^+ and K^+ occurred after exposure of rat small intestine to glucose but because of a concommitant increase in cell water the concentration of Na^+ and K^+ declined.

Table 5. Effect of 3-O-methylglucose on cell Na^+, K^+ and water

Tissue	Na^+		K^+		Water		Reference
	Control	3-O-MG	Control	3-O-MG	Control	3-O-MG	
Rat jejunum			330 ± 9	269 ± 8*			Brown and Parsons (1962)
Bullfrog intestine	55 ± 6	42 ± 4	106 ± 3	78 ± 3**	2.49	3.19**	Czaky and Esposito (1969)
Rabbit jejunum	58 ± 2	59 ± 4	145 ± 3	128 ± 4*			Koopman and Schultz (1969)
Bullfrog intestine	33 ± 7	27 ± 7	77 ± 7	54 ± 5*	3.82	4.40**	Armstrong et al. (1970)

Ion concentration in mM except for Brown and Parsons where concentration is in mmol kg^{-1} dry wt; water content in ml g^{-1} dry tissue. * $P < 0.05$; ** $P < 0.01$

Mechanisms for Stimulation of the Na Pump

Enhanced basolateral Na^+ transport would also serve to moderate any increase in cell Na^+. Greater Na^+ transport is suggested by the observed repolarization of the mucosal membrane (Fig. 4). However the stimulus for enhanced basolateral Na^+ transport is not obvious. Since a^i_{Na} was not increased by sugars the stimulus cannot be greater concentration of transported species in the cytoplasm as substrate for transport. Two ways in which basolateral Na^+ transport may be stimulated are discussed below.

Reduction of $\Delta\mu^s_{Na}$

The Na^+ pump may be activated by a reduction in the electrical driving force against which the pump must operate as proposed by Armstrong (1975). The electrogenic Na^+ pump must operate against an electrochemical gradient for Na^+ at the serosal membrane ($\Delta\mu^s_{Na}$) which is equal to

$$\Delta\mu^s_{Na} = RT\ln a^o_{Na}/a^i_{Na} + F\psi_s$$

where R is the gas constant, T is the absolute temperature and F is the Faraday constant. Since ψ_s declined to 10.6 mV (Table 4), it can be readily calculated that $\Delta\mu^s_{Na}$ is reduced about 15% after exposure to galactose. Although the rate of serosal Na^+ transport as a function of $\Delta\mu^s_{Na}$ is not known the slightly more favorable circumstance for Na^+ transport may not account for the large increase in net Na absorption.

Reduction of Cell K^+

Another possible means by which serosal Na^+ transport could be stimulated is suggested by the observations of Knight and Welt (1974) on red cells. They reported that Na^+ efflux from red cells with low intracellular Na^+ concentration (3–4 mM) is an inverse function of cellular K^+. For example reduction of cellular K^+ by the amount reported by Armstrong et al. (1970) after exposure to 3-O-methylglucose (Table 4) would increase the ouabain-sensitive Na^+ efflux in erythrocytes by 50%. The idea that cellular K^+ is a competitive inhibitor of the pump was also arrived at in a study by Garay and Garrahan (1973). In agreement with this high K has been shown to inhibit ATPase activity in red cell ghosts (Post et al. 1960) and nerve cell membranes (Skou 1957). Thus the reduction of cell K^+ that follows exposure to actively transported solutes (Table 5) may accelerate Na^+ export by relieving K^+-induced inhibition of the basolateral Na^+-K^+-ATPase.

Whether the Na^+ pump is stimulated as a result of the lowering of $\Delta\mu^s_{Na}$ or by reduction of cell K^+ the consequence is that cell Na^+ activity remains low. The stimulation of basolateral Na^+ transport without elevation of cell Na^+ permits the maintenance of the driving force for Na^+ across the luminal membrane. Thus the uphill accumulation of sugars and amino acids across the mucosal membrane can continue despite the greater influx of Na^+ without dissipation of the driving force for solute uptake (Armstrong 1975).

Equivalent Circuit of Intestinal Mucosa

In order to better understand how sugars stimulate Na^+ absorption it is helpful to develop a model in the form of an equivalent electrical circuit of the small intestinal epithelium which conforms to the observations made and which can be tested and further refined. It is proposed that the epithelium behaves like two batteries or electromotive forces in series with each other with a combined resistance which is greater than that of a parallel pathway. This is illustrated in Fig. 8. This model has been widely employed in analyzing ion transport in epithelia of low electrical resistance (Boulpaep 1967, Giebisch 1968, White and Armstrong 1971). The two electromotive forces E_m and E_s correspond to the electrical driving force at the mucosal and serosal membranes of the absorptive cell respectively. Thus they represent the transcellular pathway for ion movement. The identity of the low resistance parallel pathway is

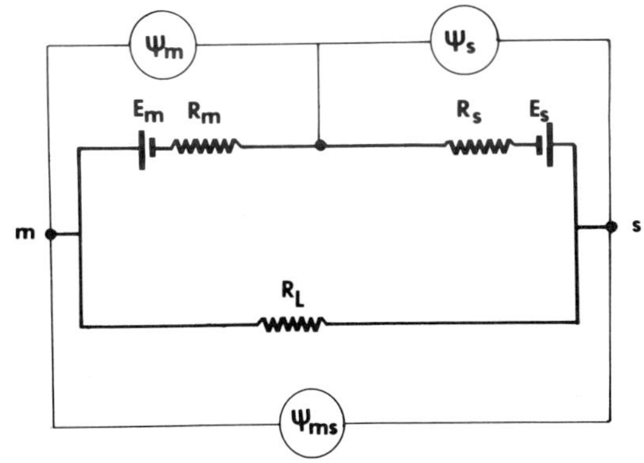

Fig. 8. Lumped equivalent electrical circuit of intestinal mucosa featuring electromotive forces *(E)* and resistances *(R)* at the mucosal *(m)* and serosal *(s)* membrane and the shunt resistance *(R_L)*

uncertain but evidence exists that the tight junctional region and lateral intracellular space between the epithelial cells is a pathway of high conductivity in epithelia (Frömter and Diamond 1972). It is the low resistance paracellular pathway which electrically couples the mucosal and serosal membranes so that changes produced in one membrane have electrical consequences in the other. The issue is to identify those electromotive forces, whether diffusion potentials or electrogenic pump potentials that, with the resistances, determine the mucosal and serosal membrane potentials (ψ_m; ψ_s) and how these are altered by transported solutes.

Luminal Membrane Permeability to K^+

Recently we have developed evidence that E_m, the equivalent electromotive force at the mucosal membrane contains a significant K^+ diffusion potential. The mucosal membrane potential should be sensitive to changes in K^+ concentration in the mucosal medium if this were the case. Using double-barreled microelectrodes K^+ activity (a^i_K) and ψ_m were measured in segments of urodele jejunum bathed on their mucosal surface with a medium containing HCO_3^- (serosal medium HCO_3^- free). When K^+ (2.5 mM) was removed from only the mucosal medium ψ_m was hyperpolarized from -35 ± 2 mV to -42 ± 2 mV ($P < 0.005$) even as intracellular K^+ declined from 85 ± 2 mM to 61 ± 4 mM. These results are shown in Fig. 9. When K^+ was removed from only the serosal medium the hyperpolarization of the mucosal membrane was even more pronounced. In the latter case the increase in ψ_m is probably due to

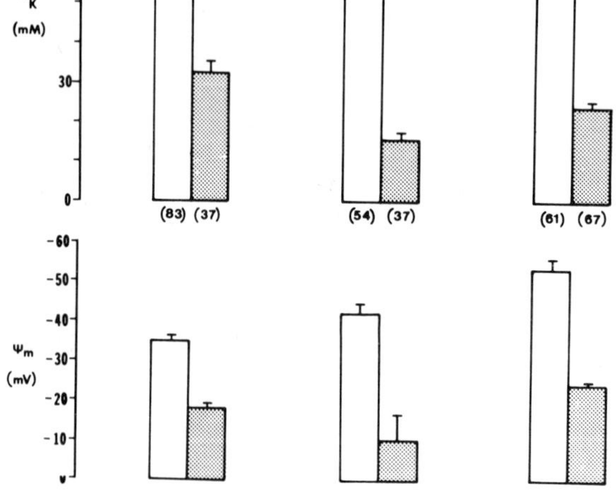

Fig. 9. Comparison of a^i_K and ψ_m before and 3–4 h after ouabain in tissues bathed bilaterally with K^+ or with K^+ absent from mucosal or serosal medium

hyperpolarization of the serosal membrane reflected in the mucosal membrane which is electrically coupled to it. In conclusion both mucosal and serosal membranes are permeable to K^+. Parenthetically, the results of Fig. 9 also indicate that the epithelial cells accumulate K^+ to higher levels when K^+ is present in the mucosal medium as if K^+ is actively accumulated across the mucosal membrane.

Evidence that the mucosal membrane is leaky to K^+ was also derived from extracellular measurements of medium K^+. A region of high K^+ activity can be detected with a K^+-specific microelectrode as it is advanced towards the brush border of the absorptive cell. The K^+ activity within 40 μm of the membrane exceeded the K^+ activity in the bulk medium by as much as 1 mM. When a K^+-free medium was employed in the mucosal compartment the K^+-activity near the membrane averaged 0.8 ± 0.1 mM. These gradients can be reduced by rapid perfusion of medium over the mucosal surface of the isolated sheet of intestine (White 1976). While it is possible that the free diffusion of K^+ ions is restricted in this region, for example by mucus, the absence of an associated electrical voltage (White 1976) rules out an electrophoretic effect and points to a leak from the cell as the source of this region of high K^+ concentration.

Luminal Membrane Permeability to Na^+

The mucosal membrane also has a finite permeability to Na^+. As seen in Fig. 10 when Na^+ was replaced with Tris ψ_m was -44 ± 1 mV which was significantly higher ($P < 0.001$) than that in control media (-35 ± 2 mV). Simultaneously a^i_K was reduced significantly ($P < 0.001$) indicating that the hyperpolarization of the mucosal membrane was not due to an increase in E_K. The decline of a^i_K would be expected if Na^+ transport normally served to maintain intracellular K^+ at a high concentration. Evidence for a second means of K^+ uptake is also seen in Fig. 10 wherein K^+ accumulation remained as high as 45 mM even 3 h after exposure to ouabain and Na^+-free medium. The retention of K^+ was not simply the consequence of an insufficient driving force for K^+ since E_K-ψ_m was -40 mV before ouabain and only declined to -38 mV after 4 h. Assuming that the changes in ψ_m do not, in this case, reflect events at the serosal membrane we conclude the mucosal membrane is normally permeable to Na^+.

Estimate of P_{Na}/P_K

An estimate of the relative permeability of the membrane to Na^+ and K^+ can be obtained from these data. At 22°C

$$\psi_m = 58.6 \text{ mV} \log \frac{[K^+]_o + \propto [Na]_o}{[K^+]_i + \propto [Na]_i} \tag{1}$$

where $\propto$ is the ratio of permeability coefficients (P_{Na}/P_K). Inserting measured values of intracellular and bath a_{Na} and a_K obtained when the tissue was bathed with a

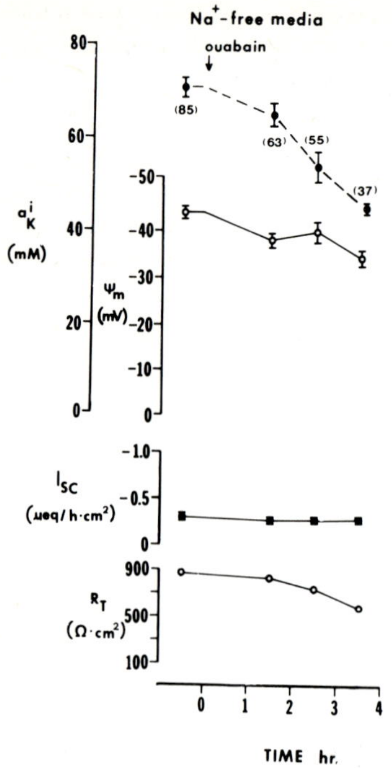

Fig. 10. Time course of change in cell K^+, ψ_m, short-circuit current (I_{sc}) and tissue resistance (R_t) before and after exposure to ouabain in 7 tissues bathed in Na^+-free media

HCO_3^--buffered glactose-free medium it was found that $P_{Na}/P_K \cong 0.17$. In comparison, Okada et al. (1975) calculated a somewhat lower value (0.07) in rat duodenum while in rabbit ileum $P_{Na} \cong P_K$ (Rose and Schultz 1971). Unfortunately Eq. (1) is valid only under certain conditions (Okada et al. 1975). For example, if there is electrogenic Cl^- or Na^+ transport which contributes to ψ_m, then ψ_m cannot be equated simply to ion diffusion potentials. Since electrogenic Na^+ transport occurs in the urodele intestine and contributes to ψ_m (Fig. 5) and since electrogenic Cl^- transport has been described in this preparation as well (White 1976), the value of relative permeability reported here must be regarded as only an approximation. Nevertheless it seems very likely that a significantly greater conductance for K^+ relative to Na^+ exists at the luminal membrane.

Estimates of R_m and R_s

Estimates of mucosal and serosal membrane resistance can be obtained from circuit analysis. Frömter (1979) and Gunter-Smith et al. (1982) relate the initial rate of hyperpolarization of ψ_{ms} ($d\psi_{ms}$) and depolarization of the mucosal membrane ($d\psi_m$) by solutes to serosal and shunt resistances by:

$$d\psi_{ms}/d\psi_m = 1/(1 + R_s/R_L). \tag{2}$$

In 7 recordings in which the initial rate of change of ψ_{ms} and ψ_m was measurable, $d\psi_{ms}/d\psi_m$ averaged 0.021. From Eq. (2), $R_s = 45.5\ R_L$; since $R_L = R_t = 126\ \Omega \cdot cm^2$ (Fig. 4) then $R_s = 5733\ \Omega \cdot cm^2$. Furthermore since R_m/R_s (the voltage divider ratio) was 0.30, then $R_m = 1720\ \Omega \cdot cm^2$.

As indicated in Table 6 the values of R_m and R_s agree remarkably well with an independent estimate obtained by an analysis of the changes in ψ_m, ψ_s and ψ_{ms} observed when K^+ was deleted from the mucosal or serosal bathing medium. The major resistance barrier in *Amphiuma* small intestine is the serosal membrane. This is also the case in rat duodenum (Okada et al. 1975) but in *Necturus* small intestine $R_m > R_s$ (Gunter-Smith et al. 1982).

Table 6. Estimates of resistance of mucosal (R_m) and serosal (R_s) membranes

Method	R_m $\Omega \cdot cm^2$	R_s $\Omega \cdot cm^2$	R_m/R_s
Response to sugar addition	1720	5733	0.30
Response to K^+ deletion	1707	6097	0.28

Summary

The equivalent electrical circuit in Fig. 11 incorporates many of the observations described in this chapter. Actively transported solutes in cotransport with Na^+ introduce a Na^+ leak into the luminal membrane which is in parallel with the normal membrane conductance for Na^+. At the basolateral membrane an electrogenic pump transfers Na^+ out in exchange for K^+. Okada et al. (1978) estimated an exchange ratio of 4 Na^+:3 K^+ in rat duodenum.

The sequence of events outlined in Fig. 12 is believed to account for observations reported here on the urodele small intestine. The mucosal membrane has a high conductivity to K^+. Upon exposure to sugars in the luminal fluid ψ_m declines away from E_K and towards E_{Na} as the luminal permeability to Na increases because of coupled

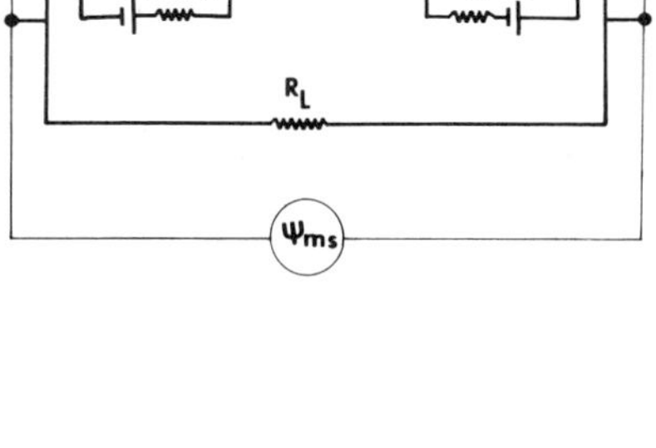

Fig. 11. Equivalent electrical circuit incorporating electromotive forces believed present at mucosal and serosal membranes. E^S_{Na} and R^S_{Na} comprise a pathway parallel to the normal Na^+ leak path that is evident in the presence of transported solutes

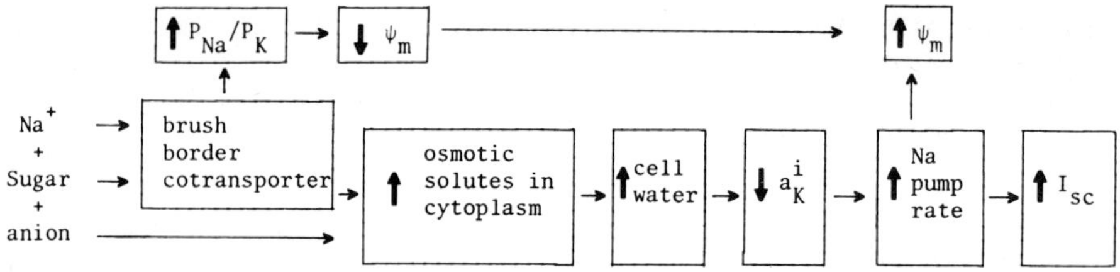

Fig. 12. Events following exposure of intestinal cells to sugar

influx of Na^+ and sugar. Eventually ψ_m repolarizes due to stimulation of serosal electrogenic Na^+ transport. The increased electrogenic Na transport is not a result of elevated cell Na and may be due to a decline in cell K^+, cell K^+ acting normally to slow the Na pump.

Acknowledgements. This work was supported by grants AM17361 and AM26870 from the National Institute of Health.

References

Armstrong WMcD (1975) Electrophysiology of sodium transport by epithelial cells of the small intestine. In: Czaky TZ (ed) Intestinal absorption and malabsorption. Raven Press, New York, p 45

Armstrong WMcD, Musselman DL, Reitzug HC (1970) Sodium, potassium and water content of isolated bullfrog small intestinal epithelia. Am J Physiol 219:1023–1026

Armstrong WMcD, Bixenman WR, Frey KF, Garcia-Diaz JF, O'Regan MG, Owens JL (1979) Energetics of coupled Na^+ and Cl^- entry into epithelial cells of bullfrog small intestine. Biochim Biophys Acta 551:207–219

Baillien M, Schoffeniels E (1961) Origin of the potential difference in the intestinal epithelium of the turtle. Nature 190:1107–1108

Barry RJC, Eggenton J (1972) Membrane potentials of epithelial cells in rat small intestine. J Physiol (Lond) 227:201–216

Barry RJC, Dikstein S, Matthews J, Smyth DH (1961) Electrical potentials in the isolated intestine. J Physiol (Lond) 17P–18P

Boulpaep EL (1967) Ion permeability of the peritubular and luminal membrane of the renal tubular cell. In: Kruck F (ed) Symposium über Transport und Funktion intracellulärer Elektrolyte. Urban & Schwarzenberg, München, p 98

Brown MM, Parsons DS (1962) Observations on the changes in the potassium content of rat jejunal mucosa during absorption. Biochim Biophys Acta 59:249–251

Cemerikic D, Giebisch G (1980) Intracellular sodium activity in *Necturus* kidney proximal tubule. Fed Proc 39:1080

Clarkson TW, Cross AC, Toole SR (1961) Dependence on substrate of the electrical potential across the isolated gut. Nature 191:501–502

Crane RK, Miller D, Bihler I (1961) The restrictions on possible mechanisms of intestinal active transport of sugars. In: Kleinzeller A, Kotyk Z (eds) Membrane transport and metabolism. Academic Press, London New York, p 439

Czaky TZ, Esposito G (1969) Osmotic swelling of intestinal epithelial cells during active sugar transport. Am J Physiol 217:753–755

Czaky TZ, Thale M (1960) Effect of ionic environment on intestinal sugar transport. J Physiol (Lond) 151:59–65

Dinno MA, Huang KC (1977) Effect of glucose and diuretics on intracellular potentials of mouse intestinal mucosa. Proc Soc Exp Biol Med 155:71–78

Frömter E (1979) Solute transport across epithelia: What can we learn from micropuncture studies on kidney tubules? J Physiol (Lond) 288:1–31

Frömter E (1982) Electrophysiological analysis of rat renal sugar and amino acid transport. 1. Basic phenomena. Pfluegers Arch 393:179–184

Frömter E, Diamond J (1972) Route of passive ion permeation in epithelia. Nature 235:9–13

Frömter E, Gessner K (1975) Effect of inhibitors and diuretics on electrical potential differences in rat kidney proximal tubule. Pfluegers Arch 357:209–224

Fujimoto M, Honda M (1980) Direct measurement of intracellular Na and K activities in the renal tubular cells with triple-barreled microelectrodes. Proc Int Congr Physiol Sci 14:119

Garay RP, Garrahan PJ (1973) The interaction of sodium and potassium with the sodium pump in red cells. J Physiol (Lond) 231:297–325

Gerencser GA, White JF (1980) Membrane potentials and chloride activities in epithelial cells of *Aplysia* intestine. Am J Physiol 239:R445–R449

Giebisch G (1968) Some electrical properties of single renal tubule cells. J Gen Physiol 51:315s

Gilles-Baillien M, Schoffeniels E (1965) Site of action of L-alanine and D-glucose on the potential difference across the intestine. Arch Int Physiol Biochim 73:355–357

Gunter-Smith PJ, Grasset E, Schultz SG (1982) Sodium-coupled amino acid and sugar transport by *Necturus* small intestine. An equivalent electrical circuit analysis of a rheogenic co-transport system. J Membr Biol 66:25–40

Kimura G, Spring KR (1979) Luminal Na^+ entry into *Necturus* proximal tubule cells. Am J Physiol 236:F295–F307

Knight AB, Welt LG (1974) Intracellular potassium. A determinant of the sodium-potassium pump rate. J Gen Physiol 63:351–373

Koopman W, Schultz SG (1969) The effect of sugars and amino acids on mucosal Na^+ and K^+ concentrations in rabbit ileum. Biochim Biophys Acta 173:338–340

Kotera K, Satake N, Honda M, Fujimoto M (1979) The measurement of intracellular sodium activities in the bullfrog by means of double-barreled sodium liquid ion-exchanger microelectrodes. Membr Biochem 2:323–338

Lee CO, Armstrong WMcD (1972) Activities of sodium and potassium ions in epithelial cells of small intestine. Science 175:1261–1264

Nagel W, Garcia-Diaz JF, Armstrong WMcD (1981) Intracellular ionic acitivities in frog skin. J Membr Biol 61:127–134

Newey H, Parsons BJ, Smyth DH (1959) The site of action of phlorizin in inhibiting intestinal absorption of glucose. J Physiol (Lond) 148:83–92

O'Doherty J, Stark RJ (1981) Transmembrane and transepithelial movement of calcium during stimulus-secretion coupling. Am J Physiol 241:G150–G158

O'Doherty J, Garcia-Diaz JF, Armstrong WMcD (1979) Sodium-selective liquid ion-exchanger microelectrodes for intracellular measurements. Science 203:1349–1351

Okada Y, Sato T, Inouye A (1975) Effects of potassium ions and sodium ions on membrane potential of epithelial cells in rat duodenum. Biochim Biophys Acta 413:104–115

Okada Y, Iramajiri A, Inouye A (1976) Intracellular ion concentrations of epithelial cells in rat small intestine. Effects of external potassium ions and uphill transports of glucose and glycine. Jpn J Physiol 26:427–440

Okada Y, Tsuchiya W, Iramajiri A, Inouye A (1977) Electrical properties and active solute transport in rat small intestine. I. Potential profile changes associated with sugar and amino acid transport. J Membr Biol 31:205–219

Okada Y, Iramajiri A, Tsuchiya W, Inouye A (1978) Contribution of an electrogenic sodium pump to the membrane potential in the intestinal epithelial cell. Jpn J Physiol 28:511–525

Post RL, Merritt CR, Konsolving CR, Albright CD (1960) Membrane adenosine triphosphatase as a participant in the active transport of sodium and potassium in the human erythrocyte. J Biol Chem 235:1796–1802

Reuss L, Weinmann SA (1979) Intracellular ionic activities and transmembrane electrochemical potential differences in gallbladder epithelium. J Membr Biol 49:345–362

Riklis E, Quastel JH (1958) Effects of cations on sugar absorption by isolated surviving guinea pig intestine. Can J Biochem Physiol 36:347–362

Rose RC, Schultz SG (1971) Studies on the electrical potential profile across rabbit ileum. Effects of sugars and amino acids on transmural and transmucosal electrical potential differences. J Gen Physiol 57:639–663

Schachter D, Britten JS (1961) Active transport of non-electrolytes and the potential gradients across intestinal segments in vitro. Fed Proc 20:137

Skou JC (1957) Influence of some cations on an adenosine triphosphatase from peripheral nerves. Biochim Biophys Acta 23:394–401

Steiner RA, Ochme M, Amman D, Simon W (1979) Neutral carrier sodium ion-selective microelectrode for intracellular studies. Anal Chem 51:351–353

White JF (1976) Intracellular potassium activities in *Amphiuma* small intestine. Am J Physiol 231:1214–1219

White JF, Armstrong WMcD (1971) Effect of transported solutes on membrane potentials in bullfrog small intestine. Am J Physiol 221:194–201

Wright EM (1966) The origin of the glucose dependent increase in the potential difference across the tortoise small intestine. J Physiol (Lond) 185:486–500

Zeuthen T, Wright EM (1981) Epithelial potassium transport: Tracer and electrophysiological studies in choroid plexus. J Membr Biol 60:105–128

Transport of Ions and Organic Molecules in the Midgut of Some Lepidopteran Larvae

S. NEDERGAARD [1]

Introduction

The animal class insects, in the phylum of arthropods, consists of families with very different dietary habits and needs, enabling them to occupy most oecological niches on the terrestrial parts of our globe. Some insects are carnivorous, living on either dead or living animals, some herbivorous and often restricted to a single plant species, while others are able to eat nearly anything. Some have very specific demands, living on a highly restricted and apparently unbalanced diet. This applies to insects living on blood or plant sap, or moth larvae living on bees' wax or keratin.

In this paper only a few species are mentioned, and only their mature larvae: the big lepidopteran larvae of *Hyalophora cecropia,* the American silkworm, *Manduca sexta,* the tobacco hornworm, *Bombyx mori,* the commercial silkworm, and a few other speices. The larvae all feed on fresh leaves, and the mature fifth instar larvae of *Hyalophora cecropia* often weigh 10 to 20 g. The digestive tract consists of a small foregut, a large midgut – which fills up about two-thirds of the body cavity, a hindgut – also of considerable size, and finally a rectum. The larval midgut functionally corresponds to the intestine of the higher animals and is the only part of the digestive system to be dealt with in this paper. The histology of the *Cecropia* midgut is described by Anderson and Harvey (1966). It onsists of an epithelium having a single cell layer with mainly two types of cells, columnar cells, the dominant type, and goblet cells. The blood side of the epithelium has a non-continuous muscle layer and thus there is free access between blood and epithelium in the living animal. The blood of these larvae is to be found not only in the arteries and veins but also surrounding the organs in the body cavity, and is therefore in direct contact with the basal side of the midgut epithelium, in contrast to the mammalian intestine.

Fresh leaves are rich in potassium, so the animal is faced with the problem of a high potassium concentration in the midgut lumen and at the same time a low potassium concentration in the blood on the other side of the epithelium. The potassium concentration in the blood is about 30 mM and the concentration is easily ten times higher in the gut lumen.

1 Institute of Biological Chemistry A, University of Copenhagen, 13 Universitetsparken, DK 2100 Copenhagen Ø, Denmark

Intestinal Transport
(ed. by M. Gilles-Baillien and R. Gilles)

Transport of Ions

The isolated midgut transports potassium actively from blood side to lumen, which creates a potential difference across the epithelium of 100–150 mV with the lumen side positive. The active potassium transport is independent of the presence of sodium in the bathing solutions but depends on oxygen supply (Harvey and Nedergaard 1964). Harvey et al. (1967) found that this active potassium transport does not increase the oxygen consumption of the tissue. This is in contrast with the observation that active sodium transport in frog skin, for instance, stimulates the metabolism, and the increase in metabolism is proportional to the amount of transported sodium.

Rubidium can be transported in place of, or together with, potassium, while caesium is transported only when no or little potassium (below 10%) is present in the bathing solutions; the transport is in an all-or-none fashion. When 5% potassium and 95% caesium are present, only caesium is transported, while with 10% potassium and 90% caesium, only potassium is transported (Zerahn 1970). Thallium is not transported by the isolated *Cecropia* midgut, but when it is added to the lumen side, it inhibits the active transport of potassium from the blood to the lumen side. The inhibition is competitive with the active potassium transport (Zerahn 1982).

If sodium is substituted for potassium in the usual bathing solution, which contains: 30 mM KCl, 2 mM $KHCO_3$, 5 mM $CaCl_2$, 5 mM $MgCl_2$, and 166 mM sucrose, the potential difference immediately drops to zero and no transport of sodium can be observed. If, however, calcium and magnesium are left out of the bathing solutions and all potassium is replaced by sodium, a potential difference persists and sodium is transported. The sodium transport amounts to about 50% of the potassium transport, and the short-circuit current, which arises when the potential difference is eliminated, is completely due to sodium. Lithium can substitute for sodium under the same conditions. The active sodium transport is not sensitive to ouabain (Harvey and Zerahn 1971).

Magnesium is transported actively across the midgut when the midgut is bathed in magnesium containing Ringer's solution, but the transport is in the opposite direction of the active potassium transport (Wood et al. 1975). Calcium is also transported by the isolated midgut (Wood and Harvey 1976).

Zerahn (1975) has found that the basolateral membrane of the midgut epithelium has a very fast potassium exchange mechanism. In 2.2 min it has exchanged 50% of the tissue potassium and it is able to exchange 90%–95% of the total amount of potassium present in the tissue.

Harvey and Zerahn (1969) found that the potassium transported actively in the midgut epithelium does not mix with the total cellular potassium and they concluded that the route of the active potassium transport from blood to lumen may be through the goblet cells, which constitute only a minor part of the total epithelial volume. Alternatively, the potassium may pass through special pathways between or through epithelial cells.

Moffett (1979) used ion-selective potassium electrodes to establish that the interior of the epithelium is negative enough to allow potassium to enter the cells passively from the blood side. Blankemeyer and Harvey (1978) used microelectrodes in order to localize the potassium pump to either the goblet or the columnar cells of the midgut epithelium. According to their interpretation, the active potassium transport takes place through the goblet cells.

In 1968 an attempt was made to find and characterize an ATPase responsible for the active potassium transport in the *Cecropia* midgut (Turbeck et al. 1968). In the presence of magnesium the ATPase discovered was acticated by potassium and rubidium, and to a lesser degree by sodium and lithium, and was not inhibited by ouabain. However, the highest activity was obtained with certain anions, bicarbonate giving about three times as high an activity as chloride.

Wolfersberger (1979) found ATPase activation by potassium and rubidium, but not by sodium and lithium, in a preparation from the midgut of *Manduca sexta*.

Mandel et al. (1980) measured the relationship between the change in short-circuit current, the change in the cytochrome oxidation, and the level of ATP in the midgut of the tabacco hornworm when the stirring gas was changed from oxygen to nitrogen. He found that ATP most likely is involved in the active potassium transport.

So far no papers on anion transport in the lepidopteran midgut have been published.

The facts agreed on so far with respect to potassium transport in the isolated midgut are summed up in Fig. 1.

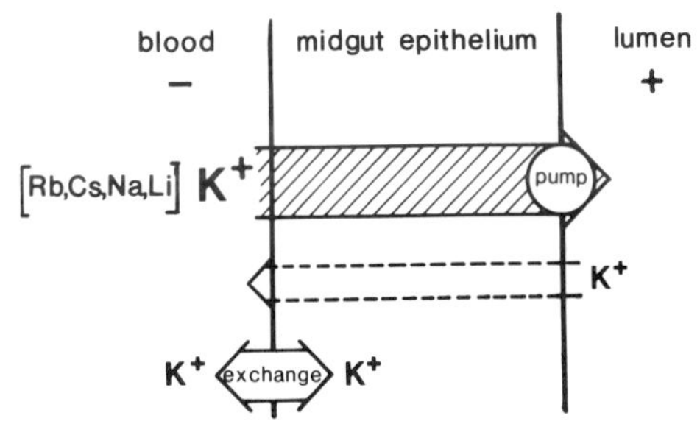

Fig. 1. Schematic drawing of the potassium transport mechanisms in the midgut epithelium. (The sizes of the *arrows* are not to be compared)

Transport of Organic Molecules

Transport of Glucose

Treherne (1958) found that the uptake of glucose by the gut wall of *Schistocerca gregaria,* the desert locust, presumably is passive because the living animal has a gradient for glucose between the gut content and the blood. The blood contains the disaccharide, trehalose, and glucose coming from the gut lumen is in the gut wall rapidly converted into the disaccharide, whereby a steep concentration gradient for glucose is created across the tissue.

Flux measurements with 3-O-methylglucose on the isolated *Cecropia* midgut showed that the fluxes in the two directions across the epithelium were equal, indicating that there is no net transport of this non-metabolized sugar, the experiments were made with the same concentration in the bathing solutions on the two sides of the gut (Nedergaard, unpublished data). This finding is in agreement with the results of Treherne.

Transport of Amino Acids

The isolated *Cecropia* midgut transports α-aminoisobutyric acid, AIB, from lumen to blood. The ratio between this uptake and the flux from blood to lumen is > 1 (Nedergaard 1972). The amino acid transport therefore seems to be an active transport as it is also dependent on oxygen supply, but the possibility must be considered that it could be a secondary transport like the amino acid and sugar transport in the intestine of higher animals. The amino acid uptake across the *Cecropia* midgut cannot be an ordinary cotransport with the actively transported ion (potassium) as it is in the vertebrate intestine, where it is a cotransport with sodium, which is transported in the same direction as the amino acids. The active potassium transport across the *Cecropia* midgut is in the opposite direction of the amino acid uptake. However, a connection must exist between the amino acid uptake and the potassium transport, because the AIB transport depends on the potential difference which is created by the active potassium transport across the epithelium from blood to lumen.

The AIB transport is not totally dependent on the potential difference, when the gut preparation is short-circuited the AIB flux from lumen to blood drops to about one third (Fig. 2). It is also seen in Fig. 2 that the effect on the AIB flux is delayed when the potential difference is eliminated. This indicates that the AIB transport mechanism is located on the luminal border of the midgut epithelium, because the AIB already accumulated in the cells by the transport mechanism will continue to leave the tissue to the blood side; if the transport mechanism were placed on the basal border, the flux would be expected to drop immediately after elimination of the driving force. The amino acid travels through the epithelium from the site of the transport mechanism in the luminal border out through the basal membrane passively (Nedergaard, unpublished data).

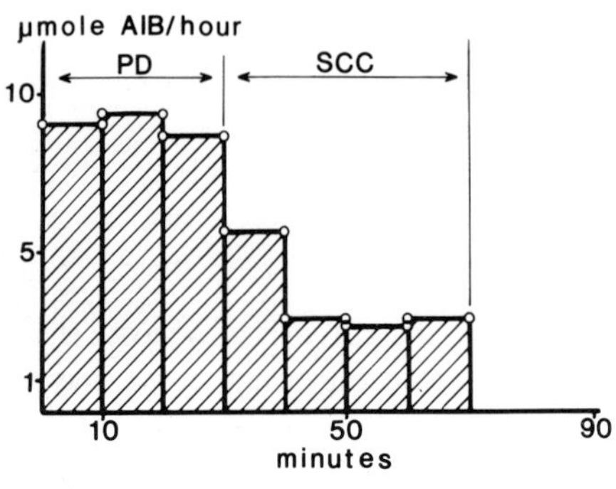

Fig. 2. AIB uptake from lumen to blood side across the midgut epithelium in μmol h^{-1}. Ten-min periods after steady state. *PD* open-circuit conditions; *SCC* short-circuit conditions

The isolated midgut from *Bombyx mori* larvae transports phenylalanine and AIB, this transport is independent of sodium and is inhibited by DNP and anoxia (Sacchi et al. 1981).

The basolateral membrane of the *Cecropia* midgut has an amino acid exchange mechanism (Nedergaard 1981), which is not involved in the exit from the cells of the amino acids being transported from lumen to blood side, and which is independent of the potassium exchange system present at the basolateral membrane (Zerahn 1975). The amino acid exchange is slow compared to the potassium exchange.

The features of the AIB transport in the *Cecropia* midgut known so far can be summarized as shown in Fig. 3.

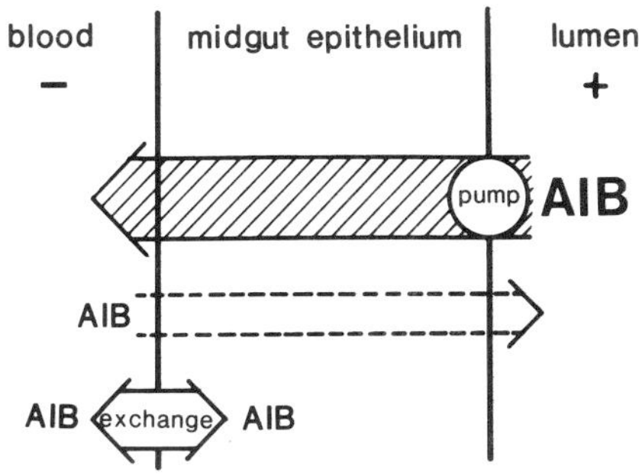

Fig. 3. Schematic drawing of the AIB transport mechanism in the *Cecropia* midgut. (The size of the *arrows* are not to be compared)

In this short review only the transports of some ions and organic molecules in the lepidioteran midgut have been considered. However, these transports are not the only functions of this epithelium: as a true intestine it produces all the necessary digestive enzymes for degradation of the food, i.e., fresh leaves. That the midgut is a very active organ can be illustrated by the rate of growth of the animal – it increases its weight about 2000 times in 6 weeks.

The Amino Acid Transport and the Potential Difference

How is the amino acid uptake dependent on the potential difference across the isolated midgut epithelium?

1. The lumen side of the midgut is positive, so if the amino acid has an overall positive charge, the amino acid flux will be down the electrochemical gradient. pH of the bathing solutions is about 8, which is close to the isoelectric point of AIB, but a small fraction on AIB molecules will have an overall negative charge at this pH.

2. A negative charge could then be essential to the transport mechanism, but experiments with lysine, both at pH 7.5 (where lysine is positively charged) and at pH 10 (lysine is neutral) showed that lysine was transported in both cases although the transport was about twice as high at pH 10 as at the low pH (Nedergaard 1973).

3. To see if potassium ions as such have anything to do with the amino acid transport, all the potassium in the bathing solutions was replaced by sodium, and as calcium and magnesium were present, sodium could not be transported and the potential difference fell to zero. An artificial potential difference was then applied across the tissue and AIB was found to be transported from the lumen to blood almost at the same rate as with the genuine potassium created potential difference (Nedergaard

1973). Potassium is therefore not necessary for the amino acid transport and sodium can be substitute.

4. The amino acid uptake in the *Cecropia* midgut could possibly be a cotransport between AIB and a passive cation flux from lumen to blood, such fluxes are reduced considerably when the potential difference across the midgut is short-circuited, as is the AIB uptake. Experiments with no potassium in the lumen side bathing solution were accordingly carried out. In these experiments potassium had to be present in the blood side bathing solution to keep up the potential difference. A continuous transport of potassium to the lumen side took therefore place during these experiments, and the lumen side was not free of potassium. The experiments were conducted in four different ways:

a) With everted guts, then potassium was transported from a volume of 5 ml to a volume of 50 ml, opposite the usual experimental set-up. The transported potassium thereby is diluted.

b) To try to decrease the effect of the transported potassium, the lumen side bathing solution was increased and stirred vigorously to keep the potassium, coming from the blood side, at a low concentration all the time. In this way possible effects of everting the gut were avoided.

c) With a reduced potassium concentration on the blood side, 10 mM instead of 32 mM KCl: the active potassium is thereby reduced by about one third but a considerable potential difference is retained across the midgut.

d) The experimental set-up was as in (b) but the lumen side bathing solution contained 32 mM of sodium instead of the low sodium concentration used in the other experiments.

The result of all four types of experiments was that the potassium concentration on the lumen side did not stop the AIB uptake completely. The amino acid uptake in the everted guts was lower, but low for both high and low lumen side potassium. That there was no effect in the experiments with sodium in the luminal bathing solution was to be expected if a cation moving from lumen to blood were cotransported with AIB, because, as mentioned earlier, a midgut bathed in sodium solution transports AIB if there is a potential difference across the epithelium.

The small effect of low potassium on the lumen side on the AIB uptake might not be a sufficient cause for discarding the cotransport hypothesis because the actively transported potassium from the blood side might be able to keep a high potassium concentration in the unstirred layer right at the luminal border between the microvilli.

An alternative way of testing a possible cotransport between the amino acid and the passive potassium flux from lumen to blood side would be to measure the AIB flux and the potassium flux in the same direction simultaneously by tracers on the same midgut preparation. If the ratio between the potassium flux and the AIB flux is constant during the experiment it could indicate a cotransport of the two compounds. As it turns out, it is not: in five experiments it varied from 1.5 to 5.0 (Table 1).

Hanozet et al. (1980) observed that the uptake of phenylalanine into vesicles prepared from the midgut of *Philosamia cynthia,* was stimulated by the presence of

Table 1. Lumen → blood fluxes of K^+ and AIB (μmol h^{-1}/100 mg)

	K^+	AIB	$\frac{K^+}{AIB}$
Open circuited conditions	18.4	5.6	3.30
	15.3	5.8	2.64
Short circuited conditions	8.7	5.5	1.58
	5.5	3.4	1.50
	3.9	3.2	1.22
	2.4	1.9	1.31

10-min periods after steady state

potassium or sodium in the medium, and they therefore suggested that the phenylalanine uptake is coupled to the potassium movement. Giordana et al. (1983) have also found that the rate of amino acid uptake into the vesicles can be saturated with respect to both external potassium and external amino acid concentration.

5. Still another possibility is that the AIB uptake is brought about by electroosmosis created by the potassium flux from lumen to blood. In order to have electroosmosis with the potassium flux it is necessary that the two potassium fluxes through the midgut tissue are separated.

Findings by Zerahn (1980) suggest that the two potassium fluxes might travel by the same route through the epithelium. However, I have obtained preliminary results with rubidium, where all potassium in both bathing solutions is replaced by rubidium, indicating that the fluxes have different routes through the midgut epithelium. This was done by using the pre-steady state flux ratio method developed by Ussing et al. (1981).

Both cotransport with potassium and electroosmosis can explain a number of the observed features of amino acid transport in the *Cecropia* midgut, and the possibility must be considered that both mechanisms contribute to the amino acid uptake. Table 1 shows that when the potential difference across the gut wall is abolished, the ratio between potassium flux and AIB flux approaches unity, which could indicate that under these conditions the two compounds are transported together. When there is a potential difference across the gut, and therefore a driving force for electroosmosis, the ratio between the two fluxes can be far from unity, indicating the possibility of a more indirect coupling between potassium flux and AIB flux.

It will, however, be necessary to perform more experiments before these problems can be solved.

References

Anderson E, Harvey WR (1966) Active transport by the *Cecropia* midgut. II. Fine structure of the midgut epithelium. J Cell Biol 31:107–134

Blankemeyer JT, Harvey WR (1978) Identification of active cell in potassium transporting epithelium. J Exp Biol 77:1–13

Giordana B, Sacchi FV, Hanozet GM (1983) Intestinal amino acid absorption in lepidopteran larvae. Biochim Biophys Acta (in press)

Hanozet GM, Giordana B, Sacchi VF (1980) K^+-dependent phenylalanine uptake in membrane vesicles isolated from the midgut of *Philosomia cynthia* larvae. Biochim Biophys Acta 596: 481–486

Harvey WR, Nedergaard S (1964) Sodium independent active transport of potassium in the isolated midgut of the *Cecropia* silkworm. Proc Natl Acad Sci USA 51:757–765

Harvey WR, Zerahn K (1969) Kinetics and route of active K-transport in the isolated midgut of *Hyalophora cecropia.* J Exp Biol 50:297–306

Harvey WR, Zerahn K (1971) Active transport of sodium by the isolated midgut of *Hyalophora cecropia.* J Exp Biol 54:269–274

Harvey WR, Haskell JA, Zerahn K (1967) Active transport of potassium and oxygen consumption in the isolated midgut of *Hyalophora cecropia.* J Exp Biol 46:235–248

Harvey WR, Cioffi M, Wolfersberger MG (1981) Portasomes as coupling factors in active ion transport and oxidative phosphorylation. Am Zool 21:775–791

Mandel LJ, Riddle TG, Storey JM (1980) Role of ATP in respiratory control and active transport in tobacco hornworm midgut. Am J Physiol 238:C10–C14

Moffett DF (1979) Potassium activity of single insect midgut cells. Am Zool 19:996

Nedergaard S (1972) Active transport of α-aminoisobutyric acid by the isolated midgut of *Hyalophora cecropia.* J Exp Biol 56:167–172

Nedergaard S (1973) Transport of amino acid in *Cecropia* midgut. In: Ussing HH, Thorn NA (eds) Transport mechanisms in epithelia. Munksgaard, Copenhagen, pp 372–381

Nedergaard S (1981) Amino acid exchange mechanism in the basolateral membrane of the midgut epithelium from the larva of *Hyalophora cecropia.* J Membr Biol 58:175–179

Sacchi VF, Cattaneo G, Carpentieri M, Giordana B (1981) L-phenylalanine active transport in the midgut of *Bombyx mori* larva. J Insect Physiol 27:211–214

Treherne JE (1958) The absorption of glucose from the alimentary canal of the locust *Schistocerca gregaria* (Forsk.). J Exp Biol 35:297–306

Turbeck BO, Nedergaard S, Kruse H (1968) An anion-stimulated adenosine triphosphatase from the potassium-transporting midgut of the larva of *Hyalophora cecropia.* Biochim Biophys Acta 163:354–361

Ussing HH, Eskesen K, Lim J (1981) The flux ratio transient as a tool for separating transport pathways in epithelia. In: McKnight ADC, Leader JP (eds) Epithelial ion and water transport. Raven Press, New York

Wolfersberger MG (1979) A potassium-modulated plasma membrane adenosine triphosphatase from midgut of *Manduca sexta* larvae. Fed Proc 38:242

Wood JL, Harvey WR (1976) Active transport of calcium across the isolated midgut of *Hyalophora cecropia.* J Exp Biol 65:347–360

Wood JL, Jungreis AM, Harvey WR (1975) Active transport of magnesium across the isolated midgut of *Hyalophora cecropia.* J Exp Biol 63:313–320

Zerahn K (1970) Active transport of caesium by the isolated short-circuited midgut of *Hyalophora cecropia.* J Exp Biol 53:641–649

Zerahn K (1975) Potassium exchange between bathing solution and midgut of *Hyalophora cecropia* and time delay for potassium flux through the midgut. J Exp Biol 63:295–300

Zerahn K (1980) Competition between potassium and rubidium ions for penetration of the midgut of *Hyalophora cecropia* larvae. J Exp Biol 86:341–344

Zerahn K (1982) Inhibition of active K transport in the isolated midgut of *Hyalophora cecropia* by Tl^+. J Exp Biol 96:307–313

Electrical Phenomena in Fish Intestine

J.A. GROOT, H. ALBUS, R. BAKKER, J. SIEGENBEEK VAN HEUKELOM
and Th. ZUIDEMA [1]

Introduction

This article is an attempt to review electrical phenomena in fish intestinal preparations. In the first two sections the various preparations and their electrical characteristics will be described. In the next section, the equivalent electrical circuit and a geometrical representation of the equations used as a solution of the equivalent circuit as introduced by Bakker (1980) will be presented. Then the electrical characteristics will be discussed on the basis of the equivalent circuit and the question of an active Cl^- transport in seawater fish will be debated. The final section will deal with sugar- and amino acid-evoked potentials.

Intestinal Preparations

In vivo Preparations

A few electrophysiological measurements in intact eel (Kirsch and Meister 1982) and tench (Buclon 1974) have been reported. The results with the tench seem to be identical to results with isolated preparations of this animal. The potential measurements in the intact eel do not correspond with results on isolated eel intestine.

In vitro Preparations

Most data are from experiments with isolated preparations.

Sac Techniques. The sac technique of Wilson and Wiseman (1954), modified by Smith (1964) and by House and Green (1965), permits transmural potential measurements on everted (Smith) or non-everted (House and Green) sacs. In this experimental set-up, transmural resistance and shortcircuit current are usually not measured. A serious drawback is that the inner side of the sac cannot be oxygenated or stirred, and

1 Department of Animal Physiology, University of Amsterdam; Kruislaan 320, NL 1098 SM Amsterdam, The Netherlands

Intestinal Transport
(ed. by M. Gilles-Baillien and R. Gilles)

because of the relatively small volume the composition will change during the experimental period so that ionic gradients (including pH gradients) may develop across the tissue. Moreover, the junction potentials of the Ringer-agar bridges may change. These difficulties have been clearly demonstrated by Ando and Kobayashi (1978) in experiments with everted and non-everted sacs.

Perfused Intestinal Segments. A technique circumventing these problems has been described by Albus and Siegenbeek van Heukelom (1976) for intact goldfish intestine. In this set-up the isolated segments can be continuously perfused with oxygenated and thermostated solutions of defined composition. Moreover, the electrode arrangement makes it possible to voltage-clamp the tissue so that transmural resistance and shortciruit current can be measured. Although the results found with this technique differ slightly from those published by Smith in 1964 and 1966, the essential finding of the first reports on fish intestinal electrophysiology are confirmed: glucose and amino acids evoke a serosa positive transmural potential.

Intestinal Sheets in Ussing Chambers. Most experiments are performed with sheets of intestine in Ussing-type chambers. With this technique it is possible to measure electrical parameters, as well as transepithelial fluxes. Results are commonly related to the area of the aperture of the mounting device. Therefore the specific resistance is given in Ωcm^2 of serosal area. In measurements of shortcircuit current values are generally corrected for the resistance of the Ringer between the potential sensing electrodes. However it is not known what correction should be introduced for the passive tissue layers underneath the mucosal cells. Even in partially stripped preparations, tissue layers with a thickness of about 100 μm remain underneath the interfold cells (Ando and Kobayashi 1978, Field et al. 1978, Cartier et al. 1979). In the stripped goldfish intestine the thickness of the remaining tissue layer is negligible (Albus et al. 1979), nevertheless its resistance is about 24% of the total transepithelial resistance (Albus and Lippens 1982).

Electrophysiological Characteristics of Fish Intestinal Preparations

In Table 1 data on transmural potentials (ψ_{ms}), mucosal membrane potentials (ψ_{mc}), shortcircuit–currents (I_{sc}) and resistances (R_{ms}) are compiled. In the first column some conditions related to acclimation of the fish and to the experimental procedures are summarized.

Transmural Potentials

The table shows the well known fact that in seawater acclimated euryhaline fish ψ_{ms} is serosa negative, whereas in fresh-water, ψ_{ms} is less negative or zero. An exception seems to be the trout *(Salmo irideus)* in which after acclimation to seawater a slightly more positive ψ_{ms} is observed. Recent in vivo measurements on seawater acclimated European eel have revealed the interesting result of a serosa positive potential across the intestine ranging from 8.5 to 38 mV. These high values for a leaky epithelium need confirmation. In stenohaline marine animals a slightly positive value is found by House and Green while negative values are found for the sea perch (*Serranus* sp.) and for the plaice *(Pleuronectes platessa)*. For freshwater fish all values are zero or positive except for *Ictalurus punctatus*.

Transmural Resistances

In the group of Pleuronectidae all values (with one exception) are about 50 Ωcm^2 for unstripped plaice and flounder. They are slightly lower in stripped winter flounder. The resistance of the eel intestine is somewhat lower in seawater-acclimated fish. This may correspond with the observed decrease in height of the folds after seawater acclimation (Hirano et al. 1976). The range of resistances stated for freshwater fish is much greater: from 26 Ωcm^2 in stripped goldfish intestine to 221 Ωcm^2 in non-stripped *Ictalurus punctatus*.

It is tempting to speculate that the difference in resistance between unstripped intestines of goldfish and Pleuronectidae has its morphological correlate in the difference in cell renewal and in the inhomogeneity in tight junction structure (Trier and Moxey 1980, Madara et al. 1981).

Mucosal Membrane Potentials

The only ψ_{mc} measurements have been reported for Pleuronectidae (–45 to –65 mV) and for the goldfish (–45 to –54 mV).

Electrical Equivalent Circuit

Circuit Description

In epithelia the electromotive forces (emf) across the barriers constituting the epithelial structure are coupled by resistances. The emf's across the mucosal and serosal membrane are coupled by an extracellular pathway which especially in leaky epithelia has a very high conductance (Diamond 1974). Therefore the electrical potential difference across a membrane is in general not equal to its electromotive force.

Simplified equivalent electrical circuits by means of which electrical potential measurements can be analysed were introduced for intestinal mucosal epithelia by White and Armstrong (1971) and by Rose and Schultz (1971). In Fig. 1 such an equivalent circuit is drawn.

In contrast to common practice potential differences are referred to the cell interior which is taken as zero ($\psi_{mc} = -\psi_{cm}$). This is to facilitate the drawing of the figures.

Table 1. Transmural potentials (ψ_{ms}), mucosal membrane potentials (ψ_{mc}), shortcircuit currents

Fish	Conditions	Glc	pH	Buffer	Gas	Temp.	Accl.
Cottus scorpius	sac, ne	2.8	7.2	2.5	air	18°C	
Serranus sp.	–						
Pleuronects platessa	Ussing, ns, Ala 2, Ac.10	10	7.4	25	5%	20°C	
Pseudopleuronectes americanus	Ussing, ps, s	20	8.2	20	1%	15°C	
	Ussing, ps		7.2	20	5%	15°C	
	Ussing, ps		7.2	5	1%	15°C	
	Ussing, ps		8.2	20	1%	15°C	15% SW
	Ussing, ns	22	7.2	25	5%	26°C	
Platichthys flesus	Ussing, ns	5.5	7.2	25	5%	23°C	SW
	Ussing, ns						FW
	Perf. segm., ns	0	7.2	25	5%	21°C	SW
	Perf. segm., ns						FW
Anguilla japonica	Ussing, post, ns	5	7.2	25	5%	20°C	FW
	Ussing, post, ns						SW
	sac, ne, post, ps	5	7.2	25	5%	20°C	FW
	sac, ne, post, ps						SW
Anguilla anguilla	in vivo – Ringer in lumen						SW
Salmo irideus	–						FW
	in vivo – Ringer in lumen						SW
Salmo gairdnerii	Ussing, ns	5	7.2	25	5%	20°C	FW
	Ussing, ns						SW
Ictalurus punctatus	Ussing, ns	5.5	7.2	25	5%	24°C	
Ictalurus nebulosus	Perf. segm., ns	14	7.2	2.3	air	–	
Tinca tinca	Ussing, ps	0	7.4	Pi	air	–	
Carassius auratus	Ussing						
	sac, ns	28	7.3	25	5%	20°C	
	Perf. segm., ns	28	7.3	25	5%	23°C	
	Perf. segm., ns	0					
	Ussing, str	28	7.3	25	5%	23°C	
	Ussing, str	0					

The sign of the potentials refers to the mucosal solution. Positive values for I_{sc} denote net positive electrical charge flux from mucosa to serosa. Abbreviations used in description of conditions: Ussing = sheets in Ussing-type chamber, Perf. segm. = perfused segments, sac = canulated sac adapted for ψ_{ms} measurement, ne = non everted, ns = non stripped, ps = partially stripped, str = stripped, post = posterior intestine, s = serosally. Ala = alanine, Ac. = acetate followed by concen-

(I_{sc}) and resistances (R_{ms}) in different fish

ψ_{ms} (mV)	ψ_{mc} (mV)	I_{sc} ($\mu A/cm^2$)	R_{ms} ($\Omega\ cm^2$)	References
+ 0.6	–	–	–	House and Green (1965)
– 0.5 to – 4.0	–	–	–	Lahlou (1976)
– 5.5	– 45 p.s.	– 107	58	Ramos and Ellory (1981), Katz et al. (1982)
(– 4.5)	– 65	– 118	38	Smith et al. (1981)
(– 2.8)	– 44	– 68	41	Smith et al. (1981)
(– 1.9)	–	– 36	54	Field et al. (1978)
(– 1.9)	–	– 40	47	Field et al. (1980)
– 3.4	–	– 22	(153)	Huang and Chen (1971)
– 1.9	–	– 45	(42)	Smith et al. (1975)
– 1.2	–	– 18	(68)	Smith et al. (1975) (fish acclimated to FW)
– 7 (– 3.4 to – 9)	–	(135)	52	Groot (unpubl. obs.)
– 5 (– 3.4 to – 6.7)	–	(111)	45	Groot (unpubl. obs.) (fish from lake Ijselmeer)
+ 0.1	–	+ 1.0	76	Ando et al. (1975)
– 2.8	–	– 82	36	Ando et al. (1975)
– 2.7	–	–	–	Ando and Kobayashi (1978)
– 7.8	–	–	–	Ando and Kobayashi (1978)
+ 3.3 to 38	–	–	–	Kirsch and Meister (1982)
+ 0.8	–	–	–	Crénesse, quoted by Lahlou (1976)
+ 1.1	–	–	–	Crénesse, quoted by Lahlou (1976)
+ 0.2	–	+ 1.8	108	Ando et al. (1975)
– 0.9	–	– 25	43	Ando et al. (1975)
– 1.7	–	– 7.7	(221)	Chen and Huang (1972)
+ 2.5	–	–	–	Datta and Savage (1968)
+ 1.5	–	–	(59)*	Buclon (1974)
+ 2.1 to 4.5	– 45	15–30	(145)	Ellory et al. (1972)
+ 4	–	–	–	Smith (1964)
+ 2.8	–	(+ 26)	106	Albus and Siegenbeek van Heukelom (1976)
+ 0.1	–	(+ 0.9)	106	Albus and Siegenbeek van Heukelom (1976)
+ 2.5	– 48	(96)	26	Albus and Lippens (1982)
– 0.1	– 54	(– 3.8)	26	Albus and Lippens (1982)

tration in mM. In the column "Glc" the glucose concentration is given in mM. In the column "buffer" the HCO_3 concentration is given. Pi = phosphate buffer. In the column "gas" percentage CO_2 is given when O_2/CO_2 mixtures were used otherwise air. SW = seawater acclimated, FW = freshwater acclimated. * = calculated from $(\Delta\psi_{ms})_{max}$ and $(I_{sc})_{max}$ induced by glycine

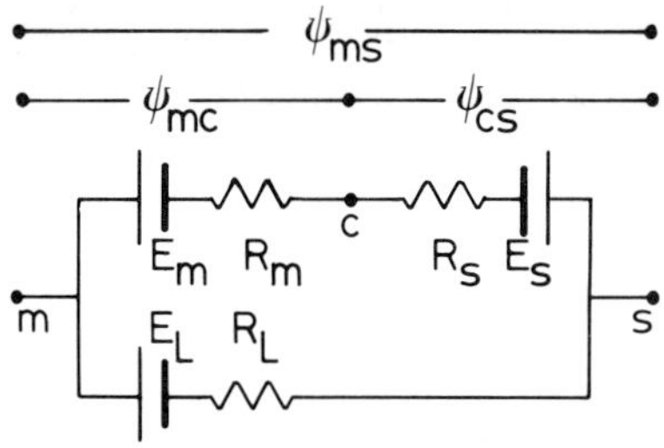

Fig. 1. Equivalent circuit for intestinal mucosa. *m* mucosal; *c* cellular; *s* serosal compartment. R_m, R_s and R_L lumped resistances of the mucosal and serosal membrane and the extracellular pathway respectively. E_m, E_s, and E_L lumped electromotive forces across the respective membranes and across the extracellular pathway. They respresent the potential differences in the absence of a leak current. ψ_{cm} (= $-\psi_{mc}$) and ψ_{cs} are the potential differences actually measured across the respective membranes and ψ_{ms} is the potential difference across the mucosa

The solutions for this circuit for ψ_{cm} and ψ_{cs} are:

$$\psi_{cm} = [R_m (E_s + E_L) + (R_s + R_L) E_m]/R_T \quad (1)$$

where $R_T = R_m + R_s + R_L$

$$\psi_{cs} = [(R_m + R_L) E_s + R_s (E_m - E_L)]/R_T \quad (2)$$

and for ψ_{ms}, the potential difference across the whole epithelium with reference to the mucosal solution:

$$\psi_{ms} = \psi_{cs} - \psi_{cm} = [R_L (E_s - E_m) - E_L (R_m + R_s)]/R_T \quad (3)$$

$R_{ms} = R_L (R_m + R_s)/R_T$ is the transepithelial resistance. (4)

In these equations the direction of the E's is as illustrated in Fig. 1 so that a positive value indicates that the emf is orientated as shown. To calculate the three resistances R_m, R_s and R_L three independent measurements are needed.

a) R_{ms} can be found from $\Delta\psi_{ms}$ induced by transepithelial current pulses. Corrections should be made for resistances of the solutions and the tissue resistance in series with the epithelial cell layer.

b) In addition, with a microelectrode in the epithelial cell the voltage deflections $\Delta\psi_{cm}$ and $\Delta\psi_{ms}$ caused by a transepithelial current pulse can be measured.
From $\Delta\psi_{cm}/(\Delta\psi_{ms} + \Delta\psi_{cm})$ the ratio R_m/R_s can be calculated. Again: $\Delta\psi_{ms}$ depends on the non-epithelial tissue resistance.

c) A third equation can be derived using an agent that exclusively provokes a change in only one of the E's, for instance E_m. In this case from Eqs. (1) and (3)

$$\Delta\psi_{cm} / \Delta\psi_{ms} = -(R_s + R_L)/R_L.$$

As will be discussed later mucosal addition of glucose, just after ouabain addition on the serosal side, is thought to be such an agent in goldfish intestinal mucosa (Bakker and

Albus (1982). Resistances, thus calculated for stripped goldfish intestinal epithelium and related to cm^2 serosal area are: $R_L = 20.7\ \Omega cm^2$, $R_m = R_s = 197\ \Omega cm^2$ while the non-epithelial tissue resistance was $6.2\ \Omega cm^2$ (Bakker and Albus 1982).

A Geometrical Representation

The Eqs. (1)–(3) can be represented geometrically as introduced by Bakker (1980). Figures 2 and 3 are based on the following rewriting of Eqs. (1) and (2):

Since $(R_s + R_L)/R_T = (R_T - R_m)/R_T = 1 - R_m/R_T$, Eq. (1) becomes

$$\psi_{cm} = E_m + (E_s - E_m)\, R_m/R_T + E_L R_m/R_T \tag{5}$$

Since $(R_m + R_L)/R_T = (R_T - R_s)/R_T = 1 - R_s/R_T$, Eq. (2) becomes

$$\psi_{cs} = E_m + (E_s - E_m)(R_m + R_L)/R_T - E_L R_s/R_T \tag{6}$$

If, for a start, E_L is assumed to be zero Eqs. (5) and (6) reduce to

$$\psi_{cm} = E_m + (E_s - E_m)\, R_m/R_T \tag{5a}$$

$$\text{and } \psi_{cs} = E_m + (E_s - E_m)(R_m + R_L)/R_T \tag{6a}$$

These relations are depicted as shown in Fig. 2 (proof is by simple geometry) in which the ordinate is in mV and the x-axis is in Ωcm^2. In practice ψ_{cm} and ψ_{ms} as well as R_m/R_s, R_{ms} and $(R_s + R_L)/R_L$ can be measured so that the figure can be constructed. From the origin to the right R_m, R_L and R_s are depicted with consecutive line segments. ψ_{cm} is drawn perpendicularly at R_m and ψ_{cs} at $R_m + R_L$. Extrapolation of the line connecting ψ_{cm} and ψ_{cs} to the ordinate gives E_m and to the vertical through $R_m + R_s + R_L$ gives E_s. The assumption of $E_L = 0$ (Rose and Schultz 1971, White and Armstrong 1971, Frömter 1982) is questionable because the basolateral Na pump is thought to increase the salt concentration within the lateral intercellular spaces (lis) so that by osmosis water is transported from mucosa to serosa (Curran 1960). This NaCl gradient from lis to serosa induces a diffusion potential, serosa negative. Moreover most epithelia have cation-selective tight junctions so that across these, diffusion potentials are generated with the same orientation: serosa

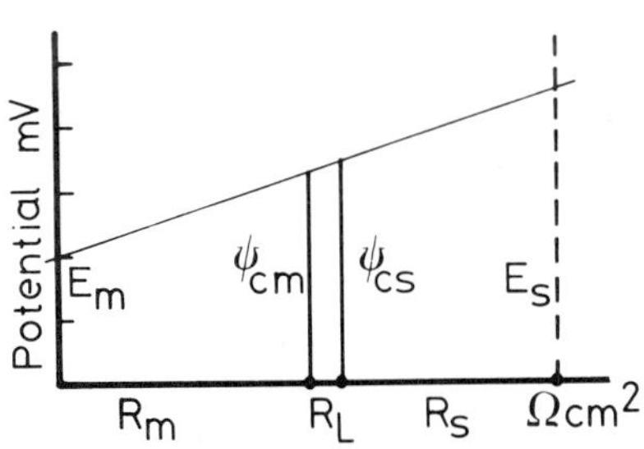

Fig. 2. Pictorial representation of Eqs. (5a) and (6a). $E_L = 0$. (For explanations see text)

negative. When E_L is not zero extrapolation of the line $\psi_{cm} - \psi_{cs}$ would intersect the ordinate at a value of $E_m + E_L R_m/R_L$ and the vertical through $R_m + R_L + R_s$ at a value of $E_s - E_L R_s/R_L$. In leaky epithelia R_m/R_L and R_s/R_L are $\gg 1$ so that it is necessary to know E_L in order to obtain reliable values for E_m and E_s.

To illustrate the effect of a finite serosa negative value of E_L on ψ_{ms} we retain E_m and E_s from Fig. 2 and add $E_L R_m/R_L$ to E_m on the ordinate and substract $E_L R_s/R_L$ from E_s on the vertical through $R_m + R_L + R_s$ (Fig. 3). The line connecting these points intersects the vertical through R_m and $R_m + R_L$ at ψ_{cm} and ψ_{cs}. It is clear that now the transepithelial potential is serosa negative without a change in E_m or E_s. This can also be seen from Eq. (3) but in our experience the pictorial representation of the equations contributes to a clear apprehension of the matter.

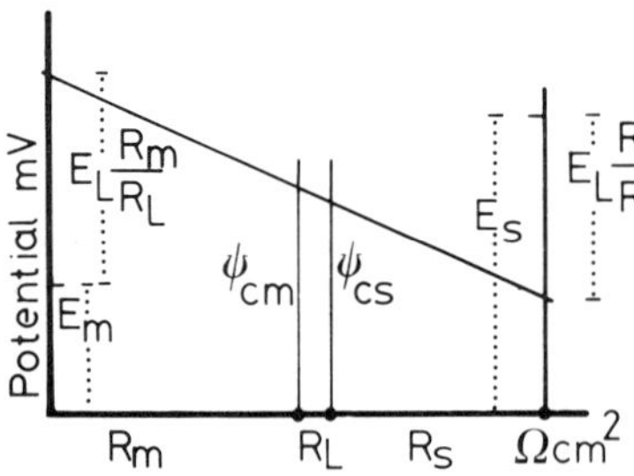

Fig. 3. Pictorial representation of Eqs. (5) and (6). The effect of a finite value for E_L in addition to E_m and E_s of Fig. 2

The Polarity of the Transepithelial Potential Difference

Introduction

It has been suggested that the tolerance of euryhaline fish to high salinity is determined by the capacity of the intestine to absorb NaCl and thereby to induce a solute-linked water flow from lumen to blood. The other parameter of importance is the osmotic permeability. Both the active NaCl transport and the osmotic permeability, increase when euryhaline fish are transferred from fresh- to seawater. Cortisol seems to be the important hormone in seawater acclimation (Skadhauge 1974, Hirano et al. 1976).

Cl^- Pump

The serosa negative transepithelial potentials, the effects of ion substitutions and the greater net Cl^- transport under short-circuit conditions have led to the hypothesis of an electrogenic, Na^+-independent Cl^- pump (Ando 1975, Ando et al. 1975). However, this idea has been challenged by the findings of Ando and Kobayashi (1978). They showed that stripped intestinal preparations do not increase in serosa negativity when bathed in choline-Ringer. The fact that ψ_{ms} decreases to zero in Cl^- free Ringer, in low-Na^+ Ringer and in Ringer with ouabain, is now considered as evidence for an interdependence of Na^+ and Cl^- transport across the eel intestine (Ando 1980, 1981).

The Role of the Cation-Selective Extracellular Pathway and the NaCl Cotransport Mechanism

As an alternative to the Cl^- pump hypothesis Field et al. (1978) proposed a coupled NaCl entry across the mucosal membrane. Net Cl^- transport and the serosa negative potential would be a consequence of the cation selectivity of the tight junctions and of a greater resistance to Na^+ diffusion than to Cl^- from the lis to the serosal side. This allows more Na^+ than Cl^- to recycle back to the mucosal solution. In other words, this model re-introduces a finite value for E_L in the equivalent electrical circuit (see Fig. 3). There are, however, a number of observations that do not fit readily into this framework.

The Cation Selectivity of the Extracellular Pathway

1. The occurrence of a potential change (streaming potential) induced by an osmotic gradient across an epithelium reflects the ion selectivity of the extracellular pathway. However, there is no correlation between streaming potentials and the sign of the transmural potential. When mucosal solutions made hypertonic with mannitol are applied, the following streaming potentials in mV/Osm across intestinal preparations were observed: rat 64 (Barry and Eggenton 1972), goldfish 29 (Albus and Siegenbeek van Heukelom 1976), tench 54 (Buclon 1974), seawater eel 26 (Ando et al. 1975) and European flounder 37 (Groot, unpublished data).

2. In Field's model the serosa to mucosa fluxes of Na^+ and Cl^- are assumed to be extracellular. The ratio of these fluxes (J_{sm}^{Na}/J_{sm}^{Cl}) is about 4 in the winter flounder and reflects the cation selectivity of this pathway. In seawater acclimated (SW) eel and in freshwater (FW) eel a ratio of 3.0 and 3.1 respectively is found. However, the ψ_{ms} in FW eel is 0 mV while it is -2.5 in SW eel (Ando et al. 1975).

3. In addition, it is assumed that 2,4,6-triaminopyrimidine (TAP) blocks the cation selective pathways in the tight junctions of leaky epithelia (Moreno 1975). Therefore TAP should lead to an increase in transmural resistance and to a diminished serosa negative potential. Mackay and Lahlou (1980) indeed observed an increase in R_{ms} from 34 Ωcm^2 to 76 Ωcm^2 but ψ_{ms} became more negative.

In conclusion, these findings do not support the idea that the cation selectivity of the extracellular pathway is the main reason for the serosa negative transmural potential.

NaCl Cotransport. With ion selective microelectrodes it has been shown that Cl^- is electrochemically accumulated in fish enterocytes (Duffey et al. 1979, Zeuthen et al. 1978, Zuidema et al. 1982). There are a number of observations indicating that the conductance of the enterocyte membrane for Cl^- is negligible. Cl^- substitutions consistently induce a hyperpolarization instead of a depolarization (Zuidema et al. 1982a, Helman and Beyenbach 1979, Halm et al. 1982, Zeuthen et al. 1978, Katz et al. 1982). Therefore electroneutral Cl^- transport across the membranes should be present. Uphill transport of Cl^- coupled to downhill Na^+ transport was postulated in the mucosal membrane by Field et al. (1978) while in the serosal membrane a KCl cotransport was proposed by Stewart et al. (1981). In goldfish intestine no evidence could be found of a coupled NaCl transport (Zuidema et al. 1982a,b). The obligatory coupling between Cl^- and Na^+ fluxes as observed in gallbladder (Cremaschi and Henin 1975) is not found in intestinal epithelia. Moreover, experiments with brush-border vesicles from rat intestine (Liedtke and Hopfer 1982) and from winter flounder intestine (Rao et al. 1982) did not reveal a NaCl cotransport either.

The Double Exchange Mechanism for Coupled NaCl Transport

This model (Turnberg et al. 1970) consists of a Na^+/H^+ and a Cl^-/HCO_3^- or Cl^-/OH^- exchange. These two mechanisms are mutually coupled by the HCO_3^--CO_2 reaction and the pH. Because of the observed pH and HCO_3^-/CO_2 dependence (Hirano et al. 1976, Ramos and Ellory 1981, Field et al. 1980, Smith et al. 1980, 1981), this is an attractive alternative. However, there remains much to be learned about the exact role of HCO_3^- and pH. We will briefly summarize the present knowledge available.

Winter Flounder: Serosal Acidification Inhibits Cl^- Transport. Table 2 shows the qualitative results of the winter flounder obtained by Field and co-workers. For quantitative results the references given in the legend of the table can be consulted. The authors conclude from these results that the inhibition of Cl^- transport is due to acidification of the serosal side. This suggests that one component in the Cl^-transport chain is a HCO_3^- pathway in the serosal membrane. No firm conclusion can be drawn about the mucosal uptake mechanism. However, the increase in J_{me}^{Cl} and the concomitant increase in cellular Cl^- (Smith et al. 1981) when the Ringer is gassed by 5% CO_2 instead of 1% CO_2 suggests an increase in Cl^- uptake in exchange for HCO_3^- or OH^- from the cell.

Goldfish: Cl^- Replacement Induces Cellular Alkalinization and Hyperpolarization. From our work with goldfish intestine we know that replacement of Cl^- by "less permeable" anions induces an alkalinization of the cell interior (Groot et al. 1982). It seems that cellular Cl^- exchanges with extracellular HCO_3^- or OH^-. Alkalinization of the cell interior by varying the HCO_3^-/CO_2 or HEPES/Tris ratio in the Ringer induces a hyperpolarization of the membrane (Zuidema et al. 1982a,b). We suggest that the hyperpolarization caused by mucosal or bilateral Cl^- replacement is related to or is a direct consequence of the cell alkalinization. However, when at pH 7.4 HCO_3^-/CO_2 is replaced by HEPES/Tris cell alkalinization and depolarization was observed (Zuidema, unpublished data). This suggests that HCO_3^-/CO_2 has a direct effect on the membrane potential.

Goldfish, Plaice and European Flounder: Inhibition of Cl^- Transport by Absence of HCO_3^-/CO_2. In goldfish intestine net mucosa to serosa flux of Cl^- is nullified when HCO_3^-/CO_2 is replaced by HEPES/Tris at pH 7.4 (Bakker et al. 1982). In the plaice (Ramos and Ellory 1981) HCO_3^-/CO_2 free Ringer causes a significant reduction of I_{sc} at pH 7.4. At pH 7.4, in the European flounder, HCO_3^-/CO_2 free Ringer reduces ψ_{ms} and acetazolamide reduces it further. Subsequent addition of ouabain has no further effect on ψ_{ms} (Groot, unpublished data). It is likely that in the winter flounder at pH 8 (Smith et al. 1980) endogenous CO_2 production provides sufficient HCO_3^- to maintain Cl^- transport.

Table 2. Effects of CO_2/HCO_3^- variations and pH on Cl fluxes, resistance and transepithelial potential in the winter flounder intestine

MES or HEPES/Tris	CO_2		HCO_3^-		pH		J_{me}^{Cl}	J_{ms}^{Cl}	I^{sc} or J_{net}^{Cl}	R_{ms}	ψ_{ms}
	m	s	m	s	m	s					
Effect of mucosal or serosal decrease of HCO_3^- and pH											
–	1%	1%	↓	20	↓	8	–	= [a]	= [a]	= [a]	= [c]
–	1%	1%	20	↓	8	↓	= [a]	↓ [a]	↓ [a]	↑ [a]	–
Effect of bilateral pH decrease by decrease in HCO_3^- or increase in CO_2											
–	1%	1%	↓	↓	↓	↓	–	↓ [e]	↓ [e]	= [e]	dep [e]
–	↑	↑	20	20	↓	↓	↑ [b]	↓ [b]	↓ [b]	= [b]	dep [c]
Effect of serosal pH decrease by CO_2 or buffer											
–	1%	↑	20	20	8	↓	–	↓ [d]	–	–	dep [d]
MES	–	–	–	–	8	↓	–	↓ *[d]	–	–	= *[d]
Effect of HCO_3^-/CO_2 free Ringer											
HEPES	–	–	–	–	8	8	–	= **[b]	= **[b]	= **[b]	–

↑ and ↓ denote increase or decrease with respect to normal (20 mM HCO_3^-/1% CO_2, pH = 8), = denotes no change, dep = deplarization, m = mucosal, s = serosal, J_{me}^{Cl} = chloride uptake across mucosal barrier, J_{ms}^{Cl} = unidirectional flux mucosa to serosa, J_{net}^{Cl} = net flux, normally mucosa to serosa, * = as compared with pH_s = 8 (MES), ** = as compared with pH 8 HCO_3^-/CO_2

[a] Field et al. (1979), [b] Smith et al. (1980), [c] Smith et al. (1981), [d] Halm et al. (1982), [e] Field et al. (1978)

The Role of Potassium

The observation of Stewart et al. (1981) that the serosal membrane potential does not respond to changes in K^+ or Cl^- concentration in the serosal solution suggests that the serosal membrane has a low electro-diffusive permeability for K^+ and Cl^-. With respect to the K^+ permeability this is in contrast with observations in the goldfish (Bakker, unpublished data) and it is of great interest in relation to the polarity of the transmural potential. To explain serosal Cl^- permeation a neutral KCl cotransport is postulated (Stewart et al. 1981).

Recently the existence of a rheogenic $Na^+ K^+ Cl^-$ cotransport in the brush border membrane of the winter flounder has been suggested (Musch et al. 1982, Field et al. 1982).

It is clear that the role of potassium is not yet fully understood and further investigations may bring exciting new findings.

The Polarity of ψ_{ms}

The finding of Stewart et al. (1981) implies that E_s is mainly determined by E_{Na} (Equilibrium potential of Na^+). This makes the sign of E_s serosa negative with respect to the cell interior. From a pictorial representation like Fig. 3 it can readily be seen that with values for ψ_{mc} and ψ_{ms} as found in pleuronectidae (see Table 1) $E_s - E_L R_s/R_L$ becomes negative when R_s/R_L is about 10. This is a ratio that is not unusual for leaky epithelia. A serosa positive orientation of E_s, if determined by E_K, would require a rather high value for $E_L R_s/R_L$. However, high values for E_L are questionable (Simon et al. 1981, Spring et al. 1981) and the highest R_s/R_L so far reported is 32 (Gunter-Smith et al. 1982). A serosa negative orientation of E_s due to impermeability to K^+ obviates this difficulty. Moreover it makes superfluous the assumption that a rheogenic Cl^- pump in the serosal membrane explains the negavitity of $E_s - E_L R_s/R_L$. Among other reasons the idea for a rheogenic Cl^- pump has emerged because Cl^- replacement reduces ψ_{ms} to zero or even reverses its sign. However, recent observations that Cl^- substitution causes alkalinization of the cell and hyperpolarization of the membrane should be taken into account. It is not known whether under this condition the serosal membrane remains impermeable to K^+, and whether changes in membrane resistances are significant. Therefore mucosal Cl^- substitution induces at least three changes in the electrical equivalent circuit: E_L changes by addition of the anion diffusion potential (serosa positive), R_L increases and E_m increases. ΔE_L and ΔE_m are in opposite directions. When, by inhibition of the NaK pump the ion gradients across the membranes are diminished, the full effect of ΔE_L should appear. We suggest that the results of ion substitution experiments with eel intestine as reported by Ando (1981) can be interpreted in this way.

In conclusion we suggest that the serosa negative transmural potential across intestines of SW fish is mainly a reflection of the very low electrodiffusive permeability of the serosal membrane for K^+. In FW fish, as observed in the goldfish, the serosal membrane potential depolarizes when serosal K^+ concentration is increased. In these cells therefore E_s would be serosa positive with respect to cell interior.

Sugar and Amino Acid Transport

Introduction

In a great number of fish (SW, FW and euryhaline) it has been shown that D-sugars and L-amino acids can be transported uphill from the mucosal to the serosal solution by a Na-dependent system. Net sugar transport can be inhibited by phlorizin, a rather specific inhibitor of Na^+-sugar cotransport systems in all sorts of animals. Results have been obtained with sacs, rings and sheets and with in vitro and in vivo perfusion techniques and most recently with brushborder membrane vesicles (Bogé and Rigal 1981). For references articles by Cartier et al. (1979) and Farmanfarmaian et al. (1972) can be consulted. In some fish the serosal tissue layers seem to constitute a barrier to in vitro transmural transport of glucose (Carlisky and Huang 1962, Stokes and Fromm 1964). In winter flounder no glucose transport could be detected by Rout et al. (1965), but recently Naftalin and Kleinzeller (1981) have reported the existence of two sugar transport systems one of which seems to be the universal Na-sugar cotransport system.

In contrast with the broad spectrum of fish in which tracer or chemical methods have been used for transport studies, electrophysiological responses to sugars and amino acids have only been published for cyprinids (goldfish and tench, for references see Table 1, Smith 1966, Mepham and Smith 1966) and for the winter flounder (Huang and Chen 1971, Naftalin et al. 1982).

Analysis of Electrical Responses to Glucose in the Goldfish

In this section contributions by Albus and Lippens (1982), Bakker and Albus (1982) and Albus et al. (1983a,b) will be examined briefly.

The responses of ψ_{mc} and ψ_{ms} to the substitution of non-transported mannitol by glucose at the mucosal side of stripped epithelium are shown in Fig. 4.

Glucose evokes a rapid depolarization of ψ_{mc} which reaches 90% of its maximum value within approximately 15 s. This is followed by a slower repolarization to a new steady state. During the depolarization the ratio of R_m/R_s decreases slightly with less than 20% but returns to its pre-glucose value within a few minutes. R_{ms} does not change. In our approximation we assume that the resistances are not altered significantly and therefore changes in ψ_{mc} and ψ_{ms} are exclusively caused by changes in the emf's. Thus Eqs. (1) and (3) can be rewritten with $\Delta\psi$'s and ΔE's and solved for ΔE_m, ΔE_s and ΔE_L. In the presence of ouabain at the serosal side, glucose again induces a depolarization but no repolarization occurs. This is evidence that the repolarization depends on the NaK-pump. We assume that in the presence of ouabain glucose does not induce a change in E_s or E_L. In the short period after ouabain addition no changes in R_{ms} or R_m/R_s could be detected so that $(R_s + R_L)/R_L$ can be calculated from $\Delta\psi_{mc}/\Delta\psi_{ms}$ induced by glucose. Now, R_m, R_s and R_L can be calculated. At the time of maximum depolarization of ψ_{mc} the change in ψ_{ms} is only about 70% of its ultimate change.

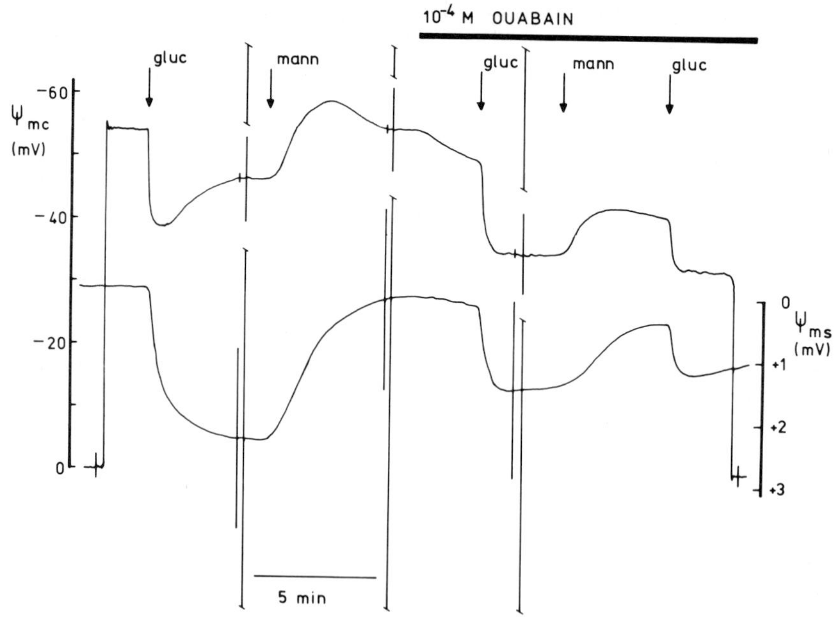

Fig. 4. Redrawing of original recording of ψ_{mc} *(upper trace* and *left voltage scale)* and ψ_{ms} *(lower trace* and *right voltage scale). Arrows* indicate the mucosal substitutions of glucose and mannitol (27.8 mM). *Heavy line* indicates the presence of oubain in the serosal solution. Voltage deflections are from transepithelial current pulses of + and − 10 μA and + and − 100 μA

Further analyses can be performed when it is assumed that during the depolarization phase of ψ_{mc}, $\Delta E_L = 0$, while during the repolarization of ψ_{mc}, $\Delta E_m = 0$. Therefore at the time of maximum depolarization of ψ_{mc}, ΔE_m and ΔE_s^1 can be calculated and from the ensuing repolarization of ψ_{mc} and the concomitant further change in ψ_{ms}, values for ΔE_s^2 and ΔE_L can be calculated. ΔE_s^1 and ΔE_s^2 are the changes in E_s during depolarization and repolarization of ψ_{mc} respectively.

Analysis of the relation between ΔE_m and mucosal glucose concentrations from 0.2 to 28 mM have revealed two rheogenic glucose transport systems with a low- and a high affinity, respectively. A similar analysis of alanine-induced changes in E_m suggests only one rheogenic alanine system. Recently in rabbit proximal tubules a low- and a high affinity Na^+ dependent D-glucose system has been demonstrated (Turner and Morran 1982).

Analysis of ΔE_m caused by separate and simultaneous addition of alanine and glucose (3.5 to 14 mM) shows that simultaneous addition evokes a greater change than the maximum change induced by saturating concentrations of either glucose or alanine separately. However, as shown in Table 3, ΔE_m, caused by simultaneous addition, is only 90% of the sum of the ΔE_m's induced by separate addition, suggesting a slight mutual inhibition. In contrast, when both substrates are applied together ΔE_s^1 is much larger than the sum of changes induced by individual addition. This suggests that the serosal mechanism is activated more intensely during this period.

Table 3. Changes in E_m and E_s during depolarization (ΔE_s^1) and repolarization (ΔE_s^2) of ψ_{mc} induced gy glucose + alanine as percentage of the sum of the changes induced by glucose or alanine separately

Concentration mM Alanine and glucose each	ΔE_m %	ΔE_s^1 %	ΔE_s^2 %	$\Delta E_s^1 + \Delta E_s^2$ %
3.5	91	179	74	92
7	91	192	71	93
14	86	166	71	94

These findings corroborate the conclusion of Read (1967) drawn 15 years ago concerning competitive effects in sugar and amino acid transport.

Recently Frömter (1982) and Gunter-Smith et al. (1982) have reported similar types of analyses on electrical responses in rat proximal tubules and in *Necturus* intestine. Frömter succeeded in measuring the response to fast concentration steps and thereby proved that the initial potential change can only be brought about by a rheogenic solute-Na cotransport system. In contrast with observations in the goldfish intestine, perfusion of the tubule lumen with glucose + phenylalanine induced a striking decrease in R_m/R_s to 52.8 ± 12% of its control value. This suggests a decrease of R_m by opening of a solute-sodium channel. Frömter observed no repolarization in the tubules, but this may be due to the short time that solutes were present. A metabolism dependent repolarization of 16–18 mV after an initial depolarization of 20–48 mV was observed in *Necturus* intestine. In this tissue R_m/R_s decreases at the peak of the potential change to about 25% of the pre-galactose value. In conclusion these electrophysiological data strongly support the Crane (1962)-Schultz and Zalusky (1964)-model for brush border Na-solute transport in fish enterocytes.

Regarding ΔE_s, two related and additive effects should be considered. First, the NaK-pump may be rheogenic and secondly the activation of the NaK-pump may alter the K^+ activity in the lis and therefore E_K at the serosal side (Zuidema et al. 1982b). The latter effect will not occur when the serosal membrane is impermeable to K^+ as is found in winter flounder.

Dependence on Oxygen

It is well known that functions of mammalian enterocytes are impeded by anoxic conditions. Cartier et al. (1979) suggest that oxygenation of intestinal preparations of fish is superfluous. In goldfish intestine the glucose evoked $\Delta\psi_{ms}$ under anoxic conditions is about 30% lower than in the presence of oxygen (Groot, unpublished data). After an anoxic period, readmission of O_2 brings the glucose evoked $\Delta\psi_{ms}$ back to the value in oxygen. The oxygen evoked potential change occurs only when substrates for the cotransport system are present at the mucosal side and can be blocked by serosal ouabain, or when sugars are present, by phlorizin. These results suggest that the oxygen evoked increase in ψ_{ms} is caused by stimulation of the NaK-pump which can occur only when the Na-solute cotransport system in the brush border membrane is active. The phlorizin sensitive 3-oxymethyl-D-glucose (3-OMG) flux from mucosa to serosa remains constant for the first hour in N_2 but decreases to approximately 50% in the second hour. The change in tissue sodium content induced by anoxia in the presence of 3-OMG is shown in Fig. 5. The much slower increase in the presence of phlorizin indicates the stimulation of sodium influx induced by 3-OMG.

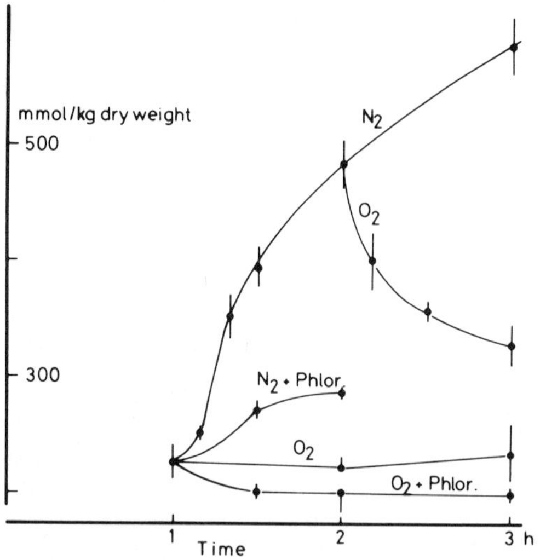

Fig. 5. Sodium content in mmol kg^{-1} dry weight in stripped intestinal epithelium. After 1 h oxygenated Ringer containing 27.8 mM 3-OMG part of the strips were further incubated under anoxic conditions in the presence or absence of phlorizin (5.10^{-5} M). As a control strips were also incubated in oxygenated Ringer containing phlorizin. *Points* are mean ± SEM of at least 4 measurements

However, from the relative constancy in R_m/R_s before and after application of glucose it is tentatively concluded that the conductance of the sodium-solute cotransport system in the mucosal membrane is shunted by a much larger conductance that is not influenced by cotransport-substrates.

In conclusion, analysis of the electrical phenomena evoked by glucose, assuming a constant Na-glucose stoichiometry over a glucose range of 0.2–28 mM, has revealed a low-affinity and a high-affinity Na-glucose cotransport system.

Moreover the electrophysiological experiments present evidence for independent sugar and amino acid transport systems (although both depend on the electrochemical gradient for sodium).

The transepithelial glucose evoked potential is partially dependent on oxygen. This oxygen-dependent increase can be blocked by ouabain.

Acknowledgment. We thank Drs. Kirsch and Meister for making their experimental data available to us.

References

Albus H, Siegenbeek van Heukelom J (1976) The electrophysiological characteristics of glucose absorption of the goldfish intestine as compared to mammalian intestines. Comp Biochem Physiol 54A:113–119

Albus H, Lippens F (1982) Electrophysiological changes induced by sugars and alanine in the isolated intestinal epithelium of the goldfish. 4th meeting of the EITG. Gastroenterol Clin Biol (in press)

Albus H, Groot JA, Siegenbeek van Heukelom J (1979) Effects of glucose and ouabain on transepithelial electrical resistance and cell volume in stripped and unstripped goldfish intestine. Pfluegers Arch 383:55–66

Albus H, Bakker R, Siegenbeek van Heukelom J (1983a) Circuit analysis of membrane potential changes due to electrogenic sodium dependent sugar transport in goldfish intestinal epithelium. Pfluegers Arch (in press)

Albus H, Lippens F, Siegenbeek van Heukelom J (1983b) Sodium dependent sugar and amino acid transport in isolated goldfish intestinal epithelium: Electrophysiological evidence against direct interactions at the carrier level

Ando M (1975) Intestinal water transport and chloride pump in relation to sea-water adaptation of the eel *Anguilla japonica*. Comp Biochem Physiol 52A:229–233

Ando M (1980) Chloride-dependent sodium and water transport in the seawater eel intestine. J Comp Physiol 138:87–91

Ando M (1981) Effects of ouabain on chloride movements across the seawater eel intestine. J Comp Physiol 145:73–79

Ando M, Kobayashi M (1978) Effects of stripping of the outer layers of the eel intestine on salt and water transport. Comp Biochem Physiol 61A:497–501

Ando M, Utida S, Nagahama H (1975) Active transport of chloride in eel intestine with special reference to sea water adaptation. Comp Biochem Physiol 51A:27–32

Bakker R (1980) A geometrical representation as an expedient for the evaluation of electrical measurements in epithelial tissues. Proc Int Union Physiol Sci XIV: Abstr No 801

Bakker R, Albus H (1982) Analysis of electrophysiological phenomena upon glucose addition to the mucosal side of isolated intestinal mucosa of the goldfish. 4th Meet EITG. Gastroenterol Clin Biol (in press)

Bakker R, Dekker K, Zuidema Th, Groot JA (1982) Transepithelial Cl^- transport in goldfish *Carassius auratus* intestinal mucosa and the effect of theophylline on fluxes and electrophysiology. In: European Society for Comparative Physiology and Biochemistry (ed) Intestinal transport: Fundamental and comparative aspects. Abstr Vol 4th Comf, pp 123–124

Barry RJC, Eggenton J (1972) Membrane potentials of epithelial cells in rat small intestine. J Physiol (Lond) 227:201–216

Bogé G, Rigal A (1981) A chloride requirement for Na^+-dependent amino-acid transport by brush border membrane vesicles isolated from the intestine of a mediterranean teleost *(Boops salpa)*. Biochim Biophys Acta 649:455–461

Buclon M (1974) Bioelectric potentials and the transfer of amino acids across the digestive epithelium of the tench (*Tinca tinca* L.). J Physiol (Lond) 68:157–180

Carlisky NJ, Huang KC (1962) Glucose transport by the intestinal mucosa of the dogfish. Proc Soc Exp Biol Med 109:405–408

Cartier M, Buclon M, Robinson JWL (1979) Preliminary studies on the characteristics of phenylalanine and β-methyl-glucoside transport in the tench intestine in vitro. Comp Biochem Physiol 62A:363–370

Chen TST, Huang KC (1972) Structural specificity in the intestinal transport of hexoses, tyrosine derivatives and electrolytes in freshwater catfish. J Pharmacol Exp Ther 180:777–783

Crane RK (1962) Hypothesis for mechanism of intestinal active transport of sugars. Fed Proc 21: 891–895

Cremashi D, Henin S (1975) Na and Cl transepithelial routes in rabbit gallbladder. Tracer analysis of the transports. Pfluegers Arch 361:33–41

Curran PF (1960) Na, Cl and water transport by rat ileum in vitro. J Gen Physiol 43:1137–1148

Datta S, Savage NB (1968) Investigation of the electrical potential of the isolated midgut of the freshwater teleost, *Ameirus nebulosus.* Experientia 24:572–574

Diamond JM (1974) Tight and leaky junctions of epithelia: A perspective on kisses in the dark. Fed Proc 33:2220–2224

Duffey ME, Thompson SM, Frizzell RA, Schultz SG (1979) Intracellular chloride activities and active chloride absorption in the intestinal epithelium of the winter flounder. J Membr Biol 50:331–341

Ellory JC, Evans HM, Heal JW, Smith MW (1972) Intracellular and transepithelial potential recording from the goldfish intestine. J Physiol (Lond) 226:29–30P

Farmanfarmaian A, Ross A, Mazal D (1972) In vivo intestinal absorption of sugar in the toadfish (marine teleost, *Opsanus tau*). Biol Bull 142:427–445

Field M, Karnaky KJ Jr, Smith PL, Bolton JE, Kinter WB (1978) Ion transport across the isolated intestinal mucosa of the winter flounder *Pseudopleuronectes americanus:* I. Functional and structural properties of cellular and paracellular pathways for Na and Cl. J Membr Biol 41: 265–293

Field M, Smith PL, Clayton DC, Frizzell RA (1979) Role of HCO_3 in the regulation of Cl transport by flounder intestine. Bull MDIBL 18:44–45

Field M, Smith PL, Bolton JE (1980) Ion transport across isolated intestinal mucosa of the winter flounder *Pseudopleuronectes americanus:* II. Effects of cyclic AMP. J Membr Biol 55:157–163

Field M, Kimberg LS, Orellana SA, Frizzell RA (1982) Potassium dependence of chloride transport in the intestine of the flounder, *Pseudopleuronectes americanus.* Bull MDIBL 21:93–95

Frömter E (1982) Electrophysiological analysis of rat renal sugar and amino acid transport. I. Basic phenomena. Pfluegers Arch 393:179–189

Groot JA, Dekker K, van Riel J, Zuidema Th (1982) Intracellular ion concentrations and pH in stripped mucosa of goldfish *Carassius auratus* intestine in relation to Cl^- transport. In: European Society for Comparative Physiology and Biochemistry (ed) Intestinal transport: Fundamental and comparative aspects. Abstr Vol 4th Conf, pp 132–133

Gunter-Smith PJ, Grasset E, Schultz SG (1982) Sodium-coupled amino acid and sugar transport by *Necturus* small intestine. An equivalent electrical circuit analysis of a rheogenic co-transport system. J Membr Biol 66:25–39

Halm D, Krasny E, Frizzell R (1982) Apical membrane potassium conductance in flounder intestine: Relation to chloride absorption. Bull MDIBL 21:88–93

Helman SI, Beyenbach KW (1979) Electrophysiological study of the apical barrier of the intestinal epithelium of winter flounder *(Pseudopleuronectes americanus).* Bull MDIBL 18:51–54

Hirano T, Morisawa M, Ando M, Utida S (1976) Adaptive changes in ion and water transport mechanism in the eel intestine. In: Robinson JWL (ed) Intestinal ion transport. MTP Press, Lancaster, pp 301–317

House CR, Green K (1965) Ion and water transport in isolated intestine of the marine teleost, *Cottus scorpius.* J Exp Biol 42:177–189

Huang KC, Chen TST (1971) Ion transport across intestinal mucosa of winter flounder *Pseudopleuronectes americanus.* Am J Physiol 220:1734–1738

Katz U, Lau KR, Ramos MMP, Ellory JC (1982) Thiocyanate transport across fish intestine *(Pleuronectes platessa).* J Membr Biol 66:9–14

Kirsch R, Meister MF (1982) Progressive processing of ingested water in the gut of sea-water teleosts. J Exp Biol (in press)

Lahlou B (1976) Ionic permeability of fish intestinal mucosa in relation to hypophysectomy and salt adaptation. In: Robinson JWL (ed) Intestinal ion transport. MTP Press, Lancaster, pp 318–328

Liedtke CM, Hopfer U (1982) Mechanism of Cl^- translocation across small intestinal brush-border membrane. I. Absence of Na^+-Cl^- cotransport. Am J Physiol 242:G263–G271

Mackay WC, Lahlou B (1980) Relationships between Na^+ and Cl^- fluxes in the intestine of the European flounder, *Platichthys flesus.* In: Lahlou B (ed) Epithelial transport in the lower vertebrates. Cambridge Univ Press, Cambridge, pp 151–162

Madara JL, Burtis RL, Lindem NM, Trier JS (1981) Junctional complex structure of small intestinal epithelial cells in winter flounder *(Pseudopleuronectes americanus)*. Bull MDIBL 20: 14–17

Mepham TB, Smith MW (1966) Amino acid transport in the goldfish intestine. J Physiol (Lond) 184:673–684

Moreno JH (1975) Blockade of gallbaldder tight junction cation-selective channels by 2,4,6-triaminopyrimidinium (TAP). J Gen Physiol 66:97–115

Musch MW, Field M, Frizzell RA (1982) Active K^+ transport by the intestine of the flounder, *Pseudopleuronectes americanus:* Evidence for cotransport with Na and Cl. Bull MDIBL 21: 95–99

Naftalin RJ, Kleinzeller A (1981) Sugar absorption and secretion by winter flounder intestine. Am J Physiol 240:G392–G400

Naftalin RJ, Thompson K, Kleinzeller A (1982) Sugar absorption and secretion by the intestine of the winter flounder *(Pseudopleuronectes americanus)*. Bull MDIBL 21:62–64

Ramos MMP, Ellory JC (1981) Na and Cl transport across the isolated anterior intestine of the plaice *Pleuronectes platessa*. J Exp Biol 90:123–142

Rao MC, Dubinsky W, Vosburg E, Field M, Frizzell RA (1982) Sodium proton antiport in intestinal brush border vesicles of the flounder, *Pseudopleuronectes americanus*. Bull MDIBL 21: 99–103

Read CP (1967) Studies on membrane transport. 1. A common transport system for sugars and amino acids? Biol Bull 133:630–642

Rose RC, Schultz SG (1971) Studies on the electrical potential profile across rabbit ileum. J Gen Physiol 57:639–663

Rout WR, Lin DST, Juang KC (1965) Intestinal transport of amino acids and glucose in flounder fish. Proc Soc Exp Biol Med 118:933–938

Schultz SG, Zalusky R (1964) Ion transport in isolated rabbit ileum. II. The interaction between active sodium and active sugar. J Gen Physiol 47:1043–1059

Simon M, Curci S, Gebler B, Frömter E (1981) Attempts to determine the ion concentrations in the lateral spaces between the cells of *Necturus* gallbladder epithelium with microelectrodes. In: Ussing HH, Bindslev N, Lassen NA, Sten-Knudsen O (eds) Water transport across epithelia. Munksgaard, Copenhagen, pp 52–63

Skadhauge E (1974) Coupling of transmural flows of NaCl and water in the intestine of the eel *(Anguilla anguilla)*. J Exp Biol 60:535–546

Smith MW (1964) Electrical properties and glucose transfer in the goldfish intestine. Experientia 20:613–614

Smith MW (1966) Sodium-glucose interactions in the goldfish intestine. J Physiol (Lond) 182: 559–573

Smith MW, Ellory JC, Lahlou B (1975) Sodium and chloride transport by the intestine of the European flounder *Platichthys flesus* adapted to fresh or sea water. Pfluegers Arch 357:303–312

Smith PL, Orellana S, Field M (1980) Role of the medium pH in the regulation of chloride transport in the intestine of the winter flounder, *Pseudopleuronectes americanus*. Bull MDIBL 19: 24–27

Smith PL, Welsh MJ, Stewart CP, Frizzell RA, Orellana SA, Field M (1981) Chloride absorption by the intestine of the winter flounder *Pseudopleuronectes americanus:* Mechanism of inhibition by reduced pH. Bull MDIBL 20:96–101

Spring KR, Hope A, Persson BE (1981) Quantitative light microscopic studies of epithelial fluid transport. In: Ussing HH, Bindslev N, Lassen NA, Sten-Knudsen O (eds) Water transport across epithelia. Munksgaard, Copenhagen, pp 190–200

Stewart CP, Smith PL, Welch MJ, Frizzell RA, Musch MW, Field M (1981) Potassium transport by the intestine of the winter flounder, *Pseudopleuronectes americanus:* Evidence for KCl cotransport. Bull MDIBL 20:92–96

Stokes RM, Fromm PO (1964) Glucose absorption and metabolism by the gut of rainbow trout. Comp Biochem Physiol 13:53–69

Trier JS, Moxey PC (1980) Epithelial cell proliferation in the intestine of the winter flounder, *Pseudopleuronectes americanus.* Cell Tissue Res 206:379–385

Turnberg LA, Bieberdorf FA, Morawski SG, Fordtran JS (1970) Interrelationships of chloride, bicarbonate, sodium and hydrogen transport in human ileum. J Clin Invest 49:557–567

Turner RJ, Moran A (1982) Heterogeneity of sodium-dependent D-glucose transport sites along the proximal tubule: evidence from vesicle studies. Am J Physiol 242:F406–F414

White JF, McD Armstrong W (1971) Effect of transported solutes on membrane potentials in bullfrog small intestine. Am J Physiol 221:194–201

Wilson TH, Wiseman G (1954) The use of sacs of everted small intestine for the study of the transference of substances from the mucosal to the serosal surface. J Physiol 123:116–125

Zeuthen T, Ramos M, Ellory JC (1978) Inhibition of active chloride transport by piretanide. Nature 273:678–680

Zuidema T, Groot JA, Siegenbeek van Heukelom J (1982a) Chloride transport in goldfish intestinal epithelium. 4th meeting of the EITG. Gastroenterol Clin Biol (in press)

Zuidema T, Siegenbeek van Heukelom J, Kamermans M, Groot JA (1982b) The transmembrane glucose evoked potential in goldfish *Carassius auratus* intestinal epithelium and the role of the extracellular space. In: European Society for Comparative Physiology and Biochemistry (ed) Intestinal transport: Fundamental and comparative aspects. Abstr Vol 4th Conf, pp 152–153

Intestinal Transport and Osmoregulation in Fishes

B. LAHLOU[1]

Introduction

The general pattern of osmoregulation in fishes was outlined long ago by H. Smith (1930) and has since been confirmed by many investigators. The most significant additional contribution has been the use of radioisotopes which allow the measurement of ionic fluxes and transport across epithelia and has been developed mainly by J. Maetz and his collaborators.

Until recently, most of the experimental work performed on this subject has been devoted for obvious reasons to the study of gill function. The branchial epithelium plays a major role in the transport of ions and water. This has moreover proved to be of general interest in membrane biology as it is the only transporting tissue to come between the blood and the environmental medium. In euryhaline species at least, by altering the external concentrations, it is possible to set up large osmotic, chemical and electrical gradients in either direction across the tissue in physiologically perfect conditions. This study is limited, however, by the complex structure of the gill, the small number of salt-secreting cells (the chloride cells) and its folded anatomy, making it impossible to apply to this tissue the rewarding methods designed for flat epithelia.

The other relevant organs are the kidney and the intestine.

The importance of the kidney is obvious in freshwater (FW) as in this medium urinary excretion is the only way in which the fish gets rid of the excess water entering through the outer surfaces. Its role is less clear in sea water (SW) because, despite the established fact that the kidney excretes nitrogen and divalent ions, it is now known that the gills are permeable to Ca^{2+}, Mg^{2+}, and organic substances.

The intestine has proved to be easier to study, thanks to in vitro techniques. Its physiological function may be described as opposite to that of the kidney. The intestine is essential in hypertonic media, such as SW, to enable water absorption to take place and compensate for osmotic losses through the gills. In FW and in other largely hypotonic media, there is no need in principle to absorb water through the gut wall, therefore drinking may occasionally take place under normal conditions.

Because of the part it plays in the osmoregulatory processes of fishes and because its function varies considerably from one adaptation medium to the other, the intestine

1 Laboratoire de Physiologie cellulaire et comparée et ERA CNRS 943, Faculté des Sciences et des Techniques, Parc Valrose, F-06034 Nice Cedex, France

Intestinal Transport
(ed. by M. Gilles-Baillien and R. Gilles)

has been the subject of much interest in recent years. In the present paper we propose to discuss some of the results obtained by newly introduced methods. Most of these concern teleostean fishes but progress has also been made using elasmobranchs.

Water Transport

This is the main intestinal function associated with osmoregulation. Drinking rates have been measured in vivo in many species under the assumption that they reflect the actual absorption rates of water. For this purpose, non-absorbed markers (e.g., colloidal radioactive gold) were included in the outer medium and the amount of water ingested calculated from the intestinal contents at autopsy. A more complicated approach, more suitable for analysis of the mechanisms involved, consists of inserting cannulae at various points along the gut lumen and causing to circulate through them a known volume of fluid of determined composition, the changes of which were followed subsequently (Skadhauge 1969, Kirsch and Laurent 1975).

The results have established firmly that in osmoregulatory teleosts, drinking and absorption rates are low (but not nil), and increase with the external salinity, even when this goes beyond full-strength SW. In sea water, where urine output of water is small and may be discounted, these rates are equated with the gill osmotic loss and are therefore used to calculate the osmotic permeability of the branchial epithelium. The opposite holds for freshwater, where the urine flow is used to evaluate this permeability.

Early measurements on gut contents revealed that in SW the luminal fluid is dilute compared to the external medium and that the dilution process has already taken place in the stomach. Thus, in *Anguilla anguilla,* Na concentration is 174 mM as against 500 outside (Sharratt et al. 1964).

This led several researchers to investigate the ionic and water permeability of the esophagus in order to localize steps of dilution. In their work on the eel, Kirsch and Laurent (1975) and Hirano and Mayer-Gostan (1976) made the striking observation that the esophagus wall is readily permeable to ions and virtually impermeable to water in SW-adapted animals. In other words, when SW is ingested, Na^+ and Cl^- are driven from the lumen into the blood owing to their chemical gradients, with little opposite movement of water. Moreover, the esophagus of FW-adapted eels is impermeable to both ions and water. Finally, transfer from FW to SW results in structural modifications of the esophagus wall. The epithelium becomes monolayered, made up of columnar cells and free of mucous cells, while large blood spaces develop beneath (Laurent and Kirsch 1975, Yamamoto and Hirano 1978).

These permeability properties are undoubtedly unusual, but their relevance to osmoregulation is difficult to assess. During adaptation from FW to SW the epithelium becomes "leaky" in such a way that the reflection coefficient of the major ions must be very small since almost no osmotic flux is driven. Further in vitro studies, involving measurements of the electrical properties of the tissue as well as of transport are necessary. It is claimed that early desalting of sea water in the esophagus prevents or limits dilution in the posterior parts of the gut at the expense of body fluids.

However, this results in salt overload which must be disposed of by the gills. Therefore the question remains: which is less costly in energy expenditure, gill secretion of ions or intestinal absorption of ions and water?

Fish intestine has a relatively simple anatomy. It may be almost uniform along its full length or divided into separate regions differing from each other by various features such as folds in the mucosa. Unlike mammalian gut, there are no crypts of Lieberkuhn (Field et al. 1978). In vitro techniques, such as the everted sac, also reveal differences in transport abilities. Thus, in goldfish, the absorption rate is relatively constant along the intestine (Smith 1964) while the anterior intestine of the trout (Bensahla-Talet et al. 1974) and the middle intestine of the eel (Ando 1980) transport more.

It is believed that the classic double-membrane model (Curran and MacIntosh 1962) and standing-gradient hypothesis (Diamond and Bossert 1967) designed for mammalian gut and gallbladders are equally valid for fish intestine. Water is driven across the mucosa as a solute-linked flow, osmotic equilibration taking place within the tissue. In vitro transport of water necessitates the presence of sodium in the lumen and is suppressed when ouabain is added to the serosal side. However, ionic influences other than Na^+ active transport have been suggested. Thus in the Japanese eel, Ando (1980) found that water absorption requires the presence of chloride as accompanying anion. Net water flux was largely reduced when Cl^- was replaced by NO_3^- or an organic anion. It has been suggested that water transport is linked to the "coupled" transfer of Na^+ and Cl^-, meaning that both ions are needed in this process. However, this does not rule out the basic preeminence of sodium pump.

In vivo experiments performed on perfused eel intestine (Skadhauge 1969) revealed an interesting adaptive feature. When a saline solution hypertonic to blood is introduced into the lumen, dilution takes place before the water and ions are absorbed. The resulting osmotic concentration for which water movement is zero (the turning-point) is higher than plasma osmolality by 126 mOsm kg^{-1} in SW animals and by 73 mOsm kg^{-1} in FW fish. Consequently, marine fish appear to gain water with higher efficiency (from about half-strength SW).

This result together with the observations reported for the esophagus, may indicate that how to save water is the first problem euryhaline teleosts must solve when they are transferred to hypertonic salt media. This explanation may also be valid for isosmotic organisms (marine Cyclostomes and Elasmobranchs, stenohaline freshwater teleosts such as carp and goldfish adapted to high salinity) which, in fact, maintain an osmotic gradient favouring a positive water balance for their body.

Diamond's model for fluid absorption requires that osmotic equilibration takes place within the intercellular spaces which are thus distended in the transporting epithelia. Fish enterocytes are high and narrow (e.g., 60 × 3.5 μm in American flounder, Field et al. 1978). However, in fish intestine, lateral space dilatations may be rare, of reduced width, or even absent. A reasonable, but not definitive, explanation, as suggested by Field et al. is that fluid accumulates and hydrostatic pressure builds up in unstripped preparations only. In fish intestine which has been cleared of its muscular layers, intercellular volume will be reduced as transported liquid moves more freely towards the serosal side. This, however, should impair the recycling of sodium across the tight-junctions towards the lumen which is presented by Field et al. as an essential component of the transport mechanism (see below).

Ion Transport

Sodium and chloride are the major inorganic ions in extracellular fluids. Little transport of these electrolytes is involved at intestinal level in freshwater fishes. In sea water, they provide the driving force necessary for water to cross the gut wall. Various techniques have been used recently, mostly in vitro, in order to analyze the basic cellular phenomena and the mechanisms of adaptation in relation to salinity. These methods include:

- the everted sac technique (Wilson and Wiseman 1954),
- the measurement of the potential difference (PD) and short-circuit current (SCC) on unstripped or stripped mucosa and of ion fluxes in Ussing chambers,
- short-term uptake across the luminal border (Schultz et al. 1967),
- the determination of intracellular potentials and ion activities using non-selective or selective microelectrodes,
- kinetics on brush-border vesicles,
- the biochemical evaluation of enzyme activities (ATPase, alkaline-phosphatase).

A major advantage of fish intestine is its functional stability at room temperature, allowing for in vitro measurements of long duration (several hours).

Transepithelial Properties

In Ussing chambers, fish intestine behaves as a "leaky" epithelium, presenting a low PD (a few mV) and low electrical resistance (less than 100 $\Omega \cdot cm^2$ in most cases). Stripping the muscular layers is feasible for some species (eel, flounder, plaice, tench) but impossible for others (trout), yet this treatment is essential if we are to obtain more reliable results by the removal of undesirable diffusion barriers. In *Anguilla japonica,* the PD is multiplied four-fold after stripping (Ando and Kobayashi 1978) whereas we ourselves have observed that in trout stable isotopic fluxes cannot be recorded until 30 min have elapsed to allow for equilibration within the entire tissue. The peculiarity of fish tissue (as compared to mammalian gut) is that it generates a positive or a negative PD (serosa vs. mucosa) depending on the species and on the medium to which the fishes are adapted. Table 1 presents a number of data including those obtained by Ando et al. (1975).

A serosa-positive PD is frequently observed in FW fishes, whereas in marine fish the PD is generally serosa-negative.

Is There a Chloride Pump?

Occurrence of negative PD's has led researchers to postulate that fishes possess an electrogenic, Na^+-independent chloride pump (Huang and Chen 1971, Ando 1975, Hirano et al. 1976, Smith et al. 1975). Finally, Cl^- transport may be observed in the absence of Na^+ transport. Thus, in winter flounder, Huang and Chen (1971) found that chloride transfer was maintained in the presence of a Ringer solution containing

Table 1. Electrical properties of fish intestine (data collected from the literature) PD in mV, SCC in $\mu A\ cm^{-2}$, means ± S.E.

Species	PD	SCC
Fresh water		
Cyprinus carpio	+ 1.41 ± 0.34	+ 30.3 ± 6.4
Carassius auratus	+ 1.6 ± 0.3	+ 25 ± 5
Channa argus	+ 0.56 ± 0.16	+ 7.2 ± 0.1
Parasilurus asotus	+ 1.07	+ 24.7
Salmo gairdneri	+ 0.23 ± 0.12	+ 1.8 ± 1.3
Anguilla japonica	− 0.53 ± 0.25	− 4.2 ± 2.2
Patichthys flesus	− 1.24 ± 0.14	− 18.2 ± 3.6
Intermediate salinity		
Carassius auratus (in 190 mM NaCl)	− 1.2 ± 0.3	− 33 ± 7
Sea water		
Cottus scorpius	nil	nil
Prionurus microlepidotus	+ 0.06 ± 0.26	+ 0.4 ± 2.1
Astroconger myriaster	− 3.25	− 50.1
Goniistius zonatus	− 2.86 ± 0.38	− 75.8 ± 35.1
Salmo gairdneri	− 0.94 ± 0.30	− 25.0 ± 10.8
Anguilla japonica	− 2.46 ± 0.28	− 42.3 ± 4.7
Platichthys flesus	− 1.94 ± 0.14	− 45 ± 5
Pseudopleuronectes americanus	− 3.35 ± 1.46	− 26.5 ± 13.3
Pleuronectus platessa	− 5.20 ± 0.21	− 90.2 ± 4.3

25 mM Na^+. Although this is not a sodium-free medium, we may at least assume that there is presumably no chemical gradient to drive Na^+ into the cells. More compelling evidence comes from the work of Ando et al. (1975) on *Anguilla japonica.* In this species, a large and sustained (30 min or more) negative PD and SCC is observed in the presence of choline-Cl Ringer or tetrathylammononium-Cl Ringer, that is, in the total absence of sodium.

Also, in European flounder adapted to FW (Smith et al. 1975) and in plaice in SW (Ramos and Ellory 1981), the SCC may be equated with chloride netflux.

Thus, when we consider in addition that a positive PD means that a Na^+-pump is operating as in higher vertebrates, we may conclude that a dual active transport of ions is disclosed in fish intestine, one component or the other being predominant.

However, other explanations for this situation have been put forward. First, in *Cottus scorpius* intestine, there is no transepithelial PD and Na^+ and Cl^- are transported at the same rate. Therefore House and Green (1965) have suggested that these ions cross the tissue by the electrically neutral NaCl absorption mechanism described for the gallbladder (Diamond 1962). However, further studies on epithelia have demonstrated that two steps (entry and exit) must be considered separately.

More recently Field et al. (1978) have proposed a novel interpretation which has the great advantage of explaining by a single mechanism all the situations encountered. Their model is inspired by that described previously (Machen and Diamond 1969) to account for the negative electrical potential observed in rabbit gallbladder. Briefly, the sole active process is the sodium pumping located at the basolateral side of the cells. This leads to the accumulation of Na^+ in the intercellular spaces, with Cl^- following passively. Na^+ subsequently refluxes back to the mucosal solution across the cation-selective tight-junctions, while Cl^- ions are repelled and move forward to the serosal side owing to their chemical gradient. This results in an excess of negative charges on this side and a build-up of negative potential. In other words, the transepithelial PD, whether positive or negative, is determined solely by Na^+ movements across the tight junctions and there is no need to postulate the presence of an independent chloride pump.

This model requires that sodium ions meet less resistance in the tight junctions than in the lateral space. This may cause excessive limitation of solute accumulation and water movement between the cells. Also in the presence of 2,4,6,-triaminopyrimidine (TAP) in the mucosal solution, the negative PD is increased in flounder intestine and not diminished as might be expected (Mackay and Lahlou 1980). The basic argument is that TAP blocks the cation-selective paracellular pathways in epithelia and consequently hinders backflux of sodium. However, it is known that this substance may also alter transcellular properties and that its action is therefore complex.

Sodium and Chloride Coupling at the Apical Membrane

Under current in vitro conditions, where a tissue is bathed on both sides by an identical Ringer solution, the electrochemical gradient favours sodium but not chloride entry into the cells across the luminal membrane.

Experiments on gallbladders (e.g., Diamond 1962) strongly suggest that Na^+ and Cl^- are absorbed in equal amounts by a compulsory one-for-one transport mechanism. Major evidence of this is that replacement of one of these ions by an non-permeant one in the bathing solutions suppresses the netflux of the other. This process, described as coupled transport, has since been found in various epithelia such as rabbit ileum (Nellans et al. 1973), amphibian proximal renal tubule (Spring and Kimura 1978), trout urinary bladder (Fossat and Lahlou 1979) and flounder intestine (Frizzell et al. 1979), to name but a few.

A direct demonstration of this mechanism necessitates some experimental access to the apical cell membrane.

In fish intestine, this has been made possible by the measurement of short-term uptake of ions following the technique of Schultz et al. (1967). In these experiments, a labelled non-permeant substance (inulin, choline, polyethylene glycol) was used as extracellular marker. Despite this correction, it was found in fish as in mammalian intestine that the apical entry of both Na^+ and Cl^- was far greater than the transepithelial fluxes. This discrepancy was explained by the presence of ion recycling across the mucosal membrane by such processes as exchange-diffusion or even active transport. Backflux of sodium through the apical junctions does not seem to be a satisfactory answer since this difference in flux amplitude also involves chloride. Finally, there is a possibility in these experiments that uptakes are overestimated because small ions may diffuse readily in unstirred extracellular layers (mucus, glycocalyx) less accessible to the extracellular markers. This ion retention cannot be of electrical origin since positive and negative particles are equally affected. In trout urinary bladder, a flat epithelium, the apical fluxes measured in conditions similar to those for intestine are basically identical to transepithelial fluxes except for an exchange-diffusion component for chloride (Fossat and Lahlou 1979).

With this technique, it has been possible to demonstrate sodium-chloride coupled entry by simultaneous use of ionic substitutions and/or specific pharmacological inhibitors. In the flounder *Pseudopleuronectes americanus,* uptake of Na^+ and Cl^- are reduced to the same extent by replacing the counter ion or by addition of furosemide to the mucosal solution (Frizzell et al. 1979). In the plaice, *Pleuronectes platessa,* piretanide, another loop diuretic, suppresses the coupled entry which accounts for about 20–30% of the uptake (Zeuthen et al. 1978, Ramos and Ellory 1981). It is interesting to note that substances acting on membrane band 3, such as SITS, are ineffective (Ramos and Ellory 1981) and that drugs such as amiloride, which interfere with sodium sites where this cation is not coupled with chloride, have no effect on fish intestine (Lahlou, unpublished data, Ramos and Ellory 1981). In the European flounder, *Platichthys flesus,* the same coupling has been observed. However, it has been found that it corresponds to the net flux of sodium and to only a third of the net flux of chloride (Mackay and Lahlou 1980).

Intracellular Recordings

Measurements of intracellular potential and ion activity by the use of conventional and selective microelectrodes have been made recently despite the small size of fish enterocytes: in plaice (Zeuthen et al. 1978, Ellory et al. 1978, Lau, pers. commun., Katz et al. 1982) and in winter flounder (Duffey et al. 1979).

It has been found that chloride accumulates in the cells, its intracellular activity being about three times that predicted for an equilibrium distribution across the apical membrane. In other words, chloride entry from the mucosal Ringer solution represents an active step requiring some supply of energy. In every tissue, removal of sodium in the bathing media leads to a reversible decrease of cellular chloride activity towards its electrochemical equilibrium value. Thus, as in other epithelia, it may be concluded that the downhill movement of sodium across the apical membrane provides energy for the uphill accumulation of chloride by the cell.

Subsequently, chloride will move downhill to the serosal side across the basolateral membrane. Little is known concerning the ion associated at this step. It has been suggested that this may involve a Cl^-/HCO_3^- exchange on the grounds that PD, SCC and transport are increased when the pH of the bathing solutions is raised above 8 (Oide 1973, Hirano et al. 1976, Duffey et al. 1979, Ramos and Ellory 1981) while the absence of bicarbonate reduces SCC without changing the uptake (Ramos and Ellory 1981).

Brush-Border Membrane Vesicles

This technique has been extended successfully to several fish epithelia: intestinal spiral valve of dogfish (Crane et al. 1979), intestine of teleosts (Bogé et al. 1982, Bogé and Rigal 1981), rectal gland of the spiny dogfish *Squalus acanthias* (Eveloff et al. 1978).

It aimed first at studying the dependence of non-electrolyte transport on mucosal sodium. The discovery that in the teleost *Boops salpa* chloride is required in addition to sodium for the uptake of amino acids (glycine and 2-aminoisobutyric acid) (Bogé and Rigal 1981) is somewhat intriguing.

This technique also permitted researchers to confirm the presence of Na^+ and Cl^- coupling at the cellular entry step and its strong inhibition by loop diuretics (Eveloff et al. 1978).

Effects of Amphotericin B: Sodium and Potassium Channels in Brush Border

In order to make clear further relationships between Na^+ and Cl^-, we have investigated the effects of Amphotericin B. This polyene antibiotic has been employed in both artificial and biological membranes to induce cationic permeability. Following studies on several epithelia (namely toad bladder, frog skin, mammalian gallbladders and colon, fish bladder) this substance is believed to act on the apical cellular membrane primarily to produce channels for sodium entry, this leading to the increase (or appearance) of serosa-positive PD (see Reuss 1981).

Since Na^+ is of prime importance in the transport processes of these tissues, little attention has been paid to Cl^- movements. Because of the peculiar status of chloride in fish intestine, as discussed above, we have analyzed the effects of Amphotoricin B on both ions in plaice (Ellory et al. 1982) and in trout (Lahlou, unpublished data).

Tables 2 and 3 summarize some results with plaice. With trout intestine only small effects were observed. This may be due to poor mixing of the drug in the thick layer of mucus on this tissue or adverse effect in unstripped mucosa or the influence of chemical composition of the cellular membranes (critical levels of cholesterol), or it may be a consequence of the small positive PD and high leakiness displayed by this epithelium when at rest.

For plaice, the PD is reversed (i.e., it becomes serosa-positive) and tissue conductance is increased. As may be expected, there is a net absorption flux for Na^+ following enhanced entry across the brush-border side. However, chloride also is sensitive to Amphotericin B. Its netflux is largely reduced on account of increased serosa-to-

Table 2. Effects of Amphotericin B (20 $\mu g\ ml^{-1}$) on Na and Cl fluxes in plaice intestine. Fluxes in $\mu mol\ cm^{-2}\ h^{-1}$, means ± S.E. Control and test periods: 40 min each. (In collaboration with J.C. Ellory)

	Control	Amphotericin	n
Na			
J_{ms} (mucosa-to-serosa)	11.6 ± 0.4	14.4 ± 0.6	22
J_{sm} (serosa-to-mucosa)	9.0 ± 0.3	9.7 ± 0.3	
J_n (netflux)	1.5 ± 0.5	2.6 ± 0.4	
Cl			
J_{ms} (mucosa-to-serosa)	8.6 ± 0.7	8.1 ± 0.6	7
J_{sm} (serosa-to-mucosa)	2.7 ± 0.6	5.3 ± 0.6	6

Table 3. Effects of Amphotericin B (20 μg ml^{-1}) on electrical parameters in plaice intestine. Means ± S.E. (n = 24). Control and test periods: 40 min each. (In collaboration with J.C. Ellory)

	Control	Amphotericin
Short-circuit current (SCC, μA cm^{-2})	– 158 ± 8	+ 67 ± 11
Transepithelial potential (PD, mV)	– 5.8 ± 0.3	+ 1.9 ± 0.3
Conductance (G, mS)	28 ± 1	38 ± 2

mucosa flux. Intracellular recordings (Lau, unpublished data) show that the apical membrane is depolarized and chloride becomes distributed passively, while its intracellular accumulation is suppressed. This may result from diminished selectivity against chloride either of the cellular membrane or of the tight-junction. However, the situation is obscured by a decrease in K^+ selectivity of the cell apical membrane. It is claimed that potassium leaks out and contributes to the induced PD by refluxing through the junctions. These K^+ channels are revealed through the use of Ba^{2+} salts, which prevent the development of the PD.

As a matter of fact, K^+ movements across the mucosa, notably its secretion through the brush-border, are currently represented as a component of the ionic exchanges in fish intestine (Stewart et al. 1980).

Adaptive Regulatory Mechanisms

With regard to osmoregulation in low and high salinities, some attempt has been made to determine the cellular steps which control or limit adaptation. By comparing apical and transepithelial ion fluxes, it has been suggested that brush-border and basolateral membranes react separately when fish are transferred from one medium to the other (Lahlou 1976). This may help us to differentiate euryhaline from stenohaline osmoregulatory capacities.

The main feature responsible for increased transport in high salt as compared to low salt media appears to be the speeding up of the basolateral sodium pump which is controlled by cortisol (Ellory et al. 1972) and mediated by Na^+/K^+-ATPase (Oide 1967). It has also been suggested that alkaline-phosphatase, which increases parallel to water and ion transport in eel intestine, plays a part in osmoregulation, at least indirectly (Oide 1973). As this enzyme is located in the apical membranes where Cl^- uptake takes place by a process which is "active" (as described above) and saturable (but of low affinity) it is clear that we must look for the presence of an anion-sensitive ATPase in the brush border. However, unlike other chloride-transporting epithelia including fish gills, fish intestine cannot be said definitely to contain this enzyme.

Fish mucosa presents adenyl-cyclase activity and cyclic AMP which is enhanced by theophylline. This indicates that hormones with short-term action are likely to modulate intestinal function. In isolated mucosa of the European flounder (Mackay et al. 1978) and winter flounder (Field et al. 1980), exogenous cyclic AMP inhibits chloride absorption but does not reverse the netflux as it does in mammals. It has been argued that this reduction is associated with acid-base balance in the fish (Mackay et al. 1978) while the failure to induce secretion has been attributed to the absence of crypts in fish intestine (Field et al. 1980). It may appear surprising that up to now few hormonal actions have been revealed in this tissue. This however is true for other epithelia carrying solute-linked water in association with Na-Cl coupled transport (e.g., renal proximal tubule of gallbladder, compared to collecting duct or amphibian skin and bladder).

In conclusion, the recent developments summarized in this paper have shed a certain amount of light on the cellular mechanisms involved in fish intestinal transport, without changing the functional significance discovered more than 50 years ago. We believe that the role of the intestine is primarily to maintain water balance. NaCl transport serves not for ionoregulation, but to provide the driving force necessary for other absorptions. In the domain of membrane physiology, several points remain to be clarified, namely: the relative amplitude of Na-Cl coupling compared with other permeation processes (this mechanism is not flexible enough to serve several types of regulation), the possible existence of an independent chloride pump (despite the strong attraction of Field's model), the mechanism of high apical ion exchanges. As regards monovalent ions, study of potassium transport is just now opening as a new fashion and its scope is as yet unknown.

In all these studies, a major aspect has always been overlooked or neglected, to wit the presence or absence of nutriments in the lumen for a protracted period.

In some species, at least – trout for example – fish feeding or fasting is of great consequence for the maintenance of intestinal structure, biochemical composition and function. This condition (see Leray and Florentz, this vol.) cannot be ignored in a presentation of reliable data on intestinal transport.

Summary

Since the early contributions of H. Smith (ca. 1930), the functional significance of the intestine in fish osmoregulation has been clearly defined. Most of the recent work on the subject deals with the mechanisms involved in ion transport and its hormonal regulation.

Available in vivo studies on the subject are not numerous. They have been mainly concerned with the measurement of drinking rates in media of various salinities and with the solute-solvent relationships along the gut lumen.

In vitro investigations have been made using all the major techniques currently available in epithelial physiology.

The electrical properties of the intestine as determined in Ussing chambers are those of a leaky tissue. Unlike that observed in higher vertebrates, the transepithelial potential difference may be serosa-positive (in most freshwater fishes) but it is frequently serosa-negative in sea water fishes. The latter feature has been explained either by the presence of an independent chloride pump or else (Field et al. 1978) by sodium refluxing across the cation-selective apical junctions.

Ionic replacements and measurements of apical uptake by the short-time incubation method devised by Schultz et al. (1967) revealed a high rate of ion turnover and the presence of a partial Na^+/Cl^- coupling at this entry step. Impalements with chloride-sensitive microelectrodes have shown that chloride accumulates against its electrochemical gradient owing to the downhill entry of sodium into the cells. Chloride transport is inhibited by furosemide and piretanide and other loop diuretics (Ramos and Ellory 1981). Sodium channels and serosa-positive potential may be induced by Amphotericin B. There is now a tendency to consider that potassium secretion at the apical border plays a part in membrane polarisation. Some of these results have been confirmed by using brush-border vesicles.

Adaptation to sea water results in a large increase in the intestinal transport of water and ions compared with dilute media. During this process, the permeation properties of the basolateral membrane (where Na^+/K^+-ATPase is located) and of the brush border adjust separately.

With regard to endocrine control, in addition to the stimulating effects of cortisol, short-term acting hormones appear to play a part since exogenous cyclic AMP diminishes chloride transport (without inducing secretion as it does in mammals).

The membrane properties of fish intestinal mucosa and the convenience of using this tissue for in vitro studies explains why it is attracting growing interest among physiologists.

References

Ando M (1975) Intestinal water transport and chloride pump in relation to seawater adaptation of the eel *(Anguilla japonica)*. Comp Biochem Physiol 52A:229–233

Ando M (1980) Chloride-dependent sodium and water transport in the sea water eel intestine. J Comp Physiol B138:87–91

Ando M, Kobayashi M (1978) Effects of stripping of the outer layers of the eel intestine on salt and water transport. Comp Biochem Physiol 61A:497–503

Ando M, Utida S, Nagahama H (1975) Active transport of chloride in eel intestine with special reference to sea water adaptation. Comp Biochem Physiol 51A:27–32

Bensahla-Talet A, Porthé-Nibelle J, Lahlou B (1974) Le transport de l'eau et du sodium par l'intestin isolé de la truite *Salmo irideus* au cours de l'adaptation à l'eau de mer. CR Acad Sci 278:2541–2544

Bogé G, Rigal A (1981) A chloride requirement for Na^+ dependent amino-acid transport by brush-border membrane vesicles isolated from the intestine of a mediterranean teleost *(Boops salpa)*. Biochim Biophys Acta 649:455–461

Bogé G, Rigal A, Pérès G (1982) The use of intestinal brush border membrane vesicles for comparative studies of glucose and 2-amino-isobutyric acid transport by four species of marine teleosts. Comp Biochem Physiol 72A:85–89

Crane RK, Bogé G, Rigal A (1979) Isolation of the brush border membranes in vesicular form from the intestinal spiral valve of the small dogfish *(Scyliorhinus canicula)*. Biochim Biophys Acta 554:264–267

Curran PF, MacIntosh JR (1962) A model system for biological water transport. Nature 193: 347–348

Diamond J (1962) The mechanism of solute transport by the gallbladder. J Physiol (Lond) 161: 474–502

Diamond JM, Bossert WH (1967) Standing gradient osmotic flow. J Gen Physiol 50:2061–2083

Duffey ME, Thompson SM, Frizzell RA, Schultz SG (1979) Intracellular chloride activities and chloride absorption in the intestinal epithelium of the winter flounder *Pseudopleuronectes americanus.* J Membr Biol 50:331–342

Ellory JC, Lahlou B, Smith MW (1972) Changes in the intestinal transport of sodium induced by exposure of goldfish to a saline environment. J Physiol (Lond) 222:497–509

Ellory JC, Ramos MMP, Zeuthen T (1978) Cl^- accumulation in the plaice intestinal epithelium. J Physiol (Lond) 287:12–13

Ellory JC, Lahlou B, Ramos MMP (1982) The effect of Amphotericin B on NaCl transport in plaice intestine. J Physiol (Lond) 325:166

Eveloff J, Kinne R, Kinne-Saffran E, Murer H, Silva P, Epstein FH, Stoff J, Kinter WB (1978) Coupled sodium and chloride transport into plasma membrane vesicles prepared from dogfish rectal gland. Pfluegers Arch 378:87–92

Field M, Karnaky KJ, Smith PL, Bolton JE, Kinter WB (1978) Ion transport across the isolated intestinal mucosa of the winter flounder *Pseudopleuronectes americanus:* 1. Functional and structural properties of cellular and paracellular pathways for Na and Cl. J Membr Biol 41: 265–293

Field M, Smith PL, Bolton JE (1980) Ion transport across the isolated intestinal mucosa of the winter flounder *Pseudopleuronectes americanus:* 2. Effects of cyclic AMP. J Membr Biol 55: 157–164

Fossat B, Lahlou B (1979) The mechanism of coupled transport of sodium and chloride in isolated urinary bladder of the trout. J Physiol (Lond) 294:211–222

Frizzell RA, Smith PL, Vosburgh E, Field M (1979) Coupled sodium-chloride influx across brush border of flounder intestine. J Membr Biol 46:27–39

Hirano T, Mayer-Gostan N (1976) Eel esophagus as an osmoregulatory organ. Proc Natl Acad Sci USA 73:1348–1350

Hirano T, Morisawa M, Ando M, Utida S (1976) Adaptative changes in ion and water transport mechanisms in the eel intestine. In: Robinson JWL (ed) Intestinal ion transport. MTP, London, pp 301–317

House CR, Green K (1965) Ion and water transport in isolated intestine of the marine teleost *Cottus scorpius.* J Exp Biol 42:177–189

Huang KC, Chen TST (1971) Ion transport across intestinal mucosa of winter flounder, *Pseudopleuronectes americanus.* Am J Physiol 220:1734–1738

Katz U, Lau KR, Ramos MMP, Ellory JC (1982) Thiocyanate transport across fish intestine *(Pleuronectes platessa).* J Membr Biol 66:9–14

Kirsch R, Laurent P (1975) L'oesophage, organe effecteur de l'osmorégulation chez un Téléostéen euryhalin, l'anguilla *(Anguilla anguilla* L.). CR Acad Sci 280:2013–2015

Lahlou B (1976) Ionic permeability of fish intestinal mucosa in relation to hypophysectomy and salt adaptation. In: Robinson JWL (ed) Intestinal ion transport. MTP, London, pp 318–328

Laurent P, Kirsch R (1975) Modifications structurales de l'oesophage liées à l'osmorégulation de l'anguille. CR Acad Sci 280:2227–2229

Machen TE, Diamond JM (1969) An estimate of the salt concentration in the lateral intercellular spaces of rabbit gallbladder during maximal fluid transport. J Membr Biol 1:194

Mackay WC, Lahlou B (1980) Relationships between Na^+ and Cl^- fluxes in the intestine of the European flounder *Platichthys flesus.* In: Lahlou B (ed) Epithelial transport in the lower vertebrates. CUP, London, pp 151–162

Mackay WC, Lahlou B, Porthé-Nibelle J (1978) AMP cyclique et contrôle des échanges ioniques au niveau de l'intestin de poisson. CR Acad Sci 287:1239–1242

Nellans HN, Frizzell RA, Schultz SG (1973) Coupled sodium-chloride influx across the brush border of rabbit ileum. Am J Physiol 225:467–475

Oide M (1967) Effects of inhibitors on transport of water and ions in isolated intestine and Na^+-K^+-ATPase in intestinal mucosa of the eel. Annot Zool Jpn 40:130–135

Oide M (1973) Role of alkaline phosphatase in intestinal water absorption by eels adapted to sea water. Comp Biochem Physiol 46A:639–645

Ramos MMP, Ellory JC (1981) Na and Cl transport across the isolated intestine of the plaice *Pleuronectes platessa.* J Exp Biol 81:123–142

Reuss L (1981) Mechanisms of the mucosa-negative transepithelial potential produced by Amphotericin B in gallbladder epithelium. Fed Proc 40:2206–2212

Schultz SG, Curran PF, Chez RA, Fuisz RE (1967) Alanine and sodium fluxes across mucosal border of rabbit ileum. J Gen Physiol 50:1241–1260

Sharratt BM, Bellamy D, Chester-Jones I (1964) Adaptation of the silver eel *(Anguilla anguilla* L.) to sea water and to artificial media together with observations on the role of the gut. Comp Biochem Physiol 11:19–30

Skadhauge E (1969) The mechanisms of salt and water absorption in the intestine of the eel *(Anguilla anguilla)* adapted to water of various salinities. J Physiol (Lond) 204:135–158

Smith H (1930) The absorption and excretion of water and salts by marine fishes. Am J Physiol 93:480–505

Smith MW (1964) The in vitro absorption of water and solutes from the intestine of goldfish *Carassius auratus.* J Physiol (Lond) 175:38–49

Smith MW, Ellory JC, Lahlou B (1975) Sodium and chloride transport by the intestine of the European flounder *Platichthys flesus* adapted to fresh of sea water. Pfluegers Arch 357:303–312

Spring KA, Kimura G (1978) Chloride reabsorption by renal proximal tubules of *Necturus.* J Membr Biol 38:233–254

Stewart CP, Smith PL, Welsh MJ, Frizzell RA, Musch MW, Field M (1980) Potassium transport by the intestine of the winter flounder *Pseudopleuronectes americanus.* Bull Mt Dsert Isl Biol Lab 20:92–95

Wilson TH, Wisemann G (1954) The use of sacs of everted small intestine for the study of the transference of substances from the mucosal to the serosal surface. J Physiol (Lond) 123: 116–125

Yamamoto M, Hirano T (1978) Morphological changes in the esophageal epithelium of the eel, *Anguilla japonica,* during adaptation to sea water. Cell Tissue Res 192:25–38

Zeuthen T, Ramos MMP, Ellory JC (1978) Inhibition of chloride transport by piretanide. Nature 273:678–680

Biochemical Adaptation of Trout Intestine Related to Its Ion Transport Properties. Influences of Dietary Salt and Fatty Acids, and Environmental Salinity

C. LERAY and A. FLORENTZ [1]

Introduction

It is well known that intestinal absorption of salt and water is an important component of the osmoregulatory processes taking place in fish living in hyperosmotic environment (Shehadeh and Gordon 1969, Sharratt et al. 1964, Skadhauge 1969, Lahlou 1976, Hirano et al. 1976). If stenohaline species, either from freshwater (FW) (Smith 1964, Ellory et al. 1972) or from seawater (SW) (House and Green 1965, Aull 1966), were sometimes taken as experimental material, euryhaline species are most frequently considered due to the suitability of their osmotic adjustments. Although the nature of the adaptative mechanism is not exactly known, it seems that regulation of both pumping efficiency and passive permeability can occur. If global studies have emphasized the passive role played by the intestine in the osmoregulatory aspect of salinity adaptation (Shehadeh and Gordon 1969, Sharratt et al. 1964), more precise studies revealed a general enhancement of active NaCl absorption (Skadhauge 1969, Lahlou 1976, Hirano et al. 1976). This event was shown in some cases to be linked to adaptative changes in enzymes related to ion transport (Hirano et al. 1976), but some discrepancy remains when the chronology of the phenomena is considered (MacKay and Janicki 1979).

Nevertheless, an inversed evolution in ion transport was recently demonstrated in the tilapia, a FW euryhaline teleost, during its adaptation to seawater (Mainoya 1982). This strategy can be also interpreted as a compensatory mechanism limiting salt invasion, the water absorption remaining curiously unmodified in that situation.

The passive role of the intestine in ionic regulation is of obvious importance in limiting entry of magnesium and sulfate into the blood as stated by Shehadeh and Gordon (1969) but it has been also demonstrated to be efficient in limiting NaCl entry (Lahlou 1976). This regulation was shown to take place at the apical part of the enterocytes of poor osmoregulators such as goldfish or trout compared with flounder, a perfect euryhaline fish which displays a constant mucosal ion entry (Lahlou 1976).

Thus it appears that, according to the efficiency of the gill to cope with a large sodium influx, different strategies at the intestinal level were adopted in euryhaline teleosts to ensure the maintenance of their water balance.

1 Laboratoire de Physiologie comparée des régulations, BP 20 CR-67037 Strasbourg Cedex, France

Intestinal Transport
(ed. by M. Gilles-Baillien and R. Gilles)

The various events involved in the regulation of ion movement across the mucosal cells point out the importance of the two steps of the transepithelial transport which are materialized by the two cellular poles: brush border and basolateral membranes. On the serosal side, the active transfer of Na^+ and K^+ linked to Na^+, K^+-ATPase is thought to play a key role in the adaptation of the intestinal epithelium to salt load (Hirano et al. 1976, Oide 1967, MacKay and Janicki 1979). A HCO_3^- activated ATP-ase was also proposed to be related to Cl^- intestinal transport in the eel (Morisawa and Utida 1976, Hirano et al. 1976) but its cellular distribution has not yet been elucidated. At the luminal border, the involvement of alkaline phosphatase in salt and water absorption was suggested considering its evolution during SW adaptation (Utida 1967, Utida and Isono 1967), its properties (Utida et al. 1968) and the action of some inhibitors (Oide 1973).

If the apical membrane Cl^- conductance was shown to be modulated by cyclic AMP in some euryhaline flatfish (MacKay et al. 1978, Field et al. 1980), the mechanism is unknown and no relation with membrane structure can yet be made.

It is well known that the mediation of ion transport either by vectorial enzymes or by defined protein carriers can be finely regulated by membrane lipid fluidity (Green et al. 1980, Freedman 1981, Yuli et al. 1981) which is itself determined by the lipid composition and assembly. This aspect does not appear to be documented in fish. The adaptive changes in ion transport process in the intestine of euryhaline fish are particularly suitable to the study of the correlation between function and molecular mechanisms. We used the rainbow trout, *Salmo gairdneri*, as experimental material since this fish can tolerate changes in environmental salinity (Leray et al. 1981, Lahlou et al. 1975) and can be reared easily with semi-synthetic food of controlled composition. In this paper, the transport properties of the brush border membranes (BBM) of trout enterocytes following rapid transfer in SW are considered in relation to their lipid composition and to some membrane bound enzyme activities. Salt supplementation of the diet was also tried to test the potential activation of osmoregulatory mechanisms and its possible use as a method of pre-conditioning rainbow trout before SW transfer.

Part of this work was presented to the third congress of the European Society of Comparative Physiology and Biochemistry (Leray and Di Costanzo 1981).

Influence of the Fatty Acid Composition of the Diet on Intestinal Membrane Properties in Freshwater Adapted Trout

Extensive studies have shown that dietary lipids affect largely the fatty acid (FA) composition of trout phospholipids (Castell et al. 1972a, Yu and Sinnhuber 1975, Castledine and Buckley 1980, Leger and Fremont 1981). Concurrently, it has been shown that fish cannot synthetize FA of (n-6) or (n-3) series but only the linolenic (n-3) series FA can be considered as essential fatty acids (EFA) in trout nutrition. The tenacity with which these EFA are retained in body or liver phospholipids (PL) during dietary deprivation (Castledine and Buckley 1980, 1982) or thermal acclimation (Hazel 1979) emphasizes their role in maintaining membrane integrity and functions. It should be observed that no comparable studies can be found on organs actively involved in osmoregulation.

Thus, we have investigated the influence of the FA quality of the PL on the ion permeability of the BBM isolated from the anterior intestine which is the only efficient osmoregulatory part (Lahlou et al. 1975). During 6 weeks trout were fed three diets prepared according to Castell et al. (1972a) and containing (Table 1) either EFA-rich cod liver oil (CLO), EFA-deficient grape seed oil (GSO) or being lipid free (D).

Table 1. Fatty acid composition of the diets (as mol %)

FA	Cod liver oil (CLO)	Grape seed oil (GSO)
sat	4.6	8.5
n-9	52.5	13.6
n-6	0.6	77.6
n-3	29.2	0.4
n-3/insat	0.30	0.004
n-3/n-6	0.48	0.005

BBM was prepared according to a suitable method modified from Schmitz et al. (1973), PL distribution and their FA composition according to recommended procedures (Di Costanzo et al, unpublished data).

Table 2 gives the proportions of the main PL fractions and cholesterol content in the three experimental groups. EFA-deficient diet induces a low PL/protein ratio while the proportion of cholesterol increases. Furthermore, this diet is able to induce higher amounts of phosphatidylcholine (PC) and phosphatidylinositol (PI) than those found with the two other diets. Phosphatidylethanolamine (PE) and sphyngomyelin (SP) being relatively stable. It must be noticed that the proportion of SP among choline containing PL is constant (15%–16%) and equivalent to that found in gill of marine fish (Zwingelstein 1981). When the FA compositions of the total PL fraction are considered (Table 3) it is clear that they roughly reflect the diet lipid composition. Thus, BBM from EFA repleted trout (CLO) are (n-3) FA rich, in opposition with the abundance of (n-6) FA in EFA depleted trout (GSO). BBM from trout fed a delipited diet (D) are comparatively depleted in (n-3) FA but enriched in saturated FA. As was

Table 2. Phospholipid and cholesterol composition of trout brush border membrane in the three experimental groups

Diet	$\frac{\text{PL}}{\text{Proteins}}$	$\frac{\text{CHOL}}{\text{PL}}$	PE	PC	PI	SP
CLO	0.50	0.98	28.5	34.6	19.6	6.5
GSO	0.27	1.58	26.9	40.2	15.1	6.9
D	0.38	0.90	33.7	33.3	16.7	6.2

$\frac{\text{PL}}{\text{Proteins}}$ on a weight basis, $\frac{\text{CHOL}}{\text{PL}}$ on a molar basis, PE, PC, PI and SP as % total lipid phosphorus.
Data obtained from pools of 10 intestinal samples

Table 3. Fatty acid composition of total phospholipid fraction from trout brush-border membrane in the three experimental groups

Diet	FA composition (as mol %)				n-3/n-6	n-3/unsat	PUFA	unsat/sat	DB/sat	DB/unsat
	sat	n-9	n-6	n-3						
CLO	39.6	11.9	10.5	34.0	3.24	0.56	38.2	1.52	6.3	4.1
GSO	38.9	14.5	37.2	9.4	0.25	0.15	34.3	1.57	4.8	3.1
D	55.1	17.1	14.4	11.4	0.79	0.25	21.6	0.81	2.4	3.0

PUFA: Polyunsaturated fatty acids (C 20, C 22); DB: double bound. Data obtained from pools of 10 intestinal samples

expected, a higher proportion of (n-9) and (n-6) FA are found when compared to CLO trout. Thus, considering the proportion of (n-3) FA (global proportion, (n-3)/unsat. or (n-3)/(n-6) ratio), the three membranes can be sorted according to the decreasing sequence CLO > D > GSO.

When the sodium apical uptake (J_{mc} Na^+) is measured on the anterior intestines of the same pretreated trout using the technique of Schultz et al. (1967) slightly modified by Florentz (1982), a significant decrease (– 36%) is observed in EFA-deficient trout when compared to control trout (Table 4). In D trout, the mucosal entry is also diminished (– 23%) but not significantly, due to the scatter of data. Comparable alterations of Cl^- apical entry were observed in the same preparations (unpublished results). Thus, the same sequence as for (n-3) FA proportion can be applied to Na apical uptake and justified the hypothesis of an important role for EFA in membrane PL in maintaining proper ion transport. A fluorescence study has recently revealed that paralleled alterations of membrane fluidity are induced by the same dietary lipid modifications (Duportail et al, unpublished data). We are unable to detect modifications in enzyme activities at the whole epithelium level (Table 5) in EFA-deficient trout, even focusing on a well-known marker of brush-border membrane, alkaline phosphatase. This enzyme was estimated in the presence of an optimal concentration of NaCl (200 mM) which was shown to be an activator as potent as $MgSO_4$ used at its optimal concentration (20 mM). Similar properties were described by Utida (1967) in the same biological material.

We therefore conclude that a decrease in EFA constituting the PL of the apical membrane of trout enterocytes is linked with a decrease in membrane fluidity which

Table 4. Na apical uptake (J_{mc} Na^+) in the trout anterior intestine in the three experimental groups

	CLO	GSO	D
n	5	5	4
J_{mc} Na^+ (n Eq mn^{-1} cm^{-2})	365.0 ± 46.2	232.7 ± 36.2	282.4 ± 40.5
		$p < 0.05$	n.s.

Table 5. Enzyme activities in anterior intestinal mucosa in control and EFA-deficient trout

Diet	Na^+, K^+-ATPase	Ouabain-binding	$\frac{Na^+, K^+\text{-ATPase}}{\text{Ouabain-binding}}$	HCO_3^--ATPase	Alkaline phosphatase (NaCl)
CLO	33.4 ± 5.6	6.1 ± 0.9	5.6 ± 0.6	129.5 ± 12.6	160.9 ± 25.6
GSO	36.4 ± 4.8	5.9 ± 0.8	6.6 ± 1.1	113.2 ± 6.1	185.0 ± 39.3

Enzyme activities are expressed as nmol mn^{-1} mg^{-1} protein. Ouabain binding as pmol mg^{-1} protein according to Hossler et al. (1978). (n = 5)

is associated with a lower NaCl permeability. This result emphasized the fundamental role of EFA at the membrane level showing that the qualitative aspect of these constituants prevails the quantitative one. These specific functions of membrane EFA were frequently suggested considering their metabolism and distribution (Castell 1979, Leger and Fremont 1981) and a partial confirmation of their role in membrane restructuring following thermal acclimation was recently given in trout by Hazel (1979).

Influence of the NaCl Content of the Diet on Intestinal Membrane Properties in FW Trout

Salt supplementation of the diet of salmonids was tried several times because of its possible use in fish culture as a method of improving food intake and conversion (MacLeod 1978) or SW adaptation (Zaugg and MacLain 1970, Basulto 1976). In this later case, salt supplementation can be expected to activate some osmoregulatory mechanisms and thus would be of interest in pre-conditioning fish before transfer to SW.

We have therefore examined the biochemical and functional condition of the intestinal membrane in FW trout fed 2 months on CLO pellets with an added NaCl content of 10%.

Na^+ apical uptake and the number of molecules of Na^+, K^+-ATPase (ouabain-binding sites) in enterocytes significantly decrease with NaCl oral intake, while Na^+, K^+-ATPase activity remains constant (Table 6). The NaCl activated alkaline phosphatase

Table 6. Na apical uptake (J_{mc} Na^+) and enzyme activities in anterior intestinal mucosa of trout fed control and NaCl supplemented diets

Diet	n	J_{mc} Na^+	Na^+, K^+-ATPase	Ouabain-binding	Alkaline phosphatase (NaCl)
CLO	10	469 ± 17	13.7 ± 2.3	2.9 ± 0.2	91.4 ± 13.7
CLO + NaCl	9	414 ± 9	12.3 ± 1.3	2.1 ± 0.2	64.4 ± 9.5
		$p < 0.05$		$p < 0.05$	

J_{mc} Na^+ expressed as nEq mn^{-1} cm^{-2}. For enzyme activities: same legends as in Table 5

activity is only slightly modified. In spite of the steady level of Na^+, K^+-ATPase activity, which probably does not reflect the in vivo activity, all modifications induced by oral salt load can be interpreted as the result of a compensatory mechanism leading to a reduction of ionic absorption across the gut wall. This likely results from changes in both passive permeability and pumping efficiency. The lipid composition of the BBM prepared from the intestine of salt-supplemented trout (Table 7) and their FA composition (Table 8) reveal some important modifications when compared to control trout. If the decrease in cholesterol content accompanying the increase in SP proportion is consistent with a low ionic permeability (Barenholz and Thomson 1980), the observed increase in (n-3) EFA content of the PL fraction would presumably outweigh this last effect. In the absence of experiments on liposomes, the extent of the antagonism between the respective lipid constituants cannot be estimated. Results of the present study indicate the ability of the salt fraction of the diet to induce modifications in the composition of specialized membranes of the intestine altering their functional properties.

Table 7. Phospholipid and cholesterol composition of brush-border membrane in trout fed control (CLO) and NaCl-supplemented (CLO + NaCl) diets

Diet	$\frac{\text{CHOL}}{\text{PL}}$	PE	PC	PI	SP
CLO	0.98	28.5	34.6	19.6	6.5
CLO + NaCl	0.75	26.6	32.5	20.5	11.8

Same legends as in Table 2

Table 8. Fatty acid composition of total phospholipid fraction from brush-border membrane in trout fed control (CLO) and NaCl-supplemented (CLO + NaCl) diets

Diet	FA composition (as mol %)				$\frac{\text{n-3}}{\text{n-6}}$	$\frac{\text{n-3}}{\text{unsat}}$	PUFA	$\frac{\text{unsat}}{\text{sat}}$	$\frac{\text{DB}}{\text{sat}}$	$\frac{\text{DB}}{\text{unsat}}$
	sat	n-9	n-6	n-3						
CLO	39.6	11.9	10.5	34.0	3.24	0.56	38.2	1.52	6.3	4.1
CLO + NaCl	43.8	7.5	7.5	39.4	5.23	0.70	44.9	1.28	6.2	4.8

Same legends as in Table 3

Further investigations would be necessary to evaluate the transepithelial NaCl movements in such pre-adapted fish and the time course of the functional and biochemical adaptation after shifting to a salt diet. As expected, no modification of the number of ouabain binding sites could be observed at the gill level, the NaCl excretion operating easily along the salt concentration gradient. This is in contrast with the increase of Na^+, K^+-ATPase activity observed by Zaugg and MacLain (1970) in Coho salmon. It is noteworthy that the number of ouabain-binding sites is about ten times more imprtant in intestinal mucosa than in gills.

Effect of Seawater Adaptation on the Properties of Diet-Modified Intestinal Membranes

Changes in Membrane Lipid Composition

During preliminary experiments we have observed different modifications in the FA composition of the main PL fractions prepared from the whole scrapped mucosa (Di Costanzo and Leray 1981) soon after the transfer of trout in SW. The time course of these PL alterations is diet-dependent and emphasized the importance of the (n-3) FA (Table 9). It can be seen that whatever the importance of the (n-3) FA in the diet, there was a saturation of all PL after 2 days in SW, delay corresponding to the physiological "crisis" and the onset of the mortality (Leray et al. 1981). This rapid PL alteration is well characterized by an increase in the level of saturated FA concomitant with a decrease in the level of (n-6) and (n-3) FA. The best index of these structural modifications seems to be the unsaturation index (number of double bonds per mole of saturated fatty acid). After 7 days in SW, the initial state is recovered with a diet rich in EFA (CLO), while a progressive saturation was observed with an EFA deficient diet. Among the PL analyzed, PE is more acutely affected, in contrast with the conservative mechanism which characterized this PL in EFA deficiency (Castledine and Buckley 1982) or cold exposure (Hazel 1979) in trout.

Although these analyses were processed on the PL of the whole mucosal membranes, it could be expected that they reflect the evolution of plasma membrane. Thus different transport properties must accompany these rapid cellular perturbations during SW adaptation.

Table 9. PC, PE and PI fatty acid composition in trout intestinal mucosa during seawater adaptation

		sat.		n-6		n-3		DB/sat		DB/unsat	
		CLO	GSO	CLO	GSO	CLO	GSO	CLO	GSO	CLO	GSO
PC	FW	39.3	37.4	6.5	41.1	43.3	11.4	7.3	6.2	4.7	3.7
	SW2d	47.0	42.0	1.1	41.2	41.2	7.8	5.4	4.9	4.8	3.6
	SW7d	44.0	44.6	5.8	37.5	42.9	11.5	6.2	4.8	4.9	3.9
PE	FW	29.3	23.0	5.7	58.0	52.9	15.0	11.4	13.5	4.7	4.0
	SW2d	35.2	32.5	3.5	49.6	46.1	13.9	8.2	8.5	4.5	4.1
	SW7d	30.7	47.0	3.5	32.5	61.4	13.6	12.0	4.7	5.3	4.2
PI	FW	48.6	49.4	18.1	33.8	32.7	16.8	5.2	4.9	4.9	4.7
	SW2d	63.2	46.8	13.6	37.1	23.2	16.1	3.0	4.9	5.2	4.3
	SW7d	50.8	55.3	10.4	30.1	38.8	11.7	5.2	3.5	5.3	4.4

Fatty acid compositions are expressed as mol %. Data obtained from pools of 5 intestinal samples

Changes in Membrane Transport Function

Effect of EFA Deficiency. The rapid increase of the plasma Na content (Fig. 1) is largely influenced by the EFA content of the pelleted food previously distributed to FW trout. The high plasma Na increase observed after one day in SW in EFA-repleted trout can be correlated with their higher capacity of intestinal ion apical uptake as seen previously. Later, the evolution of the plasma ion concentration is similar in both EFA-repleted and deficient experimental groups. These effects may result from differential efficiency in gill extrusion rate but connections with intestinal absorption rate must be considered since it depends on the FA nutritional state (Fig. 2). An EFA deficiency induces a progressive permeability increase, significant after 2 days in SW, while the reverse is observed with control fish fed an EFA-supplemented diet. These results are thus consistent with the fact that EFA play an important role in the permeability as well as the plasticity of membranes. This specific role was postulated as one of the factors accounting for differences in FA content between fresh-water and marine fish and between warm-water and cold-water fish (Castell 1979). The demonstration of a role for (n-3) FA in salt uptake by trout enterocytes may be correlated to their influence on the swelling rate of trout liver mitochondria (Castell et al. 1972b).

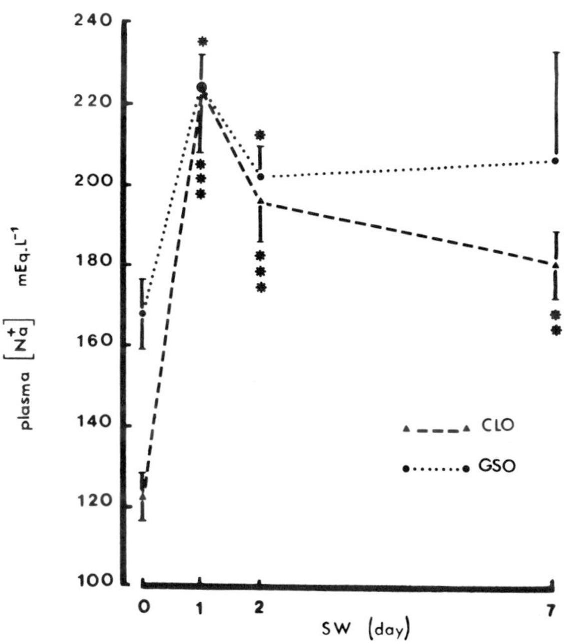

Fig. 1. Plasma Na^+ content in trout fed control (CLO) or EFA deficient (GSO) diet during seawater adaptation. Comparison with FW value: * $P < 0.05$, ** $P < 0.01$, *** $P < 0.001$

The modification in transport function observed at the BBM level soon after the SW transfer might be transitory since we could determine that the apical Na^+ uptake in trout kept one month with CLO diet in SW (355 ± 13 nEq min^{-1} cm^{-2}) is similar to that in FW (365 ± 46). Thus, compensatory changes detected during SW adaptation

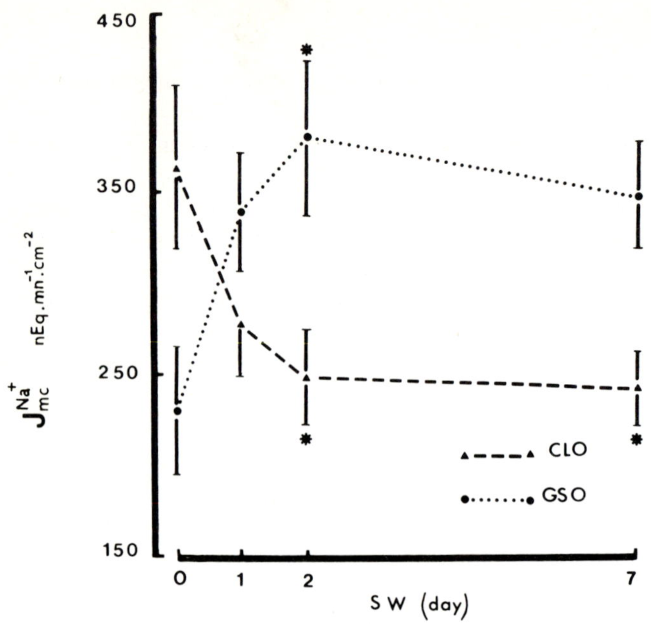

Fig. 2. Na^+ apical uptake (J_{mc} Na^+) in the anterior intestinal mucosa of trout fed control (CLO) or EFA-deficient (GSO) diet during seawater adaptation. Comparison with FW value: * $P < 0.05$

at the functional level as well as at the structural level of enterocytes consist of relatively rapid events which are complete after a few days. This early adaptation was shown to include several physiological and biochemical responses at the plasma, gill and muscle levels (Leray et al. 1981).

Two days after the transfer in SW, the changes in membrane permeability are not associated with changes neither in Na^+, K^+-ATPase activity nor in ouabain binding sites number (Figs. 3 and 4). After 7 days in SW, only the catalytic activity of this enzyme slightly increases in control as in EFA-deficient fish. MacKay and Janicki (1979) report a time course of Na^+, K^+-ATPase activity in *Anguilla rostrata* which is similar to our results. If we postulate that, in trout, salt absorption increases as soon as SW transfer is completed, as in eel (MacKay and Janicki 1979), it can be hypothetized that ion transport and Na^+, K^+-ATPase activity are not always stringently linked. The urgent mechanisms essential for the animal survival must be independent of protein synthesis and the results of the present study suggest rather membrane restructuation through PL composition to limit salt invasion.

Later, slowly occurring events include an elevation in Na^+, K^+-ATPase activity linked to increased ion and fluid absorption by intestinal mucosa (Lahlou et al. 1975), the only efficient way to maintain osmotic balance. The level of Na^+, K^+-ATPase activity and the number of ouabain binding sites measured after a one month adaptation period in seawater are identical to those measured in freshwater, this probably reflects the lack of correlation between in vitro measurements and in vivo activity of the enzyme. This was already postulated by MacKay and Janicki (1979) studying changes in the eel intestine during seawater adaptation.

Oide (1973) and Utida (1967) suggested that alkaline phosphatase rather than Na^+, K^+-ATPase is responsible for NaCl and fluid absorption by the intestine of euryhaline fish. Recently, Morisawa (1978) stated that HCO_3^--ATPase plays also an important role in active transport of Cl^- across the eel intestine. All these conclusions were mainly drawn from the observation of an increase in enzyme activities during the first week following FW to SW transfer.

Our investigations in trout do not confirm these enzyme variations, since HCO_3^--ATPase (Fig. 5) and alkaline phosphatase (Table 10) activities decrease significantly the first day after the transfer. It seems unlikely that these enzymes might be directly implicated in the mechanism of trout intestinal adaptation. It would now be very valuable to understand the function of these membrane bound enzymes and their precise role in the intestinal transport function. Cellular localization studies and subcellular systems such as vesicles might be useful to solve this important problem.

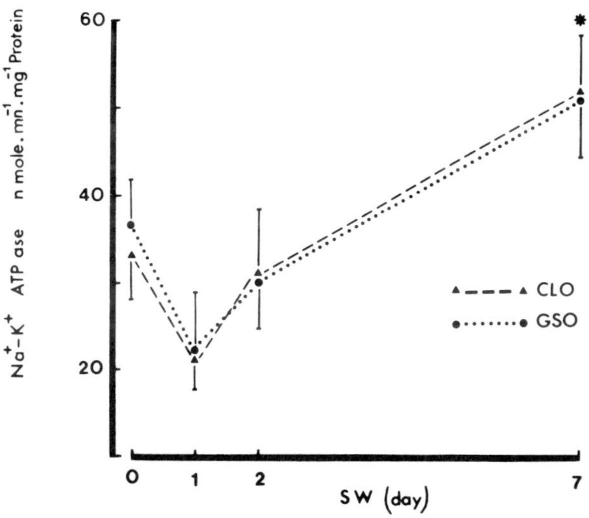

Fig. 3. Na^+, K^+-ATPase activity in the anterior intestinal mucosa of trout fed control (CLO) or EFA-deficient (GSO) diet during seawater adaptation. Comparison with FW value: * $P < 0.05$

Influence of Diet NaCl Content. We have verified that addition of NaCl (10%) in a control diet (CLO) containing 0.16% NaCl improves survival of FW trout when abruptly transferred in SW (Fig. 6). The median lethal time increases from 18 h to 66 h when fish (average weight: 70.5 ± 2.1 g, n = 100) are fed a 10% NaCl supplemented diet two months before the experiment. The size of the fish was chosen such as they could not survive direct transfer to SW in opposition with the other experiments reported here during which larger fish were used (200–250 g). In a preliminary experiment using 250 g trout fed a control diet for 2 months where all NaCl fraction was omitted, we demonstrated they could not survive in 50% SW. The abrupt transfer of control fish at this salinity is never harmful. Thus, these studies demonstrate that any salt content in the diet has an influence on SW adaptation and that supplementation has beneficial effects.

The possibility of increased enzyme activity associated with the salt pre-conditioning of trout before and after transfer in SW was examined at the intestinal mucosa level.

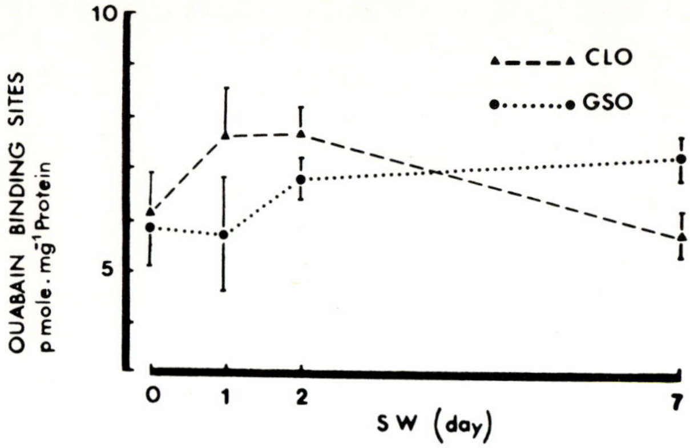

Fig. 4. Number of ouabain-binding sites in the anterior intestinal mucosa of trout fed control (CLO) or EFA-deficient (GSO) diet during seawater adaptation

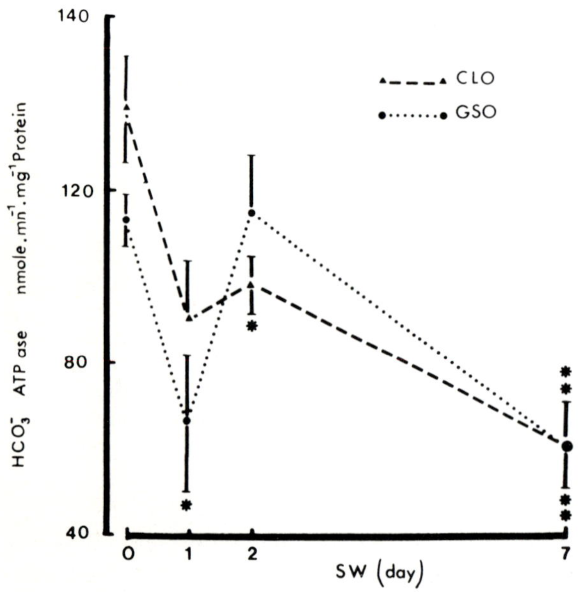

Fig. 5. HCO_3^--ATPase activity in the anterior intestinal mucosa of trout fed control (CLO) or EFA-deficient (GSO) diet during seawater adaptation. Comparison with FW value: * $P < 0.05$, ** $P < 0.01$

The results (Table 10) show that Na^+, K^+-ATPase activity and the number of ouabain-binding sites are less important in salt supplemented fish than in control in both environments. Curiously, trout held in SW for 1–2 days have a transient decrease in these parameters suggesting a decrease in ion pumping efficiency. Thus it can be inferred that dietary NaCl tends to reduce both active and passive transport in the anterior part of the trout intestine so that subsequent SW exposure would cause lower stress in the animal. Further studies should attempt to link enzyme modifications with other perturbations at the level of plasma, cell and membrane parameters. The use of deprivation or supplementation of other components of the salt mixture would be fruitful to detect any beneficial effect on SW adaptation of young trout. The possibility of an interaction between salt and EFA supplementation cannot be excluded.

Table 10. Enzyme activities in anterior intestinal mucosa of trout fed control (CLO) and NaCl-supplemented (CLO + NaCl) diets during seawater adaptation

	Na^+, K^+-ATPase		Ouabain-binding sites		Alkaline phosphatase (NaCl)	
	CLO	CLO + NaCl	CLO	CLO + NaCl	CLO	CLO + NaCl
FW	13.7 ± 2.3	12.3 ± 1.3	2.9 ± 0.2	2.1 * ± 0.2	91.4 ± 13.7	64.4 ± 9.5
SW 1d	12.4 ± 2.7	5.7 ++ ± 1.1	2.3 ± 0.4	1.3 ± 0.3	39.7 + ± 15.3	47.2 ± 6.5
SW 2d	17.0 ± 0.3	12.1 ± 0.5	2.7 ± 0.4	1.7 ± 0.2	35.2 + ± 10.6	29.9 + ± 2.2
SW 7d	17.9 ± 2.8	17.2 ± 2.3	3.3 ± 0.3	3.3 ++ ± 0.3	38.5 ++ ± 8.4	49.3 ± 7.3

For enzyme activities same legends as in Table 5. Data are obtained from 9 (FW and SW 7d) or 4 (SW 1d and SW 2d) trout

* $P < 0.05$ significance of the difference between diets

\+ $P < 0.05$ significance of the difference between FW and SW

++ $P < 0.01$ significance of the difference between FW and SW

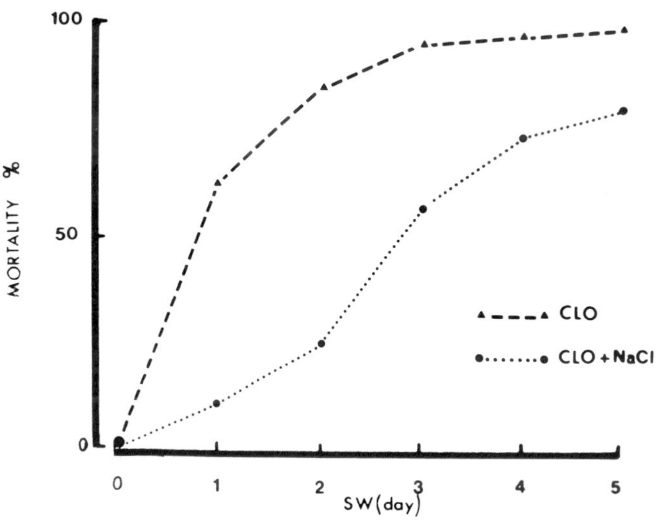

Fig. 6. Effect of seawater adaptation on the survival of young trout fed control (CLO) or salt-supplemented (CLO + NaCl) diet. 100 trouts in each group

Conclusions

The results of the present study indicate the value of examining the time course of compensation following a marked perturbation of external salinity. This protocol gave us previously (Leray et al. 1981) the opportunity to describe rapid adjustments followed by slowly occurring adaptations.

Clearly, EFA play an important role in maintaining flexibility of membrane structures which keep the possibility of molecular reorganization under the influence of environmental salinity.

The mechanisms which mediate the action of EFA at the membrane level are presently unknown, but they can modify either the permeability of the lipid phase itself or the properties of vectorial enzymes implicated directly in ion transport. It can be expected that only slight modifications of the kinetic properties of enzyme systems would alter their functional activity in vivo without any detectable changes in in vitro assays. Unfortunately, no work has been made in fish on the influence of the polyunsaturated fatty acids in the relationships between phospholipids and Na^+, K^+-ATPase molecules. Furthermore, a specific association between this enzyme and any phospholipid fraction has not yet been described. Our investigations on the biochemical adaptation of trout intestine during SW adaptation show clearly that during the "crisis" following immediately the transfer in an hypertonic environment important alterations of the lipid part of membranes occur. These structural changes lead to a decrease in permeability and pumping efficiency, thus allowing the excess of ingested salt to be eliminated at the gill level. A proper diet composition would help the fish to carry on these rapid changes before enzyme modifications happen.

Thus, it appears that comparative informations about the precise nutritional requirements of fish in FW and SW would allow to select nutrients essential for maintaining an adequate water balance. The efficiency of preconditioning FW trout for improving their SW survival by the use of salt supplemented pellets is a guide for future experiments designed to study the role played by the intestine in the osmotic adjustment of euryhaline fish.

Acknowledgements. We are indebted to Mrs L. Debellemaniere, G. Gutbier and Mr R. Meens for their technical help. This work was supported by the Centre National de la Recherche Scientifique et le Centre National d'Exploitation des Océans.

References

Aull F (1966) Absorption of fluid from isolated intestine of the toadfish, *Opsanus tau*. Comp Biochem Physiol 17:867–870

Barenholz J, Thomson TE (1980) Sphingomyelins in bilayers of biological membrane. Biochim Biophys Acta 604:129–158

Basulto S (1976) Induced saltwater tolerance in connection with inorganic salts in the feeding of atlantic salmon *(Salmo salar* L.). Aquaculture 8:45–55

Castell JD (1979) Review of lipid requirement of finfish. In: Halver JE, Tiews K (eds) Proceedings of the world symposium on finfish nutrition and fishfeed technology, vol 1. Heeneman, Berlin, p 59

Castell JD, Lee DJ, Sinnhuber RO (1972a) Essential fatty acids in the diet of rainbow trout *(Salmo gairdneri):* lipid metabolism and fatty acid composition. J Nutr 102:93–100

Castell JD, Sinnhuber RO, Lee DJ, Wales JH (1972b) Essential fatty acids in the diet of rainbow trout *(Salmo gairdneri):* physiological symptoms of EFA deficiency. J Nutr 102:87–92

Castledine AJ, Buckley JT (1980) Distribution and mobility of ω 3 fatty acids in rainbow trout fed varying levels and types of dietary lipid. J Nutr 110:675–685

Castledine AJ, Buckley JT (1982) Incorporation and turnover of essential fatty acids in phospholipids and neutral lipids of rainbow trout. Comp Biochem Physiol 71B:119–126

Di Costanzo G, Leray C (1981) Importance of the essential fatty acids in the phospholipids of the rainbow trout enterocytes: relationships between diet composition and water salinity. 23rd Conf ICBL, Nyborg

Ellory JC, Lahlou B, Smith MW (1972) Changes in the intestinal transport of sodium induced by exposure of goldfish to a saline environment. J Physiol (Lond) 222:497–509

Field M, Smith PL, Bolton JE (1980) Ion transport across the isolated intestinal mucosa of the winter flounder, *Pseudopleuronectes americanus:* II. Effects of cyclic AMP. J Membr Biol 55: 157–163

Florentz A (1982) Relations acides gras essentiels et échanges ioniques transmembranaires au niveau de l'intestin de la truite arc en ciel lors de son transfert en eau de mer. Thèse 3e cycle, Strasbourg

Freedman RB (1981) Membrane-bound enzymes. In: Finean JB, Michell RH (eds) Membrane structure. Elsevier, Amsterdam, p 161

Green DE, Fry M, Blondin GA (1980) Phospholipids as the molecular instruments of ion and solute transport in biological membranes. Proc Natl Acad Sci USA 77:257–261

Hazel JR (1979) Influence of thermal acclimation on membrane lipid composition of rainbow trout liver. Ann J Physiol 261:R91–R101

Hirano T, Morisawa M, Ando M, Utida S (1976) Adaptive changes in ion and water transport mechanism in the eel intestine. In: Robinson JW (ed) Intestinal ion transport. MTP Press, Lancaster, p 301

Hossler FE, Sarras MP, Barrnett RJ (1978) Ouabain binding during plasma membrane biogenesis in duck salt gland. J Cell Sci 31:179–197

House CR, Green K (1965) Ion and water transport in isolated intestine of the marine teleost, *Cottus scorpius.* J Exp Biol 42:177–189

Lahlou B (1976) Ionic permeability of fish intestinal mucosa in relation to hypophysectomy and salt adaptation. In: Robinson JW (ed) Intestinal ion transport. MTP Press, Lancaster, p 318

Lahlou B, Crenesse D, Bensahla-Talet A, Porthe-Nibelle J (1975) Adaptation de la truite d'élevage à l'eau de mer. Effets sur les concentrations plasmatiques, les échanges branchiaux et le transport intestinal du sodium. J Physiol (Paris) 70:593–603

Leger C, Fremont L (1981) Métabolisme des acides gras et des lipides neutres. In: Fontaine M (ed) Nutrition des poissons. CNRS, Paris, p 215

Leray C, Di Costanzo G (1981) Time course of phospholipids and enzymes modification in the intestinal mucosa of rainbow trout following abrupt changes in external salinity. 3rd Conf ESCPB, Noordwijkerhout

Leray C, Colin DA, Florentz A (1981) Time course of osmotic adaptation and gill energetics of rainbow trout *(Salmo gairdneri* R.) following abrupt changes in external salinity. J Comp Physiol 144:175–181

MacKay WC, Janicki R (1979) Changes in the eel intestine during seawater adaptation. Comp Biochem Physiol 62A:757–761

MacKay WC, Lahlou B, Porthe-Nibelle J (1978) AMP cyclique et contrôle des échanges ioniques au niveau de l'intestin de poisson. CR Acad Sci 287:1239–1242

MacLeod MG (1978) Relationships between dietary sodium chloride, food intake and food conversion in the rainbow trout. J Fish Biol 13:73–78

Mainoya JR (1982) Water and NaCl absorption by the intestine of the tilapia *Sarotherodon mossambicus* adapted to fresh water or seawater and the possible role of prolactin and cortisol. J Comp Physiol 146:1–7

Morisawa M (1978) Changes in activity of HCO_3^--activated adenosine triphosphatase in intestinal mucosa of the eel during seawater adaptation. Annot Zool Jpn 51:5–9

Morisawa M, Utida S (1976) HCO_3^--activated adenosine triphosphatase in intestinal mucosa of the eel. Biochim Biophys Acta 445:458–463

Oide M (1967) Effects of inhibitors on transport of water and ion in isolated intestine and Na^+, K^+-ATPase in intestinal mucosa of the eel. Annot Zool Jpn 40:130–135

Oide M (1973) Role of alkaline phosphatase in intestinal water absorption by eels adapted to seawater. Comp Biochem Physiol 46A:639–645

Schmitz J, Preiser H, Maestracci D, Ghosh BK, Cerda JJ, Crane RK (1973) Purification of the human intestinal brush border membrane. Biochim Biophys Acta 323:98–112

Schultz SG, Curran PF, Chez RA, Fuisz RE (1967) Alanine and sodium fluxes across mucosal border of rabbit ileum. J Gen Physiol 50:1241–1260

Sharratt BM, Bellamy D, Jones IC (1964) Adaptation of the silver eel *(Anguilla anguilla* L.) to sea water and to artificial media together with observations on the role of the gut. Comp Biochem Physiol 11:19–30

Shehadeh ZH, Gordon MS (1969) The role of the intestine in salinity adaptation of the rainbow trout, *Salmo gairdneri.* Comp Biochem Physiol 30:397–418

Skadhauge E (1969) The mechanism of salt and water absorption in the intestine of the eel *(Anguilla anguilla)* adapted to waters of various salinities. J Physiol (Lond) 204:135–158

Smith MW (1964) The in vitro absorption of water and solutes from the intestine of goldfish, *Carassius auratus.* J Physiol (Lond) 175:38–49

Utida S (1967) Effect of sodium chloride on alkaline phosphatase activity in intestinal mucosa of the rainbow trout. Proc Jpn Acad 43:783–788

Utida S, Isono N (1967) Alkaline phosphatase activity in intestinal mucosa of the eel adapted to fresh water or sea water. Proc Jpn Acad 43:789–792

Utida S, Oide M, Oide H (1968) Ionic effects on alkaline phosphatase activity in intestinal mucosa with special reference to seawater adaptation of the japanese eel *Anguilla japonica.* Comp Biochem Physiol 27:239–249

Yu TC, Sinnhuber RO (1975) Effect of dietary linolenic and linoleic acids upon growth and lipid metabolism of rainbow trout *(Salmo gairdneri).* Lipids 10:63–66

Yuli I, Wilbrandt W, Shinitzky M (1981) Glucose transport through cell membranes of modified lipid fluidity. Biochemistry 20:4250–4256

Zaugg WS, MacLain LR (1970) Adenosine triphosphatase activity in gills of salmonids: seasonal variations and salt water influence in Coho salmon, *Oncorhynchus kisutch.* Comp Biochem Physiol 35:587–596

Zwingelstein G (1981) Les lipides amphiphiles du poisson. In: Fontaine M (ed) Nutrition des poissons. CNRS, Paris, p 249

Subject Index

E. Skadhauge

Osmoregulation in Birds

1981. 42 figures. X, 203 pages
(Zoophysiology, Volume 12)
ISBN 3-540-10546-8

Contents: Introduction. - Intake of Water and Sodium Chloride. - Uptake Through the Gut. - Evaporation. - Function of the Kidney. - Function of the Cloaca. - Function of the Salt Gland. - Interaction Among the Excretory Organs. - A Brief Survey of Hormones and Osmoregulation. - Problems of Life in the Desert, of Migration, and of Egg-Laying. - References. - Systematic and Species Index. - Subject Index.

Structural and Functional Aspects of Enzyme Catalysis

32. Colloquium der Gesellschaft für Biologische Chemie
23.-25. April 1981 in Mosbach/Baden

Editors: **H. Eggerer, R. Huber**
1981. 116 figures. IX, 216 pages
ISBN 3-540-11110-7

Contents: Mechanism of Enzyme Action. - Dynamics of Molecular Recognition. - Function of Metals in Enzymes: Thermophilic Enzymes. - Biological and Chemical Modifications of Enzymes. - Selected Topics of Enzyme Catalysis. - Subject Index.

Biochemistry of Differentiation and Morphogenesis

33. Colloquium der Gesellschaft für Biologische Chemie
25.-27. März 1982 in Mosbach/Baden

Editor: **L. Jaenicke**
1982. 158 figures. XI, 301 pages.
ISBN 3-540-12010-6

Contents: Gene Expression. - Transfer of Genes. - Cell Differentiation. - Cell Recognition. - Morphogenesis. - Subject Index.

"Differentiation" and "Morphogenesis" have been of interest for many decades not only to developmental biologists, but also to biochemists who have long been aware that this field holds much potential for their skills. Early studies concentraded mainly on the search for diffusible factors that induce developmental events, while more recently research has focused in molecular biology, with remarkable success.

The aim of the 33rd Mosbach Colloquium was to summarize and comment on these achievements. This volume contains the proceedings of this meeting, with 27 contributions by experts from different countries, providing the reader with a wealth of information presented from different and sometimes unusual angles. It also shows the broad overlaps and indentations that make biochemistry such a strong link between the physical and biological sciences.

Springer-Verlag
Berlin
Heidelberg
New York
Tokyo